C0-ATW-978

Computer Techniques in Neuroanatomy

CONTRIBUTORS

Ellen M. Johnson
Department of Physiology
The University of North Carolina
Chapel Hill, North Carolina

R. Ranney Mize
Department of Anatomy and Neurobiology
University of Tennessee Health Science Center
Memphis, Tennessee

David G. Tieman
Department of Biological Sciences
State University of New York
Albany, New York

Harry B. M. Uylings, Jaap van Pelt,
Ronald W. H. Verwer, and Patricia McConnell
Netherlands Institute for Brain Research
Amsterdam, The Netherlands

Computer Techniques in Neuroanatomy

Joseph J. Capowski

Eutectic Electronics, Inc.
Raleigh, North Carolina
and The University of North Carolina
Chapel Hill, North Carolina

Plenum Press • New York and London

Library of Congress Cataloging in Publication Data

Capowski, Joseph J.
Computer techniques in neuroanatomy / Joseph J. Capowski.
p. cm.
Bibliography: p.
Includes index.
ISBN 0-306-43263-3
1. Neuroanatomy—Data processing. I. Title.
[DNLM: 1. Computer Systems. 2. Neuroanatomy. WL 101 C245c]
QM451.C36 1989
591.4′8′0285—dc20
DNLM/DLC 89-16046
for Library of Congress CIP

Cover illustration: HRP-filled spinocervical tract neuron from the spinal cord of an adult cat. The cell was identified and stained by L. M. Mendell and M. J. Sedivec of the Department of Anatomy and Neurobiology, SUNY, Stony Brook, New York. The reconstruction was performed by J. J. Capowski of the Department of Physiology, The University of North Carolina, Chapel Hill, North Carolina. Reproduced with permission from Oxford University Press, New York, New York.

A Division of Plenum Publishing Corporation
233 Spring Street, New York, N.Y. 10013

Printed in the United States of America

Preface

This book is the story of the marriage of a new technology, computers, with an old problem, the study of neuroanatomical structures using the light microscope. It is aimed toward you, the neuroanatomist, who until now have used computers primarily for word processing but now wish to use them also to collect and analyze your laboratory data. After reading the book, you will be better equipped to use a computer system for data collection and analysis, to employ a programmer who might develop a system for you, or to evaluate the systems available in the marketplace.

To start toward this goal, a glossary first presents commonly used terms in computer-assisted neuroanatomy. This, on its own, will aid you as it merges the jargon of the two different fields. Then, Chapter 1 presents a historical review to describe the manual tasks involved in presenting and measuring anatomic structures. This review lays a base line of the tasks that were done before computers and the amount of skill and time needed to perform the tasks.

In Chapters 2 and 3, you will find basic information about laboratory computers and programs to the depth required for you to use the machines easily and talk with some fluency to computer engineers, programmers, and salesmen.

Chapters 4, 5, and 6 present the use of computers to reconstruct anatomic structures, i.e., to enter them into a computer memory, where they are later displayed and analyzed. The data may be entered from tissue directly mounted in a microscope or from drawings, photographs, or televised images of the tissue. Errors are made during data entry, both by the person entering the data and because of optical constraints of the microscope. Techniques to correct these errors are found in Chapter 7. Chapter 8 presents techniques for building displays of the data that are superior to the original microscope image. These displays form one of the two major reasons for using a computer to assist in the study of neuroanatomy.

The second major reason for using a computer in neuroanatomy is to summarize statistically a structure, perhaps to compare one population of structures to another. These statistical methods are covered in Chapters 9, 10, and 11 by Harry Uylings, his colleagues, and me.

In Chapter 12, Ellen Johnson describes functions that should be included in a computerized data collection system to make it easy to use. This chapter especially will help you design your own programs, instruct your own programmer more intelligently, or survey the marketplace with more insight.

Imaging systems that combine computers and television devices offer you new ways to look at structures. Dave Tieman in Chapter 13 explains their basics, and Ranney Mize follows in Chapter 14 with a complete treatment of their use in the analysis of immunocytochemically treated tissue. Television also offers the potential to automate the tedious data entry process, especially when used in conjunction with the laser-scanning microscope. Chapter 15 closes the text portion of the book by describing the current capabilities and the future of such automation.

The final chapter compiles commercially available products for computer-assisted neuroanatomy. These are presented side by side, as a shopping mall.

Why do you need to know the material in this book? In the last 20 years, computers have been applied successfully to many fields, reducing tedium and increasing knowledge and productivity. Often the initiative for applying this technology comes as much from the computer manufacturers as it does from the users. When the marketplace is large, computer companies are willing to invest their people and financial resources in solving its problems. Unfortunately, the neuroscience community does not form a large market, and the neuroanatomy marketplace is even smaller. So the burden of applying the new technology falls primarily on you with some help from smaller companies.

Many journal articles that describe the application of computers and ancillary hardware (e.g., television cameras, video digitizers, computer graphics display devices) have been published in the last two decades, though scattered among neuroscience research journals, neuroscience techniques journals, microscope journals, and computer journals. Very few books have consolidated this information. This one does, and in it techniques and principles are presented broadly enough so that the book will maintain its relevance well past the lifespan of any one computer system.

We use the book at UNC as a text in the course "The Computer in the Laboratory," but it is more likely that you will read it to learn more about the subject. You will probably not read the book from cover to cover; rather, you will read the individual chapters that discuss the particular problem that interests you. I have attempted to make each chapter stand alone, even though this means that some repetition does occur.

The marriage of computers and neuroanatomy, like all marriages, does have some strings attached, however. Using or programming computers can be a fascinating and captivating hobby with never-ending challenges. My 16-year-old stepson spends uncounted hours in front of our home computer programming video games; he won't come to the dinner table in the same way that young Pete Rose wouldn't come in from the baseball diamond for lunch. The neuroscientist who discovers computers and doesn't use them effectively can also miss lunch, that is, his neuroscience. It takes discipline to remain a neuroscientist and not become a technologist interested more in the means of solving a neuroscience problem than in its solution.

I came to the Physiology Department of the University of North Carolina as a computer design engineer in 1972. The department was a leader in computer use in neuroscience at that time, although its hardware consisted only of a PDP-11 computer, which many people signed up to use. It was attached to a large IBM 370 computer owned by our university. With the help of many people, I designed hardware and software for neuron reconstruction, serial section reconstruction, silver grain counting, and other now standard neuroanatomical measurement tasks. Since that time, many improvements have been made to all these tasks, and new ones have been added. The work was not done in a vacuum; many neuroscientists at UNC and at other schools contributed their suggestions, and several computer engineers who had neuroscience interest at other institutions have also been very helpful.

I must personally thank Edward R. Perl, the chairman of the UNC Department of Physiology, for his moral support over 12 years. Miklos Réthelyi, 2nd Department of Anatomy, Semmelweis University Medical School, Budapest, and a frequent visitor to our department, has contributed much time and expertise in critiquing my efforts. William L. R. Cruce, Department of Neurobiology, Northeastern Ohio Universities College of Medicine, has been most helpful and has been a cheerleader. Ellen M. Johnson and C. William Davis, both of our Department of Physiology, have been wonderful backboards for technical discussions on how to solve problems.

The writing effort of this book has also been shared. I want to thank my wife Carolyn W. Capowski, who is an instructional developer, and her colleagues, technical editors David Carozza, Judy Clark, Connie Cowell, Craig Henkle, and Kim Miller for their efforts, especially in the face of my unreasonable demands.

I am delighted to acknowledge multiyear financial support from a National Institutes of Health program project grant on the effects of spinal cord injury, number NS-14899. Most recently, I am indebted to Eutectic Electronics, Inc., Raleigh, NC, for seeing enough potential marketplace for a neuron-tracing system and a serial section reconstruction system to provide me laboratory grant support for their continued development.

I hope that the book helps you. Please call or write me if there are any issues you wish to discuss. Good luck with your reading.

Joseph J. Capowski

Chapel Hill, North Carolina

Contents

Chapter 3. Software in the Neuroanatomy Laboratory

Chapter 4. Semiautomatic Entry of Neuron Trees from the Microscope

Chapter 7. Intermediate Computations of Reconstructed Data

Chapter 8. Three-Dimensional Displays and Plots of Anatomic Structures

Chapter 11. Statistical Analysis of Neuronal Populations

HARRY B. M. UYLINGS, JAAP VAN PELT, RONALD W. H. VERWER, and PATRICIA McCONNELL

Chapter 12. Controlling the Computer System: The User Interface

ELLEN M. JOHNSON

Chapter 15. Fully Automatic Neuron Tracing

Chapter 16. Commercially Available Computer Systems for Neuroanatomy

Color Plates

FIGURE 8-1. A three-dimensional reconstruction of the soma of a cortical pyramidal neuron from an Alzheimer's patient. The tissue containing this soma was cut into 100 sections, and the soma, nucleus, and nucleolus outlines were traced on a data tablet. Tiling triangles were generated to connect the sections, resulting in about 35,000 polygons in the complete image. It was rendered on a Cray supercomputer in about 1.5 min and presented on a Silicon Graphics workstation. Reproduced from Frenkel (1988) with permission.

FIGURE 8-20. A view of a rat brain reconstructed from serial sections of an atlas. The hippocampus may be seen through the translucent cortical surface. Produced at the UCLA Laboratory of Neuro Imaging and provided courtesy of A. W. Toga.

FIGURE 13-8. Three pseudocolor representations of the image in Fig. 13-3C. In each panel, the gray scale has been represented in the upper part. The color continuum has been applied within the gray region from 128 to 248. Hence, the color images are analogous to the images in Fig. 13-3. Values below 128 were set to the same color as 128, and the values above 248 were set to black. A shows the color continuum in Fig. 13-9A. B shows the color continuum in Fig. 13-9C. C shows the color continuum in Fig. 13-9B.

FIGURE 14-7. Immunocytochemically labeled tissue section. (A) Section through the cat superior colliculus labeled with an antibody to BSA-conjugated γ-aminobutyric acid. The chromagen is diaminobenzidine, which produces a brown reaction product. (B) Pseudo (attributed) color digital image of the section. Colors are assigned to different gray levels. Black–blue, densest immunoreactivity (lowest gray-level values); red–gray, least dense immunoreactivity (highest gray-level values).

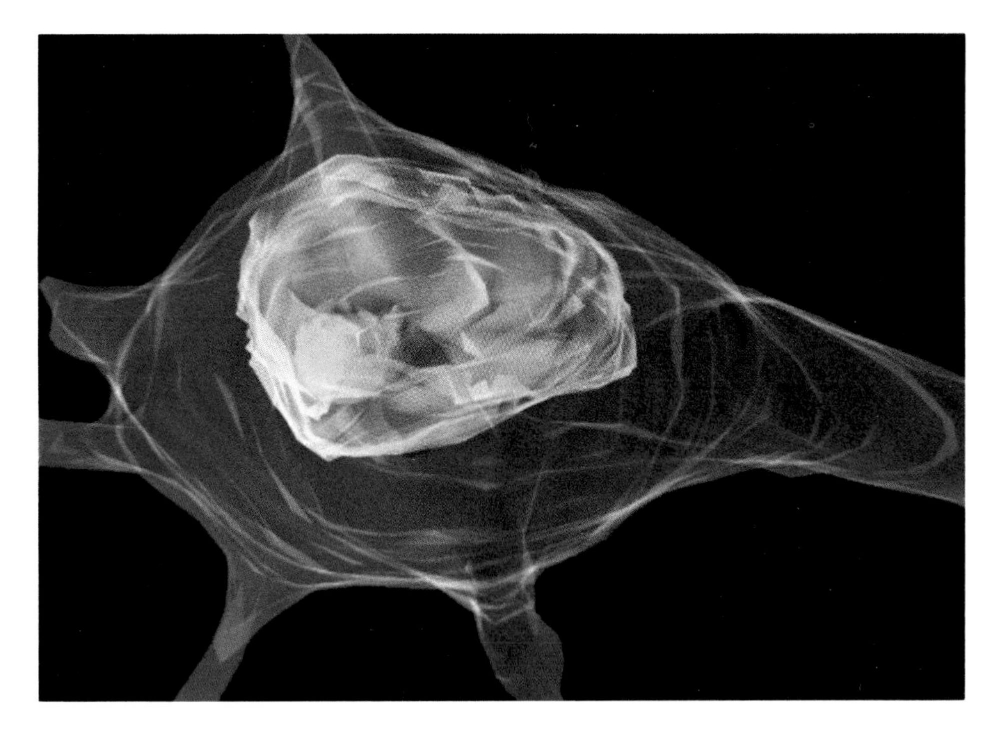

FIGURE 8-1

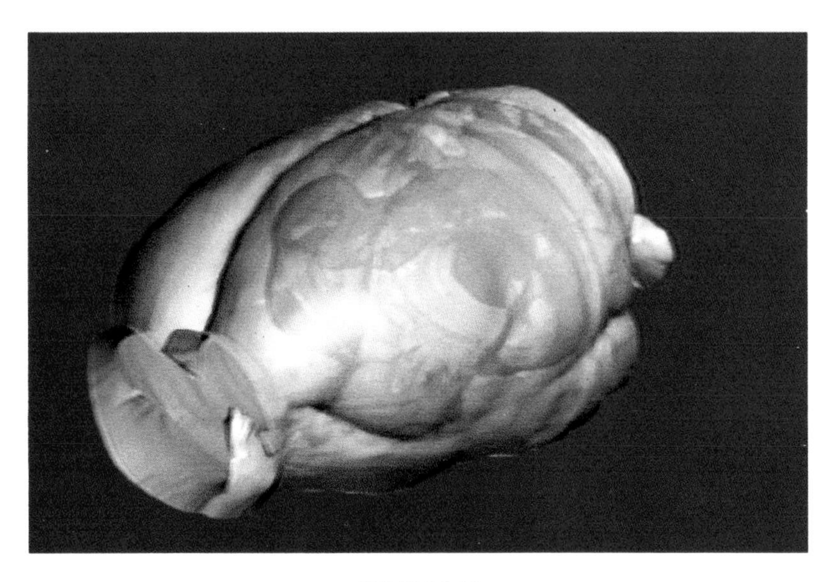

FIGURE 8-20

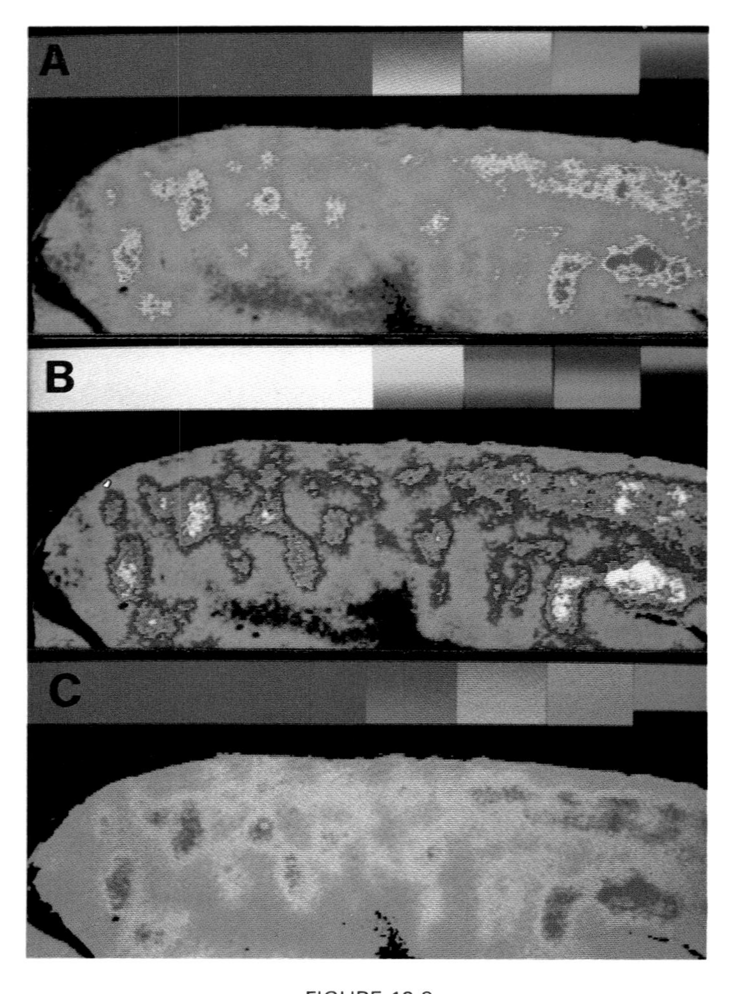

FIGURE 13-8

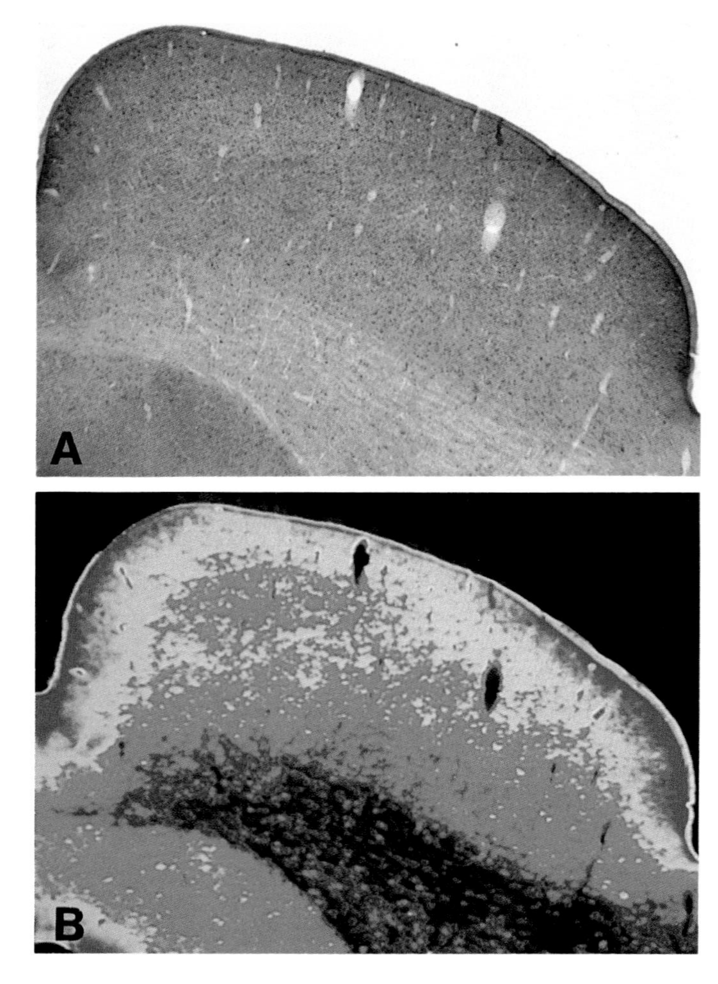

FIGURE 14-7

Glossary

The terms in this book are commonly used in the fields of computers, neuroanatomy, and video image processing, all of which are covered in this book.

This glossary can be either scanned or used as a dictionary. Its main function is to ease you through some of the jargon so that you can talk intelligently with colleagues and others about the subjects. A knowledge of the language is particularly useful when you are dealing with sales representatives.

68000 Series A series of CPU integrated circuits manufactured by the Motorola Corporation. Numbers in the series include 68000, 68010, 68020, and 68030. They are fast and use 32-bit arithmetic. The Apple MacIntosh line of computers uses these CPUs.

8086 Series A series of CPU integrated circuits originally manufactured by the Intel Corporation, consisting of numbers 8086, 80186, 80286, and 80386. These are used in the IBM PC, XT, AT, and PS/2 models as well as their clones.

8087 Series A series of math coprocessor integrated circuits originally manufactured by the Intel Corporation and consisting of the numbers 8087, 80187, 80287, and 80387. See Math Copressor.

ADA A high-level programming language used in projects for the U.S. Department of Defense. Named after Lord Byron's daughter, who was possibly the first computer programmer.

ADC See Analog-to-Digital Converter.

Address A location, usually in a computer memory, where data are stored.

AGC See Automatic Gain Control.

ALGOL Acronym for algorithm, a high-level programming language popular in Europe.

Algorithm A series of step-by-step instructions that are followed to solve a problem. Computer programming consists of two steps: (1) writing the program so that the problem is defined, and (2) coding the program into a series of instructions in a language that the computer can understand. ''Programming'' is used loosely to describe the whole process.

Aliasing An error made when sampling an analog function (such as an electrical waveform or a video image) into a series of digital values when the sampling rate is not high enough. A classic example occurs in drawing a nearly horizontal straight line on a television screen. The line appears as a staircase, or, in slang, it has the ''jaggies.''

ALU See Arithmetic and Logical Unit.

Ambilateral Type See Topological Type.

Analog A range of numeric values or voltages that is continuous, like a volume control on a radio.

Analog-to-Digital Converter (ADC or A/D) A computer card that converts analog voltages into digital numbers. As an example, input voltage range might run from −10 V through +10 V, and these will be converted into digital numbers ranging from −2048 through +2047.

AOI See Area of Interest.

Aperture A little hole. In optics, a hole through which light may pass; used to limit the photons of light entering a photomultiplier tube to only a narrow beam.

Area of Interest (AOI) See Region of Interest.

Arithmetic and Logical Unit (ALU) That part of a digital computer where arithmetic operations, such as "adds" and "subtracts," and logical operations, such as "ands" or "ors," are performed on numeric data.

Arithmetic Operations Mathematical operations such as "add" or "subtract" that are performed inside the ALU. Contrast to Logical Operations.

Array A group of data in a computer program accessed by a collective name and indexed by a number. For example, an array of size 100 refers to elements whose names are $x(1)$ through $x(100)$. An array may have two or more dimensions, e.g., $x(512,512)$, and a particular element may be $x(334,128)$.

ASCII Acronym for "American Standard Code for Information Interchange," a code used to designate numeric values for characters. A maximum of 256 characters are available. This code is used in all personal computers and in all large computers except IBM computers. In contemporary personal computers, ASCII has become a synonym for character.

ASCII File A data file stored on a computer disk consisting of ASCII characters, or text. These can be printed directly onto a printer.

Aspect Ratio The ratio of width to height for a television picture. In the United States, the aspect ratio is 4 to 3; that is, it is $1\frac{1}{3}$ times as wide as it is high.

Assembler A computer program whose function is to translate an assembler language program into machine language.

Assembler Language A low-level computer language in which each instruction, such as an "add" or a "subtract," is translated into one machine-level instruction that the computer can execute.

Assembly The process of translating an assembler language program into machine language.

Asynchronous Adapter See Serial Port.

Asynchronous Communication Synonymous with RS-232 communication. It is serial transmission of data over a single wire without an external device to synchronize the transmitting and receiving devices.

Automatic Gain Control (AGC) An electronic circuit, frequently found in a television camera, that automatically adjusts the contrast level of the video signal to some preset level;

therefore, a darker signal will be amplified more, and a lighter signal, or a whiter signal, will be amplified less.

Average Density (IOD/area) The total amount of darkness (integrated optical density) per unit area in an image.

Axonal Field See Field (of a neuron).

Background Image In video microscopy, a digitized image of the microscope field with no specimen. This image captures the unevenness in illumination and flaws in the optical system; it will be subtracted from later digitized images of specimens to cancel the effects of imperfect illumination.

Background Subtraction The procedure in digital image processing for subtracting a background image, usually with no data, from the captured image to cancel any effects of uneven illumination, dirt, or misalignment in the optical system. It is sometimes used to determine if there is motion in the image by subtracting a previous image.

BASIC (for Beginner's All-Purpose Symbolic Instruction Code) A simple programming language, usually interpreted. Easy to learn, it is used in many fields to write small programs.

Basic Input/Output System (BIOS) That part of the operating system, in an IBM-type computer, for handling the details of input/output. This is software that is burned into read-only memory integrated circuits.

Baud The rate of information sent over a serial line measured in bits per seconds; named after the French engineer, Emile Baudot.

Bifurcation Point See Branch Point.

Binary Having two parts or two states. Computers store and calculate numbers using a binary or base 2 system; i.e., each digit has the value of 0 or 1.

Binary Tree A data storage structure that defines a tree. It consists of points that are origins of trees, branch points, and endings. There must be one origin and one more ending than branch points.

BIOS Acronym for Basic Input/Output System.

Bit A contraction for "binary digit." It is the smallest unit of information in a digital computer and has two possible values, usually designated 0 and 1.

Blooming The tendency of a dot on the screen of a CRT to appear slightly larger than its real size. It is caused by too high intensity of the electron beam.

Blur A smeared image seen on a CRT, caused by selection of a long-persistence phosphor. The display of a moving image will blur, and you will see remnants of old images as the current one moves.

Board A card on which integrated circuits are mounted.

Boolean Operations Logical operations, e.g., "and," "or," performed on binary numbers. Named after the logician Boole.

Boot The process of starting the operating system when the computer is either started or restarted. It is so called because the procedure is analogous to someone who pulls himself up "by his bootstraps." A certain amount of the program is stored in read-only memory, and this program is used to read in the rest of the operating system from a storage device such as a disk.

Bootstrap See Boot.

Branch Angle The angle between the two distal segments of a branch point. It may also be the angle between the distal segment and the imaginary extension of the proximal segment that enters the branch point.

Branch Order A measure of the amount of branching that occurs between the soma and any point on a tree. It is 1 for the primary segment of the branch—the most proximal segment of a dendrite—and increases by 1 with each branch. Contrast Centrifugal Order with Centripetal Order.

Branch Point The point at which a dendrite or axon splits into two smaller branches. A bifurcation is a two-way branch; a trifurcation is a three-way branch.

Branch Segment See Segment.

Bug A problem in computer hardware or software. The name originates with a 1947 incident at Harvard. An insect flew into the Mark II computer and blocked the closing of a mechanical relay, thereby causing a machine failure. Also used as a synonym for integrated circuit.

Burn In (CRT or camera sense) A hole burned in the phosphor coating on the surface of the CRT or camera tube. Caused by the continuous impinging of the electron beam on the same spot in the phosphor, which generates too much heat at that spot.

Burn In (general electronic equipment sense) The process of running a just-manufactured piece of equipment for a period of time (perhaps 24 consecutive hours) to see if a failure will occur.

Bus A collection of wires through which information is sent from component to component in a computer system. Individual cards are plugged into a bus.

Byte A group of 8 binary digits or 8 bits. A byte may have 256 possible values, usually numbered 0 through 255. Memory capacity is usually expressed in a number of bytes.

C A very popular computer language, especially on the IBM personal computer line. Designed by Bell Laboratories, it is a high-level compiled language.

C-mount The standard method for mounting television cameras to other hardware or mounting lenses on television cameras. The camera optical input has a 1-inch-diameter hole that is threaded with 32 threads per inch. C-mount adapters are available for most microscopes.

Camera Tube A vacuum tube in a television camera that converts the image incident onto its face into an electronic or video signal.

Card See Board.

Cathode Ray Tube (CRT) A vacuum tube usually shaped like a bottle. Electrons are emitted from the neck of the bottle and accelerated toward the tube face, which is coated with phosphor. When the electrons strike the front face of the CRT, photons of light are emitted. On television sets and oscilloscopes, the front surface of the CRT forms the viewing screen.

CCD See Charge-Coupled Device.

CCIR A standard color television transmission format used in Europe and in several other countries. The name comes from the French, Comité Consultatif International des Radio Communications.

CCTV See Closed-Circuit Television.

Central Processing Unit (CPU) The part of a computer that fetches instructions from a memory, executes them, and returns the calculated data to memory. Usually this is considered the heart of a computer.

Centrifugal Ordering (of a neuron tree) The assigning of a value or order to each branch segment of a tree, starting from the tree's root. The value for each segment is the number of segments between this segment and the tree's root. Contrast to Centripetal Ordering.

Centripetal Ordering (of a neuron tree) The assigning of a value or order to each branch segment of a tree, starting from the tree's terminals. The value for each more proximal segment is increased according to one of several standard criteria. Contrast to Centrifugal Ordering.

CFF See Critical Flicker Frequency.

CGA See Color Graphics Adapter.

Charge-Coupled Device (CCD) An integrated circuit used in small television cameras instead of a video tube. It consists of an array of approximately 500 by 500 elements on a wafer approximately two-thirds of an inch square. Each element stores an electrical charge proportional to the amount of light incident on it. This charge may then be read from the device to form the electrical video signal.

Chip Synonym for integrated circuit; so named because the electrical components are chemically grown on a small chip of silicon.

Clone A computer designed to operate exactly like one from a different manufacturer.

Closed-Circuit Television A television system consisting of a TV camera and a TV monitor. The TV monitor always presents the view that the television camera sees.

Closed-Loop System A system designed to perform a task in which the input device can sense what happens to the output and can make corrections if necessary. Contrast to an Open-Loop System.

Closing (in image processing) A dilation followed by an erosion, used to eliminate individual points outside a structure.

Closing (of data files) To finish writing data into a file on a disk and to record the existence of the file on the disk directory.

COBOL Acronym for "Common Business-Oriented Language," a programming language commonly used for financial and business applications on all computers.

Coding The process of reducing an algorithm to machine-understandable language.

Color Graphics Adapter (CGA) A computer board and a standard for color graphics display on an IBM personal computer and clones. Its resolution is 320 × 200 pixels, each with four possible colors.

Command A function to be executed by the computer, e.g., copy.

Command Language The rules for using possible commands, e.g., the various commands that DOS can execute.

Compatibility The ability of one piece of hardware or computer software to interface with another. See also Upward Compatibility.

Compiler A computer program that translates a program written in high-level language to machine language to make the program directly executable by the computer hardware.

Composite Video Signal The complete television signal consisting of video information, vertical synchronization pulses, and horizontal synchronization pulses, all composed in one electrical signal on one wire.

Concentric (concerning microscope lenses) Two microscope objective lenses are concentric if, when you switch from one lens to the next, something that is in the center of the field in one lens remains in the center of the field in the second lens.

Contrast (in video) The difference in gray scale or color intensity between two regions in an image.

Convolution A mathematical procedure performed on a digitized image, substituting for each pixel a combination of itself and several of its neighbors. Doing so can enhance an image, for example, make edges show up more clearly.

Core Memory The main or random-access memory of a computer. A holdover from the time when memory was composed of tiny doughnuts of iron called magnetic cores.

CPU See Central Processing Unit.

Crash An unexpected termination of a computer program, resulting in the loss of its memory-stored data and returning to the operating system. So called because of the sound the terminal makes as the frustrated operator throws it to the floor.

Critical Flicker Frequency A frequency, expressed in the number of frames per second or hertz, above which a CRT image will not appear to flicker.

CRT See Cathode Ray Tube.

Cursor An indicator, such as a cross, on a computer screen that is moved over an image to indicate a location of the image. Also, a device that may be moved over the surface of a data tablet to point to, or trace items from, an image that is placed or projected onto the data tablet. This latter device is sometimes called a "hand cursor" or "puck."

Cylinder (in data storage on a disk) The cylindrically shaped region formed by a circular track from several disk surfaces that are stacked one on top of another.

DAC See Digital-to-Analog Converter.

Data Bits The seven or eight discrete bits of information that make up a byte transmitted through a serial port using the RS-232 format.

Data Tablet A device for converting the position of a hand-held cursor to digital values that the computer can sense. The cursor is moved by the user across the flat surface of the tablet.

Data Base A collection of data stored in a computer and manipulated by computer programs.

Data-Base Management Manipulation of data stored in a computer by a computer program. Also, a specific type of program for sophisticated means of accessing and formatting data.

Daughter Angle See Branch Angle.

Daughter Fiber One of the two segments of an axon or dendrite exiting from a branch point.

dB See Decibel.

Debug To search for and remove problems, or bugs, in a computer program or in computer hardware.

Decibel A unit for comparing the magnitude of two signals; it is 20 times the log of the ratio of the signal voltage. For example, if a voltage is reduced to half its original value, it is said to be 6 dB down.

Decimal The base-10 numbering system, using digits 0 to 9. This is not a convenient way for computers to store information in scientific work, but it is the way most humans are used to making calculations.

Deflection The movement of the electron beam in a CRT in the up-and-down and left-and-right directions.

Degree of a Tree (in topology) The number of terminal segments or terminal tips in a tree.

Dendritic Field See Field.

Density (in video images) See Optical Density.

Depth of Focus The distance along the focus axis in a microscope over which the image remains crisp.

Digital A value that can occur only in one of predefined, distinct, states. Digital computers use only these predefined values. Contrast to analog.

Digital-to-Analog Converter (DAC) A computer circuit used to convert a digital number to an analog voltage. For example, a digital number of range −2048 through +2047 will be converted and presented on its outputs as an analog voltage of range −10 V through +10 V.

Digitizer A device for converting some continuous movement or voltage to digital values. This is used in two ways: see Data Tablet and Video Digitizer.

Dilation In image processing, an operation that expands the outlines of an object by adding pixels at its perimeter. It is the opposite of Erosion.

Direct Memory Access (DMA) A method of reading or writing data to memory. A DMA device in a computer can read and write information into the computer memory directly—without requiring the involvement of the CPU. It is frequently used in analog-to-digital conversion to dump a large amount of data into memory very quickly.

Directory A list of the contents of a disk. Also a portion of a disk that can be set aside and accessed by name. This is frequently used with large hard disks to allow the establishment of separate areas for individual purposes or users.

Disk A magnetic or optical storage medium within a computer for storing a large amount of information. See Hard Disk, Floppy Disk, or Optical Disk.

Disk Drive The device that reads and writes data to and from a disk but generally does not include the medium on which the data are stored.

Disk Operating System (DOS) A computer program read from a disk to oversee the operation of a computer. It frequently includes utility programs to perform common tasks, e.g., disk copying. Commonly known by its acronym, DOS, it is used in the IBM microcomputer environment to refer specifically to the operating system developed by the MicroSoft Corporation.

Diskette See Floppy Disk.

DMA See Direct Memory Access.

DOS See Disk Operating System.

Dot-Matrix Printer A type of printer that prints information as individual dots. Characters or graphics are generated by optionally printing the individual dots within a two-dimensional matrix. A common and inexpensive form of printer for small computers.

Double-Sided Disk A disk on which information can be stored magnetically on both sides.

Drum Scanner A device for scanning an image into a computer. The specimen, e.g., a photo, is mounted on a rotating drum, and a light sensor moves parallel to the drum as it rotates, thus covering the entire specimen in a raster pattern. The computer records the light intensity at each point in the raster.

EBCDIC Abbreviation for "Extended Binary-Coded Decimal Information Code." A standard system of encoding characters for computers, most frequently used in large IBM-manufactured computers. It is an alternative to the ASCII coding system.

Edge Detection In digital image processing, a process to record the boundaries of a digitized image by some computer algorithm.

Edge Enhancement In digital image processing, a computer algorithm for rendering the edges of a previously digitized structure more visible.

Editor A computer program designed to change data in some way. For example, text editors are used to change ASCII files.

EGA See Enhanced Graphics Adapter.

EIA Abbreviation for "Electronic Industries Association," a U.S. committee for setting electronic equipment standards.

Electron Gun The part of a CRT that shoots electrons out of the neck of the glass bottle and aims them toward the front face of the tube.

End User A person who uses a computer; specifically, the person for whom a computer is eventually acquired for performing some task.

Ending In neurons, any of the points that represent the end of a dendrite or axon.

Enhanced Graphics Adapter (EGA) A computer board and a standard for color graphics display on an IBM personal computer and clones. Its resolution is 640×350 pixels, each with 16 possible colors.

EPROM Acronym (often pronounced ee-prom) for "Erasable Programmable Read-Only Memory," an integrated circuit (IC) that stores numeric values even when power is off. EPROMs can be erased by exposure to strong ultraviolet light.

Erosion In digital image processing, an operation to shrink the boundary of an object by eliminating pixels at its perimeter. It is the opposite of dilation.

Expansion Slot An area in a computer into which a board may be plugged to expand the capability of a computer.

Extended Binary-Coded Decimal Information Code See EBCDIC.

Felt-Tip Pen Plotter A computer output device that draws on a sheet of paper with a felt-tip pen by moving the pen around under the control of the computer.

Field (in television) One of the two scans of the electron beam across the face of a CRT. The first field, called the primary field, starts in the upper left corner of the picture. In U.S. television, it contains $262\frac{1}{2}$ lines, finishes in the bottom center of the field, and is drawn in 16.7 msec. The second field, called the interlace field, starts in the top center of the picture, also contains $262\frac{1}{2}$ lines, finishes in the bottom right of the picture, and also requires 16.7 msec. The lines of the interlace field are drawn between those of the primary field.

Field (of a neuron) The region of tissue into which the neuron sends its dendrites or axons.

File A collection of data stored on a disk and accessed by name; it may be textual information, a program, or numeric data.

File Extension The last part of a file name, most often used to declare what type of file it is, and usually preceded by a ".'' For example, the file name "joe.txt" describes a file named "joe" that contains text.

Firmware Computer software that has been burned into a read-only memory and exists permanently within a computer or a computerlike device.

Flatbed Scanner A computer input device for scanning a specimen. The specimen is laid on a flat surface, and a light sensor is passed, either electrically or mechanically, over the specimen. The computer records the light intensity at each point in the specimen.

Flexible Diskette See Floppy Disk.

Flicker A phenomenon that occurs on a CRT screen when an image is not drawn frequently enough for the viewer to think that he is seeing a continuous image; the image appears to blink on and off.

Floppy Disk A disk made of thin mylar and coated with iron oxide and enclosed in a paper or plastic envelope. It is placed in a computer's floppy disk drive that rotates it inside its envelope. The computer may read or write data from or to the disk. This is the most common storage medium currently used on personal computers. Also called Flexible Diskette.

Format Generally, the organization of any data. Specifically applied to computer disks, the writing of basic timing and spatial information on a disk so that computer programs know where to write their information onto the disk. A disk must be formatted before it is used.

FORTRAN Acronym for "FORmula TRANslation," a very popular scientific programming language in use on many machines. It is a high-level language and is generally translated into machine language by a compiler.

Fourier Transform A mathematical transformation of recurring events in a waveform or in an image into the frequency domain, i.e., description of these events by sinusoidal components of different frequencies. The components may then be separately filtered, and an inverse transformation may be applied to generate the original events or images with selected frequencies enhanced or reduced.

Frame A complete video picture. In U.S. television, a frame is composed of 525 lines on a TV screen that take 33 msec to draw.

Frame Buffer A computer memory for storing an image in a pixel-by-pixel format of, for example, 512 × 512 pixels, each pixel having eight bits or one of 256 possible gray values or brightness levels.

Frame Grabber A device for capturing a video image and storing it in a digital frame buffer memory in one frame time (33 msec).

Gamma (γ) The slope of the light transfer curve of a video device. For a camera, an expression of the relationship of the output video voltage from the camera to the input light. A γ of 1 means that the camera is linear; i.e., for twice as much light you get twice as much voltage output.

Ghost See Blur.

GPIB Abbreviation for "General Purpose Instrument Bus." See IEEE-488.

Grab To capture a frame or a television image in a frame buffer memory, usually in one frame time or 33 msec.

Gray Level The brightness of pixels in a digitized video image. This usually ranges from 0 through 255, with 0 being the blackest value and 255 the whitest.

Gray-Level Histogram A commonly used graph that displays the numbers of pixels in an image containing each gray level.

Handler A piece of software used to pass data to or from a device.

Hand Cursor See Puck.

Hard Disk An inflexible rotating disk with iron oxide coating in a computer for magnetically storing large amounts of information. It cannot, as a general rule, be removed from the computer. Generally uses a recording procedure called "Winchester technology" and may be called a Winchester disk.

Hardware In the computer field, the electrical and mechanical components of the computer. Contrast with Software or Firmware.

HDTV See High-Density Television.

Hercules Monochrome Graphics Adapter (MGA) A computer board and a standard for monochrome text and graphics display on the IBM PC line of computers. It was defined and first manufactured by the Hercules Corporation. Its resolution is 720 × 348 pixels, each either on or off.

Hexadecimal A base-16 numbering system using the digits 0–9 and A through F, used most frequently to express the numbers and addresses in IBM computers.

High-Density Television (HDTV) A newer television standard using approximately 1025 lines to draw a picture. This results in a higher-resolution image than current consumer television. Picture and video waveform standards are still in embryonic form.

High-Level Language A computer language in which each statement translates into several machine language instructions.

Horizontal Scan Line In a television image, one horizontal movement of the electron beam across the face of the CRT. This requires 63.5 μsec in U.S. television.

Horizontal Sync Pulse A pulse that tells the television monitor or television camera to start a horizontal scan of the electron beam.

Hot Spot A bright region in an image that is usually caused by uneven illumination of the specimen.

HPIB Abbreviation for "Hewlett–Packard Instrument Bus." See IEEE-488.

Hue The wavelengths of light that cause the sensation of color.

IC See Integrated Circuit.

Icon A pictoral or nontextual character, e.g., a happy face, used in computer graphics.

IEEE See Institute of Electrical and Electronic Engineers.

IEEE-488 A standard computer bus for connecting instruments. It contains eight data lines, five address lines, and several miscellaneous lines. Designed originally by the Hewlett–Packard Corporation and called the Hewlett–Packard Instrument Bus (HPIB); later renamed General-Purpose Instrument Bus (GPIB).

Image An electronically reproducible scene.

Image Analysis The process of extracting numeric information from an image by computer.

Image Averaging The process of averaging, on a pixel-by-pixel basis, several images together to reduce noise or to increase the signal-to-noise ratio.

Image Dissector A television cameralike device to convert light levels into voltage levels, except that its electron beam does not sweep a fixed raster; rather, it moves to a position on the screen controlled by a computer. Sometimes called a vidissector.

Image Enhancement A procedure for changing a digitized image to improve the picture.

Image Histogram See Gray-Level Histogram.

Image Processing Applying a computer algorithm to an image to change it in some way.

Inflection Point (in neurons) A place where a dendrite or axon bends. This is more general than the analytical geometry definition referring to the point at which a curve changes from concave to convex or where its second derivative changes sign.

Institute of Electrical and Electronic Engineers (IEEE) The U.S. association for electrical engineering. Among other things, it establishes industrial standards for electrical equipment.

Integrated Circuit (IC) The smallest manufactured portion of a computer, consisting of transistors that have been grown onto a silicon chip. An integrated circuit may consist of 5000 transistors that may store information and perform logical functions on that information. Sometimes called Chip or Bug.

Integrated Optical Density (IOD) The sum of the gray levels of pixels in a region of a digitized image, representing the total amount of "darkness" in a region.

Interactive An adjective describing a computer system that allows the user to participate in its actions.

Interlace Field (in television) See Field.

Intermediate Segment A portion of a dendritic or axonal tree between two consecutive branch points.

Interpreter A computer program that translates and executes a program written in a high-level language one statement at a time.

Interrupt A hardware construction and method of programming in which an infrequently used device signals the computer that, for a brief period of time, it would like the computer to run a program to service it. For example, when you press a key on a computer keyboard, you interrupt the computer from the program it is running; the computer executes a program that determines which key you have typed and places the value of that key in a location in memory.

IOD See Integrated Optical Density.

Jaggies A slang term for the staircaselike effect created when a nearly horizontal line is drawn on a raster CRT.

Joystick A vertical stick in a ball-and-socket joint that a user can deflect in the *X* and *Y* directions. In a three-dimensional joystick, the user can also twist it along its axis, and the computer can sense the deflection as well as the twisting of the stick.

Kbytes An expression of memory storage. K means kilo or 1024. A typical memory size might be 512 kbytes, i.e., 512 times 1024 bytes.

Keyboard The peripheral device on a computer on which the user types, as on a typewriter keyboard.

Kilobytes See Kbytes.

Laser Printer A type of printer using a laser and a xerography process to produce its output. The computer inside the printer directs a laser beam over a heated drum to write the pattern of the characters or graphics. Powdered ink is then spread over the drum and adheres to the written areas. Paper rolls over the drum and receives the image. Laser printers are more expensive than dot-matrix printers, but they produce work of excellent quality.

LED See Light-Emitting Diode.

Letter-Quality Printer A type of printer that prints each character by a separate metallic instrument. Contrast this to a Dot-Matrix Printer that uses a two-dimensional matrix of dots. Although output quality is better, letter quality printers are generally slower and more expensive than dot-matrix printers.

Light Pen A penlike device with which a user can point to a part of a display on a CRT. Inside the pen is a light sensor whose output can be read by the computer. As the CRT's electron beam passes under the sensor, the computer is interrupted. Since the computer knows which part of the image is being drawn at that instant, it knows to what part of the display the user is pointing.

Light-Emitting Diode (LED) A small solid-state device that converts a low voltage into a light. It is frequently used to indicate when power is "on" in an electronic device.

Logical Operations Mathematical operations such as "and" or "or" that are performed inside the ALU. Contrast to Arithmetic Operations.

Look-up Table (LUT) A special memory in an image processor. Each pixel gray value is digitized from an image and is used as an address into this memory. This memory contains other gray levels that are placed in the frame buffer. This is called an "input look-up table." An "output look-up table" takes the gray-level values from the frame buffer, uses them as an address in a memory, the look-up table, and sends that element of the look-up table's contents to the television monitor. Look-up tables can be filled in order to change contrast, to threshold a picture, to apply pseudocolor, and to do other image-processing applications.

Low-Level Language A computer language that translates closely to machine language instructions.

LUT See Look-up Table.

Machine Language The actual program that a computer can execute, consisting of a series of numbers that specify the instructions to be executed.

Macro A group of commands issued to perform a sophisticated function repeatedly. For example, in text editing, a macro may be defined to find all of the instances of the word "excellent" and change them to read "superb." Also called a Script.

Main Memory See Random-Access Memory.

Mainframe A large computer, frequently shared by many users who work from terminals at some distance from the mainframe.

Math Coprocessor A special-purpose integrated circuit optionally included in a personal computer to perform floating-point arithmetic quickly and to assist in the fast calculation of transcendental mathematical functions such as sines and cosines.

Mbytes Abbreviation for millions of bytes (M standing for mega). It is used to express the capacity for memory storage, either on disk or in the integrated circuit memory.

MDA See Monochrome Display Adapter.

Megabytes See Mbytes.

Memory Capacity The amount of room available for storing information in a memory or on a disk, usually measured in thousands or millions of bytes.

Menu A list of command possibilities presented on the computer terminal. The user may choose from them as he would choose items from a restaurant menu.

MGA See Hercules Monochrome Graphics Adapter.

Microcomputer A very small computer, costing a few thousand dollars or less. Frequently called a personal computer.

Microprocessor A very small computing device tailored to perform a certain task, frequently packaged in one integrated circuit. Often a microprocessor forms the CPU of a microcomputer.

Minicomputer A computer costing between $5000 and $15,000 with a large memory capacity and a faster speed than a microcomputer.

Modem Acronym for Modulator–Demodulator, a computer device that converts numbers into acoustic sounds sent over telephone lines—this is the modulation function—and to convert acoustic sounds back into data for the computer—this is the demodulation function. A modem enables a computer to pass information over a telephone line.

Modulator–Demodulator See Modem.

Module In software, a unit of a program that performs a specific task. If it performs a task initiated by another program, it is a synonym for Subroutine. In hardware, see Board.

Monitor A televisionlike device that converts a video signal, e.g., from a camera, into a television picture. It does not include radio-frequency decoding circuitry to enable it to receive a standard television broadcast station.

Monochrome Display Adapter (MDA) A computer board and a standard for a monochrome text display on an IBM personal computer and clones. Its resolution is 720 × 350 pixels, forming 25 lines of 80 characters each.

Mother Board In a small computer, the main board on which integrated circuits are mounted, frequently having connectors for attaching other boards.

Mother Fiber A portion of a dendrite that enters a branch point from the proximal direction.

Motor Controller An electronic device to interface a computer with motors. It converts computer commands to move the motors into the form necessary to drive the motors.

Mouse A small interactive device that may be rolled across a flat surface on a pair of perpendicular wheels. The computer senses the wheel rotation and thus the X,Y motion of the mouse. It is frequently used to move a cursor around a screen.

Multiplex To share a scarce resource. For example, the lines on a bus may be time-multiplexed so that sometimes they are carrying addresses and other times they are carrying data.

Multisync Monitor A television monitor that can be driven by several standard forms of video signal.

NA See Numeric Aperture.

Nonvolatile Memory A computer memory that does not lose its contents when the power is turned off. For example, it may be powered by a battery during the times when the computer is not turned on. A disk is also non-volatile, for it retains its contents when the computer is off.

Noninterlace A format in television in which the entire face of the television is scanned in one pass. It is not used in standard American television; however, some contemporary graphics systems use a noninterlace raster. Also called a one-to-one interlace.

NTSC Abbreviation for "National Television Systems Committee," a U.S. broadcast engineering group. The initials refer to the standard 525-line, 60-field-per-second color video adopted in the United States in 1953.

Numeric Aperture In a microscope, the half angle of the cone of light accepted by the objective lens times the refractive index of the medium between the specimen and the lens. The resolution of a microscope is proportional to the numeric aperture of the objective lens and the numeric aperture of the condenser lens.

Nyquist Frequency The sampling frequency necessary to capture a given frequency component of an analog signal. Analog signals may contain many frequency components; when sampled only those components less than half the sampling frequency will be recorded.

Octal A base-8 numbering system, using the digits 0 through 7. This system is used often to express the data and addresses calculated by the Digital Equipment Corporation PDP-11 microcomputers.

OEM See Original Equipment Manufacturer.

Open-Loop System A group of components designed to perform a task without feedback. The controlling part of the system assumes that the system's output follows its wishes reliably. For example, if you use a computer to drive a stepping motor and the computer says, "move to a certain position," the computer has no way of knowing whether the motor actually worked. The integrity of the system is based solely on the perfect functioning of each component.

Opening (in image processing) The process of erosion followed by the process of dilation. Used to include individual pixels in a structure.

Opening (of a data file) The procedure of establishing a file on a disk for writing of data to the file or for locating a file for reading data from the file.

Operand A piece of data on which an operation will be performed. For example, if you add 2 and 3 to obtain 5, 2 and 3 are the operands.

Operating System See Disk Operating System.

Optical Density The gray value or darkness of a pixel of a digitized image.

Optical Disk A rotating computer disk on which data are written using a laser and read with an optical means.

Optical Sectioning Using a high-numeric-aperture microscope objective to achieve a shallow depth of field for looking at objects in a short range of focus. This is contrasted to tissue sectioning in which the actual tissue is cut into very narrow slices.

Order See Branch Order.

Origin See either Slide Origin or Tree Origin.

Original Equipment Manufacturer (OEM) A manufacturer makes a product, e.g., a disk drive, and sells it to another manufacturer who incorporates it in his product, e.g., a computer, and sells it to the end user. The OEM is the second manufacturer.

Orthogonality (relating to human–computer interaction) A design criterion for human–computer interaction in which changing one value does not affect any other value. For example, in controlling an automobile, if one "steps on the gas," the car's steering is not affected. This means that the speed control and the steering control are orthogonal.

Orthogonality (relating to CRTs and TV monitors) A control knob on a TV monitor or camera to adjust the perpendicularity of the X and Y deflections.

PAL Acronym for Phase Alternating Line System, a color TV standard format used in some countries in Europe and the rest of the world, excluding North America.

Palette In a computer graphics display, the total number of colors that the device may present, though usually not at one time.

Pan On a CRT system, to move an image left or right to present a different portion of the image. Contrast to scroll and zoom.

Pantograph The instrument formed by linking a microscope and an analog X,Y plotter. When the user moves the stage, the plotter senses the stage position via potentiometers and drives the plotter pen to the stage position. Neurons and other structures are drawn on paper in this fashion.

Parallel Port An electrical connector and its associated electronics out of which computer information can be sent over many wires simultaneously in parallel. Sometimes referred to as a Printer Port when it is in the standard form and used to drive a printer.

Parameter A numeric value stored in a program.

Parent Fiber See Mother Fiber.

Parfocal (as relates to microscope objective lenses) An adjective describing the position of two or more lenses when they are inserted into the viewing path of an item and the same item remains in focus.

Parity An additional bit sent to indicate the number of data bits transmitted over a serial port. Parity may be odd or even; when it is odd, an extra bit is sent to make the total number of bits sent odd. If parity is even, then an extra bit is sent to make the total number of bits sent even.

PASCAL A high-level compiled computer language used in the scientific field. PASCAL is named for the French mathematician Blaise Pascal.

PEL See Picture Element.

Peripherals Short term used for peripheral devices (e.g., disk drive, printer, plotter) that are attached to a computer.

Personal Computer (PC) A small computer designed to be used by one person.

PC See Personal Computer.

Phosphor A chemical compound spread on the inside of the front surface of a CRT. When struck by electrons, the phosphor emits photons of light. Many different compounds are now used, some not containing the element phosphorus at all, yet the name is still used.

Phosphor Persistence The amount of time that a spot of phosphor will emit photons after it is struck by an electron beam.

Photomultiplier A device for multiplying photons, thereby increasing the amplitude of an image. It has two forms. In its image intensifier configuration, it is an electronic device on whose cathode a dim image is focused. Electrons are emitted from the cathode toward a phosphor screen on which an amplified form of the original image is presented. In its counting configuration, the electrons are counted rather than displayed, so that a computer can sense the number of photons or the light intensity incident on it.

Picture Element One element of a picture that is stored in digital form; the term Pixel is used by most computer manufacturers except IBM, which uses PEL.

PIO See Programmed Input/Output.

Pipeline A series of computer processors that act on data, the output of one processor feeding the input of the next. The first processor operates on one piece of data; the next one operates on a previous piece of data; the third one operates on data that are two previous, etc., so that all the processors are working at once even though they are working on data that are serial.

Pixel See Picture Element.

PL/1 A high-level, general-purpose compiled language used primarily on large IBM computers.

Polling A method of computer programming in which the computer program asks each device, "Do you have any information for me?" Contrast to Interrupt.

Pop-up Menu A menu that is presented on a CRT at the position of the cursor when the program's user requests it by pressing a button or a key.

Port A connector through which the computer can make electrical connection to other devices.

Potentiometer A resistor whose resistance is variable, usually controlled by a knob. A computer may sense its value; thus, the potentiometer may be used interactively to control a program.

Primary Field (in television) See Field.

Principal Axis of a Tree The axis for which the sum of the squared distances of all the tree points is minimal, i.e., the axis that is perpendicular to the plane of the tree.

Principal Plane of a Tree For a reasonably flat neuron, it is that plane that contains the neuron.

Printer A peripheral device attached to a computer for making paper, or "hardcopy," output of information, text or graphics.

Printer Port See Parallel Port.

Program A series of instructions that the computer executes to perform a task.

Program System An interrelated group of computer programs to do a task.

Programmable Read-Only Memory (PROM) An integrated circuit (IC) containing data that remain even after power is turned off. The data may be burned into the read-only memory by the user with a device called a ROM programmer.

Programmed Input/Output (PIO) A method of sending information from the computer to some device by having the computer's central processing unit (CPU) read the data one piece at a time from the computer memory and send it to the device. Contrast with Direct Memory Access.

Programmer A person who writes computer programs.

Programming The process by which a person reduces a complicated problem into a series of step-by-step instructions, i.e., the generation of an algorithm. The term is loosely used to refer to the whole process of generating and coding the algorithm for the computer.

PROM See Programmable Read-Only Memory.

Pseudocolor In digital image processing, the process of assigning colors for output onto a color television monitor according to gray levels stored in the frame buffer. This may be used to emphasize small differences in gray levels.

Puck The device that the user slides over the surface of a data tablet. Sometimes called a "Hand Cursor" or "Cursor."

Pull-Down Menu A menu presented on a CRT when a program's user moves a cursor to the top of the screen and presses a button or a key. The menu appears to be pulled down from the screen top like a window shade.

RAM See Random-Access Memory.

RAM Disk An array of random-access memory ICs, usually mounted on a single board, designed to appear to the computer program as though it were a disk drive. Also refers to using any random-access memory to store data in disk-stored format.

Random-Access Memory (RAM) A computer memory for storing information in which each element of the memory is equally accessible. Also refers to the Main Memory of a computer.

Raster A sequential scanning pattern used in video. It is the pattern that the electron beam sweeps over the front face of the CRT, independent of whatever information is displayed on the CRT.

RBG/TTL Monitor A television monitor that can be driven either by red, green, and blue analog signals or by computer digital values.

Read-Only Memory An IC that stores preprogrammed information. The information remains in the memory even when the power is turned off. Generally, one cannot write information into a read-only memory; it is preprogrammed at the factory.

Real Time A computer process that is performed in its normal time course, without any computer-caused delay. For example, a video frame will be grabbed in its actual frame time—33 msec.

Red, Green, Blue (RGB) A color video standard in which the red, green, and blue signals that modulate the electron beams generating the red, green, and blue light intensities are sent down separate wires to a television monitor. The resolution is higher than in an encoded signal such as NTSC, in which the red, green, and blue signals and synchronization pulses are encoded in a single waveform.

Refresh To redraw an image continually on a CRT. With each drawing, the image persists for a few milliseconds. If the image is refreshed often enough (usually about 35 times per second), the viewer will interpret it as a continuous image.

Region of Interest (ROI) In image processing, a region of an image that a program will manipulate. In simple image-processing systems, the region may be rectangular; in more complex ones, the region may be irregular, as traced by the user.

RGB See Red, Green, Blue.

RGB Input The connectors on a color television monitor for receiving the red, green, and blue analog video signals.

Robust A computer program that gives predictable results for all possible data inputs, that is, one that will not "crash" even when it is used improperly.

ROI See Region of Interest.

ROM See Read-Only Memory.

Root Directory The lowest-level directory or section of a disk, from which a computer will boot the operating system.

Root Segment The most proximal or initial segment of a dendrite; the part of the tree between the soma and the first branch point.

RS Abbreviation for Recommended Standards, a prefix for various electronic standards established by the Electronics Industries Association.

RS-170 A standard U.S. format (525 lines, 60 Hz per field, two-to-one interlace) for transmitting monochrome video in a closed-circuit TV system.

RS-232C A standard for serial communications between computers. It involves transmitting individual bits over one wire through a serial port.

RS-330 A standard format for transmitting monochrome video information in a closed-circuit television system. It has slight detailed improvements over the RS-170 format.

S/N See Signal-to-Noise Ratio; also Serial Number.

Scan Line The movement of an electron beam horizontally across the face of a CRT. In U.S. television, the horizontal scan takes 63.5 μsec.

Scanning Stage A flat, movable platform, frequently used on a microscope, and driven by DC motors. It may be moved at high speeds to scan large areas of a specimen. Contrast to Stepping Stage.

Script See Macro.

Scroll On a CRT screen, to move an image up or down to present a different portion of the image. Contrast to Pan and Zoom.

SECAM Acronym for "Système Electronique Couleur Avec Mémoire." A color TV standard format used in some countries in Europe and the rest of the world, excluding North America.

Sector In storage of data on a disk, a part of a circular track, i.e., an arc.

Segment A portion of a dendrite between branch points, from the soma to the most proximal branch point, or from a branch point to an ending.

Segmental Growth Model A mathematical model expressing the growth of a tree in which all segments have an equal probability to branch. Contrast to Terminal Growth Model.

Serial Number A unique number defining a piece of hardware or software.

Serial Port A computer card containing a connector through which computer information in RS-232 format may be sent or received. The data are sent on one line in serial format. Contrast with Parallel Port.

Signal-to-Noise Ratio The ratio of the amplitude of the signal to the amplitude of unwanted noise, usually expressed in decibels. The higher the signal-to-noise ratio, the better the quality of the signal.

Single-Sided Disk A disk on which information can be stored magnetically on only one side.

Slide Origin The 0,0,0 coordinate in a piece of tissue from which all distances are measured.

Slot See Expansion Slot.

Software A synonym for a computer program. It is the instructions that one writes to solve a problem.

Stepping Motor An electric motor that rotates in discrete steps every time an appropriate pulse is received. It is easily driven by a computer, allowing the computer precise control over the motor's position.

Stepping Stage A flat, movable platform, frequently used on a microscope, and driven by stepping motors. It may be driven by a computer with low to modest speed at high precision. Contrast to Scanning Stage.

Stop Bits One or two bits sent at the end of each byte to signify that this is the end of data for a byte. Used when transmitting data in RS-232 format from a serial port.

Stylus A device, similar to a ball-point pen, that the user moves around the surface of a data tablet. One way a stylus might be used would be to trace outlines on a photograph placed on the data tablet.

Submenu A menu presented on a CRT, selected by the program's user from a previous menu.

Subprogram A computer program to perform some task that is initiated by a different program module.

Subroutine Synonym for Subprogram.

Swim A phenomenon in vector graphics in which the entire picture appears to move slowly and slightly up and down. This is caused by electromagnetic interference (usually 60 Hz) beating against the refresh frequency rate of the image (perhaps 57 Hz), resulting in a 3 Hz movement, or "swimming," of the image.

Sync See Synchronization.

Synchronization The assurance that two or more things happen in proper relative time one to another. In video, synchronization pulses make sure that the electron beam of the monitor moves in a position and time corresponding to the electron beam in the television camera. Thus, the light intensity from one place on the television camera is copied to the same place on the television monitor.

System A group of hardware and software components to do a task. Loosely used to mean any combination of items.

Terminal A televisionlike display device for seeing computer output.

Terminal Growth Model A mathematical model describing the growth of a neuron tree in which only the end (terminal) segments have an equal probability to branch. Contrast with Segmental Growth Model.

Terminal Point See Ending.

Terminal Segment The portion of a dendritic tree between a branch point and an ending.

Threshold A certain value over which something happens. In digital image processing, a threshold is usually specified as a gray level below which all pixels are set to black and above which all pixels are set to white. This allows one to distinguish between objects of interest and background.

Tissue Origin See Slide Origin.

Toolbox A package of subroutines for performing standard functions in a field, such as image processing, on a particular piece of computer hardware. With a toolbox, writing a program to do a task is much easier, because the programmer does not have to be concerned with the lowest-level details of the task.

Topological Diameter (of a tree) The path length (in number of segments) from the tree's root to its most distant terminal tip.

Topological Path Length (of a terminal or branch point in a tree) The number of segments between that point and the root of the tree.

Topological Size (of a neuron tree) The number of segments of a tree.

Topological Tree Types (of a neuron tree) The finite number of different patterns that a binary tree may take from a given number of terminal tips and branch points.

Topology The branch of mathematics concerned with the schematic shape of structures, regardless of their physical size. For example, both a coffee cup and a donut are structures with a surface and one hole.

Track On a floppy disk, one circular line of data extending around the surface of the disk.

Trackball A computer input device consisting of a ball partially enclosed in a cabinet. The user can rotate the ball around any axis, and the computer will sense the motion.

Transistor–Transistor Logic (TTL) The standard used in contemporary computers for representing binary numbers by voltages. The digit 0 is represented by a range of 0.0 through 0.8 V, and the digit 1 is represented by a voltage range of 2.4 through 5.0 V. Named for the pair of transistors that amplify the signal to these voltages.

Transmittance The amount of light passing through an optical system. Usually this is the opposite of optical density.

Tree Origin The point that is the beginning of a tree, possibly a dendrite or a piece of a dendrite.

Trifurcation Point See Branch Point.

TTL See Transistor–Transistor Logic.

TTL Input The connector on the back of a TV monitor for accepting computer-generated TTL pulses that drive the monitor.

Turnkey System A system, composed of a computer, peripheral devices, software, and documentation, that is ready to use without any programming. The term stems from automobiles that are completely ready for someone to drive—all the buyer has to do is turn the key to start it.

Underscan In a TV monitor, the process of reducing the raster size so that the entire raster, including its edges, may be seen on the screen. It is useful for adjusting the monitor accurately.

Upward Compatibility The characteristic in computer software that data written by one version of a program can be used by a later version. In computer hardware, older devices may be used with newer ones.

User A person who uses a computer. See End User.

User Friendly An adjective that applies to computer programs and systems, meaning easy to use and easy to learn.

User Module In a system of programs, a program to do a specific task that is written by the end user, not by the company that provided the majority of the software.

Variable A named mathematical value stored in a computer program, accessed by its name.

Vertical Sync Pulse The pulse in a video signal that instructs the monitor to begin a vertical sweep of a field.

VGA See Video Graphics Array.

Vidissector See Image Dissector.

Video The picture portion of a television signal, contrasted to the synchronization and audio portions. It carries only the gray-level information. Sometimes the term is used loosely to mean the entire signal.

Video Digitizer An electronic device for capturing a video image and storing it in a memory composed of picture elements, where each element has a gray level.

Video Graphics Array (VGA) A computer board and a standard for color graphics display on the IBM Personal System 2 computer. Its resolution is 640 × 480 pixels, each with 16 possible colors.

Video Microscopy The use of video in conjunction with a microscope to televise a microscope image, possibly to enhance it or to analyze it.

VME Bus Abbreviation for ''Versabus Module European.'' A standard computer bus structure defined by the Motorola Corporation and frequently used to connect 68000-series computer components together.

Volatile Memory A computer memory that loses its contents if the power is turned off. Normal random-access memories are volatile.

Winchester Disk See Hard Disk.

Yoke An assembly of coils placed over the neck of a CRT bottle that either focuses the electron beam into a sharp point when it hits the front face of the screen or deflects the electron beam to sweep it to a certain position on the face of the screen.

Zoom On a CRT screen, to magnify an image to see a small portion of the image with greater detail. Contrast to Pan and Scroll.

1

The History of Quantitative Neuroanatomy

1.1. INTRODUCTION

For several centuries neuroanatomists have been looking at tissue in the light microscope, describing what they see in words, and drawing pictures of the structures that they see contained in the tissue. Microscope equipment and viewing techniques have developed over the decades, as have histological procedures for preparing the tissue for the researcher's examination. Independently, the computer field has been blossoming, but at a much more rapid pace. The pressures of World War II caused the first digital computers to be developed for practical use. Through the 1950s and 1960s, large computers were developed for military, banking, and census work. These machines were far too expensive and cantankerous for a biological research laboratory. In the late 1960s, however, the laboratory computer began to appear at a price within the range of well-funded research laboratories. At this time, the two fields, computers and biological research, began to merge. And in the last 20 years, the laboratory computer has been applied to almost every laboratory task. This book describes the computer's application to neuroanatomy and its role in collecting and analyzing the shapes and functions of anatomic structures.

In this chapter we present the historical perspective for the marriage of these two technologies, describing the precomputer methods of neuroanatomical data collection and analysis and indicating the dire need for the data-processing capabilities that the computer may provide.

1.2. EARLY HISTORY OF DRAWING NEURONS

Santiago Ramon y Cajal, possibly the most famous neuronanatomist of all time, is shown in Fig. 1-1, drawing neurons at his microscope. As the figure shows, without using any drawing aid, he had to look repeatedly from the microscope to the paper to draw a picture of a neuron. Figure 1-2 shows an example of one of his hand drawings. Many other drawings appear in his classic textbooks (e.g., Cajal, 1906, 1909).

Cajal, as did subsequent researchers, faced two major disadvantages when trying to

FIGURE 1-1. Ramon y Cajal drawing by hand at the microscope. Reproduced from Cajal (1984) with permission.

describe neuronal material he saw in the microscope. First, the optical system of the microscope seriously handicapped him. Second, when he viewed the image in the microscope, and even if he drew it on paper, he still had only an image of the neuron. To obtain any sort of data from the image, he had to resort to many additional, almost heroic, measures. Furthermore, because data were collected from a hand-drawn facsimile of the tissue sample, margin for error was high.

Throughout the later chapters of this book, other authors and I attempt to explain how

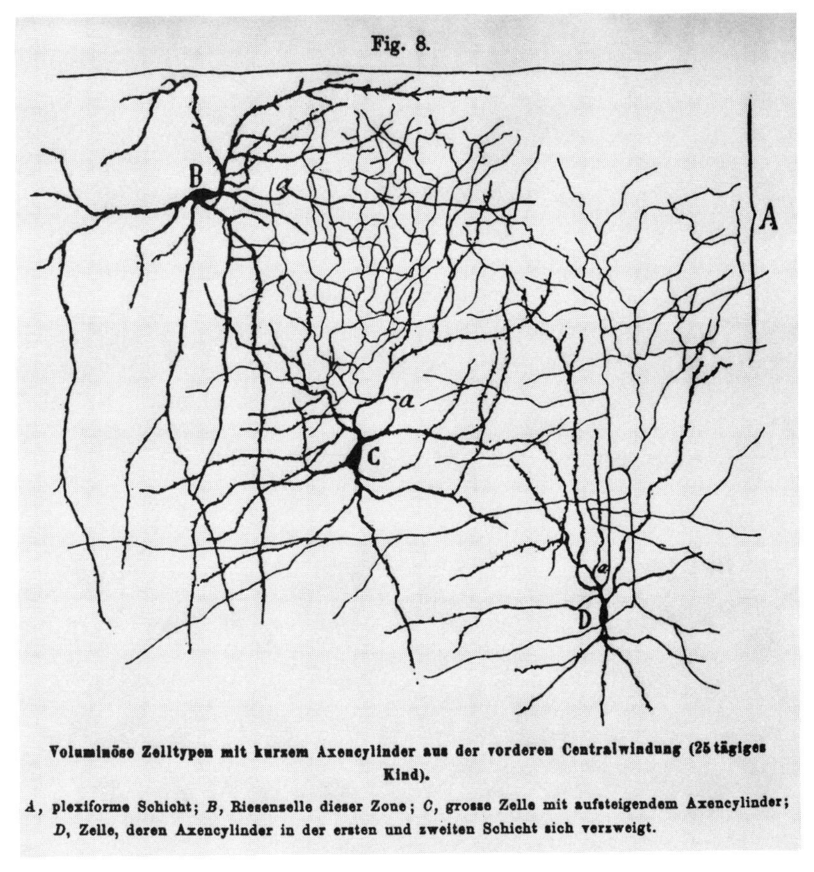

FIGURE 1-2. Neurons hand drawn by Cajal from a microscope without the benefit of any other drawing device. Reproduced from Cajal (1906) with permission.

a computer and ancillary equipment may be attached to a microscope to help overcome these disadvantages.

1.3. HOW THE MICROSCOPE HANDICAPS THE USER

The constraints that you feel when you use a microscope are easily described by comparing the view of a sycamore tree in winter to a microscopic neuronal tree, both shown in Fig. 1-3. When looking at the two images, consider how the two different viewing devices, the naked eye in the first case and the microscope in the second, would affect someone who wished to study the images scientifically. You will notice how the microscope imposes five major constraints on the viewer: (1) medium, (2) perspective, (3) magnification, (4) contrast, and (5) measurement.

1.3.1. Medium

First, consider the medium. A sycamore tree exists in a transparent medium, air. Thus, you can see the entire depth of the sycamore tree at once. On the other hand, a

FIGURE 1-3. Left: A sycamore tree in winter. Right: An HRP-stained neuron from the spinal cord of a cat. The problems in viewing these two types of trees are contrasted in the text. Sycamore reproduced from Argus (1981) with permission. Neuron modified from Capowski et al. (1986) with permission.

neuron exists in a translucent medium, tissue. You can look into the tissue only to a depth of 100 to 200 μm. If the neuron extends deeper into the tissue than 200 μm, you must cut the tissue into additional sections and look at each section individually. Then, because the tissue has been sectioned, you must contrive some means to reattach the sections to view the intact neuronal tree.

1.3.2. Perspective

The perspective of the microscope view is also limited. In most cases, unless the tree is at the edge of a cliff, for example, you may look at a sycamore tree from one angle and then easily walk around it to view the sycamore from any other direction. The perspective of the microscope view is, however, defined rigidly at the time that the tissue is sectioned. This is because the tissue can only be mounted on a microscope slide horizontally. Thus, it is very hard to get a side view of a neuron through a microscope.

1.3.3. Magnification

Although you certainly know that magnification is an issue when viewing a neuron through a microscope, you may not have considered that you are working with magnification when you look at a sycamore tree. For instance, you can look at the sycamore by standing far back from it to notice its global shape, and you can walk up to a small twig to inspect it in detail. With each step you take toward the tree, you are examining it through a different magnification. Additionally, you can switch your magnifier back and forth easily by taking steps backward or forward. When looking at a neuron in a microscope, however, it is possible to change the magnification, but not as easily and not continuously. Furthermore, because of the sectioning process, you frequently cannot, even by switching magnifications, see the entire neuron tree at once.

1.3.4. Contrast

Contrast is also different between the two images. A sycamore tree, as it exists in nature, is opaque within its transparent medium, air; a neuron tree is translucent within its translucent medium, tissue. Therefore, the neuron tree must be stained to render it visible. Typical staining techniques, Golgi impregnation and horseradish peroxidase injection, do render a neuron visible, but they also stain unwanted elements of the tissue, other neurons included. Moreover, there are questions as to whether the staining is complete and whether it changes the structure of the neuron.

1.3.5. Measurement

The final way viewing the two trees differs relates to measurement. If, for example, a forester wants to measure an environmental effect on one year's growth of the sycamore tree, he can choose a point on its trunk, place a tape measure around the trunk, and measure its circumference. Then a year later, he can go back to that same tree, place a tape measure around the same part of the trunk, and determine the effect on the growth by simple subtraction. With a little extra effort in the test design phase, he can also control for rainfall, fertilization, the felling of neighboring trees, and other environmental perturbations.

By comparison, when you look at a neuron tree, you have destroyed its source—it is not possible to go back the next year, or even the next day, to examine environmental changes on the neuron. Indeed, this is the classic destructive test process, such as occurs when testing camera flashbulbs. Consequently, the only way to assess the effects of environmental changes on neurons is to measure large samples of two populations and statistically compare them. This means that there is a great amount of labor involved. And, without some kind of automatic data-processing assistance, the task is almost impossible.

1.4. DRAWING WITH THE CAMERA LUCIDA

Such optical handicaps as mentioned above restricted tracing to people who had artistic skill and a great deal of patience. Indeed, Cajal and his colleagues spent countless hours drawing pictures of neurons and related structures from the microscope, as shown in Figs. 1-1 and 1-2. Since they had to look from the microscope to the paper and draw by short-term memory, a significant amount of artistic talent was required to generate drawings with the quality and complexity of the one shown in Fig. 1-2. Clearly some sort of device was needed to make the drawing possible by people with less artistic skill.

In 1807, a British optical expert, William Hyde Wollaston, invented a device called a camera lucida. It consisted of a tube and a prism and was used to project an external image down onto a sheet of paper. Wollaston's original camera lucida is on display in the Smithsonian Museum of American History in Washington, D.C.; a diagram of this device is shown in Fig. 1-4. Since the original model was invented, the device has been used as an ''artist's assistant'' in many fields.

The German optical expert Ernst Abbe modified Wollaston's camera lucida during

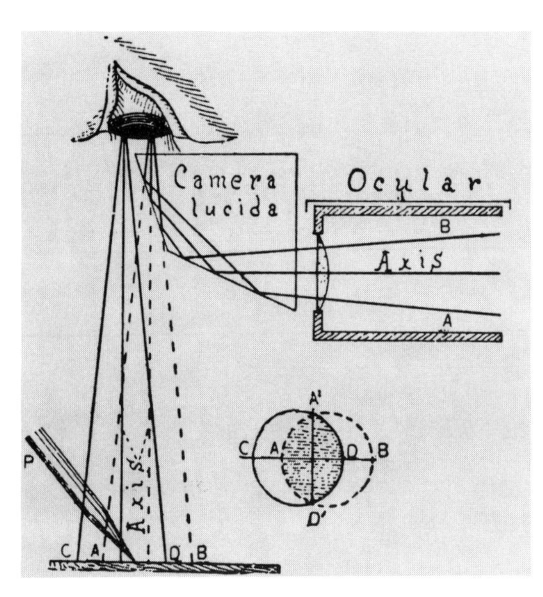

FIGURE 1-4. Wollaston's camera lucida. The microscope image enters from the ocular on the right and is deflected by the prism to the user's eye. He sees the microscopic image overlaid on top of the image of his pencil and paper so that he may trace accurately. Reproduced from Gage (1941) with permission.

the 1880s into more or less the device we have today. The modern camera lucida, when attached to a microscope, combines an external image with the image of the specimen and presents the two images superimposed in the microscope eyepieces. Thus, you can view an external image at the same time you are looking into the microscope, without removing your eyes from the microscope. Figure 1-5 shows a person using a microscope with a

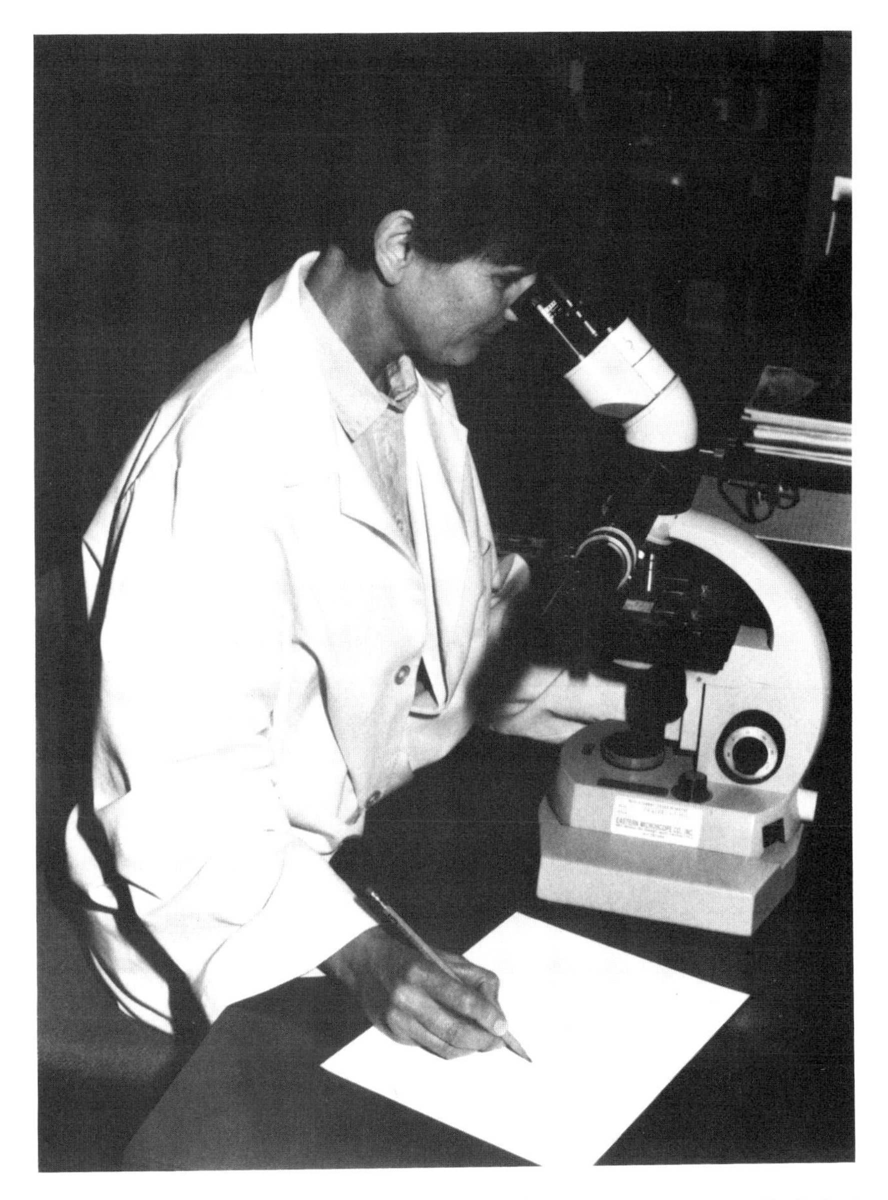

FIGURE 1-5. A researcher using a microscope and camera lucida to trace a structure. She looks into the microscope while drawing with a pencil on a sheet of paper. The camera lucida mixes the pencil's image with the specimen's image so she may accurately trace the neuron.

FIGURE 1-6. The microscope view while drawing a neuron with the aid of a camera lucida. The hand and pencil of the person drawing the neuron may be seen overlaid on the image of the neuron. He is currently drawing the dendrite that extends toward the right side of the photograph.

camera lucida to draw a neuron. The camera lucida, loosely called a "drawing tube," is the horizontal pipe extending to the right from the microscope. It has an optical input port that views a piece of paper on which the person is drawing. Although a moderate amount of artistic talent is needed to trace with a camera lucida, thousands of neurons have been drawn this way by many individuals in neuroscience laboratories.

When you look into the microscope eyepieces, you see the image of the neuron; you also see the image of the sheet of paper, the pencil, and whatever part of the neuron you have already drawn on the paper. These are all shown in Fig. 1-6. With your pencil, you may draw a realistic two-dimensional tracing of the neuron. If the neuron extends over more than one field of view of the microscope, you may shift the paper and the specimen in order to continue tracing over a wide area. You may use several sheets of paper to trace a large neuron at a high magnification.

A limitation of camera lucida tracing is that the final drawing is only a two-dimensional projection of the neuron. All the focus-axis information is lost, and the only view possible is that of the plane of sectioning. Another disadvantage of this technique is that when you have completed the drawing, the only end product is a drawing—you cannot statistically summarize the drawing without somehow measuring it. Figure 1-7 shows a complex neuron traced with a camera lucida.

Literally thousands of camera-lucida-drawn neurons have been presented in the literature. The drawings of Mannen (1975) and Mannen and Sugiura (1975) deserve a special mention because of their quality, accuracy, and beauty.

1.5. THE PANTOGRAPH: A PLOTTER FOR THE MICROSCOPE

A more sophisticated instrument for manual tracing of neuroanatomical structures is called a pantograph or microscope plotter. An early one is shown in Fig. 1-8. With a

FIGURE 1-7. A camera lucida tracing of a branching axon in cat lumbar spinal cord. The axon extended through 20 sections, and the drawing required about 25 hr to complete. Reproduced from Casale (1988) with permission.

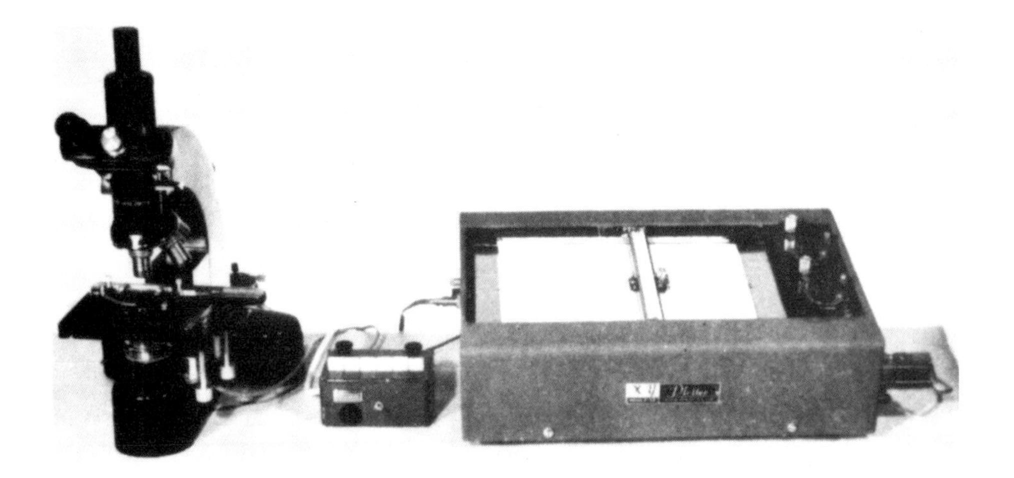

FIGURE 1-8. The classic microscope pantograph. A researcher views a neuron or other structure in a microscope (left) and manipulates the stage and focus knobs in order to pass the structure under a cross hairs that he has inserted in one of the eyepieces. Potentiometers are mounted on the stage and focus controls, and analog electronics (center) pass the stage coordinate to an analog *XY* plotter (right). The plotter thus tracks the portion of the tissue that is under the cross hairs. With a foot switch (not shown) the researcher may lower the plotter's pen onto the paper. Reproduced from Boivie et al. (1968) with permission.

pantograph, potentiometers are mounted on the knobs that you use to control the stage movement of the microscope. To use the pantograph, you place a cross-hairs reticulum in one of the eyepieces of the microscope. Then you may follow the dendrites or structure borders by moving the stage to pass the course of the structure under the cross hairs. The potentiometers present two voltage outputs indicating the X,Y coordinates of the point in the tissue that is under the cross hairs. If the voltages are delivered to an analog X,Y plotter, as shown in Fig. 1-8, then the neuron's structure will be plotted onto a sheet of paper that has been placed on the plotter's bed.

Although not necessary, a footswitch is usually attached to a pantograph. With it, you may control when the plotter's pen is lifted off of or placed on the paper. A structure traced with a pantograph is shown in Fig. 1-9. The pantograph, like the camera lucida, allows a nonartist to plot a microscopic structure on a sheet of paper. So the pantograph plot possesses the same disadvantages that the camera lucida drawing does: it is a two-dimensional drawing on a sheet of paper limited to the plane of sectioning. Thus, for statistical summarization, the plot must still be measured.

1.6. PHYSICAL MODEL BUILDING

To overcome the two-dimensional limitation of drawing on paper, numerous researchers have chosen a different type of system for visualizing neuronal structures; they chose to build physical models of them. This, however, requires a great deal of skill and patience. Mannen (1975) deserves special mention here for his construction of complex and realistic models of neurons made from cardboard and wire. Figure 1-10 shows one of his physical models.

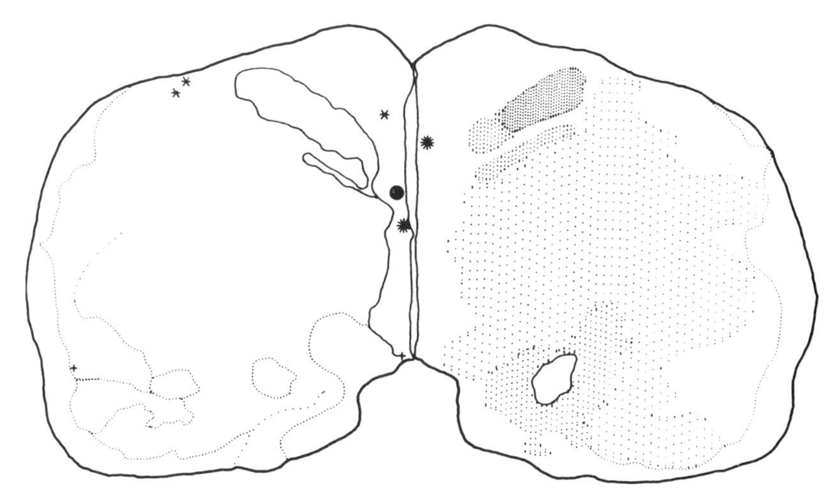

FIGURE 1-9. A cross section of cat medulla oblongata plotted with a pantograph system similar to the system shown in Fig. 1-8. Any microscopic structure may be reproduced in this fashion. Reproduced from Williams and Elde (1982) with permission.

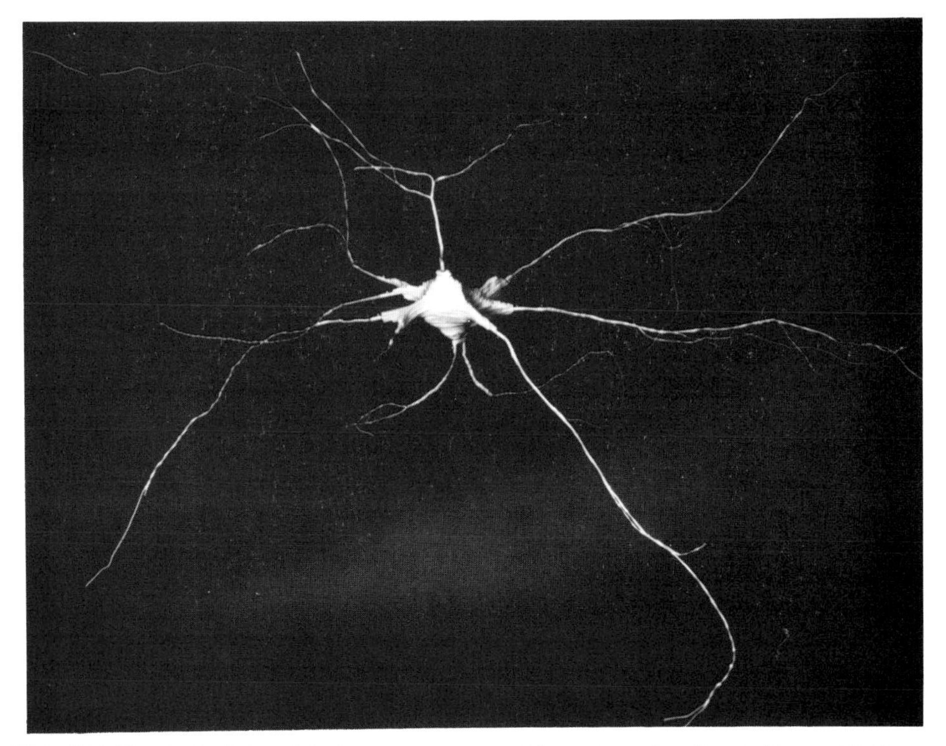

FIGURE 1-10. A physical model of a neuron constructed from wire and cardboard. Mannen (1966) constructed numerous similar models. Reproduced from Capowski (1985) with permission.

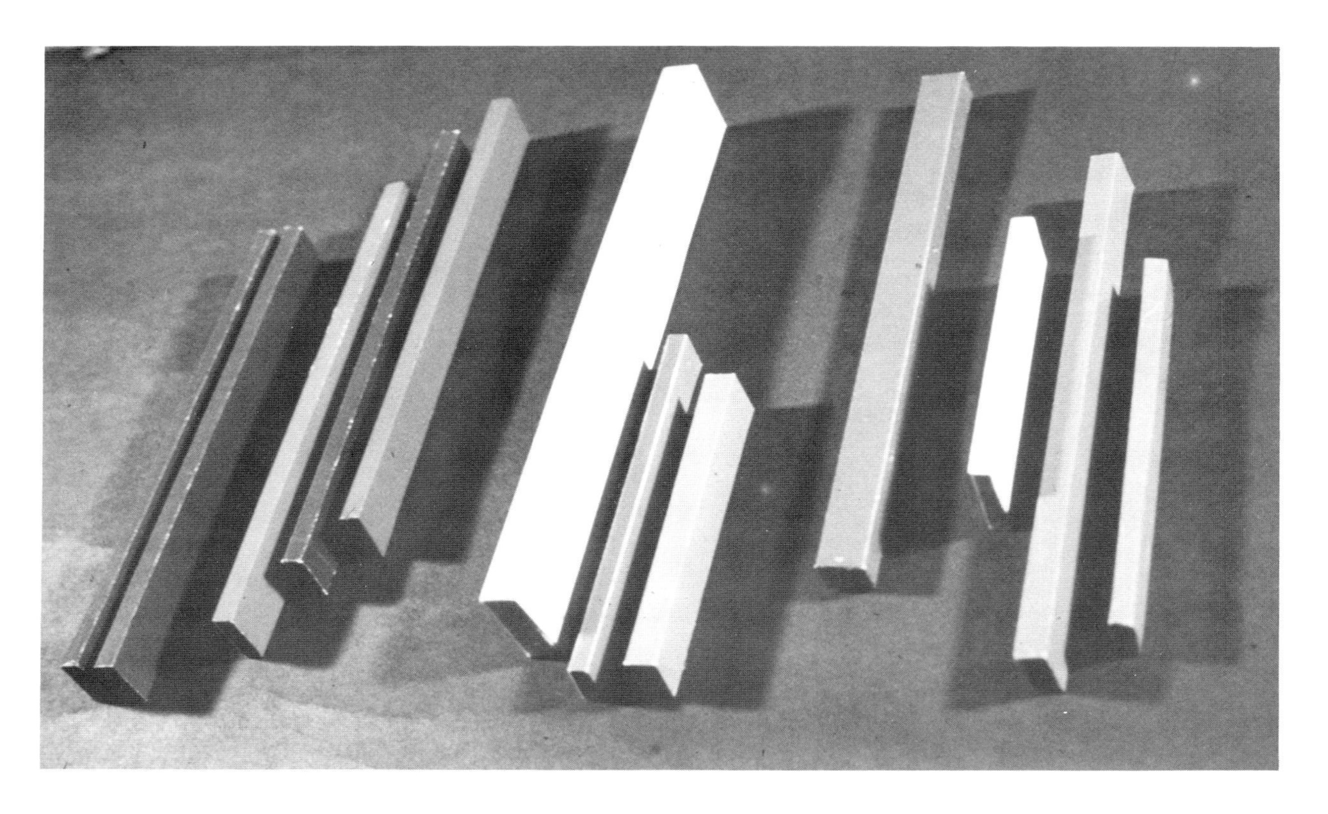

FIGURE 1-11. Wooden blocks cut to size proportional to the dendritic extent of several neurons. Here a neuron's size may be easily represented. Reproduced from Réthelyi (1981) with permission.

A simpler version of physical model building is depicted in Fig. 1-11. Here, a rectangular wooden block is used to model the size of a neuron. The block size is proportional to the extent of the neuron along the three axes. The neurons are painted various colors according to the type of neuron or their experimental population. This elementary form of model building is also a form of mathematical summarization.

It should be realized, however, that any physical model, no matter whether it is complex (Fig. 1-10) or simple (Fig. 1-11), still has the disadvantage that it must be measured manually. Thus, for characterizing a neuron model or comparing it to another model, a more sophisticated form of statistical summarization is needed.

1.7. EARLY ATTEMPTS AT STATISTICAL SUMMARIES

In his 1956 book, *The Organization of the Cerebral Cortex,* Donald A. Sholl first presented statistical summaries designed to characterize the structure of neurons. He did manual drawings of neurons and made measurements from the drawings for use in his graphs. Two graphs, which bear his name and are still in common use today, are illustrated in Figs. 1-12 and 1-13. Sholl's work, of course, predated laboratory computers, so all the summarizations were performed manually.

1.8. HOW THE COMPUTER HELPS VISUALIZING AND SUMMARIZING

To summarize the chapter up to this point, there are two basic problems in using a microscope to look at and to analyze the structure of neurons. First, the microscope constrains your view in several ways, and second, the process of statistically summarizing a drawing is so tedious that you will rarely do it. By attaching a computer to a microscope and reconstructing the neuron in the computer memory, both of these problems may be minimized.

The computer is a superior bookkeeper, easily and accurately managing lists of numbers, whereas you are a superior pattern recognizer. So if your pattern recognition skills are blended with the computer's bookkeeping skills, neuroanatomical tasks that

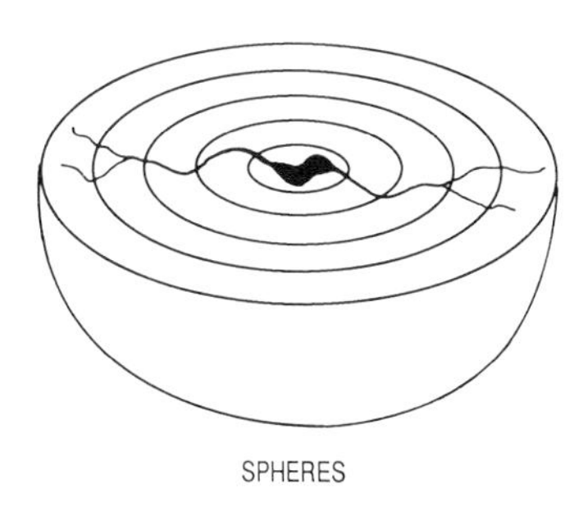

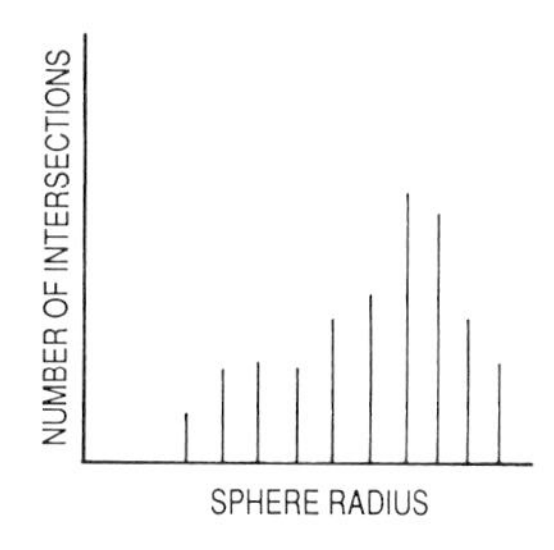

FIGURE 1-12. A Sholl "concentric sphere" diagram. Left: A series of imaginary concentric spheres is drawn, centered at the neuron's soma. Right: The number of intersections with each sphere is counted and plotted versus the radius of the sphere. This describes the number of branches that exist at various distances from the soma.

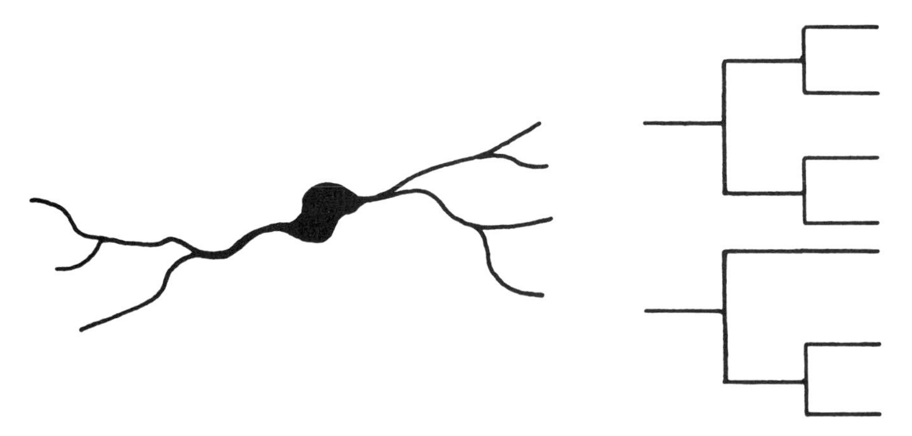

FIGURE 1-13. A Sholl schematic or ''tennis tournament'' diagram. Left: A drawing of a simple neuron. Note the simple branching pattern of its dendrites. Right: The schematic diagram representing the branching of this neuron. The upper schematic diagram represents the right-hand dendrite.

have in the past been too tedious may become reasonable. If a computer can monitor the X,Y,Z coordinates of the tissue that is under a cross hairs, then as the viewer passes the tissue under the cross hairs, the computer will record a series of coordinates in its memory that forms a mathematical model of the structure. In other words, the structure in the tissue is disassembled into a series of coordinates that are passed, one at a time, into the computer memory, where the structure or, more properly speaking, a mathematical model of it is reassembled or ''reconstructed.''

The model stored in computer memory, unlike the original structure stored in the tissue, may be easily accessed, because computer memory is ''tractable.'' That is, data within the memory may easily be read and mathematically manipulated, and new data can be written into the computer memory.

From the mathematical model of the neuron structure stored in computer memory, computer graphics techniques may be used to generate pictures of the model from any viewpoint, at any magnification, and as realistically as computer programs may allow. To whet your appetite, Fig. 1-14 presents a neuron in its tissue and a computer plot of its reconstruction. You will see that the computer plot is superior to the original because the entire neuron that extends through a dozen tissue sections can be seen at one time and because small details (e.g., swellings and spines) that can only be seen in a local high-power view can be seen here over the entire neuron.

Figure 8-1 presents an enriched view of a neuronal structure. It is possible with current technology to present a better-than-life view of a structure to allow you to visualize some particular aspects of a structure particularly well.

When the model of a structure has been stored in the computer memory, statistically summarizing its structure becomes a more manageable task. For now our only need is to write programs that instruct the computer how to summarize the large amounts of data, rather than dealing with the actual coordinates ourselves. So the computer forms a bookkeeping aid to the neuroanatomist, rendering possible those tasks that previously could not be practically performed because of the tedium involved.

When the computer, equipped with a television camera, can monitor the brightness

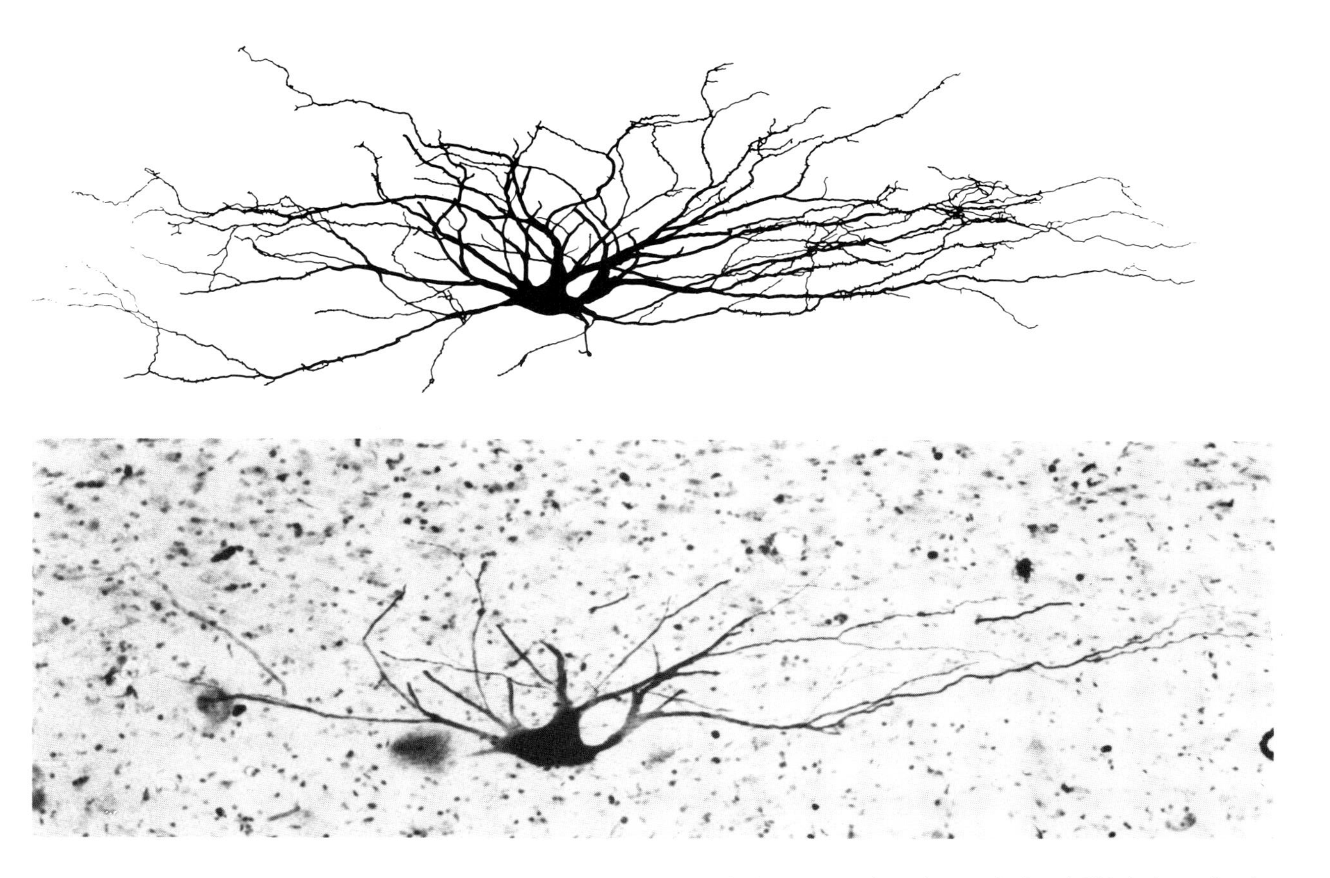

FIGURE 1-14. Lower: The tissue section containing an HRP-stained spinocervical tract neuron from the cat spinal cord. This is the section that contains the soma. The neuron's dendrites extend through about 12 tissue sections. Upper: The same neuron reconstructed with the UNC neuron-tracing system. All the dendrites may now be seen, and spines and swellings are also visible. Reproduced from Capowski et al. (1986) with permission.

of portions of a microscope image, two new categories of applications in neuroanatomy become available. The first category of applications uses optical density or "darkness" of the image to measure the number of structures in an area (e.g., the location of labeled neurons) or uses the darkness to measure the metabolic rate of a region in the brain. The second category of applications uses the computer's ability to scan the gray level of portions of the image to automate some task. In this situation, for example, particles representing cells may be counted and their size measured by the computer. And in the future, it may be possible to automate some more complex task, perhaps to follow the branching of dendrites automatically and with a high level of accuracy.

1.9. FOR FURTHER READING

McKenzie and Vogt (1976) report the use of a microscope, camera lucida, and a mechanical device to draw two perpendicular views of a neuron simultaneously.

Glenn and Burke (1981) describe a noncomputer method for drawing stereo pairs of a neuron, and Haberly and Bower (1982) describe a noncomputer method for rotating a camera lucida drawing to see a different view of the neuron from that in the plane of sectioning.

1.10. WHAT THIS BOOK PRESENTS

Most neuroanatomists have been trained qualitatively with few mathematical course requirements and even fewer computer science requirements. Frequently they have used computers for word processing only. This book is targeted toward those neuroanatomists who are now starting to use computers for the collection and analysis of laboratory data. As computers become prevalent in the laboratory, their training may change.

Chapters that follow first describe how the computer works and then explain some of its applications in neuroanatomy. The methods of entering anatomic data, processing it internally, for example to compensate for errors made during entry, displaying the stored structure, and summarizing it are presented in detail.

2

Laboratory Computer Hardware

2.1. INTRODUCTION

So that you can make the most intelligent use of computers to collect, display, and analyze your neuronanatomical data, you need a certain level of knowledge about computing hardware and software. This chapter focuses on computing hardware, the electronic and mechanical equipment that is used to perform laboratory tasks. Chapter 3 offers an overview of typical computer programs, called "software," that give the hardware its operating instructions. Both chapters provide information about computers and related laboratory equipment to the level of detail you require. Once you understand the material presented in this chapter and in Chapter 3, your expectations of what computers can do for you in the laboratory should be realistic. The information presented will also enable you to talk intelligently with the people who market computer systems for the laboratory.

2.2. OVERVIEW OF KEY COMPONENTS

A computer is an electronic and mechanical device for manipulating and storing textual and numeric information. It can operate on these data by executing a predetermined series of instructions known as a "program." Recently computers have been widely accepted in laboratories (among other places) primarily for two reasons: (1) programs have been written that attack standard laboratory problems, and (2) the cost of the machines and programs has fallen to where they are easily within typical laboratory budgets. Two common laboratory computers are shown in Figs. 2-1 and 2-2.

Figure 2-3 is a block diagram of a typical laboratory computer. Sections 2.2.1 through 2.2.10 describe its components in a low level of detail, to provide you with background information. Basic computer concepts are provided in Section 2.3, followed by more detailed hardware descriptions in Section 2.4.

2.2.1. Bus

A bus is a collection of wires that link the various components of a computer. It is important to note that in the design of computers, there is little direct connection from one

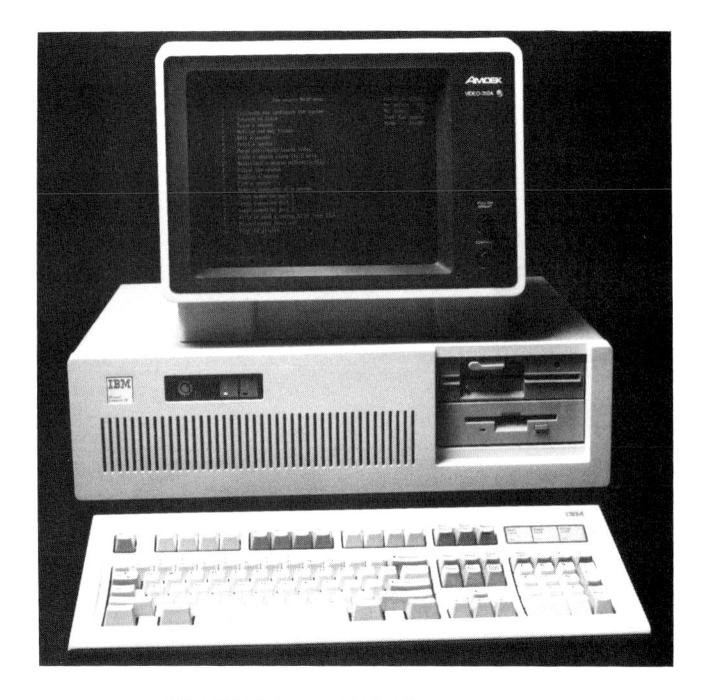

FIGURE 2-1. An IBM AT computer.

internal component to another. All the information travels by the bus; thus, it is possible to design and install computer components independently.

2.2.2. Central Processing Unit

The central processing unit or CPU is the heart of the computer. Here, arithmetic and logical operations are performed on data that are stored in the computer's memory. The data may be numeric, as when you are using a scientific program, or it may be textual, as when you are using a word-processing program. Since the computer may only deal with numbers, letters and other characters are always encoded as numeric values.

2.2.3. Memory

Memory (sometimes called "random access memory" or "RAM") is used to store a numeric form of the computer programs as well as the data on which the programs operate. Memory consists of a large number of locations or "addresses," each of which contains an electronically stored number.

FIGURE 2-2. An Apple Macintosh Computer.

2.2.4. Disks

Two magnetic media for storing data are commonly used today: hard disks and floppy disks. A hard disk is a rigid mylar disk coated with iron oxide into which data are written electrically and then stored magnetically until changed. Such a disk has high

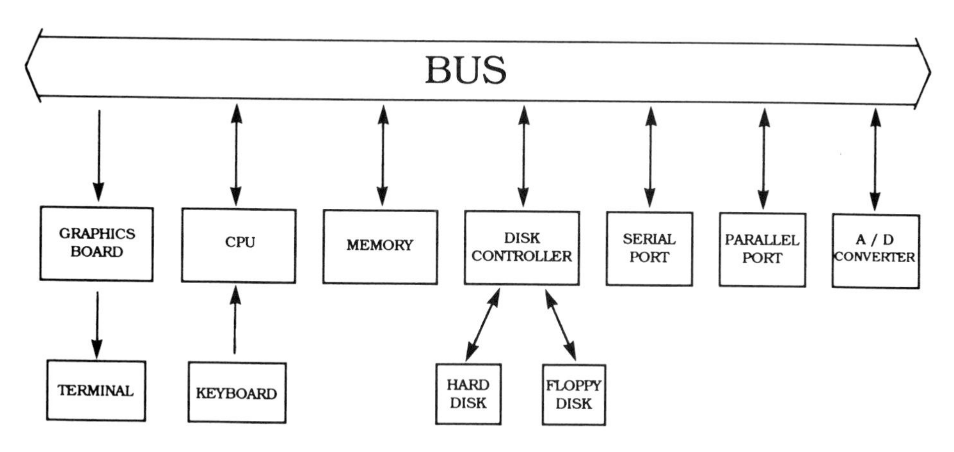

FIGURE 2-3. A block diagram of the internal components of a laboratory computer. Details are given in the text.

capacity but may not be removed from the computer. A floppy disk is paper thin and rotates within either a paper or rigid plastic envelope. It may be removed from the computer for storage on a shelf.

2.2.5. Disk Drives

The device that reads and writes data to either kind of disk is called a disk drive. It contains a motor that rotates the disk and an arm that extends out over the surface of the disk. On the end of the arm is a head that generates electrical pulses to write the data to the disk and senses the magnetic field generated by the data to read it from the disk. Disk drives for both hard and floppy disks are linked to the computer bus by a disk controller card.

2.2.6. Keyboard

The operator of the computer may enter text and numeric information to a program on a typewriterlike keyboard.

2.2.7. Terminal

While you are typing on the keyboard, you may view text and numeric responses from the computer on a televisionlike screen called a terminal. Computer programs may display their output in numeric, textual, and graphics forms on the terminal.

2.2.8. Graphics Board

A graphics board presents computer-stored information on a televisionlike screen. The information may be textual, such as characters that you type on the keyboard, or it may be picturelike, composed of many thousands of dots.

2.2.9. Ports

A port is an electrical connector on the back of the computer cabinet that you may use to connect the computer to external devices. Two types, serial and parallel, are now in very common use. Serial ports transfer data one bit (the atomic unit of information, defined in Section 2.3.1.2) at a time over a single wire. Information thus flows one bit after another, or serially, down the wire; hence the name. Parallel ports transfer a number of bits of information simultaneously, each along a separate wire.

2.2.10. Analog-to-Digital Converter

An analog-to-digital converter, also called an A/D converter or an ADC, is a device that samples continuous voltages presented to its input and converts them into numeric form so that the computer program may operate on them.

2.3. CONCEPTS AND DEFINITIONS

In Section 2.2, you saw a brief introduction to the most common components of a general-purpose computer. In Section 2.4, you will see a more rigorous presentation of these common components and those less common ones that are needed for laboratory computing work. But in order to understand the components, you must first understand some concepts.

2.3.1. Forms of Data

Several forms of numeric and textual data have evolved in the computer field. These data may be presented to the computer by you or by other laboratory devices, manipulated within the machine, and may be returned to you in some new form.

2.3.2. Analog and Digital Values

An analog value is a continuous value; it has no discrete steps. In neuroscience-related computing, the term analog is most often applied to voltage waveforms. Actually the term comes from analog computers, which are electronic circuits that use continuous voltages to model (or to build an analog of) the movement of mechanical systems (e.g., the suspension system of a car). The term analog grew to mean a continuous value.

Digital is the opposite of analog. It refers to data that exist in discrete states, for example, as the digits 0 through 9. Almost all laboratory computers are digital and thus deal with discrete values. A specific case of a digital value is the binary value, described below.

2.3.3. Storage of Digital Data

Bit, the basic unit of storage, is a contraction of binary digit. "Binary" means having two states; thus, a binary digit can have two states, a one or a zero, or an "on" or an "off." Bits are clustered together into bytes, eight binary digits. Some manufacturers also define a word, but there is no industry standard; for instance, words on IBM computers have 32 bits, whereas words on computers designed by Digital Equipment Corporation have 16 bits.

A bit may take one of two possible values, typically called a zero and a one. Two bits may take one of four possible values, usually called zero, one, two, and three. Eight bits may take one of 256 possible values, which are usually assigned the values zero through 255. The 256 possible values may alternatively be assigned the range −128 through +127 in order to represent both negative and positive numbers. A group of 16 bits may take 65,536 possible values and is usually assigned the range −32,768 through +32,767. Such a representation is called a "two's complement" representation. Generally, n bits may take 2^n possible values.

Three kinds of data normally reside in a computer: integers, floating point numbers, and characters. Integers typically use 16 bits and range from −32,768 up to +32,767, with no fractional values. It is necessary to perform the various arithmetic processes on these integers inside the CPU.

The second type of storage is floating point, or scientific notation (e.g., 1.234×10^6), which has a certain amount of accuracy in the values stored; unlike integers, however, there is not a perfect representation. The range of values possible in scientific notation is much greater than that of integers and can also be much less than 1, so that fractional values (e.g., 1.234×10^{-6}) can be accommodated. For the computer to add together two numbers with different magnitudes (e.g., $1.234 \times 10^3 + 4.567 \times 10^6$), the following operations must be performed. First, scale the values as necessary to make their exponents identical; then add; then, normalize so the fractions are between 1 and 10. This can take many steps in a general-purpose CPU. To speed up the process, people frequently build (or add) a math coprocessor, described below, to the computer.

The third type of frequently used data is known as textual or character data. At the very lowest level, the computer may only store and manipulate groups of bits that form numeric values, so characters must be encoded into numeric values. A standard code called ASCII (for American Standard Code for Information Interchange) has evolved for storing text characters in most laboratory computers. Each character is represented by a one-byte numeric value; the letter "A," for example, is encoded into the value 65.

2.4. COMPUTING HARDWARE DESCRIBED IN SOME DEPTH

In this section, a more detailed description of the hardware components of a computer configured for use in the laboratory is presented. Some of the components were introduced in Section 2.2.

2.4.1. Central Processing Unit

The central processing unit (CPU) is the heart of a computer—the part that executes the instructions, controls the processing of the data, and actually does the arithmetic and logical operations on the data. Typical arithmetic operations are SUB (subtract), MULT (multiply), DIV (divide), and NEG (negate or change the sign of a number). Logical operations are also performed by the CPU; examples of these are AND, OR, and NOT. These logical operations (also called "Boolean" operations after the logician George Boole) combine two numeric values, bit by bit. The part of the CPU where these operations are performed is referred to as the "arithmetic and logical unit" or ALU.

In most laboratory computers, the CPU is a single integrated circuit (see Section 2.5). To illustrate the CPU's operation, a typical instruction, ADD, is executed by fetching the instruction from memory into the CPU, fetching two data values, and then, in the CPU, performing the addition. Finally, the sum is sent from the CPU to be stored in the memory.

2.4.2. Math Coprocessors

A math coprocessor is a special purpose processor, usually constructed of one integrated circuit, that can perform the standard mathematical operations on floating point numbers in a single step. If a computer program has many mathematical operations, a math coprocessor may greatly reduce its execution time.

2.4.3. Random Access Memory

The function of random access computer memory is to store numeric information. Information is stored in RAM as individual bits, each in a capacitive "cell" that holds a static electrical charge representing an on or off state until changed. The storage capacity is usually specified in kilobytes (kbytes), or megabytes (Mbytes), although some computer manufacturers specify "words." Since the size of a word varies, if a person says that a memory size is 32k, ask for a specific unit. Note here that the metric abbreviation "k" (for "kilo") in the computer field means 2^{10} or 1024. Similarly, "M" (for "Mega") means exactly 2^{20} or 1,048,576.

2.4.3a. Addressing. You may recall from Section 2.3.3 that a 16-bit number can take on 65,536 possible values. You may therefore select one of 65,536 possible things with this 16-bit number. If the thing you are selecting is a particular location in memory, then you are "addressing" 65,536 possible locations with a 16-bit value. If each location of memory contains a byte of data, then you may address 65,536 bytes of memory with a 16-bit word. The 16-bit value that specifies the location in memory is called the address. A 32-bit word can access approximately 4 gigabytes, or four billion bytes of memory, which, at least in 1988, is more memory than most computers can handle. Generally, n bits can address 2^n possible locations. Further information on this concept of addressing is presented in Section 2.4.8.

2.4.3b. Random Access. A memory in which you can access one particular location as easily as any other location is called a "random access memory" or "RAM." A constant time is required to generate any memory address and to read or write data to this address in memory. Contrast this idea with a reel of magnetic tape—it may be relatively quick to access one location at the near end of the tape, but it may take 30 sec to reach the other end of the tape. A tape is therefore a linear rather than a random access memory. A disk drive is semilinear, because a recording head is moved to a track, as a phonograph needle is placed in a specific record groove. Thus, when you use a disk you must wait for the proper location within the track to spin into position underneath the recording head. So the disk memory access takes a varying amount of time depending on where the recording head is and where the data to be read are positioned on the disk. Disk manufacturers specify an "average access time" to indicate the speed of access to the data on a disk. It is a statistical summary of the amount of time required to locate disk-stored data in various situations.

2.4.3c. Volatility. Another memory concept is that of volatility. In a volatile computer memory, the data are stored as electronic charges. When the power is turned off, the electrons dissipate, and the data stored in the memory are lost. In a nonvolatile computer memory, such as that on a disk or on a tape, the data are stored by magnetic fields aligned in an iron oxide coating. Those data are stored permanently or at least until rewritten. They are not lost when the power goes off.

2.4.3d. Read-Only Memory. The final memory concept is that of a ROM, or "read-only memory." This is a memory composed of integrated circuits (Section 2.5) into which data are written permanently and, in most cases, may not be changed. Data are said to be "burned" into a ROM because each data bit is written by melting fuse links within the ROM integrated circuits. Once written, the data are always available, even if the power is interrupted.

Frequently, software to perform general-purpose tasks is provided by the computer manufacturer. This software is burned into a ROM that is mounted inside the computer. When software is burned into a ROM, it is given the name "firmware."

2.4.4. Disks

Two kinds of disks may be found in contemporary small computers. The first kind is a floppy or flexible disk or diskette. The second is a hard or Winchester disk. A flexible diskette, made usually of very thin mylar, is most often $3\frac{1}{2}$, $5\frac{1}{4}$, or 8 inches in diameter. It is encased in a paper or sometimes plastic envelope that is inserted in a disk drive, and the data are recorded in a series of circular tracks. On an IBM AT disk, for example, there are 77 tracks. A recording head moves in and out over the disk like a phonograph needle and positions itself on one of the tracks. Once it is positioned, it reads or writes information from or onto the track. As of 1988, the disk drive costs between $100 and $200, and the disk itself costs about $2. The storage capacities of floppy disks vary between 140 kB and 1.4 MB.

The other type of computer disk is the hard disk. Instead of flexible mylar, the disk is a rigid plastic platter coated with iron oxide. A high storage density is made possible by a recording technology called Winchester technology. A disadvantage of this technology is that the recording head must be very close to but not touch the actual magnetic medium. Therefore, any particles of dirt between the disk and the recording head can seriously compromise the disk. As a consequence, these disks must be manufactured in clean rooms, and they are not generally removable from the computer. Hard disks, however, are much faster than floppy disks. The storage capacity is also much higher, between 10 and 100 megabytes of information.

2.4.5. Magnetic Tapes

A reel of tape is another form of magnetic medium. In the computer industry, two tape formats find some use, although neither approaches the ubiquity of disks. Older computer-industry standard tapes use reels 10 inches in diameter that contain about 2400 feet of tape and can store about 40 million bytes of data. Contemporary tape cartridges come in many sizes but typically store about 20 million bytes of data. Because seconds or even minutes may be required to reach the data on a particular part of the tape they are now used primarily as backup devices on which to store a second copy of infrequently used data.

2.4.6. Graphics Display Cards

Most laboratory computers have a televisionlike computer terminal on which you can see the output from computer programs. The computer terminal is driven by a card, usually called a graphics display card, that plugs into the computer. There are two kinds of these cards and computer terminals; the simpler ones display only text, and the other, more complicated, display bit-map graphics. A text-only card usually displays 25 lines of text on a screen, where each line can contain 80 characters.

A bit-map graphics display card is used to draw pictures on the terminal. Bit-map graphics is so-called because of a memory on the card that contains the picture. The picture is stored as a matrix of, typically, 500 × 500 picture elements or ''pixels.'' The computer simply turns on bits in the memory, and then, independent from that, the contents of the memory are drawn or ''mapped'' onto the computer terminal one pixel at a time. Depending on the complexity of the card, each picture element may be a bit that is turned on or off, or it may be one of a number of gray levels, or it may be one of a number of colors.

Several common standards for graphics cards have evolved in the personal computer industry, mainly because of the marketing power of the IBM Corporation. The IBM monochrome display adapter (MDA) is a text-only display with a resolution of 25 lines and 80 characters per line. The Hercules monochrome graphics adapter (MGA) is a bit-mapped graphics display of 720 × 340 pixels. Each pixel can be either on or off. The IBM color graphics adapter (CGA) is also bit-mapped, with a resolution of 320 × 200 pixels, where each pixel may take one of four colors. IBM then provided an advanced version of the CGA, called an enhanced graphics adapter (EGA), with 640 × 350 pixels, each with one of 16 colors. Most recently, a card called the video graphics array (VGA) is available with a spatial resolution of 640 × 480 pixels, each with one of 16 colors.

2.4.7. Computer Speeds

Specifications of a computer system may state, for example, that the computer has an 8-MHz clock. What does that mean? Inside the computer is a crystal-controlled oscillator that is generating eight million pulses per second. Each pulse steps the CPU through part of the execution of an instruction. The instruction to add two numbers together and store the result in memory takes a number of steps. The first one is to fetch the instruction from memory. The next one is to calculate the address for the first piece of data, send that to memory, and get the data back from memory into the CPU. Then the address of the next piece of data is calculated and sent to memory, and data are returned from memory into the CPU. The sum is calculated, the memory address for the sum is calculated, and finally the sum is placed in memory. This procedure takes approximately ten steps and therefore ten clock-pulses or ten times the period of an 8-MHz clock to execute an instruction.

A computer is faster if it has a higher clock speed. However, it is misleading to compare the speeds of incompatible computers by simply comparing their clock speeds. Different instructions do different amounts of processing; i.e., the power of each instruction varies. More powerful instructions executed on a slower-clock-speed computer may actually process more scientific data per second than less powerful instructions executed on a faster computer. You should really perform a ''benchmark'' experiment; that is, you should run a program relevant to your work on the various computer systems and compare the times required to do the entire task. Fortunately, however, within a general price range, the speed of a computer is not the most important consideration for purchase. Software availability (see Chapter 3) usually is.

2.4.8. Bus

The bus consists of three parts: (1) wires that carry data from device to device, called "data lines," (2) wires that carry addresses from device to device, "address lines," and finally, (3) the other lines or wires that perform miscellaneous functions (e.g., to specify when the data are available for one device from another, "hand-shaking lines").

The "width" of the bus (the number of address or data lines) is specified in bits (e.g., 16-bit bus). Generally, the wider a bus, the more memory it can address, and the faster the computer can operate. This occurs because more information can be sent from device to device in parallel at any instant. Besides the number of wires, each manufacturer specifies electrical characteristics of the bus as well as the type and size of connector into which cards plug. As with other devices, corporate marketplace power determines whether a bus will become a standard. This is important because many other manufacturers will make devices that plug into the bus of a standard, common computer. Later in this chapter (Section 2.6), common laboratory computers are described briefly, including the names of their buses.

In some computers there is a single bus that links all the devices in the machine. If a CPU wants to send information to a certain device (e.g., memory, disk, digital-to-analog converter), all that is required is for the CPU to provide an address and to provide the data. The receiving device recognizes the address and accepts the data; the sending device, in this case, the CPU, does not have to know where the data are going.

Alternatively, there is a two-bus computer. The two-bus system is composed of a memory bus that connects the CPU to the memory and an input–output bus (I/O bus) that connects the CPU to input–output devices such as disk drives and digital-to-analog converters. Two-bus computers are useful because writing data into and out of memory is simpler and faster than writing data to, for example, a disk. Therefore, the memory bus and the software instructions that read and write data to memory may be optimized to minimize the time to transfer this data. IBM has always used two separate buses in their computers.

In general, time sharing refers to using one resource for different purposes at different times. Applied to a bus, this concept means to share data and address lines, thus allowing both to be transferred along the same set of wires. For example, an engineer designing a machine with 16 bits of data and 16 bits of address on the bus has two options. One is to provide 32 wires. The 32 wires allow the CPU to send data to a disk drive quickly by simultaneously providing 16 bits of data and 16 bits of address. The second option is to time-share the address and the data lines. Time sharing would send, for example, address bits in the first instant and data bits in the next instant. An additional wire would be dedicated to telling the receiving device whether, in a particular instant, addresses or data are being sent. Thus, by time sharing, the engineer could save almost half the copper wire in the bus but slow the machine down by a factor of two. The slower, less-expensive computers usually time-share their addresses and data, and the faster, more expensive ones do not.

2.4.9. Plotters

A plotter is a computer output device for drawing a picture onto paper. It most often comes in the form of a felt-tip pen plotter (Fig. 2-4). Some models have a carriage that

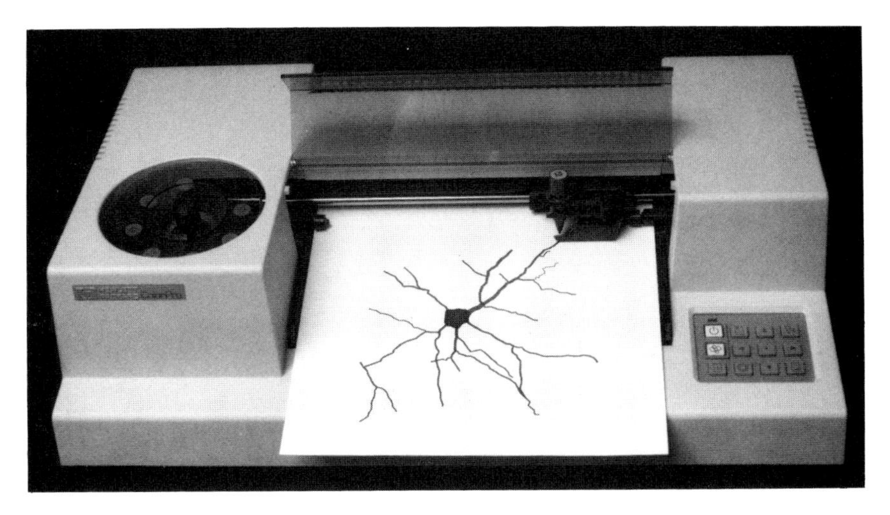

FIGURE 2-4. A felt-tip pen plotter. The computer drives the pen to the right and left and the paper in the near–far direction. Thus, a plot may be generated on the paper. Different colors may be plotted by changing pens.

moves a felt-tip pen over the paper; others move the paper under a stationary felt-tip pen; and a third style offers a combination in which on the *X* axis the pen moves over the paper and on the *Y* axis the paper is rolled back and forth under the pen. The pen can move from point to point on the paper without drawing a line, or it can draw a straight line from one point to the next. Any complex drawing is thus generated as a series of short straight lines. There may be a carousel of several pens from which the computer can select. You can put various colored pens into the stalls of the carousel to create a multiple-color plot.

2.4.10. Printers

Printers are used to provide paper output of either text or graphics from a computer. In the contemporary laboratory computer field, there are two common kinds of printers, dot matrix and laser. Laser printers are more expensive, about $1600 compared to $350 for a dot-matrix printer. However, they are faster, quieter, and produce printing of higher quality. Both of these devices can attach to any contemporary laboratory computer.

2.4.10a. Dot-Matrix Printers. A dot-matrix printer (Fig. 2-5) consists of a print head that has a series of horizontal pins $\frac{1}{50}$ inch apart. The printer is called ''dot matrix'' because it generates characters out of a matrix of dots. As its print head moves from left to right across the paper (Fig. 2-6), each wire in the print head can be thrust forward, depressing the ribbon against the paper and leaving a little round black dot. The computer and the printer direct the thrusting of each pin to generate letters.

Dot-matrix printing is generally done at low resolution and high speed. A typical speed is 250 characters per second with a matrix resolution of seven dots by nine dots. Most dot-matrix printers have a ''near-letter-quality'' mode. A true ''letter-quality'' printer is one whose output looks like that of a real typewriter, such as an IBM Selectric ''golf-ball'' typewriter. Actually, it is not actually possible to achieve this quality with a dot-matrix printer, but with high spatial resolution, near letter quality can be achieved.

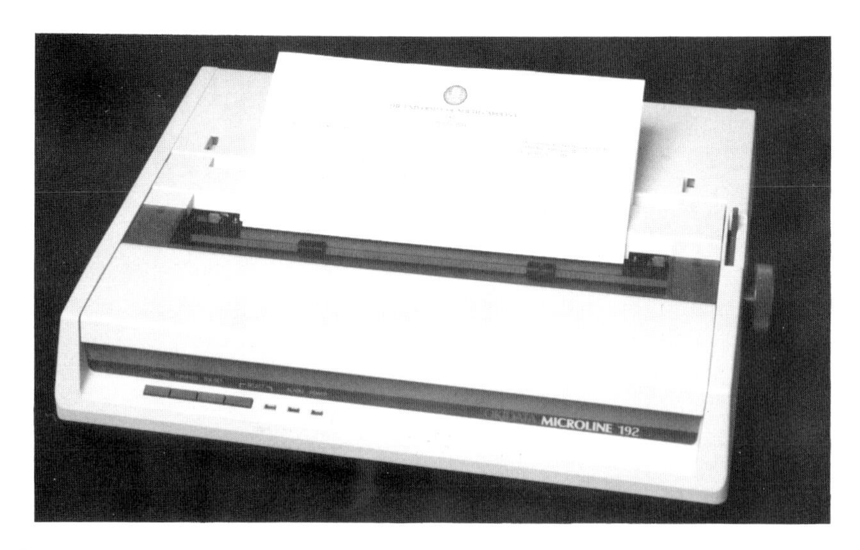

FIGURE 2-5. A dot-matrix printer. This is a common inexpensive printer used frequently with personal computers. Details of its operation are given in Fig. 2-6 and in the text.

Speed, however, is sacrificed for higher resolution because two passes are made to print a line of high-quality text. During the first pass, a portion of each letter is printed. Then the paper is advanced one half the vertical spacing between the print wires, and the second half of each character is printed. The list of neuron coordinates shown in Fig. 4-15 was made on a dot-matrix printer in near-letter-quality mode.

The dot-matrix printer can also generate pictures. Unlike the typewriter, it doesn't "think" in letters; it "thinks" in dots. Given the proper software, these printer dots can be arranged as a picture as well as a set of text.

2.4.10b. Laser Printers. A laser printer (Fig. 2-7) deflects a laser beam over the surface of a drum, and an electrostatic charge remains on the drum wherever the laser strikes it. A dry ink powder is spread over the drum and sticks where the charge is, that is, where the information was written by the laser. A blank piece of paper rolls over the drum

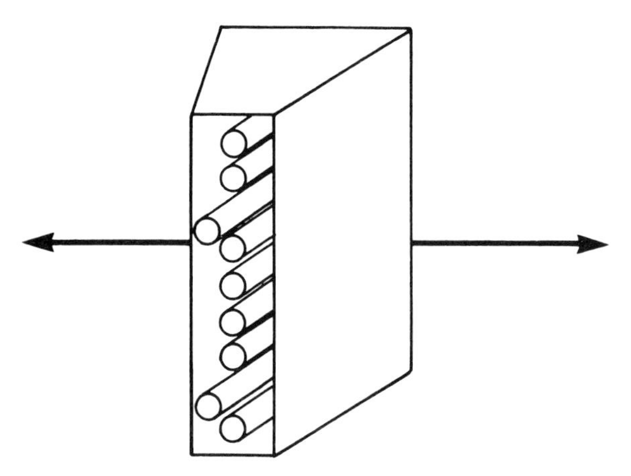

FIGURE 2-6. A simplified diagram of the print head of a dot-matrix printer. The print head is positioned against an ink ribbon and paper (not shown) on the near side of the print head. The print head is moved right and left along the paper. A solenoid sharply ejects each pin toward the ribbon, leaving a small round dot on the paper. Characters are formed by ejecting combinations of the pins coordinated with the left–right movement of the print head.

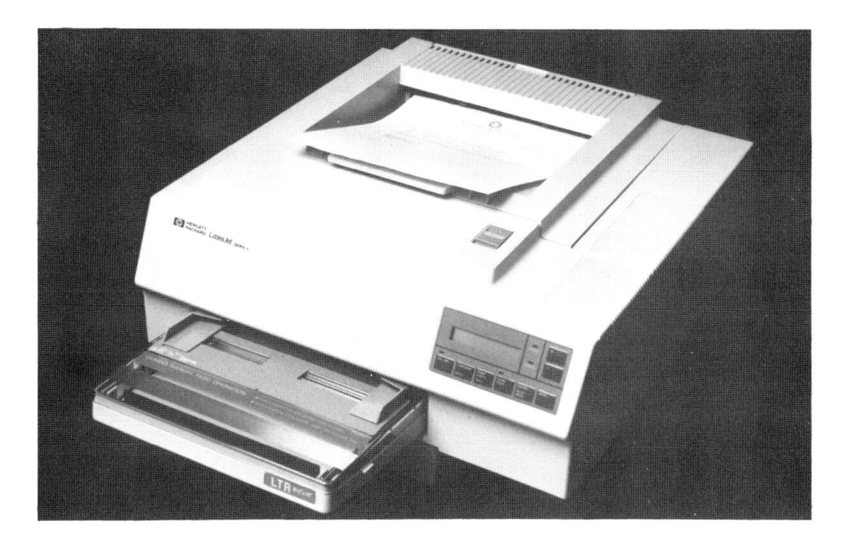

FIGURE 2-7. A laser printer. This type of printer, though it costs about $1600, generates higher-quality text and graphics output. Details of its operation are given in the text.

and absorbs that pattern of dried ink. Text or graphics can be written this way. The laser beam can be deflected to resolutions of 200 dots per inch in vertical and horizontal directions, so it is possible to make a very good-looking character set of text or to generate good-quality black-and-white graphic displays on a piece of paper. The program listings shown in Fig. 3-2 were printed on a laser printer.

2.4.11. Analog-to-Digital Converter

Analog-to-digital converter cards and digital-to-analog converter cards are common instruments that are plugged into computers to record and to generate the analog voltages that abound in neuroscience laboratories. An analog-to-digital converter card accepts analog voltages of a continuous range of perhaps −10 V through +10 V and digitizes them into numbers, for example, of −2048 through +2047.

2.4.12. Digital-to-Analog Converter

A digital-to-analog converter or DAC is the opposite of an analog-to-digital converter (see Section 2.4.11). The computer provides a numeric value of 1024, for example, and the card generates a voltage output of 5.00 V.

2.4.13. Modem

Modem is an acronym for "modulator/demodulator" and is a computer accessory for transmitting computer data over a telephone line. During transmission with a modem, each byte of information is decomposed into its bits, and the modem generates an audio tone for each bit that is sent over the telephone line. That is, the value of the bit (0 or 1) modulates (changes) the frequency of a standard tone. At the receiving end, another modem demodulates (extracts the data from the standard tone) the tones and converts each one into a bit; each 8 bits is then reassembled into a byte. Older modems are separate boxes that connect to a computer through a serial port. They convert the data from the serial port into acoustic tones. More recent ones take the form of computer cards that plug directly into the computer bus. They receive data directly from the CPU and transmit it over the phone line.

2.4.14. Graphics Display Systems

Generally, a computer may provide pictorial output onto a screen; this is called a computer graphics display, and many special devices have been designed to help with this task. Two kinds of display technologies, raster and vector, are normally used to present these images, each with advantages and disadvantages. Since the raster displays depend on television for their output, it is worthwhile to describe the operation of a TV camera and a TV monitor, the two components that comprise a closed-circuit television system.

2.4.14a. Video Systems. A video system consists of two components: a camera to sense an image and convert it to an electrical signal, and a monitor to convert the electrical signal back into an image. The heart of both devices is a glass bottle called a cathode ray tube or CRT from which all the air has been evacuated (Fig. 2-8). In the neck of the glass bottle is a negatively charged cathode from which electrons are emitted. These electrons are attracted toward positively charged electrodes so that they accelerate toward the front surface of the bottle. The electrons are focused into a beam that forms a sharp point when it reaches the front of the bottle. For a camera, the inside of the front surface of the bottle is coated with a phosphorus compound with the characteristic of variable resistance at any spot as a function of the amount of light incident on it at that location.

The electron beam is swept across a series of horizontal lines from the top to the bottom of the tube in a fixed pattern called a raster. At any point in this raster, the electron beam hits the phosphor and flows through it, forming an electrical circuit. The current that

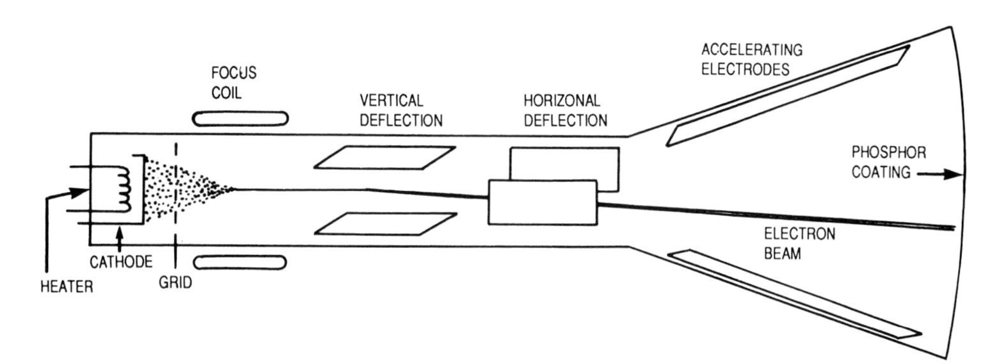

FIGURE 2-8. A stylized drawing of a cathode ray tube used as a monitor. In an evacuated bottle, electrons are boiled off a heated cathode. They accelerate toward positively charged electrodes in the wide region of the bottle. A grid modulates the number of electrons in the beam. The electrons are magnetically focused into a narrow beam and are deflected by two pairs of charged plates in order to sweep the surface of the front of the CRT. A phosphor coating on the surface emits photons of light where the electrons land. Modified from Foley and van Dam (1982) with permission.

flows through the phosphor at that point is proportional to the light incident on the phosphor at the point. The amount of current flowing is called the video signal. The signal varies as the electron beam is swept through its raster to form an electrical representation of the image on the front surface of the camera's tube. Most commonly, a video camera presents a composite video signal that consists of the video signal plus pulses to synchronize the monitor with the camera. These pulses indicate the start of each horizontal line (horizontal sync pulses), and they indicate the start of each new series of horizontal lines or each frame (vertical sync pulses).

A video monitor is similar to a video camera except that the phosphor on the CRT's front surface is different. The phosphor in the monitor emits light proportional to the number of electrons striking it. The monitor receives the synchronization signals from the camera and makes its electron beam sweep over the same fixed raster as the camera, generating light at any point in the raster proportional to the size of the video signal at that instant. Thus, the picture on the monitor is a faithful reproduction of the image incident on the camera.

The preceding description is for a black-and-white video system. A color camera and monitor are just like three black-and-white ones joined together. Remember your physics? To make all the colors of the rainbow, you need only to mix varying amounts of red, green, and blue light. Thus, a video monitor simply contains three different electron beams that are aimed at individual dots of phosphorus compounds whose chemistry is different, so that one generates red light, one generates green light, and one generates blue light. Thousands of these little dots are on the front of the screen, and the three electron beams are set up with three separate video signals—red, green, and blue—so that any color of the rainbow can be generated.

The raster pattern swept by the electron beam is called an interlaced raster. Its pattern, the one used in standard U.S. television, is shown in Fig. 2-9. The electron beam

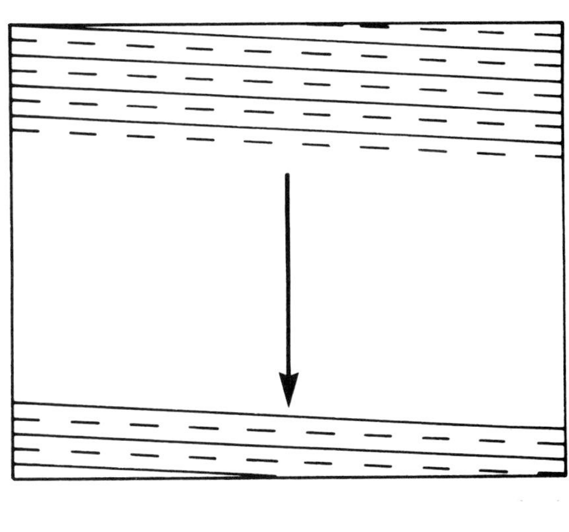

FIGURE 2-9. The interlace raster pattern in United States television. The electron beam sweeps the face of the CRT in two passes, called fields. In the first field, called the primary field, the beam scans $262\frac{1}{2}$ lines, starting at the top left and ending at the lower center. This field is shown by the solid lines and takes 63.5 μsec per line or 16.7 msec per field. Then the beam scans the interlace field, drawing its lines in between the primary field scan lines, starting at the top center and ending at the bottom right. This interlace field, shown by the dashed lines, also requires 16.7 msec. The entire frame consists of 525 lines and is drawn in 33.3 msec.

makes two passes over the front surface of the camera and monitor. In the first pass, called the primary field, each electron beam starts in the upper left and scans $262\frac{1}{2}$ lines and ends in the bottom center of the screen. This pass takes $\frac{1}{60}$ sec. The beam is then positioned in the top center of the screen and makes another scan of $262\frac{1}{2}$ lines, physically spaced between the lines of the primary field. The second pass is called the interlace field. This field also takes $\frac{1}{60}$ sec, so the total time to scan a frame is $\frac{1}{30}$ sec. Only 480 of the total 525 lines are visible on the screen; lines are lost because of the time required to reset the beam to the top of the screen to begin the next field. Each horizontal line requires 63.5 μsec, though the beam is visible for only 51.5 μsec. The remaining time is spent resetting the beam to the left edge of the screen.

Commercial television in the United States was well standardized before computers become common, so the interface hardware that allowed computers to use video devices had to be designed to existing TV standards.

2.4.14b. Television Digitizer and Display Board. A television digitizer and display board is a device that takes the incoming video signal and digitizes it into a memory that is arranged in a matrix of pixels. Electronic circuitry such as that found in a TV monitor tells the computer board into which pixel to deposit each portion of the video signal. An analog-to-digital converter converts the analog video signal at that instant into a digital number, and that digital number is stored in the memory. Thus, a television digitizer board captures a digital version of the TV image. The memory in which the image is stored in digital form is called a frame buffer, because it buffers or stores a television frame. For a color system, three frame buffers can be used, one each for the red, green, and blue components of the video signal. To generate a display of information from the frame buffer, the computer must generate the same video signal that the TV camera generates, i.e., the video signal and the synchronization pulses. The video signal is read, pixel by pixel, from the frame buffer and is converted into an analog video signal by a digital-to-analog converter. Special logic on the display board called a ''sync generator'' generates the synchronization pulses so that the monitor may sweep its electron beam through the raster.

The computer can also access the memory of the frame buffer. It can read the pixels, thereby sensing the gray levels of the TV image, and it can also write information into the frame buffer, so that computer-generated data can be presented on the TV monitor. As we will see in later chapters, this hardware can be used to achieve many different special effects for making measurements from images, enhancing images, or even synthesizing images solely from computer programs.

2.4.14c. Raster Graphics Display Systems. A raster graphics display system is a special-purpose computer consisting of a frame buffer and display hardware that will draw a computer-generated image on a TV monitor. The graphics display card (Section 2.4.6) and the computer terminal that it drives form a raster graphics display system. Because it is a raster system, the electron beam in the monitor always traverses the same pattern, regardless of the image being presented, and the computer must generate the gray levels or color values of every pixel. There are advantages and disadvantages to this type of display system. This display system may be inexpensive because it uses mass-produced television technology. It has the further advantage that complex and beautiful pictures, such as that shown in Fig. 8-1 can be drawn by generating the color value of each picture element in a scene. Another advantage is that surfaces can be shown filled, smoothed, and textured as desired. A whole computer industry exists to manufacture raster graphics systems.

Disadvantages of the raster graphics display systems are caused by the quantity of data in a picture. Consider that a picture is divided into 500 × 500 picture elements, a quarter of a million pixels! Each pixel may be one byte to specify one of 256 gray or color values. That is a lot of information for a computer to generate, especially if the picture is moving so that the computer has to generate successive frames, every $\frac{1}{30}$ sec. So raster graphics display systems are excellent for presenting realistic, static images. The complexity of the dynamic images that they can generate is limited, however, by the time required to generate all the pixels.

Raster graphics systems have one other problem, that of aliasing, an error caused by imperfect digital sampling of continuous data. If the computer presents a line that is near to horizontal on the CRT, a staircase effect (Fig. 2-10) is created where it intersects with the horizontal raster lines swept by the CRT's electron beam.

2.4.14d. Vector Graphics Display Systems. Like other graphics systems, vector graphics display systems use a cathode ray tube and an electron beam to draw an image on the front surface phosphor of the CRT; the electron beam, however, is deflected in a different manner. The electron beam is deflected only to those positions on the screen where something is drawn, rather than always deflected over the entire screen regardless of what is drawn. Thus, it is more like a laboratory oscilloscope than a television set.

The vector system also has advantages and disadvantages. It has the advantage that the memory requirements are much lower than for the raster system, because all that needs to be stored in the computer are the coordinates of the endpoints of the lines, rather than values for individual pixels. A complex image consisting of 1000 lines might take 8000

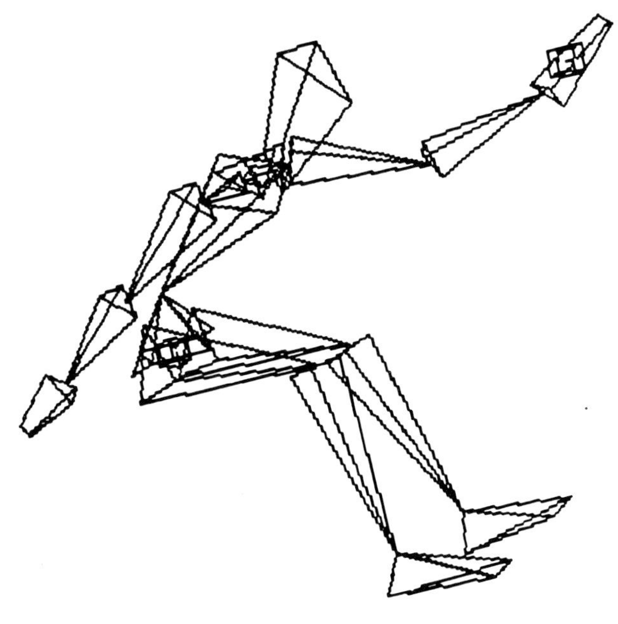

FIGURE 2-10. A wire-frame image drawn on a raster graphics system of modest resolution. The staircase effect as the image lines intersect the raster lines is clearly shown. Reproduced from Badler et al. (1987) with permission.

bytes of storage instead of a quarter million bytes. Because less data are stored, less computing is needed to generate a frame's worth of information. It is thus possible to compute individual frames fast enough so that dynamic pictures can be presented. A disadvantage of a vector system is its higher cost. This is because it cannot use the consumer marketplace of television, it can only use the smaller oscilloscope marketplace. It has a further disadvantage in that it is impossible to fill areas. Consequently, when an image is drawn, it is typically a wire-frame image composed of a series of straight lines (Fig. 2-11). More information about the advantages of both raster and vector systems for presenting anatomical data are presented in Chapter 8.

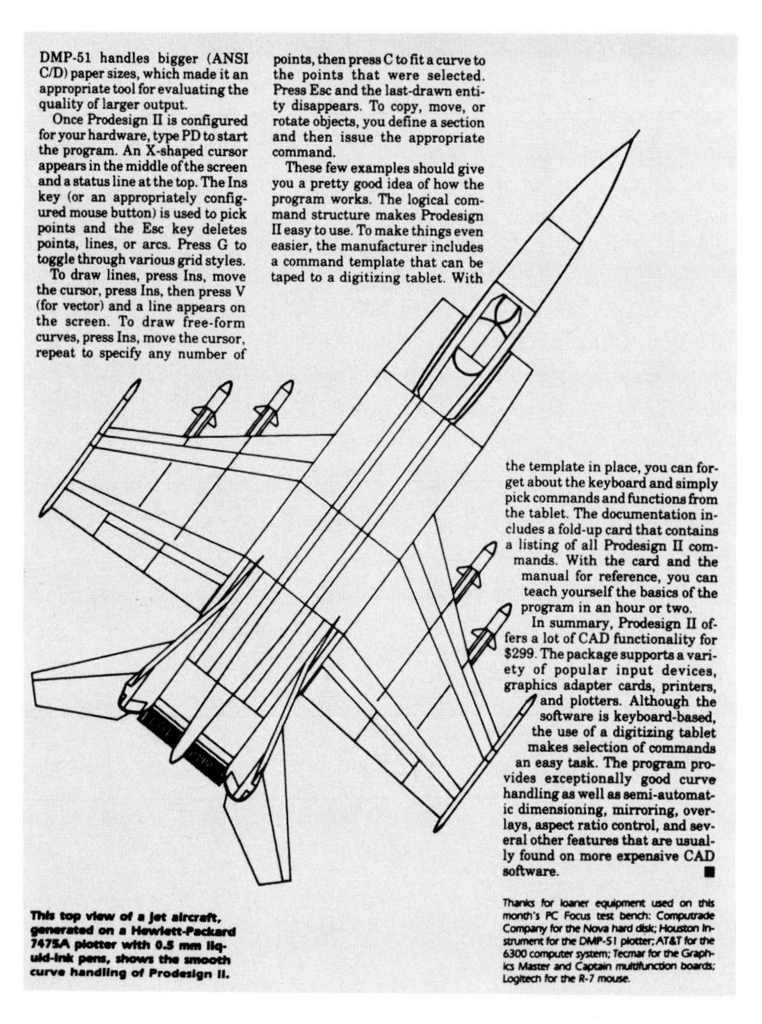

DMP-51 handles bigger (ANSI C/D) paper sizes, which made it an appropriate tool for evaluating the quality of larger output.

Once Prodesign II is configured for your hardware, type PD to start the program. An X-shaped cursor appears in the middle of the screen and a status line at the top. The Ins key (or an appropriately configured mouse button) is used to pick points and the Esc key deletes points, lines, or arcs. Press G to toggle through various grid styles.

To draw lines, press Ins, move the cursor, press Ins, then press V (for vector) and a line appears on the screen. To draw free-form curves, press Ins, move the cursor, repeat to specify any number of points, then press C to fit a curve to the points that were selected. Press Esc and the last-drawn entity disappears. To copy, move, or rotate objects, you define a section and then issue the appropriate command.

These few examples should give you a pretty good idea of how the program works. The logical command structure makes Prodesign II easy to use. To make things even easier, the manufacturer includes a command template that can be taped to a digitizing tablet. With the template in place, you can forget about the keyboard and simply pick commands and functions from the tablet. The documentation includes a fold-up card that contains a listing of all Prodesign II commands. With the card and the manual for reference, you can teach yourself the basics of the program in an hour or two.

In summary, Prodesign II offers a lot of CAD functionality for $299. The package supports a variety of popular input devices, graphics adapter cards, printers, and plotters. Although the software is keyboard-based, the use of a digitizing tablet makes selection of commands an easy task. The program provides exceptionally good curve handling as well as semi-automatic dimensioning, mirroring, overlays, aspect ratio control, and several other features that are usually found on more expensive CAD software. ■

This top view of a jet aircraft, generated on a Hewlett-Packard 7475A plotter with 0.5 mm liquid-ink pens, shows the smooth curve handling of Prodesign II.

Thanks for loaner equipment used on this month's PC Focus test bench: Computrade Company for the Nova hard disk; Houston Instrument for the DMP-51 plotter; AT&T for the 6300 computer system; Tecmar for the Graphics Master and Captain multifunction boards; Logitech for the R-7 mouse.

FIGURE 2-11. A wire-frame image drawn on a vector graphics system. Note that all its lines, even those near the horizontal, are smooth. Reproduced from McGrath (1985) with permission.

2.4.15. Data Tablets

A data tablet is a computer input device composed of a flat surface, usually made of plastic, and a cursor or puck (Fig. 2-12) that you may slide over the surface of the tablet. The computer senses the *X,Y* coordinates of the cursor by one of the methods described below.

The operation of an acoustic tablet is the easiest to explain (Fig. 2-13). The cursor looks like a ballpoint pen. At its tip is a little spark gap like the one in a spark plug of an automobile. Sparks are generated at a frequency of 10 per second, and each spark makes a popping noise. That noise radiates away from the cursor like the ripples from a stone dropped into a pond. Simultaneously with the spark generation, the computer inside the data tablet starts two counters called the *X* counter and the *Y* counter.

At both the left and the top edges of the data tablet are strip microphones. When the radiating sound wave reaches the left microphone, the *X* counter is turned off. The number in the counter is proportional to the amount of time it took for the sound wave to reach the microphone and therefore is proportional to the *X* coordinate of the cursor. Similarly, the *Y* coordinate is determined, and both values are sent to the computer. The computer can therefore sense the *X,Y* coordinates of the cursor as you move it over the tablet surface.

Contemporary data tablets generally do not use acoustic technology; rather, they use a magnetostrictive technology. Inside the data tablet is a very fine grid of copper wire, and the cursor is an antenna. Radio waves sent out from the antenna induce a stress wave, actually a physical shortening in the copper grid, that radiates out as did the acoustic wave. Instead of microphones, strain gauges sense the wave's arrival. These tablets are quiet and less susceptible to noise in their environment.

A data tablet is an absolute device. That is, if you position the cursor at a particular point on the tablet, the tablet will always provide the same *X,Y* coordinates. This is true even if you pick the cursor up from the tablet surface and replace it.

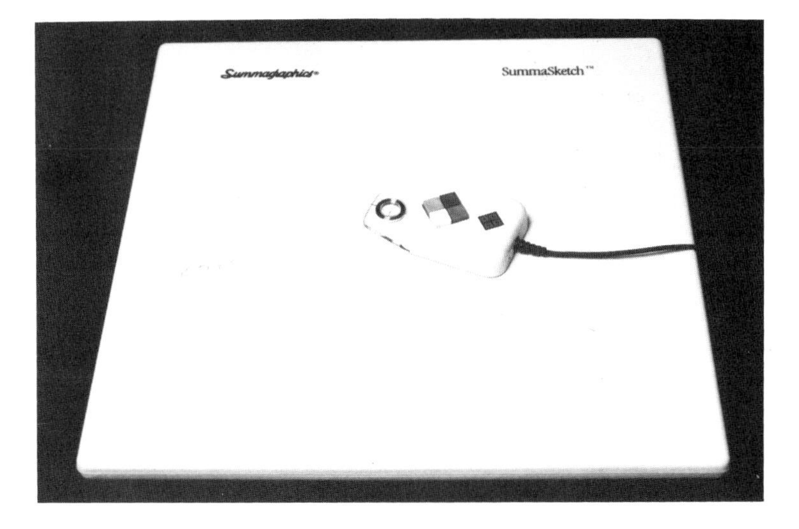

FIGURE 2-12. A data tablet. The user may slide the puck across the flat surface of the tablet, and the computer will sense the position of the cross hairs. He may also press buttons on the puck, and the computer will sense his efforts.

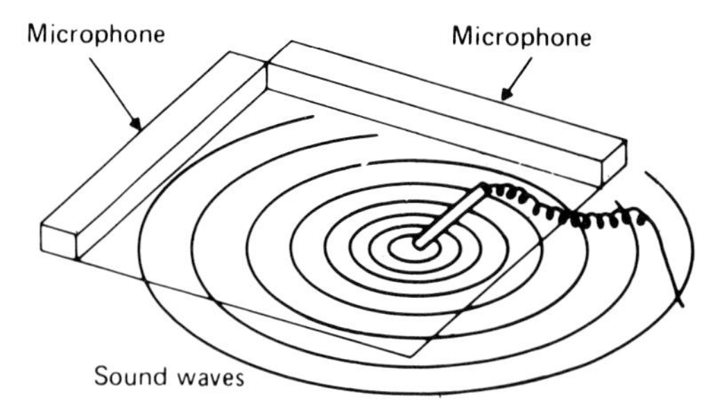

FIGURE 2-13. The functioning of a data tablet. Sound waves ripple outward from a spark gap at the tip of the stylus. The waves are sensed by microphones at two edges of the tablet. The time required for the waves to reach the microphones is proportional to the position of the stylus on the tablet. Reproduced from Foley and van Dam (1982) with permission.

2.4.16. Mouse

A mouse (Fig. 2-14), like a data tablet, is a computer input positioning device. It consists of a small plastic box resembling a real mouse that you may roll around a flat surface. On the underside of the mouse, as shown in Fig. 2-15, are two wheels oriented at right angles. The axle of each wheel is connected to an optical shaft encoder that senses how much its wheel has rotated. Thus, the computer may sense how far you have moved the mouse in the X and Y directions.

Mice cost less than data tablets. They are available for about \$100, whereas a data tablet for an IBM personal computer might cost \$500. However, the mouse is a relative positioning device. That is, if you move a mouse around a table, lift the mouse off the table, and put it down somewhere else, the computer cannot sense that you have lifted it and cannot sense where you put it back down. The computer senses only the rotation of the mouse's wheels. You must accommodate this relative positioning in computer software.

2.4.17. Joysticks

A joystick, shown in Fig. 2-16, is another device with which you can give position information to the computer. The joystick consists of a vertical stick that you may hold in

FIGURE 2-14. A mouse. The user may grasp the mouse with his hand and move it across the table. The computer will sense his motions. He may also depress the pushbuttons, and the computer will sense these efforts.

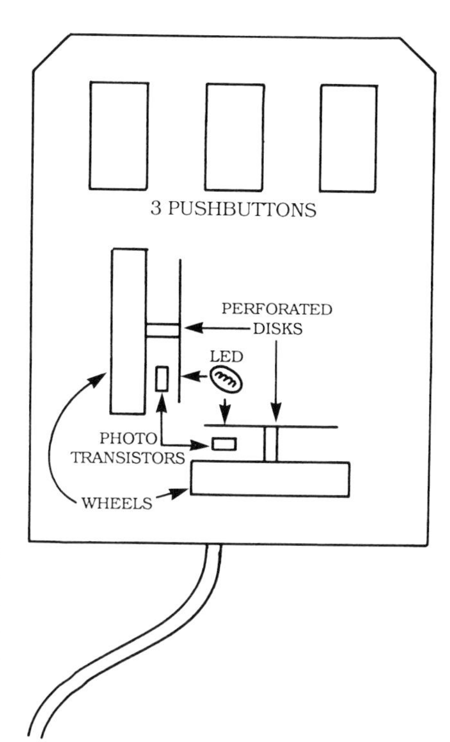

FIGURE 2-15. The gross anatomy of a mouse. Wheels oriented perpendicular to each other turn as the user rolls the mouse on the table. The wheels turn perforated disks, which repeatedly interrupt beams of light that are sensed by phototransistors. The computer may sense the number of pulses of light, thus sensing the amount that the wheels are turned. Several pushbuttons are also provided so that the user may signal the computer.

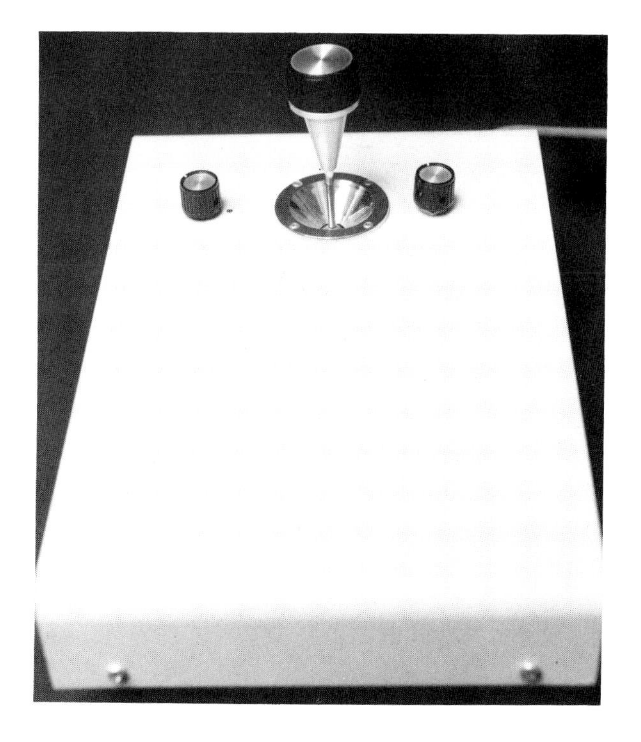

FIGURE 2-16. A three-dimensional joystick mounted in a cabinet. The stick may be deflected, and the knob on the top of the stick may be rotated. All three of these motions drive potentiometers, which the computer may read in order to sense the position of the joystick. Two other potentiometers are mounted in the cabinet, and their knobs are visible next to the joystick. Note the wide flat area of the cabinet. This provides support for the user's palm to allow him comfortable and precise use for an extended period of time.

your hand and deflect in the right–left (X) and near–far (Y) directions. At the base of the stick are a ball-and-socket joint and a pair of potentiometers. The ball-and-socket joint holds the base of the stick while you deflect it, and the potentiometer shafts rotate proportionally to the deflection that you apply. The computer may sense the deflection along both axes by reading the voltages output from the potentiometers through analog-to-digital converters.

Some joysticks are three-dimensional; that is, they have a third knob that you can twist on the top of the handle. This knob is also attached to the shaft of a potentiometer so the computer can monitor its position, thus reading X, Y, and Z coordinates from the joystick. Joysticks are absolute positioning devices. The computer always senses the same coordinate for the same position.

2.4.18. Trackball

A trackball is an input device similar in many ways to a mouse. It consists of a large ball (Fig. 2-17) set into a box so that only the top surface of the ball protrudes through the box. You can rotate the ball within its container. Inside the box are a pair of wheels perpendicular to each other that are rotated by the rotation of the trackball. Each wheel is connected to an optical shaft encoder so that the computer can sense the rotation of each wheel. Thus, the computer may sense the rotation of the ball in the X and Y directions. Both the trackball and the mouse may be used to move a cursor around a screen, but the

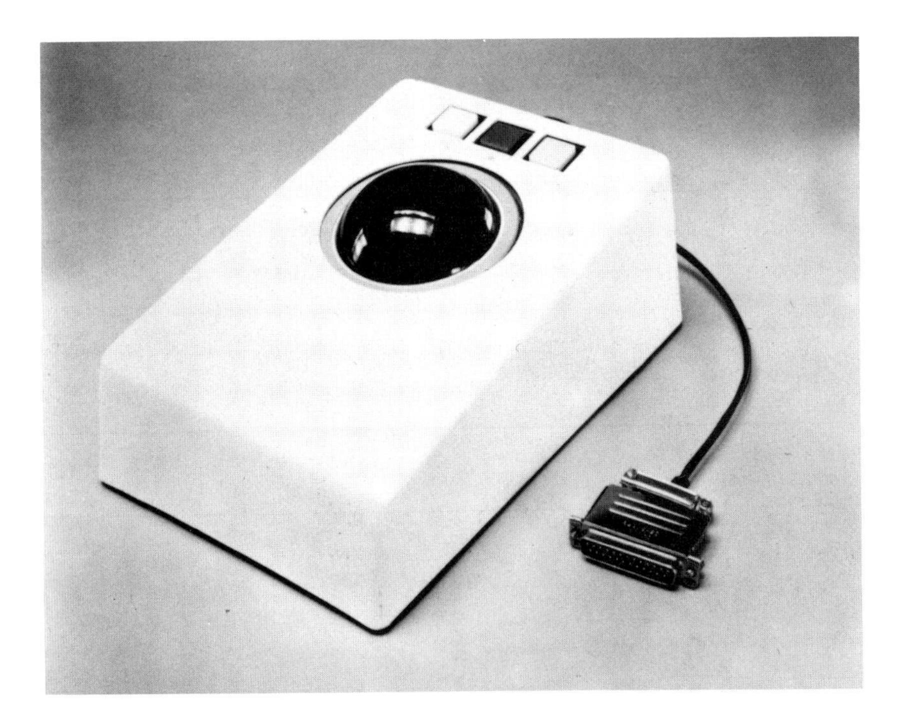

FIGURE 2-17. Trackball. A spherical ball protrudes from its cabinet. The user may rotate the ball, and the computer may sense its rotation. Courtesy of Measurement Systems, Inc., Norwalk, CT.

trackball offers the advantage of not requiring as much table space as the mouse. Except for a few specific industries (e.g., air traffic control), the trackball has not gained the popularity of the mouse.

The data tablet, the mouse, the joystick, and the trackball are all "interactive" computer input devices. You may use any of them to control or to interact with a computer program, provided that the appropriate software is available. Chapter 12 discusses their best use in the neuroscience laboratory.

2.4.19. Stepping Motors

A stepping motor is a small motor that can be easily interfaced to a computer so that the computer's program can move an object. When the motor is turned on, its torque holds its shaft at a fixed position until a pulse is received from the computer. Then the motor shaft rotates from one fixed position to the next or from step to step; hence the name. Usually the stepping motor operates in an open-loop system, allowing no feedback from the motor to the computer (i.e., the computer drives the motor and assumes that the motor actually does its motion). Typical stepping motors have resolutions of 15°, 7.5°, or 1.8° rotation per step.

In neuroanatomy, stepping motors are usually mounted in one of three ways: (1) on a microscope stage to drive the stage from one position to another, (2) on a microscope focus axis to drive the focus from one position to another, or (3) on a filter changer to change a filter in the light path of a microscope to switch quickly from one wavelength of light to another.

If a stepping motor is mounted on a microscope stage to drive the stage in the X and Y directions, you must be concerned with the resolution, or the size of each step. Typical step sizes for microscope stages vary from 0.5 μm through 100 μm.

2.4.20. Components Used for Sensing Position and Motion

This section explains the devices with which the computer can sense position and motion. These components form the heart of the interactive devices described above.

2.4.20a. Potentiometers. Potentiometers are electrical resistors with taps that you may move. You move the tap of a rotary potentiometer by rotating its shaft; you move the tap of a linear potentiometer by sliding it. If a constant voltage is applied across the potentiometer, the voltage output from the movable tap is proportional to the tap's position. The computer may sense the voltage and thus the position through an analog-to-digital converter.

2.4.20b. Shaft Encoders. A shaft encoder is another component that allows the computer to sense rotary motion. A perforated disk is attached to the shaft whose position is to be encoded. As you turn the shaft, the disk repeatedly interrupts a light path between a light bulb (or a light-emitting diode) and a phototransistor whose output can be sensed by the computer. The computer may tell whether the light beam is interrupted or passes successfully through the hole in the disk. Thus, as the shaft is rotated, the computer can count the number of times that the light beam is interrupted, sensing the movement you have applied to it.

2.4.21. Ports

A computer port is a connector for sending information from the computer to some external instrument or for receiving information into the computer from some external instrument. There are many kinds of ports that have been defined for personal and laboratory computers; the three ports that are most frequently used are described below.

2.4.21a. Parallel Printer Ports. Two decades ago, the Centronics Corporation, a manufacturer of computer printers, defined an interface between computers and their printers. Called the "Centronics port," it consisted of eight parallel wires transferring one byte at a time from the computer to the printer and a few other "hand-shaking" wires to tell the computer, for example, that the printer is turned on and that the printer is ready to accept a character. The IBM PC and other common computers adapted this parallel port and called it their "line printer port" or "parallel port." This port is most frequently used to drive a printer; however, it is also used to drive other instruments that can receive eight bits of data at a time.

2.4.21b. Serial Ports (Receive-Send 232C Port). A serial port is an electrical connector on the computer and circuitry for transmitting data, a bit at a time, along a single wire from the computer to some external device. The transmitting port disassembles each byte of information into individual bits and sends them over the wire, one after another. The serial port in the receiving device reassembles the bits into bytes to be used by this device. Data may be transmitted in the opposite direction on a second wire. The format of the data has been standardized by the Institute for Electrical and Electronics Engineers (IEEE) and given the name RS-232C. (RS stands for "recommended standards"; the terms "serial" and RS-232C are now used interchangeably.)

A serial port is also called an "asynchronous" port. It is called that because the transmitting port sends its information bit by bit and the receiving port receives and reassembles the information without the two devices being synchronized by any third device. For this concept to work, the transmitting device must send its bits in a format known by the receiving device, so that the receiver can reconstruct the data accurately. The format is specified through a number of parameters, which must be set in both ports, usually by software or by setting switches. Setting the parameters is known as "configuring the ports."

The port parameters specify the speed of transmission and details about the format of bits that are sent. The baud rate (simply put, a "baud" means bit per second, named after the French engineer Emile Baudot) must be specified to instruct the ports how rapidly to transmit the data bits. The number of data bits per character should be specified. Either eight data bits may be sent per byte or the leading bit may be omitted and seven data bits may be transferred. An error-checking scheme called "parity" may also be enabled. With this scheme, an extra bit, called the "parity bit" is sent after the data bits. The parity bit indicates whether the number of data bits sent with the value "one" is even or odd. When "odd parity" is enabled, the parity bit is transmitted with the value "one" if an odd number of data bits with the value "one" have been sent. The receiver checks whether the total number of data bits received with the value "one" is odd. If it is, then the receiver expects to receive a parity bit with the value "one." If the parity bit it receives has the value "zero," then the receiver port knows that a transmission error has occurred. Either one or two "stop" bits may be sent. Stop bits are sent to terminate the transmission of a byte of data. For most contemporary devices that are connected to the serial port of a laboratory computer, the ports are configured to 9600 baud, eight data bits, parity disabled, and one stop bit.

A special cable called a "null modem" cable is used to link a computer's serial port to the serial port of a device (e.g., a plotter). This cable is so named because there is no modem or telephone hookup in the connection.

2.4.21c. IEEE-488 Ports. The third type of computer port found in the laboratory is called an IEEE-488 port, or simply an IEEE port. This is a parallel port that may transmit 8 data bits simultaneously. Unlike the printer port (Section 2.4.21a), an IEEE port connects the computer to a bus, called the IEEE bus, which can link computers and other instruments, much as a computer bus links the components within a computer. The port may provide addresses as well as data, so that several instruments may be linked, and the computer may direct its data to a specific one. The IEEE-488 bus was designed by the Hewlett–Packard Corporation for linking together laboratory instruments; it was originally called the Hewlett–Packard Instrument Bus (HPIB) until it was adopted as a standard by the Institute for Electrical and Electronics Engineers. Plug-in cards that provide an IEEE-488 port are readily available for common laboratory computers.

2.5. PHYSICAL CONSTRUCTION OF LABORATORY COMPUTERS

This section reviews the basic elements from which a computer is manufactured. This is important for you the neuroscientist because you may want to install a new accessory in your machine. The cost of laboratory computing hardware is now low enough that even if you cannot budget for computer personnel, you still frequently will install and want to upgrade your computer. Often you will have to do the installation and upgrading yourself.

The basic unit of computer construction is the "integrated circuit" ("IC," "chip," and "bug" are synonyms), shown in Fig. 2-18. Within its plastic case is a small wafer of silicon on which the many transistors which perform the functions of the machine are grown. The silicon wafer is much smaller than its plastic case; the larger case size is necessary to allow electrical connections from the wafer to the metal connectors or "legs" of the IC.

The chips are plugged into "printed circuit boards" ("PC board" is a synonym). The PC boards are made of plastic onto which several layers of copper lines have been laid by a photography-based acid etching procedure. The procedure that prints the circuits on the boards (hence their name) mass produces them at a low unit cost. The copper lines connect the ICs together to form the more complex circuitry required in a computer and to provide power to the ICs. The PC board also provides physical mounting for the ICs.

Several boards may be connected to form a computer by plugging them into connectors in the bus. In smaller computers, such as the IBM AT shown in Fig. 2-19, the bus wires and its connectors are etched into a large PC board called the "mother board," and other cards plug into it. On the mother board is usually mounted the necessary logic to

FIGURE 2-18. Integrated circuits. The largest microprocessor chip on the left is about 3 inches long and contains about 100,000 transistors. The smallest, a small storage element, is about 1 inch long and contains about 1000 transistors.

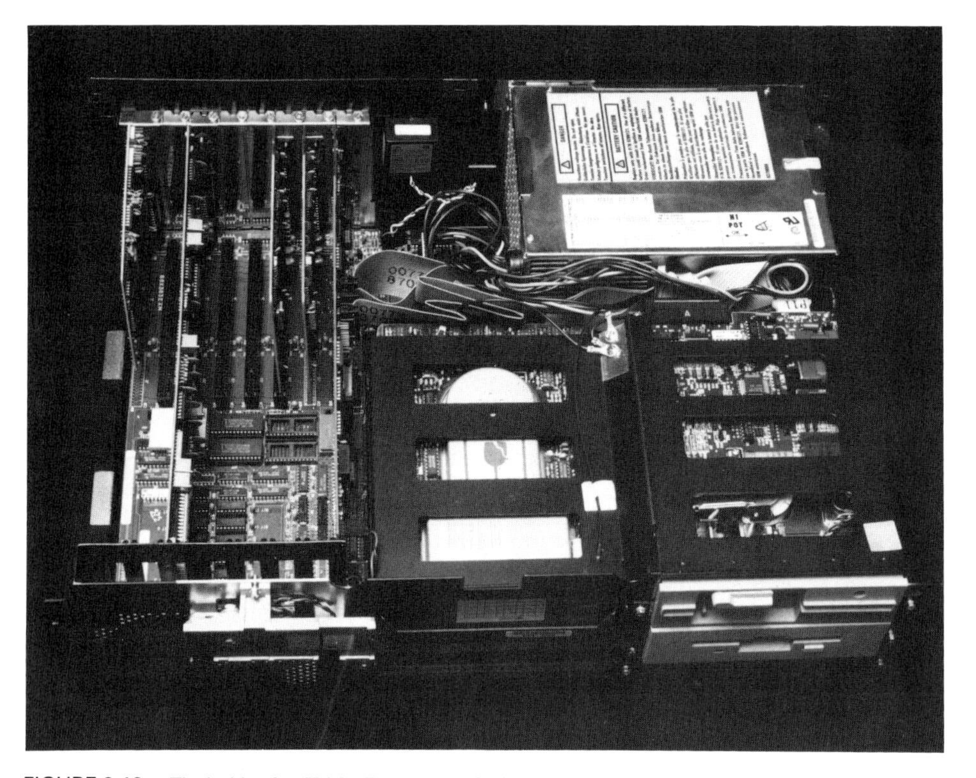

FIGURE 2-19. The inside of an IBM AT computer. Optional cards are plugged into connectors on the left side of the machine. The IBM AT bus links these connectors together. The hard disk drive is in the lower center, the floppy disks are in the lower right, and the power supply can be seen in the upper right.

make the computer function to a basic degree, e.g., the CPU and a modest amount of memory.

In larger contemporary machines and in most older machines, there is no mother board. Rather there is a "card cage" with slots into which cards are slid. At the rear of the cage are connectors that are wired together to form the bus and into which contacts on the rear of the cards are plugged. There is no electronic logic in the cage; all computing is done on the cards that plug into it.

Most contemporary integrated circuits require 5-V direct current. To provide this voltage, the computer contains a "power supply" that converts the AC power from a wall receptacle into the form for the integrated circuits. Some of the smallest machines, usually called "lap-top" computers for obvious reasons, may also be operated on batteries.

2.6. COMMON LABORATORY COMPUTERS

In this section we review some of the computers currently used in anatomic laboratories in the United States. Chapter 16 reviews commercially available systems consisting of computing hardware, computer software, and other instruments assembled to perform a laboratory data collection and analysis task. Frequently, the computers described in the next sections are included in the Chapter 16 systems.

2.6.1. IBM PC

The IBM personal computer and its later derivatives (IBM XT, IBM AT, and 386 machines, discussed below) are the computers most widely used in neuroanatomical research laboratories. When introduced in 1983, the IBM personal computer had a CPU called an 8086, manufactured by the Intel Corporation. The computer had a maximum of 640 kbytes of memory and one or two floppy disk drives, each with a capacity of 360 kbytes. No hard disk drive was included. A math coprocessor chip called the 8087 could be plugged into its mother board. The IBM PC bus contains eight data lines, which are time-shared with 20 address lines.

When it was introduced, the IBM PC sold for about $3000. Since that time, the price has fallen to below $1000, and many companies began (and continue) to manufacture them. A "clone" (see Section 2.7.1) is a copy of a computer manufactured by another company and usually sold at a lower price. Approximately 20 million IBM PCs and clones have been sold in the United States; about 85% of them used are for word processing.

2.6.2. IBM XT

In 1984, IBM introduced the IBM XT (XT stands for "extended technology"). This machine is an IBM PC but contains a hard disk, usually with a 10-Mbyte capacity.

2.6.3. IBM AT

In the following year, IBM announced the IBM AT (Figs. 2-1 and 2-19) (AT stands for "advanced technology"). It contains an Intel 80286 CPU chip and an optional 80287 math coprocessor. A floppy disk drive of 1.2-Mbyte capacity and a hard disk of approximately 30-Mbyte capacity are typically included. The bus was expanded to include 16 data lines, thereby allowing faster CPU access to many devices. This computer could execute all the programs that could run on the PC or XT but was several times faster than the PC. The computer hardware was designed to allow several people to use it simultaneously, each with his own terminal, each believing that he had his own IBM XT. However, no one developed a popular operating system that allowed several people to use it simultaneously. Consequently, most buyers purchased the AT because it has a faster speed than does the XT. IBM discontinued manufacturing the AT in 1987, yet many companies still manufacture AT clones. The IBM AT and its clones probably form the most popular computer in the laboratory today.

2.6.4. 80386 Machines

The Intel Corporation next developed a CPU chip called the 80386 and a math coprocessor called the 80387. The Compaq Corporation first packaged this processor into an IBM AT upgrade in 1987 and was followed by many other companies into the marketplace. IBM did not build an 80386 upgrade and instead concentrated on its Personal System/2 (see Section 2.6.5). The 80386 machines are similar to 80286 machines except that the bus has been expanded to 32 address lines, and the CPU is several times faster than the IBM AT.

2.6.5. IBM Personal System/2

The IBM Personal System/2 is a line of personal computers introduced by IBM in 1987. As this is written, the least-expensive version is the model 30, which is a newly packaged IBM XT containing an 8086 CPU. The models 50 and 60 are repackaged IBM ATs containing an 80286 CPU, but with two significant changes. A new bus, called "microchannel architecture" (MCA) has been included. It contains 32 data bits and 32 address bits. Also, a new computer terminal called the VGA (Section 2.4.6) provides improved graphics capabilities. The PS/2 model 80 contains an 80386 CPU, the MCA bus, and the VGA graphics terminal. It is several times faster than the AT.

All the IBM computers, current as well as old (including their clones), have a major advantage—software compatibility from one machine to the next. They all also use the same operating system (Section 3.3.1), called MS-DOS ("MS" stands for Microsoft, the company that wrote the operating system; "DOS" stands for disk operating system).

2.6.6. Apple II

The Apple Corporation introduced the Apple II line of machines in 1982. It was the first personal computer that sold successfully into the scientific laboratory, achieving this success because of its low price (at that time, approximately $2000) relative to the then very popular microcomputer, the DEC PDP-11/03 (Section 2.6.10). Its CPU was a Zilog Z-80, so the machine was incompatible with both the DEC and the IBM lines of computers. It had a low-resolution graphics screen yet became popular in video games, school systems, word processing, and laboratory work.

2.6.7. Macintosh

Introduced in 1984, also by the Apple Corporation, the Macintosh (Fig. 2-2) was the first machine designed for use by a computer novice. It used the mouse rather than the keyboard as the primary interactive device. Its CPU was the Motorola 68000, so this machine was also incompatible with the DEC and IBM lines. A number of easy-to-use programs for word processing (e.g., MacWrite) and making illustrations (e.g., MacPaint) were provided; because of these, the machine became popular. The Macintosh was designed for "bounded" tasks; that is, no provisions were made to expand the computer by plugging in new boards. This kept the price low and allowed simple-to-use software to be developed. A faster version, called the Macintosh Plus, was introduced 2 years later, followed by the Macintosh SE. These machines are still commonly found in the laboratory.

2.6.8. Macintosh II

The Apple Corporation upgraded the CPU in the Macintosh to the Motorola 68020, rendering this machine quite a powerful computer. It contains a bus, called the "Nu-bus," that contains 32 data wires and 32 address wires. This machine has promise in the laboratory, as numerous companies are manufacturing accessories and writing software for it.

2.6.9. VME-Bus Machines

The Motorola Corporation introduced a series of CPU integrated circuits called the 68000 series in 1981. They are powerful CPU chips that contain the logic for 32-bit arithmetic. They established a bus called the VME (for Versabus Module European) bus and built a series of computers using these components. The VME bus contains 32 address lines and 32 data lines that are not time-shared, so the machine is technically superior to the IBM line and is especially suited for applications that demand a large amount of memory, such as image processing. In spite of this, VME-bus machines have never been aggressively marketed into the laboratory, nor has a popular operating system ever been devised.

2.6.10. DEC PDP-11

The Digital Equipment Corporation introduced computers into the laboratory in the late 1960s with a series of small machines called the PDP-1 through the PDP-15—PDP stands for "programmable data processor." The most popular laboratory computer in the 1970s evolved; it was the PDP-11, and numerous models of it were manufactured. It contained a DEC-manufactured CPU and had two general versions, a minicomputer version and a microcomputer version. The minicomputer version, highlighted by the PDP-11/45, contained a bus called the Unibus with 16 data lines and 18 address lines. Expansion was easily done by plugging cards into the bus; DEC and other vendors made many cards that performed the standard laboratory functions such as analog-to-digital conversion. A typical price in 1975 for a PDP-11/45 was $40,000, depending on mass storage devices.

A microcomputer version of the PDP-11, called the PDP-11/03 or the LSI-11, was introduced in 1979. With it, DEC introduced a new bus, called the Q-bus, which contained 16 data lines time-shared with 22 address lines. They also provided many laboratory-related plug-in cards for this bus. The PDP-11/03 sold then for about $15,000, and since it could execute the software of older Unibus PDP-11s, it became a very popular computer in the laboratory. Faster and less expensive versions called the PDP-11/23 and PDP-11/73 were introduced during the next few years. These machines are still in wide use today.

2.6.11. DEC VAX

The Digital Equipment Corporation has elected not to compete in the personal computer marketplace; rather, they have spent their recent effort developing higher-performance machines. For the neuroscience laboratory they have made a machine called the VAX (for "virtual address extension"). VAX versions range from small machines called the MicroVAX (about $20,000) through large VAXes (about $200,000). The VAX machines are powerful, generally multiuser machines that are incompatible with the IBM and Apple lines of personal computers. Interestingly, the VAX has entered the laboratory from above, that is, by scaling down large computers. This is in contrast to IBM or Apple, whose computers started very small and to which functions and speed are continually added.

2.7. PURCHASING A COMPUTER

This section, and the ones following, give some ideas on how, where, and from whom to purchase a computer for the neuroscience laboratory. When you purchase a machine, you should keep in mind that the goal is not to buy a computer; the goal is to do a laboratory task. Indeed, with the reduction in price of computing hardware in the last decade, the cost of the computer hardware forms only a small portion of the cost of a task.

2.7.1. Clones

IBM, a large company, designed and manufactured a common laboratory computer called the IBM AT. If you purchase this machine directly from IBM, you will pay a high price for it, even though you may benefit from such aggressive marketing techniques as state contracts and university discounts. The machine will not be at the cutting edge of technology, and it will not have the highest speed. You may be reasonably confident, however, that the machine's manufacturing quality and its reliability will both be high. The vendor will provide training, customer sales and technical support, and local repair service should there be a breakdown. You may also have confidence that the vendor will exist a year after your purchase.

You may, as an alternative, purchase a clone. This is a copy of the original machine, manufactured and sold by a different company. Clones are usually significantly less expensive and may have higher performance. Clone manufacturers have several advantages over the original company that allow them to offer you a higher-performance machine at a lower price. A clone company is usually smaller and therefore has less overhead, both in cost and in the time required to bring a product to the marketplace. Clone manufacturers usually provide fewer local customer services, instead relying on mail-order firms to market and service their machines. Thus you, as a customer, may receive higher technology at a lower price.

There are some disadvantages with purchasing a clone. The mechanical quality of a clone is generally not as high; for example, the clone companies do not spend as much money on keyboards or cabinetry. If you spend 4 hr a day at a keyboard, you may want to spend the extra money for a keyboard that is well designed and feels good. Customer support, including manuals, is generally not as available. You will therefore need more knowledge to set up and start to use the machine. Finally, the financial future of the manufacturer may be unstable. Clone companies vary; those that are the most stable, reputable, and service-oriented charge more for their machines.

Software compatibility is an issue when buying a clone. There are many companies who write and sell software for common laboratory computers. When a clone company duplicates one of the common computers, it must also duplicate its operating system. The lowest-level part of the operating system is not usually provided on a floppy disk; it is burned into read-only memory ICs and thus provided as "firmware" (Section 2.4.3) by the hardware manufacturer. Most major manufacturers of laboratory computers have secured copyright protection for their firmware, so clone manufacturers cannot duplicate it exactly. As a consequence, there may be subtle differences between the way a software package executes on a common computer and the way it executes on a clone. Whether or not the clone will suffice for your task is a function of the software and the degree to which the clone emulates the original machine. Therefore, you should attempt to run the software you wish to use on the clone before you purchase it.

There is a similar problem with hardware. There are many companies that manufacture accessory boards for the common computers. If you choose to buy a clone, you may find that the accessories will not function properly because of subtle hardware differences between the clone and the common computer. Again, the answer is to try the hardware in the clone before buying it.

2.7.2. Compatibility

When you buy a computer, you must also consider with what or whom you want to be compatible. If you have a colleague within your department or at another university who has software that you would like to use and that is written for an Apple computer, you might consider buying an Apple computer. Then you can use his software or his data directly without having to translate either from one machine to another. This software compatibility is becoming ever more important, because software that is not mass-produced is now more expensive than computer hardware. And rewriting software simply to adapt it to another computer may be an expensive task, providing no new functions. Therefore, when you purchase hardware, you should think "software first." Find the software that accomplishes the task you desire and then obediently purchase the hardware on which the software runs.

2.8. FOR FURTHER READING

General articles on how to select a laboratory microcomputer have been published by Flaming (1982), Moreton (1985), and by Poler et al. (1985).

Details of specific computer-related laboratory hardware have also been published. Capowski and Schneider (1985) describe a motor controller for computer-assisted microscopy. Gdowski et al. (1987) describe an optical encoder system for the microscope focus axis and present accuracy tests of the encoder. A video column digitizer for capturing television images into the computer has been reported by Kapps and Mays (1978). Computer graphics boards for the line of IBM personal computers have been reviewed by Litzinger (1988).

A complete description of the often-confusing serial communications or RS-232 standard is available in Artwick (1980) and Putnam (1987).

Finally, general reviews of the use of computers in neuroanatomy have been done by Mize (1984) and by Capowski (1988b).

3

Software in the Neuroanatomy Laboratory

3.1. INTRODUCTION

In the early 1980s, the cost of a computer program for a typical laboratory task approached and began to exceed the cost of the computer itself. This economic event, though its exact date is impossible to define, has changed the manner of thinking by everyone involved in using, writing, buying, and selling laboratory computer products. Since this period, the cost of computers has continued to fall, and though tempered by mass production, the cost of software has continued to rise. Furthermore, there is every indication that these trends will continue. What does this mean for you, a neuroanatomist who is just starting to use computers for laboratory tasks? It means that you should think more about the computing functions you want to perform in your laboratory and less about the hardware you want to buy.

This chapter offers you information about computer software as it relates to the laboratory. The chapter presents the basic elements of software to the extent you need for talking with a computer programmer and with a computer vendor. It also describes several options that you may exercise and their pitfalls when you confront an anatomic problem that requires data processing.

As author of this chapter on software, I had to face a classic "function and form" problem. For to describe how software is translated and used requires descriptions of the software instruments involved; but to describe the instruments first requires a description of how they are used and also makes boring reading. Therefore, the chapter briefly presents the procedure of writing software and then follows with more rigorous definitions of the components of software.

3.2. HOW SOFTWARE IS WRITTEN

"Software" and "code" are both synonyms for computer program. The term software is more commonly used, as it emphasizes the contrast between the purely mental constructs, or thought stuff, of programming and the actual electrical and mechanical components of the computer, known as "hardware" (Chapter 2).

A computer programmer begins with a problem and solves it by devising an "algorithm," a series of step-by-step instructions to solve the problem. To illustrate, imagine that you must design a program to teach a drug store cashier how to make change when a customer hands him a dollar bill and makes a purchase for less than the dollar. You would need to teach the cashier how many pennies, nickels, dimes, and quarters he must return to the customer for any purchase price. You would probably write down a series of instructions such as: Calculate the change to be returned by subtracting the purchase price from one dollar; if the change to be returned is greater than 75 cents, hand the customer three quarters; and so on. This series of instructions forms the algorithm that you devise to solve the problem; it is independent of any specific computing hardware. The process of specifying the instructions of such an algorithm is called "programming."

3.2.1. Source Files

Once the algorithm for change making in the above example has been devised, it is "coded" into some computer language such as the following:

```
\* Source file for making change from a dollar */
   CHANGE = 100 - PRICE                              (3-1)
   IF (CHANGE > 75) NUMBER_OF_QUARTERS = 3
```

Here, for the first time, the programmer must get involved with a computer. He uses a text editor or word-processing program to generate an ASCII "source file" on the computer disk. "ASCII" (see Chapter 2) means characters of text, "file" means an area on the computer disk, and the source file contains these step-by-step instructions. The source file (Eq. 3-1) may be read by both a human being when he augments or changes it and by a program (Section 3.2.2) that translates it into a form that the computer may execute.

Strictly speaking, a "programmer" generates the algorithm as a computer-independent series of instructions, and a "coder" converts these instructions into the source file. Loosely, however, the term "programming" is used to cover both of these functions. Frequently the algorithm is never written down in a machine-independent form; rather the programmer works only from his mental concept of the algorithm to generate the source file.

3.2.2. Translation and Execution

The source file that the programmer has generated is in a form that he (and other people) may read but is not yet in a form that may be executed by a computer. A translation process must be first invoked, as shown in Fig. 3-1.

A computer program called a "compiler" reads the source file and translates it into an intermediate form called an "object file." This intermediate form is actually a series of calls to other program modules that are commonly used and are described in the next paragraph. Modules, sometimes called "functions" or "subroutines," are small computer programs that may be invoked from other programs.

Other object files, usually to perform general-purpose tasks, (e.g., to convert the numeric value 123 to the ASCII characters "1," "2," "3" so that they may be sent to a

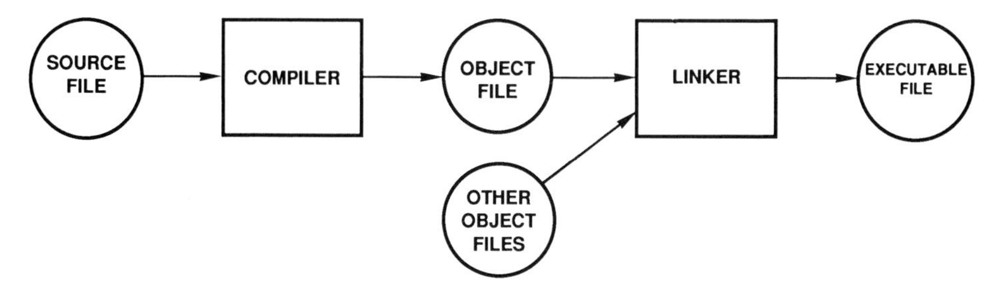

FIGURE 3-1. Translating a source file into a machine-executable file. The source file is compiled and linked with other programs to form an executable file, which the computer may run. Further details are given in the text.

printer) are provided by the company that writes the compiler program. These general-purpose modules are frequently clustered in a "library" of modules; i.e., many of them are stored in one object file.

The next step in the translation process is called "linking." Here another program called a "linker" reads the object files, combines them into one program, and translates them into machine-executable code, or machine code for short. This form of the data is stored in a disk file called an "executable file" or a "run file" or a "save" file.

An executable file contains the program now completely translated into a form that only the computer may understand. It comprises the actual program and possibly some data on which the program will operate. You may recall from Chapter 2 that a CPU executes an ADD instruction by fetching the instruction and the data from memory, doing the addition, and storing the sum back into memory. When the CPU is designed at the factory, a unique numeric code is declared for each instruction (e.g., the code for ADD is 107, the code for SUBTRACT is 223) that the CPU can execute. An executable file is a series of these numeric codes that specifies the series of instructions that the CPU must execute in order to run the program.

Once the translation is finished and the executable file has been generated, all that remains is to "run" the program, to make it perform its desired task. When you instruct the computer to run a program, the computer's operating system (a "master program," see Section 3.3) fetches the executable file from the disk and loads it into the computer's random access memory. The operating system then starts the CPU executing the first instruction in the program. The CPU executes the instructions in sequence, usually until the last one is reached, at which time, the program is terminated.

The whole translation process is harder to describe than it is to employ. Indeed, the entire translation process may be effected by simply typing a few keys. If the name of your program is "BILL" and the program is written in FORTRAN (Section 3.5.2), then you simply need to type "FORTRAN BILL" and "LINK BILL" to perform the translation. Furthermore, a series of commands, even as short as these, may be entered into a file called a "batch file" (also called a "scripting" or "command file"). Then all the commands may be executed by simply typing the name of the batch file. Nor is the translation process too time-consuming. About 2 min is required to translate a 1000-line source file on an IBM AT computer.

At any time during this entire process, either you or the translator software may discover an error in the program. An error in computer hardware or software is called a "bug," from a 1947 incident with Harvard University's Mark II computer. A moth flew in a window and lodged itself between two relay contacts in the machine, causing a failure. The process of correcting mistakes in hardware or software is called "debugging."

Because the translation process is easy and not time-consuming, great effort has not been made to modify or "patch" the executable file once it has been generated. The normal way to change a program is to edit the source file and translate it again. Programs to patch an executable file do exist but are infrequently used.

3.3. SYSTEM SOFTWARE

Computer software is loosely classified into two categories: "system software" and "application software." System software performs general computer-related tasks, such as language translation. Generally you purchase it with the computer, possibly even from the manufacturer of the computer hardware. The computer hardware and the system software form a computer "system" ready for an application; hence the name. Application software (Section 3.4) is more related to the application than to the computer. A program for neuronal morphometry forms an example of application software.

System software consists of the operating system, language translators, text editors, and utility programs that perform a variety of needed functions.

3.3.1. Operating Systems

The operating system is the "master program" that controls the operation of the computer and other programs that are executed by the computer. It is divided generally into two parts: that part which resides in a disk file and is loaded into memory as it is needed and that part which is burned into a "read only memory" (Section 2.4.3) and therefore exists in computer memory at all times. Since the majority of the operating system is stored on the disk, it is frequently and generically called a "disk operating system" or DOS.

When the computer is turned on, its CPU is designed to start executing the portion of the operating system that is stored permanently in memory. This program's first task is to fetch the rest of the operating system from the disk into memory. Such a self start is called "bootstrapping," from the concept that a person could lift himself off the ground by pulling up on his own bootstraps. The portion of the operating system that fetches the rest of the operating system is called the "bootstrap."

The operating system contains software for performing frequently needed functions, such as reading the keyboard to see if you have typed a key. Whenever you do, it displays or "echoes" the key that you have hit on the computer terminal. If you type a command to run one of your programs, the operating system's software fetches your software from disk into memory and starts the CPU executing your program. When your program is finished, the CPU again executes code within the operating system to await your next command. Note here that a computer is never idle; its CPU is always executing some program, though it may be only the part of the operating system that asks repeatedly "Have you typed a key yet?"

Many utility programs are part of operating system software and are available at your pleasure. For instance, you may list a "directory" or a "catalog" of the files on a disk to see their names. You may delete a single file or all the files on a disk. You may "format" a disk, that is, write basic information on the disk tracks that defines how later files will be stored. You may copy a file from one disk to another or from a disk to a printer.

Two operating systems for common small laboratory computers deserve mention. The line of IBM personal computers uses an operating system called MS-DOS (for Microsoft Disk Operating System). It is written by the Microsoft Corporation and is provided by IBM and clone manufacturers with their computers. DOS works in conjunction with a special section of code called BIOS (for basic input–output system), which controls the elementary functions for reading data from and writing data to disks. BIOS, provided as part of the computer hardware, is burned into ROM and resides permanently in memory.

Another common laboratory operating system is RT-11. It stands for "real time on the PDP-11" and was written by the Digital Equipment Corporation for their PDP-11 line of computers. It has been widely used for laboratory work.

3.3.2. Time-Sharing (Multiuser) Operating Systems

The operating systems just described allow one person at a time to use your computer system. This is reasonable for inexpensive computers and for tasks whose data need not be shared among several people or programs. Such an operating system is simple, requires little disk and memory storage, and, once the program is running, uses a very small amount of the CPU time to perform its ongoing tasks such as updating a clock.

The single-user concept fails, however, if the computer is too expensive for any individual to purchase or if several programs need simultaneous access to the same data. Then a multiuser operating system allocates use of the computer to several users, each running a program from his own terminal, by "time sharing." The CPU executes the program of the first user for a brief period, say 0.1 sec. Then the operating system stops this user's calculations, saves whatever intermediate results exist, and starts to execute the second user's program. After his time slice, the third user is started; eventually the first user gets another turn.

The operating system must juggle the execution of all the programs, usually keeping their programs and data separate. It must ensure that one user cannot accidently delete the data or programs of another user. When a multiuser operating system is well designed and it is running on a computer that is not overloaded, each individual user may indeed think that he has the computer all to himself. The multiuser operating system is a complex program, however, and requires space and time overhead. In fact, the operating system must use a fraction, possibly 20%, of the CPU power simply to perform its functions.

If the various programs want to share data, then the operating system becomes even more complex, for it must also resolve conflicts. If two users want to access data at exactly the same instant, the operating system must give priority to one of them. Access to data may take different forms. For example, a statistical summarization program may be allowed to read but not write data to a disk file, whereas a data-collecting program may need both capabilities. Some data may need to be secure; no one except its sole user should have access to it, even just for reading. The operating system must allocate these types of data access.

Two common multiuser operating systems deserve mention. UNIX was developed at the Bell Telephone Laboratories for the Digital Equipment Corporation PDP-11 computer line. The programming language C (Section 3.5.1.) was introduced with this operating system. UNIX has since been adapted to many common computers, large and small. OS/2 (for Operating System/2) is now being introduced by IBM for its line of PS/2 computers and will probably find greatest use in the office environment. Data files written by programs running with the OS/2 operating system will not be compatible with those written by programs running with the DOS operating system. At this time, it is difficult to predict the popularity of this operating system in the laboratory environment.

3.3.3. Language Translators

The system software provides the capability for you to write programs in various languages. In other words, you first prepare a source file for a program according to the rules of one of the standard languages described in Section 3.5. Then you may translate it into machine-executable code by running a translator program, which the computer manufacturer provides.

There are three basic methods of translation: assembling, compiling, and interpreting. These, covered in the paragraphs below, provide differing capabilities and levels of efficiency of the resulting executable program.

3.3.3a. Assemblers. From Section 3.2.2, you may recall that machine-executable code consists of a long series of numbers that form codes for arithmetic operations, numeric representations of data, and addresses where data are stored in memory. A typical machine language instruction is shown here:

32705 20443 20574 (3-2)

The first number specifies the operation "ADD." The next two values are memory addresses. And the instruction means: add the number stored in the memory location 20443 to the number stored in the memory location 20574 and store the result in the memory location 20574. A machine language program is simply a long list of similar values.

This makes perfect sense to the machine, but it is very difficult for a human. You could write a program strictly in machine language; in fact, 25 years ago, programs were commonly written this way. It is, however, a very difficult and error-prone process. So a programming language called "assembly language" and a translator program called an "assembler" were designed. Now you might write this assembly language instruction as:

ADD XDATA, YDATA (3-3)

You write the mnemonic ADD and names for the data, called "variables" to specify the addresses where the data are stored. By giving the instructions and the variables meaningful names, you will make fewer errors than if you use nonsense strings of numbers. The assembler program translates each assembler language instruction into a machine language instruction. It converts the characters "ADD" into the proper numeric code, assigns locations in memory where the data are to be stored, and places the addresses in the machine code.

An important point to note here is that each assembler language instruction translates into one machine language instruction. For this reason, assembly language programming is called "low-level" programming.

Using assembly language, a programmer may write the most flexible programs, which run in the minimum amount of time and require the least amount of memory. Assembler language programming is machine-specific, because each CPU uses unique codes to specify its instructions. Assembly language is therefore used primarily for very critical functions, such as the frequently used routines within the operating system. Because assembly language programming is the least productive (Section 3.6.2) and is unlike the common algebraic equations that scientists normally deal with, only a small amount of laboratory programming is done in this fashion.

3.3.3b. Compilers. When engineers and scientists began programming computers in the late 1950s and early 1960s, assemblers were the only translators available. But they wanted to write mathematical programs as formulas, a form more familiar to them, for example:

$$Y = Ax^2 + Bx + C \quad (3\text{-}4)$$

rather than as a series of ADDs and MULTs. So a "compiler" was designed. A compiler is a computer program that translates a source file of complex statements such as equation 3-4 into a series of machine language instructions that perform the necessary arithmetic. Thus, the programming language FORTRAN (for FORMula TRANslation, Section 3.5.2) was conceived, and formulas were translated by machine.

With compilers, the translation process is done once to create the executable file, and then the program may be executed many times using varying data. Because the values of the data are not known when the program is translated, a compiler must assign memory locations where future values will be stored. A compiled language is a high-level language, because each statement is translated into a series of machine language statements.

3.3.3c. Interpreters. Interpreters are translators that convert high-level languages into machine language, but the translation is somewhat simplified. Both translation and execution are performed one line at a time from the source file every time the program is run. This means that the values of the data are known when the translation is performed, so that the translation procedure is more simple. Interpreting a program is more time-consuming than running an already-compiled program, however, because each line has to be both translated and executed every time.

Interpreted languages are designed to make it easy to do simple programming tasks. Frequently a text editor is even included as an integral part of the interpreter, making it very easy to change and run a program repeatedly. However, interpreters make it awkward to write complex programs, so they are typically used for simple problems, whereas compiled languages are used for complex ones.

3.3.4. Text Editors (Word Processors)

A text editor or word processor is a program that allows you to enter text from the keyboard, change it, format it in various ways, store it on disk, and print it. Text editors are often provided by the computer manufacturer as part of the system software, or you may purchase them for most computers from software companies. The text that you create may take the form of a document or, if it is in the proper form for translation, the source file of a computer program.

Some text editors read and write from ASCII files. These files contain only standard ASCII characters with no additional information. Other text editors read and write files that include special characters that cannot be read by other programs. For generating program source files, you should select a text editor that writes ASCII files, so that a compiler can read the editor's files directly. Also, when you write and debug a program, you start and stop the text editor much more frequently than when you write a document. Therefore, the ideal text editor for programming should require only a little time to start or stop.

3.4. APPLICATIONS SOFTWARE

Contrasting to system software is applications software. These are programs to perform specific tasks unique to the laboratory or to the user. They may be provided by a company whose function is to write software, or you may write them yourself.

3.4.1. Specific Laboratory Tasks

In the neuroanatomy laboratory, computers are used to collect, display, and analyze data. This use of computers forms the subject matter of the rest of this book.

To preview, data from anatomic structures may be entered into the computer from a microscope or from a data tablet. Two- and three-dimensional displays may be generated from this now computer-stored data. Statistical summaries may be generated in order to characterize structures and to compare one population of them to another. Video images may be captured. These images may represent structure location, chemistry, or metabolism. Electrical waveforms may also be captured through the computer's A-to-D converters. These may represent neuronal activity or, with some transducers, animal behavior.

3.4.2. General Laboratory Tasks

A neuroanatomical laboratory can be likened to a cottage industry performing a very specific function; consequently, a neuroanatomical laboratory has the same software needs as other small businesses. A few standard ones are briefly described below.

A "spreadsheet" is a computer program for doing mathematical calculations on rows and columns of numbers. It consists of a large matrix of cells, into which you may enter numeric values. Limited by the size of your terminal screen, a certain number of these cells may be displayed at once. You may then instruct the program to combine the cells mathematically in many different ways. For example, to add rows of numbers together, compute the mean, and compare one mean to another. LOTUS 1-2-3 is probably the most famous spreadsheet program.

Statistical summarization programs are readily available for most computers. The program reads data from ASCII files and may perform numerous sophisticated calculations such as the chi-square test or the analysis of variance. Two common statistical packages are called SAS for "statistical analysis system" (SAS-PC for the IBM personal computer) and SYSTAT.

Another useful program is a "data base manager." With this program, you may enter data of many forms, scientific or clerical for example, and then access the data in many ways. I use it to store the name, address, and phone number of colleagues. Each person's entry may be accessed by name, as a card file could be, or I may ask the question, "Whom do I know in Boston?" By searching the entire data structure for the character string "B-o-s-t-o-n," the program provides me information that a card file could not.

Programs may be purchased for accounting, for writing purchasing requisitions, for generating slides or posters, and for managing journal references. When you plug a modem into your computer and attach it to the telephone line, many new capabilities appear. You may send and receive free national or international electronic mail, and you may perform searches of data bases such as those in the National Library of Medicine, the U.S. Patent Office, the yellow pages, and so on.

3.5. COMMON PROGRAMMING LANGUAGES

Many programming languages have been defined throughout the past 25 years, but only a very few have become popular. It is interesting to note that most languages have not been used at all except by the author of their language translator. This is not surprising, however, because the rules of many languages are defined and their translators are written for thesis credit in university computer science departments. Here are brief descriptions of some of the most popular languages.

3.5.1. C

C is a high-level language originally developed by the Bell Telephone Laboratories in the middle 1970s as part of the UNIX multiuser operating system for DEC's PDP-11. It is a compiled language that has become very popular recently; most scientific programming on the IBM personal computer is now done in this language. Figure 3-2 shows a small program written in the C language.

3.5.2. FORTRAN

FORTRAN is the oldest language used for engineering and scientific work. It was developed in the 1960s and is usually a compiled language. Most scientific programming is still done in FORTRAN, especially on larger computers. In 1977, a standard version called FORTRAN-77 was defined, and compilers for this version are available on almost every computer. Figure 3-2 shows a small program written in FORTRAN.

```
/*  Print data until a key is hit  */
#include <stdio.h>
main ()  {
int I;
float X;
for (I = 1; I <= 100; I++) {
if (kbhit()) break;
X = 3 * I;
printf ("The data value is %f\n",X);
    } }
```

```
10 'Print data until a key is hit
20 FOR I=1 TO 100
30 A$ = INKEY$
40 IF A$ <> "" THEN 80
50 LET X = 3*I
60 PRINT "The data value is"; X
70 NEXT I
80 END
```

```
c       Print data until a key is hit
        LOGICAL KBHIT
        DO 120 I = 1,100
        IF (KBHIT()) GO TO 200
        X = 3 * I
120     WRITE (*,121) X
121     FORMAT (' The data value is',F8.1)
200     STOP 'End of the program'
        END
```

FIGURE 3-2. The source files of a simple program written in the languages C (top), BASIC (middle), and FORTRAN (bottom). The program passes through a loop 100 times, calculating and displaying a data value that is three times the loop count. The program terminates when the looping is done or when its user types a key.

3.5.3. BASIC

BASIC (for beginner's all-purpose symbolic instruction code) is a language designed for quick solutions to simple programming problems. BASIC programs are usually interpreted, so executing them is less efficient than executing C or FORTRAN programs. BASIC interpreters are available for almost all computers; a BASIC interpreter is even burned into ROM in most contemporary personal computers. Figure 3-2 shows a small program written in BASIC.

3.5.4. Other Programming Languages

C, FORTRAN, and BASIC are the languages most often used in laboratories in the United States. Many other languages exist though, and a few are important enough that a computer-literate person should know of them.

COBOL (for common business-oriented language) is probably the most widely used computer language in the United States. It is used in banking and other financial industries and is available on almost every computer.

PASCAL, named for the French mathematician Blaise Pascal, is similar to C and enjoys moderate use in Europe and the United States. Compilers for this language may be purchased for most larger computers.

PL/I (for programming language 1) was introduced by IBM in the late 1960s, with the intention to use PL/I as a replacement for FORTRAN and COBOL. IBM wrote and aggressively marketed PL/I compilers for all of their large machines. The language achieved only modest success, however. I do not know of a PL/I compiler for any personal computer.

ALGOL (short for algorithm) was introduced in the 1960s and achieved some popularity in Europe. Its use in the laboratory is rare in the United States.

ADA (named for Augusta Ada, the daughter of Lord Byron, and considered to be the first programmer) has been defined as the new and universal language of the United States Department of Defense. ADA compilers are available for many larger computers. The language is rarely used in the laboratory; it is too early to measure its popularity elsewhere.

3.5.5. High-Level Proprietary Languages

Companies that sell computer systems sometimes define rules for proprietary very-high-level languages and provide translators for them. These languages may have very powerful instructions, e.g., "Capture 1000 samples from an A/D converter and place them in memory at location XYZ," or "Display a waveform on the CRT from the just-captured data." Such statements translate into many machine language statements and usually require special hardware, which the company provides.

Programming in such a language may be very easy, for it is often possible to do much work with only a few statements. There are two tradeoffs with high-level proprietary languages, however: inflexibility and incompatibility. Inflexibility means that it may not be possible to do something slightly different from what the manufacturer has defined in his language. So, expanding on the first example of the paragraph above, the language may not permit you to capture samples from two analog channels, each at a different frequency. Incompatibility means that your data, your program, and the programming skills you develop with such a language are not transferable to a different computer. Section 3.6.2 provides more insight on whether to use a very-high-level language.

3.6. SOFTWARE COSTS AND PRODUCTIVITY

When you are faced with a neuroanatomical problem that demands some data-processing assistance, you must answer four questions: How much money can you afford to spend on the problem? How much effort can you personally give to the problem? What computer skills do you have? And finally, how much calendar time can you let elapse before you see some results?

In this section, I present some lessons I have learned, some from personal and painful experience and some from the software engineering literature. I hope that these lessons may help you find answers to the four questions.

3.6.1. Software Costs Related to Hardware Costs

The cost of a laboratory computer has fallen dramatically during the past decade. For instance, DEC's PDP-11/03, which cost $15,000 in 1979, has been replaced by the IBM AT clone, which cost $2500 in 1988. In addition, such new personal computers have much more capability than the old laboratory computers. This dramatic decrease in price and increase in power has caused some very interesting ripples in the economics of laboratory computing.

Millions of personal computers have been sold for offices, homes, factories, and laboratories. Their mass production coupled with healthy competition has increased their speed-to-cost ratio manyfold. But it must be realized that a a computer has no value without software. The number of machines sold has generated a tremendous demand for common software (e.g., word processors) that is used in many environments. This software demand has generated a need for well-paid programmers, and universities have been happy to provide them. It is very inexpensive to mass produce a software package by copying disks, less costly than to mass produce a computer. Because of the mass production of common software, you can buy, for example, a well-refined word processing program for less than 100 dollars in spite of the large effort that went into its development.

Unfortunately, laboratories do not demand software for their needs in the same quantity that offices do. So although there is some laboratory software mass production, it does not approach that of business or clerical software. As a consequence, laboratory software costs remain fairly high in spite of the frequent laboratory use of computers. In 1981, we estimated that at the University of North Carolina, 55% of the money required to do a laboratory computer task was spent on software, even with the artificially low cost of labor in the university environment. Today, companies that sell software and hardware products into the neuroscience marketplace (see Chapter 16) charge from about $1000 through about $15,000 for software. So even this modestly mass produced software often costs significantly more than the hardware on which it runs.

Custom programming, where there is no mass production, is even more expensive. A graduate student programmer may cost $10,000 per year for half-time work, and a professional may earn a salary of $40,000 per year. And depending on its complexity, an anatomic data collection and analysis software package may take years to write.

All these facts mean that you must evaluate software carefully, because this is where the major investment lies. Section 3.6.4c describes some criteria for purchasing already-written software.

3.6.2. Software Costs Related to Software Level

As software becomes the major item purchased to perform a laboratory task, you might wonder, "How then, can I write the software efficiently to maximize the programmer's effectiveness?" As noted in Section 3.3.3, languages have a "level" such that each statement of a high-level language translates into more machine code instructions than does a statement of a low-level language.

Brooks (1975) states that "computer programmers have a productivity which is fairly constant when measured in lines of code written, independent of the language used." So if you wish to maximize the productivity of a programmer, have him select as high a level language as possible, for with it, he will be able to code the problem in a minimum number of statements. Brooks later pleads (Brooks, 1987), however, that you not look for too much improvement simply by switching languages. For in writing a computer program, the programmer faces two sets of problems: the ones inherent in the task and the ones caused by the computer and its languages. Only the latter can be improved by selecting a higher-level language.

3.6.3. Programs and Program Products

When you seek to purchase a computer program for a specific laboratory task, a salesman may demonstrate the program for you and tell you its price. Frequently you may believe that the price is high and that you could easily write the program yourself or pay someone to write it for you for much less money. But before you fall for this tantalizing alternative, consider the types of programs detailed in Fig. 3-3 and consider which type you are evaluating and which type you might write.

You may write a computer program. A program is designed to be used by its author or by someone else with the author's close guidance. Then the program's user, perhaps yourself, understands the organization of the program and will use it in the manner for which it was designed. You will not, for example, enter data that exceed some limits in the program. This is the type of software you usually consider writing when you make the comparison to commercially available packages.

You might estimate that it will take a month of your labor, and you can calculate the software's value based on your salary. Some margin for error is necessary for the time estimate, however, because as you work through the details of your program, you always encounter unforseen nuisances that extend the time required.

In contrast to a computer program, a program product is a piece of software designed to be used without the author's presence. The program must be written in a manner so that by using it you form a realistic mental image of the program's flow. Only then can you use it in the manner for which it was designed. It must be better tested and better documented, and instructions on its proper use must be included. Telephone support for installing and

PROGRAM → X3	PROGRAM SYSTEM
↓ X3 PROGRAM PRODUCT	PROGRAM SYSTEM PRODUCT

FIGURE 3-3. Four types of computer software. Writing a computer program (upper left) requires a certain amount of effort. The program is designed for use with its author present. To convert this into a product (lower left) to be used remote from the author, triple the effort is required. A system of programs that interact with one another (upper right) requires three times more effort to write than a program. To convert the system of programs into a product (lower right) to be used remote from the author requires yet another tripling of effort. Modified from Brooks (1975) with permission.

using the program must be provided. Because of these criteria, it may take three times as long to write a program product than it does to write a computer program.

A system of programs is a group of programs that interact, so that one program affects another. Because of the intercommunication, it takes approximately three times the effort to write a system of programs than it takes to write the same number of individual programs. Interfaces must be defined, and changes to one program frequently require changes in others.

Finally, Fig. 3-3 shows the system program product. This is a system of programs designed for use in the absence of the author. Such a system takes about nine times as long to write, debug, test, and document as the simple computer program requires.

3.6.4. How to Get the Job Done

With all the above characteristics of computer programs fresh in your mind, you must now decide how to solve your computer problem. Generally, you have four options: buy the computer hardware and write the software yourself; buy the computer hardware and hire a programmer; buy the computer and somehow obtain already-written software; or buy a turnkey system ready to do your task. The next sections present some ideas about each of these options.

3.6.4a. Program It Yourself. If you feel you have the time and skills, you may want to write the program yourself. Don't go into this too hastily, however, for programming may be a tar pit from which you may never be able to extricate yourself. Computer programming is a fascinating task, and when you start, it takes discipline not to become a technologist who is more interested in how to solve the problem than in its solution.

If you write programs for only your own use, then "quick and dirty" programs that would be hard for others to use may suffice. In Chapter 12 the high cost of making a program easy to use for both an expert and a novice is detailed. A novice needs menu selection and structure, whereas an expert needs flexibility and efficiency. Since you are instantly an "expert" at using your own programs, you may escape many of these costs. You may be further able to shorten your task by purchasing a software "toolbox" and writing your program using it. A toolbox is a set of program modules that perform general-purpose functions (e.g., to draw a line on a graphics screen), so that you may be free to concentrate more on the tasks specific to your problem.

3.6.4b. Hire a Programmer. You may trade money for some of your personal effort by hiring a programmer, and with some patience on your part, a good program may result. In order to write an application program, two knowledge sets are required: knowledge of the problem and knowledge of the computer. When you write the program yourself, you learn both sets, so there is no time wasted by poor communications. Hiring a programmer with no neuroanatomical experience, however, guarantees that writing the program will be slower and possibly frustrating for you while he learns the problem.

Programming consists of reducing a problem again and again until it so simple that it can be explained to a machine. Frequently, when you or a programmer writes a program, you find that the problem is not defined well enough for the machine, and you must continually redefine it. If two people are involved in this reduction and neither understands both the problem and the machine, then poor communications and wasted effort will invariably result. You will be a better employer of a programmer if you have some programming experience yourself, since you will then understand the capabilities and limitations of the machine.

Student programmers are available from many university departments, but their quality varies widely from poor to extremely high. During the academic year, their coursework rather than a programming job is highest priority, so if you can link the program to their course work, the quality of the result will frequently be raised. Students also graduate and depart, leaving you with a program and no access to its author. Then fixing errors or changing or upgrading the software becomes difficult.

Professional programmers are expensive, and their quality also varies, but when measured in productivity per dollar of salary, their results may be better. Because laboratory computers are usually interfaced to various types of equipment, people with engineering backgrounds may be more useful than those with mathematics backgrounds. The ideal laboratory programmer is an engineer who understands both the biological problem and the computer hardware and software.

3.6.4c. Find Already-Written Software. Another option for solving your problem is to find already-written software. Unfortunately, you face a harsh economic reality here. The neuroanatomical marketplace is a small one, so you should expect little attention to your problem from large companies that manufacture computers, microscopes, or other laboratory instruments. Consequently, you will have to turn to small companies to purchase the needed software. This means that you must evaluate the company as well as the product. For when you purchase a piece of software, you are also purchasing continual support to help you install, use, and upgrade the software. If the company is financially insecure, future support may not be available.

Your colleagues in other universities are another source of software. University-written software varies greatly in quality. It may range from a computer program designed to be used only by its author (Section 3.6.3) up to a quite professional product.

Regardless of the source of the software, it is important to test the program on your data rather than on demonstration data provided by the software source. Only then will you be able to tell if the program will function for you.

Although it is tempting to buy computing hardware quickly, you should think of software first, for it forms the major expense and effort! A good idea is to search the marketplace, both commercial and university, for the software that will work well on your data, and then obediently purchase the computer hardware on which the software runs. You may face a difficult decision here if the hardware on which the software runs is obsolete or uncommon.

3.6.4d. Buy a Turnkey System. A final option for solving your computer-assisted neuroanatomy problem is to purchase a "turnkey" system from a vendor. A turnkey system contains both hardware and software assembled into a package so that it is ready to be used in your laboratory. All you do to start it is to turn a key, as you would do with an automobile.

This is usually an expensive way to purchase a laboratory function, but it has some significant advantages. The responsibility for assembling the hardware and making sure that the software runs on it lies with the vendor. In systems whose hardware consists of a computer and several other devices, this can be especially important. When you purchase hardware and software independently, if the combination does not work correctly, it can be difficult to determine where the fault lies.

The same caveats for purchasing software apply to purchasing a turnkey system. It is important to test the system on your own data. Also, evaluating the company is as important as evaluating the product.

3.7. THE VENDOR'S DILEMMA

To be an intelligent shopper in the marketplace, you should recognize the dilemma a vendor faces. Most customers purchase computer software or turnkey systems after only the briefest demonstration, using test data provided by the vendor. These data are carefully selected, of course, to show off the best features of the system while minimizing its flaws.

Because of this pattern of customer activity, the vendor must decide whether to make his system easy to sell or easy to use. His temptation is to make the system easy to sell, that is, to make it easy for a novice to use on the demonstration data. This may not be the best for you in the long run, for once you purchase the machine and gain some expertise using it in your lab on your own data, your needs change. Only then do you realize whether the system is easy to use and whether it will help your research. You may counter this to a degree before you make the purchase by talking to experienced and unbiased users of the product. This may be difficult, however, because these laboratory systems continually evolve so users of the exact product that you are considering may not exist. Eventually you have to trust the company to provide for your needs.

In the very long run, it is in the vendor's best interest to manufacture a system that you will find easy to use on your own data in your own laboratory. Then you will recommend the system to your friends, and he will benefit from increased sales. However, each vendor faces a need to make short-term profits, and he participates in a race with his competitors to bring a product to market quickly. To achieve these goals, he is tempted to make a product easy to sell to a novice. Therefore, proceed cautiously.

3.8. FOR FURTHER READING

Few journal articles exist concerning laboratory software and how to select it. Two that contain general suggestions on software selection are Flaming (1985) and Sing and Salin (1985).

Where and how to obtain the personnel to perform laboratory computer work has been reviewed by Moraff (1976) and by Capowski and Cruce (1981).

Cruce and Steusse (1987) have provided a biologists' perspective from their use of computers in their neuroanatomy laboratories for more than a decade.

4

Semiautomatic Entry of Neuron Trees from the Microscope

4.1. INTRODUCTION

Chapter 1 introduced the concept of neuron tracing as a method of transferring a neuron from its original form in the tissue to a different form—either a drawing on a sheet of paper or a physical model. Although these new forms help to overcome some of the viewing constraints of the microscope, they still are not "tractable"; that is, they are still difficult to work with mathematically. If you wish to do a statistical summarization of the neuron, for example to plot its dendritic length versus thickness, you must make tedious measurements of the drawing or model to gather the data for the graph. To alleviate this problem, the neuron should be transferred (Fig. 4-1) to a computer memory. Once stored there, a computer program can work with the structure to build three-dimensional displays and plots and to generate statistical summaries of the neurons with far less effort than you could as a human.

This chapter presents contemporary techniques for tracing a neuron from its tissue into a computer memory. The techniques are semiautomatic, so that a human researcher, using his superior visual system, follows the dendrites and axons, and the computer and microscope are outfitted and programmed to do the necessary bookkeeping procedures.

In all semiautomatic neuron-tracing systems, the researcher must somehow move a cursor over the image of the neuron. The image of the cursor must therefore be mixed with the image of the neuron. Two techniques of mixing these two images are generally employed: optics and television. This chapter describes semiautomatic techniques for neuron tracing using optical methods for locating the cursor on the image. Chapter 6 discusses the use of video, with its advantages and disadvantages for performing the same task.

4.2. PRINCIPLES OF SEMIAUTOMATIC NEURON TRACING

Semiautomatic neuron tracing involves the marriage of you and a computer to record the repeatedly branching structure of a neuron. If several principles are applied when the neuron tracing system is designed, then you may enjoy an efficient tracing experience.

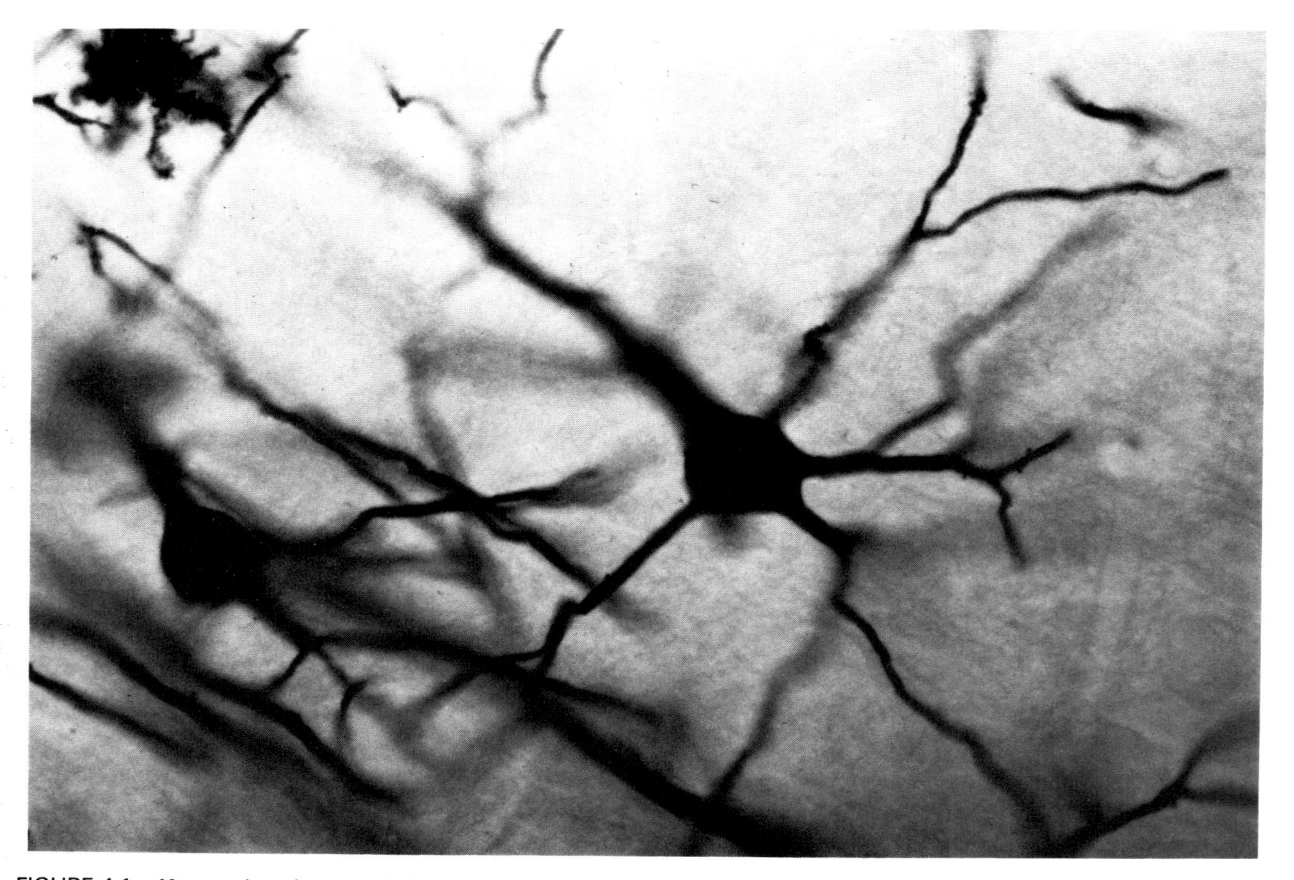

FIGURE 4-1. Neurons: how do we get their complex three-dimensional structure into the computer? Tissue courtesy of R. Weinberg, UNC, Department of Anatomy.

4.2.1. The Marriage of the Researcher to the Computer

The first principle of semiautomatic neuron tracing is to take advantage of the capabilities of you and of the computer. You, being human, are a good pattern recognizer, able to observe an image and instantly decide what is important, what to ignore, and, for neuron tracing, where the dendrites grow. You are, on the other hand, a poor bookkeeper. When you must work with lists of numbers, you are slow and error-prone. The computer is just the opposite; it is an excellent bookkeeper, quick and accurate when dealing with numbers, but poor at recognizing global patterns in an image. Ideally, a neuron-tracing system should be designed to employ the best capabilities and avoid the worst capabilities of both participants. You should recognize the dendrites and pass a cursor over them (or pass them under a fixed-position cursor), and the computer should record their spatial coordinates.

4.2.2. A Single Pass over the Data

Another principle is that you should make only one pass over the data. During this single pass, as much information as possible, or at least as much as is needed for current and foreseeable research, should be recorded. If the research study requires that the dendritic shape (X,Y,Z coordinates), thickness, and the existence of special features (e.g., dendritic spines) be recorded, the system should be designed to record all this information at once. Forcing you to make two passes over the same data is disheartening and encourages errors.

To capture the X,Y,Z coordinates and thickness in a single pass, it is necessary to design a neuron-tracing system in which you may control four analog variables simultaneously. You must control the stage or cursor movement in X, Y, and Z (the Z axis is the focus axis) and at the same time control the diameter of the cursor to match it to the diameter of the dendrite. Special features of the dendrites must be recorded. Tree origins, branch points, various kinds of endings, varicosities, and spines may be important in anatomic studies; thus, the system must be designed to record the location and size of these features for later viewing and statistical summarization.

4.2.3. Identify Different Structures in Their Environment

It is necessary to distinguish among different neurons, different dendrites or axons of a neuron, and different regions within a dendrite. So some flexible means to identify neurons, trees, and regions within trees must be designed. It may be difficult to determine in advance exactly what the needs are here, so flexibility is crucial.

In addition to just tracing the neurons, you want to locate them in their environment. You may want to trace many different neurons from the same piece of tissue, accurately recording their proper relative locations. And you want to record other features from the tissue in order to locate the neurons in their environment. Tissue borders, lamina boundaries, and other nonneuronal structures provide valuable insight to the form and function of the neurons.

4.2.4. Feedback

As you trace, you need feedback that the procedure is going well, that the computer is recording points accurately. You should receive this feedback in the form that is most convenient for you, in this case, visual. This best form of feedback in neuron tracing is a visual display of the computer-stored data overlaid on the original tissue image. Then you may continually and easily notice whether the computer has accurately recorded the neuron structure.

4.2.5. Work from the Best Image

The final principle is that you want to work from the highest-fidelity image possible; usually that means the original tissue section directly observed through a light microscope. Besides being the highest-fidelity image, the original tissue section is also the image with which you are most comfortable because you have spent much of your career looking at tissue sections in a microscope.

4.3. THE UNC NEURON-TRACING SYSTEM

In this section, a neuron-tracing system currently in use at the University of North Carolina that utilizes the basic principles that were detailed above is described. After the description its techniques are compared to techniques used by other neuron-tracing systems described in the literature. All are semiautomatic systems in which you pass a cursor over the dendrites and the computer captures a mathematical model of the traced neuron; the systems vary in the specific details of how this process is implemented.

4.3.1. Hardware of the UNC Neuron-Tracing System

Figure 4-2 shows a block diagram of the UNC system, and Fig. 4-3 shows an anatomist seated at the system. He is sitting in front of a research light microscope and

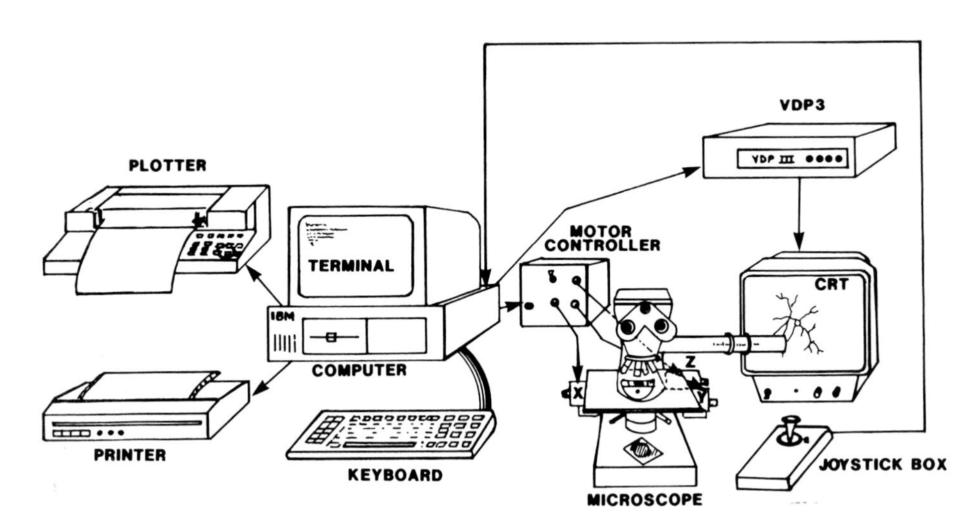

FIGURE 4-2. A block diagram of the UNC system for tracing neurons into a computer. Details are given in the text.

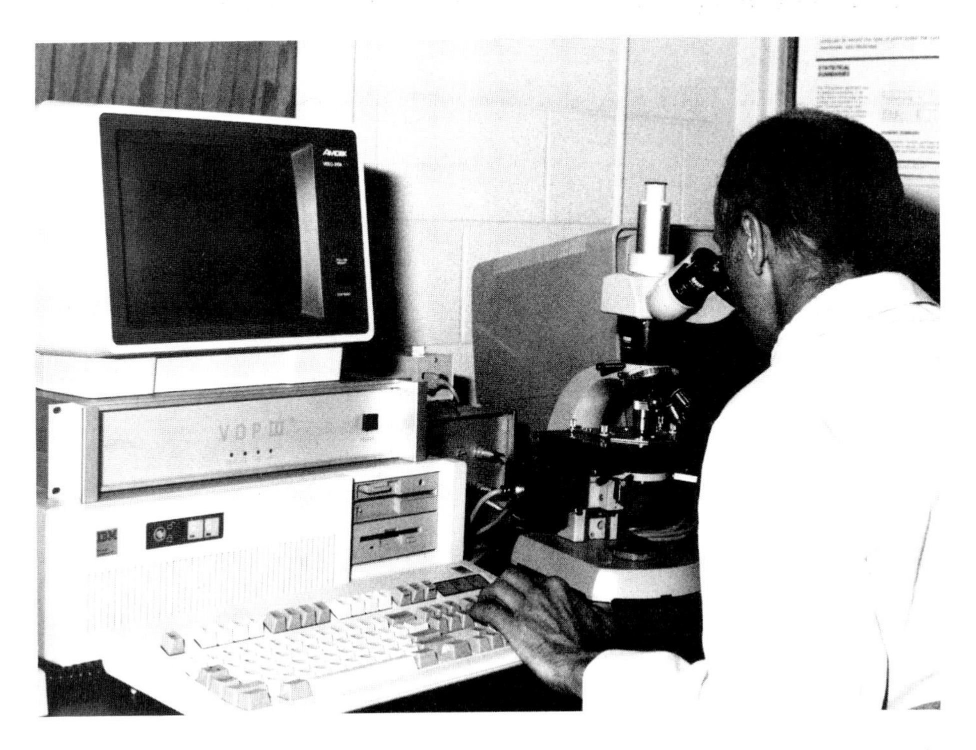

FIGURE 4-3. A researcher tracing neurons with the UNC system. While looking into the microscope, he signals the computer to record various types of points by pressing buttons on the computer keypad. Infrequently used commands are presented to him on the terminal screen above the computer.

looking into it. Immediately to his left is an IBM AT computer, a small laboratory computer that contains 512,000 bytes of memory, a hard disk with a capacity of 30 megabytes, and two flexible or floppy disk drives. Immediately above it is a vector computer graphics display box, called the vector display processor model 3 or VDP3. Above it is a computer terminal on which the computer presents information to the anatomist. During tracing, it is used to present a menu of infrequently used commands. In front of the computer is the keyboard with which the anatomist will command the computer.

The microscope is a Zeiss WL, a typical research light microscope, that has been outfitted with stepping motors on its X,Y stage and on its fine focus knob. The motors may position the stage and focus to a resolution 0.5 μm. The motors are driven by a motor controller box, which is in turn driven by a parallel port of the computer. Thus, the computer can position the microscope stage in X, Y, and Z to a resolution of 0.5 μm.

Figure 4-4 is a view from the right side of the anatomist, showing several new things. There is a large-screen CRT on which the output from the graphics display box is presented. A camera lucida or drawing tube attachment has been mounted on the microscope, and its optical input port is turned to view the CRT screen. Thus, the CRT image is overlaid on top of the image of the neuron, and both are presented to the user in the microscope eyepieces, as will be shown later in Fig. 4-7 (Section 4.3.4).

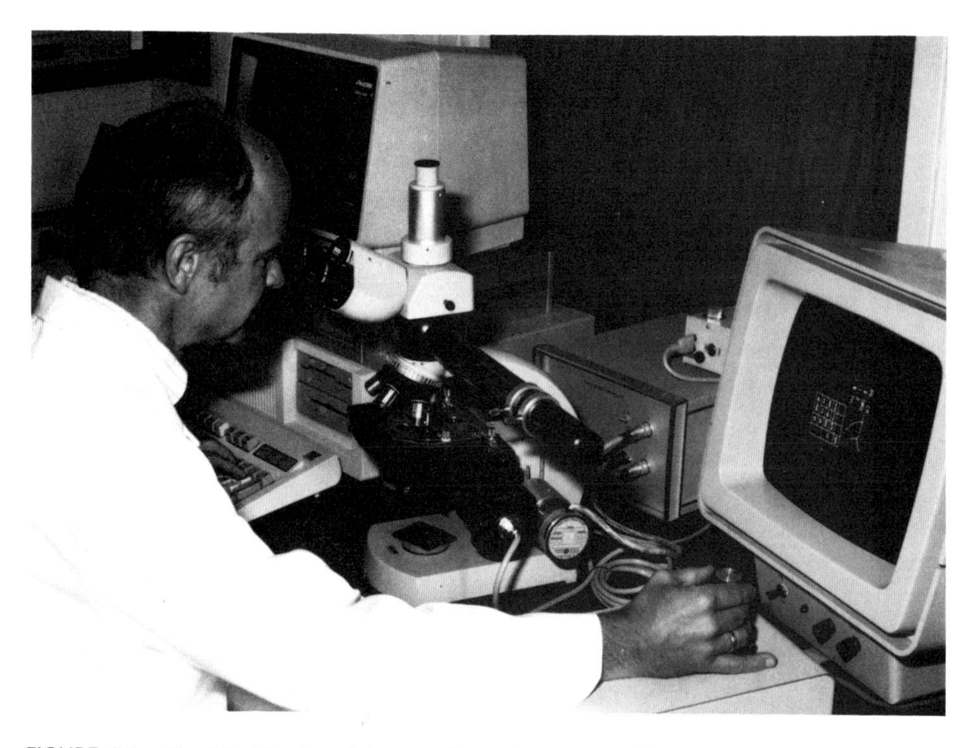

FIGURE 4-4. The right side view of the researcher tracing neurons with the UNC system. He controls the microscope's motorized stage and focus by deflecting and twisting the three-dimensional joystick with his right hand. Behind his right hand is a CRT on which is displayed a menu of point types and other information necessary for tracing. The optical input port of the camera lucida is turned toward the CRT screen and overlays this computer-generated image on the tissue image so that he sees both images in the microscope's eyepieces.

Below the camera lucida is a box containing a joystick, a vertical stick that the anatomist can deflect in the X and Y directions. The joystick is three-dimensional; that is, a potentiometer is mounted on top of the stick and may be twisted about the joystick's Z axis. On either side of the joystick are additional knobs that are attached to potentiometers readable by the computer. The joystick box is the primary device with which you control the microscope stage during tracing. Details about this are given in the next section.

The output from the joystick box is read by an analog-to-digital converter card that is mounted inside the computer, so that the computer program can monitor at all times the current position of the joystick and the two adjacent knobs.

4.3.2. Control of the Stage

Figure 4-5 shows a detailed view of this joystick box. You can grasp the knob on top of the joystick between your thumb and index finger and deflect the joystick in the left–right directions (X axis) and in the near–far (Y axis) directions. You may also twist the knob on the top of the stick (the Z axis) and simultaneously rotate the potentiometer to the

FIGURE 4-5. The use of the joystick box during tracing. With his thumb and index finger, he may deflect and twist the joystick in order to control the stage position in X, Y, and Z. Simultaneously, he may turn the adjacent potentiometer knob with his little finger in order to control the diameter of the cursor. Thus, he may control four analog variables simultaneously with one hand. The left-hand potentiometer is used only by a left-handed tracer.

right of the joystick box with the little finger of your right hand. With this scheme, you can control all four analog variables at one time.

The joystick box is used to control the stage movement and the diameter of the cursor. The position of the cursor is fixed in the middle of the image (Fig. 4-7), and its diameter varies according to the position of the knob to the right of the joystick (Fig. 4-5). The stage movement must be controlled to a resolution of less than 1 μm but must have a sufficiently long range to follow dendrites as they meander through several thousand micrometers. "Positional" joystick control of the microscope stage is used to accomplish this task; if you wish to move the cursor to the upper right (actually to have the stage move to the lower left), you simply move the cursor slightly to the upper right, and the microscope stage will follow or track your joystick motion exactly. You will be able to cover approximately 75 μm of distance with the deflection of the joystick. When you reach full deflection, a message will appear in the microscope eyepieces, instructing you to center the joystick and then type any key. The computer program will reset its stage tracking logic so you may continue tracing as if nothing had happened. With such a scheme you are able to cover long distances at high accuracy, even with a joystick whose deflection is quite limited.

4.3.3. Coordinate System and Origins in the UNC Neuron-Tracing System

The joystick control system described above must move the stage as you deflect the joystick. The problem is complicated, however, because the optical systems of microscopes vary. Some microscopes present an image to the viewer that is inverted both about the tissue's right–left (*X*) axis and about the tissue's near–far (*Y*) axis. Other microscopes contain an extra prism just below the eyepieces so that the image presented is only inverted about the tissue's *Y* axis. Some microscopes do not invert the image at all. Because of the variety of microscopes, the coordinate system has been defined in the tissue rather than in the presented image. This is shown in Fig. 4-6: *X* is to the right, *Y* is away from the user, and *Z* is down, so defined because moving in *Z* really means raising the stage in most research light microscopes.

At the center of the coordinate system is the tissue or slide origin. This is the 0,0,0 point in the tissue, and all points will be recorded relative to this origin. After you have placed a slide in the specimen holder, your first tracing task is to set this tissue origin. It should be an easily recognizable point that you may easily find at any time. If your task is to trace a neuron whose soma exists in the tissue section, then frequently you will set the slide origin in the center of the soma. You set the slide origin by positioning the cursor over the feature you wish to use as an origin and press the "O" key (for origin) on the computer keyboard.

As the computer drives the stage and focus axis, it keeps track of the current stage coordinates, i.e., the *X,Y,Z* values of the point currently in focus under the cursor. Setting the slide origin clears the current stage coordinates. Moving the stage under joystick control varies the coordinates over their range, which is microscope dependent but is typically 32,000 μm in *X* and in *Y* and 2000 μm in *Z*. The resolution of the stage coordinate is 0.5 μm in each of the three axes.

4.3.4. Outlining a Soma with the UNC System

A typical early step in tracing is to outline a soma. The view in the microscope eyepieces during this procedure is shown in Fig. 4-7. Below the cursor is a grid, each box of which has a several-letter code. That grid is a map of the keypad area, which is the

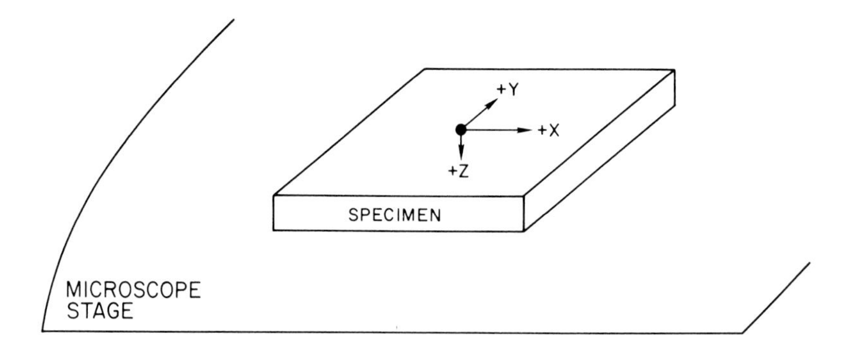

FIGURE 4-6. The tissue coordinate system used in the UNC neuron-tracing system. Reproduced from Capowski (1977) with permission.

right-hand end of the computer keyboard. Each of the boxes in the grid corresponds to a particular key on the keypad, and each of the buttons on the keypad is used to enter a point type. You start outlining a soma by moving the cursor to a point on the perimeter of the soma and then press the button called "SOS" for "soma outline start." When you press this button, the computer records several parameters. It records the point type, in this case a soma outline start; it records the "tag" for this point, an arbitrary numeric value, described in Section 4.3.5b; it records the *X,Y,Z* coordinates of the cursor; and it also records the diameter of the cursor at that moment, though the cursor diameter is not important for outlining a soma.

Next, you move the cursor a short distance around the perimeter of the soma and press the button labeled "CP" for "continuation point." The computer then records the parameters of this continuation point. A continuation point is a point along a dendrite or a point around a soma that has no particular anatomic significance. It is important to record its coordinates, though, because the computer represents a structure by a series of short line segments from point to point. So it is necessary to use a series of points to indicate any curving in a dendrite or in a soma outline.

As you move the cursor around the soma outline and record points, the just-traced

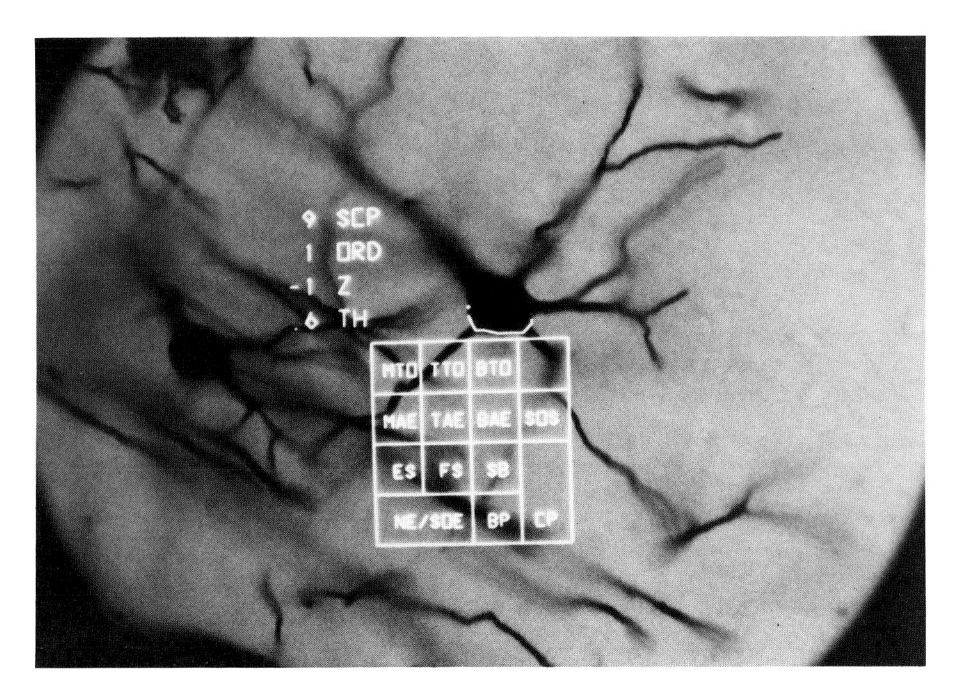

FIGURE 4-7. The view in the microscope when outlining a soma with the UNC neuron-tracing system. The circular cursor has been reduced to a dot and can be seen at the left edge of the soma. The already-traced portion of the soma is overlaid on its outline so that the user can see that tracing is proceeding correctly. He moves the joystick to pass the tissue under the cursor and records points by pressing buttons on the computer keypad. The grid below the soma is a map of the computer keypad, and the letters inside the grid indicate what type of point is recorded for each button push. To the left of the soma, other feedback is provided to the tracer. Nine points have already been recorded, and the most recent one is a soma continuation point (SCP). The current branch order is 1, the current *Z* coordinate is -1 μm, and the current diameter of the cursor is 0.6 μm.

portion of the outline is displayed by the computer overlaid on its corresponding place in the tissue. This overlay allows you to monitor that the tracing is proceeding correctly. Generating the display of the just-traced portion is difficult, since the position of the cursor is fixed and you move the tissue below the cursor. The computer program must therefore move the computer-generated outline in synchronization with the movement of the stage.

To finish the outline of a soma, you move around the soma and press the continuation point button until you have returned to the starting point. You then finish by entering a "soma outline end" (or "SOE") point type. When you have finished outlining a soma, the computer has recorded a series of points consisting of a "soma outline start," a series of "soma continuation points," and a "soma outline end," which defines the outline of a soma. The microscope view when a soma has been outlined is presented in Fig. 4-8.

4.3.5. Tracing a Dendrite with the UNC System

The next step in the tracing process is usually to trace a dendrite. To do this, you move the cursor to the place where a dendrite emanates from the cell body and focus carefully on that point. You expand the cursor to make the diameter of the cursor the same as the diameter of the dendrite at that point. You then enter a point called a "middle tree origin" (MTO) that signifies the start of a tree in the middle of a tissue section. The

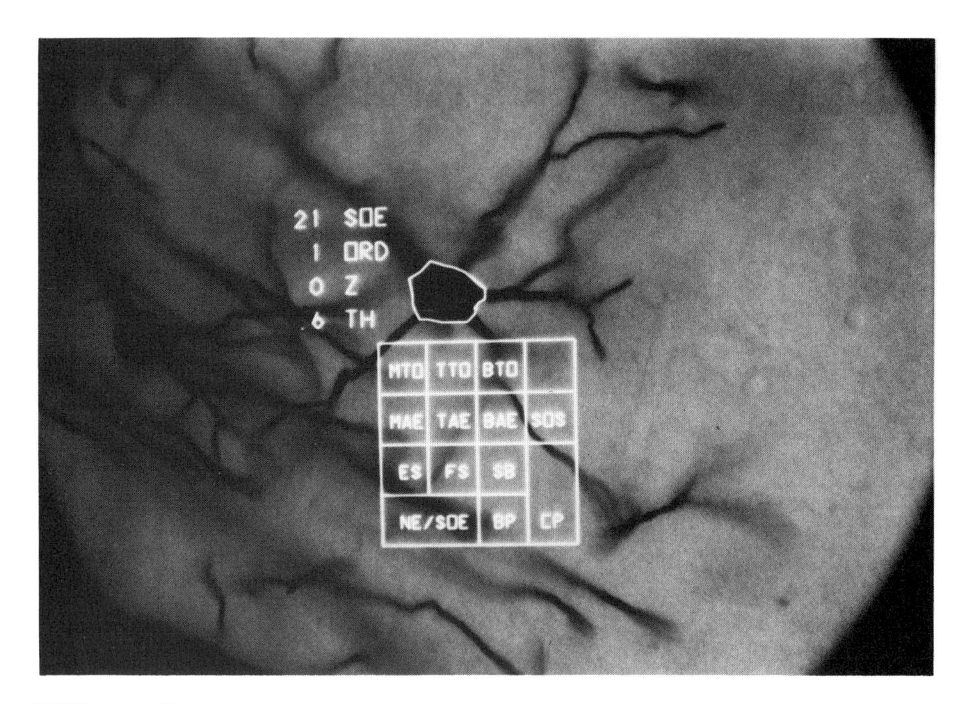

FIGURE 4-8. The view in the microscope when a soma has been outlined with the UNC neuron-tracing system. The completed outline appears on the soma so that the user can confirm that the tracing was done correctly. Twenty-one points were used to outline the soma, and the last point recorded was a soma outline end (SOE).

computer records the coordinates at that point, the thickness of the dendrite at that point, the type of point, and its tag. You move down the dendrite a short distance and focus on a point, expand or contract the cursor as necessary, and enter a "CP" for continuation point.

Figure 4-9 shows the microscope view when tracing has proceeded this far. Several features can be seen here. A stick figure of the neuron appears as you are tracing it to show you that the tracing process is proceeding normally. As you move the microscope stage, the computer-generated stick figure follows the stage motion faithfully so as to always be overlaid on the neuron.

Numeric information is also available, as shown in the upper left corner of Fig. 4-9. This information includes the number of points the computer has already recorded and the type of the most recent point, the current branching order, i.e., the number of branch points plus 1 between the soma and any particular point (covered in detail in Chapter 11), and the current Z coordinate. As you focus up through the tissue, the number of this Z coordinate decreases; as you focus down, the number increases. This feedback becomes most important at a tissue surface, for it indicates whether the surface is the top or the bottom of the tissue. Finally, the current diameter of the cursor is shown in micrometers.

When tracing, if you reach a branch point, you press the BP button, and the computer records this as a branch point. You then continue tracing down either of the two paths exiting from the branch point. When you reach an ending, you enter one of the five types

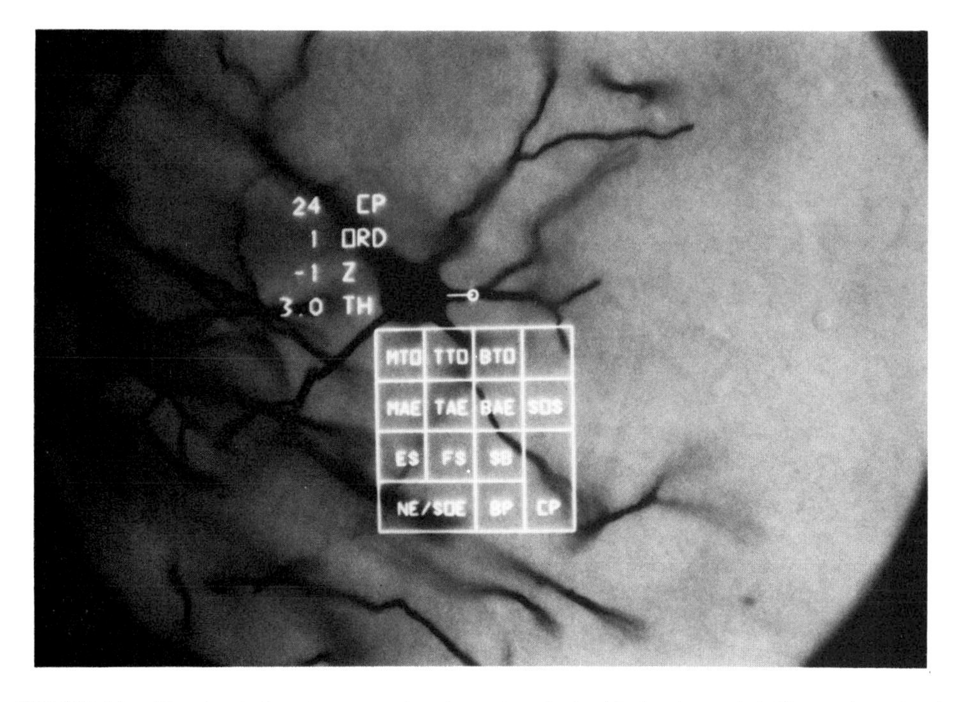

FIGURE 4-9. The view in the microscope when the tracing of a dendrite has just started. The user has entered a middle tree origin and then a continuation point (CP). Note that he has expanded the cursor to match the size of the dendrite, now 3.0 μm. Whenever he pushes a button, the thickness is recorded as well as the point type and coordinate.

of endings (described in Section 4.3.5a), and then the computer moves the microscope stage back to the most recent branch point from which you have not yet traced both exiting paths. You should then trace the other exiting path. This procedure prompts you not to miss any branch of the tree. Figure 4-10 shows the microscope view just after an ending has been recorded.

More advanced tracing of a tree is illustrated in Fig. 4-11. By this time, you have entered several branch points and the computer has generated an overlay of the just-traced dendrite. Note that the dendrite is three-dimensional; its branches meander in and out of focus. Indeed, the already-traced segment to the right of the cursor is completely invisible in the image, though its computer overlay is shown.

When you enter an ending and there are no more branch points from which both distal paths have not been traced, then the tree is finished. The computer program displays a message to indicate this. You may generate a display called the "full neuron overlay" as shown in Fig. 4-12. This is a display of the computer representation of the entire already-traced structure overlaid on top of the tissue from which the data came. This provides one more opportunity for you to see that the computer has faithfully recorded the tree structures. The "full neuron overlay" has another use, that of merging multiple section trees. This is covered in Chapter 7.

4.3.5a. Point Types to Represent a Tree. In the UNC system, there are 12 different types of points that are used to describe a tree. Three of them are origins: a "middle tree

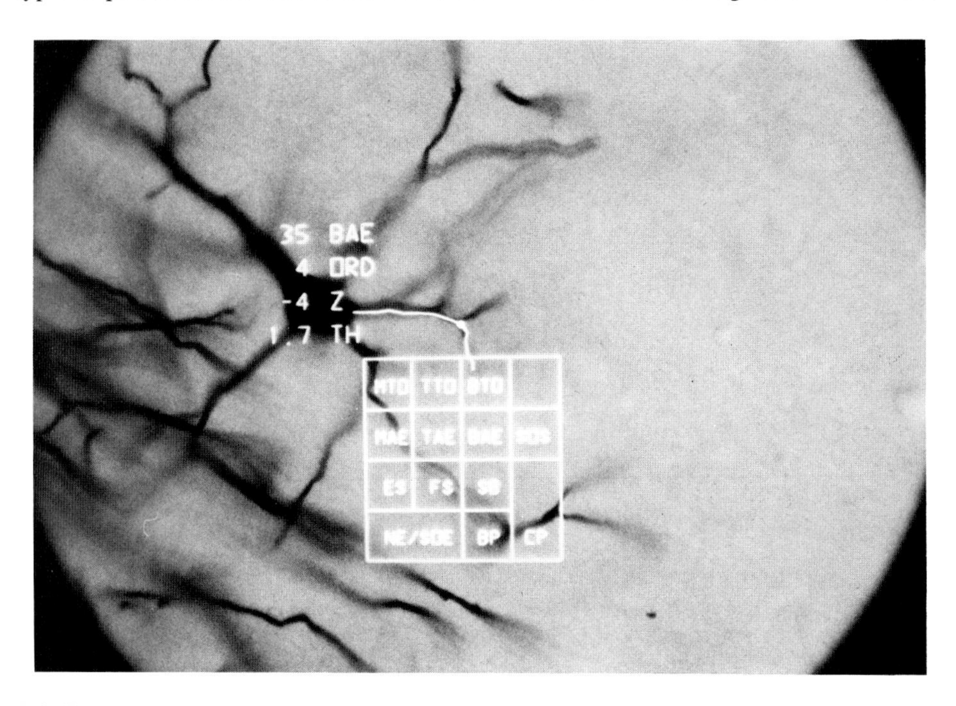

FIGURE 4-10. The view in the microscope when the user has just finished tracing a branch segment to its ending. After he enters a bottom artificial end (BAE), the computer drives the microscope stage to position the most recent branch point under the cursor. He may then trace the other path exiting from the branch point starting at the location of the cursor.

origin,'' a ''top tree origin,'' and a ''bottom tree origin.'' These are used to define the start of a tree within the tissue section, at the top of the tissue section, and at the bottom of the tissue section, respectively. Trees that start with a top tree origin are usually trees that do not emanate from somas directly but that are continued from a previous tissue section, so that within the current tissue section, only the start of a piece of the tree is available at the top tissue surface. These three origins are entered with the keypad pushbuttons labeled MTO, TTO, and BTO, respectively, shown in Figs. 4-7 through 4-11. The fourth type of point is a ''continuation point'' (entered with the CP keypad pushbutton) and is used to mark a point along a dendrite or axon that has no particular anatomic significance. A ''fiber swelling'' (FS) marks a varicosity or swelling along the course of a fiber. It looks like a snake that ate a tennis ball. A ''spine base'' (SB) marks the location, though not the length or direction, of a dendritic spine. A branch point (BP) is used to mark the bifurcation of a dendrite into two branches. Occasionally there may be a trifurcation where a dendrite splits three ways. This can be recorded by two coincident branch points.

There are five types of endings. A ''natural end'' (NE) is used to mark the tapering end of a dendrite. An ''end swelling'' (ES), looking much like a thermometer bulb, marks a varicosity or swelling at the end of a dendrite. There are three kinds of artificial ends: a ''middle artificial end'' (MAE), a ''top artificial end'' (TAE), and a ''bottom artificial end'' (BAE). A middle artificial end is used to indicate the place where the dendrite can no longer be followed within the tissue section for some artificial reason, e.g., the stain becomes faint, or it passes behind a dense silver deposit. A top artificial end indicates a place where the dendrite meanders up to the top surface of the tissue section and has been

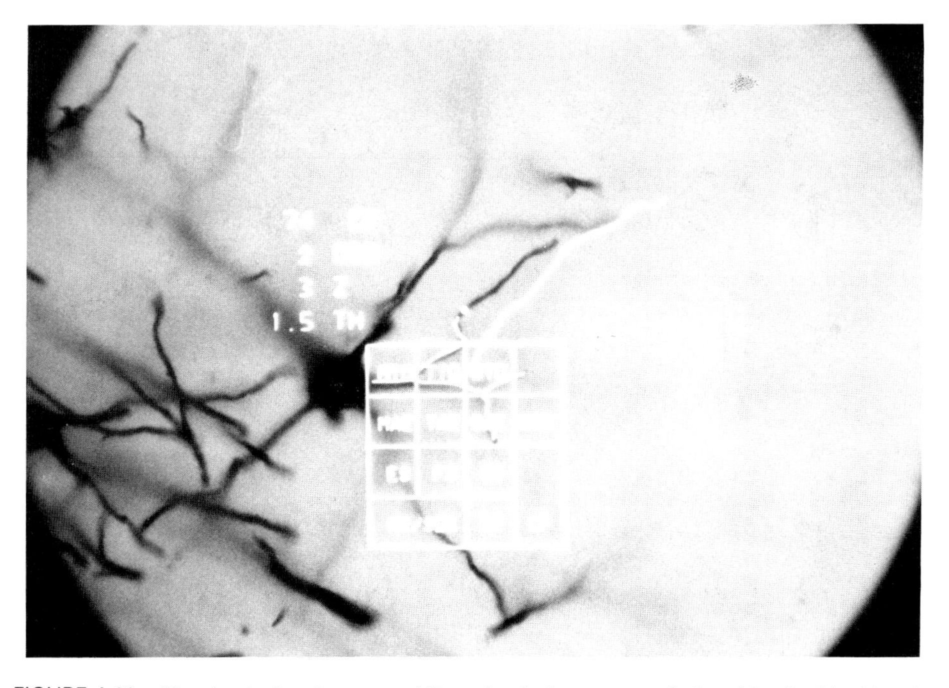

FIGURE 4-11. The view in the microscope while tracing the last segment of a branching dendrite. Note that the user has changed the focus to follow the dendrites as they meander in all three dimensions.

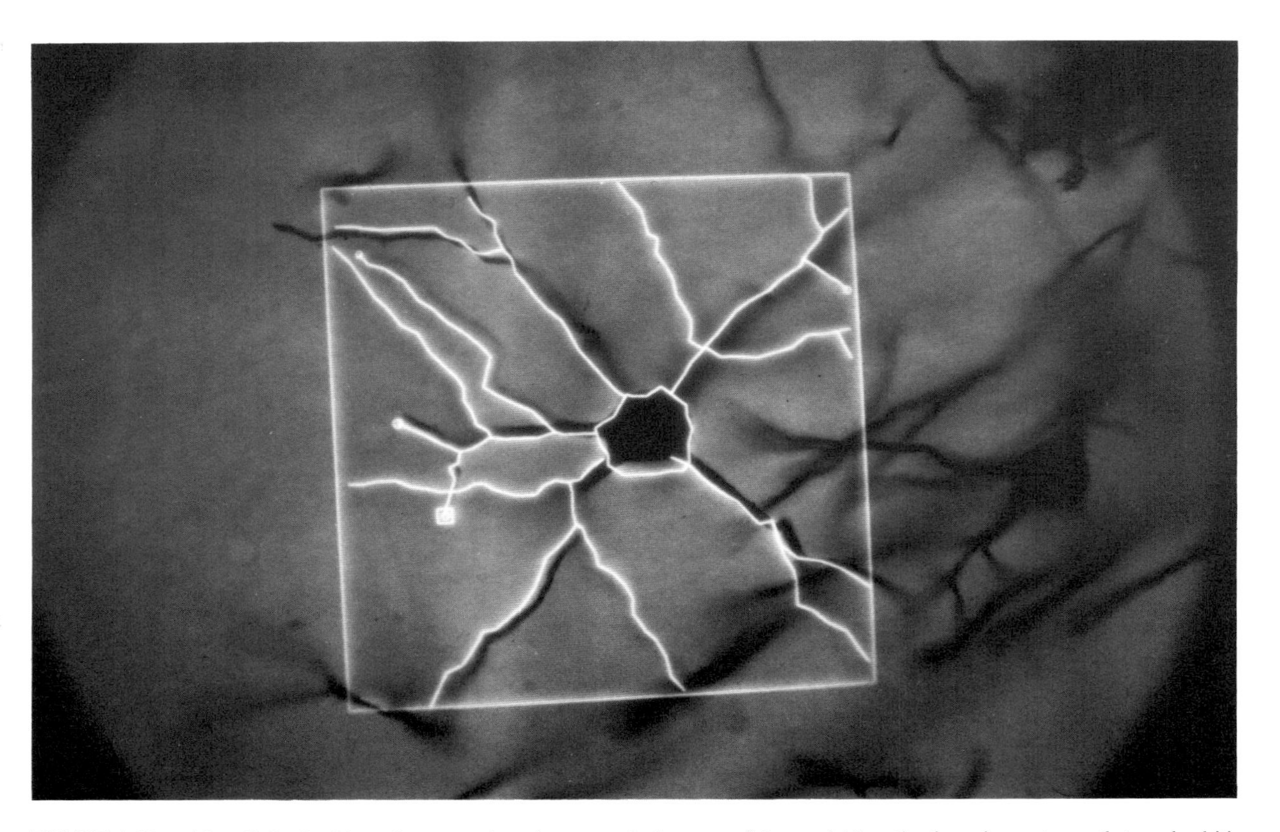

FIGURE 4-12. After all the dendrites of a neuron have been traced, they may all be overlaid on the tissue image to see that no dendritic segments were missed. The square box is 130 μm across.

cut by the microtome. Similarly, a bottom artificial end may be cut at the bottom surface of the tissue by the microtome.

4.3.5b. Distinguishing among Trees and Portions. When you have finished tracing a tree, you may give it a name that may be used later to select it from the data base for display, plotting, or statistical summarization. To assist in the selection of trees from the data base, a "wild-card" feature is frequently used, signified by an asterisk in the name. For example, if five trees, three from population *a* and two from population *b* have been entered, and they are named "treea1," "treea2," "treea3," "treeb1," and "treeb2," then it would be possible to select the three trees of population *a* by providing the name "treea*." The computer program would then display only the trees of population *a*.

In addition to specifying names to distinguish among trees, it is sometimes desired to label a portion of a tree. To do so, you can assign it a tag, which is an arbitrary numeric value. Tags can be applied to any point and are stored with the data. As an example, if you notice that a certain portion of a dendrite appears to make synaptic contact with another neuron, you may, as you trace this portion, label all the points of this portion with the same tag value. Then later, the tagged portion could be separately analyzed or perhaps plotted in a different color.

4.3.5c. Automatic Point Capturing to Speed Up Tracing. A feature of tracing called "automatic point capturing" may be used to increase tracing speed. To do this, you simply pass the cursor over the dendrite, and, at a fixed spatial interval, the computer automatically records a series of continuation points. This is a faster method of tracing, but it is more dangerous. Since the computer has no way of knowing whether you are really tracing a dendrite or not, should you deviate from the actual dendrite, the computer will simply continue taking points. For a novice, this technique generates too many errors, but for an expert, it is a good time-saving device.

4.3.6. Locating and Outlining Other Structures

Neurons reside in an environment, and you frequently wish to map this environment. More specifically, you may want to outline a tissue edge or region boundaries, and you may want to locate structures (neuronal or otherwise) in the tissue. With the UNC neuron-tracing system, there are two general methods for doing this mapping. You may use the tracing part of the neuron-tracing system and merely outline, for example, a tissue boundary by tracing it as if it were a tree with no branch points. There is, however, a much better method: to use a lower-power microscope image and a different section of the neuron-tracing program, called OUTLINING AND MAPPING TISSUE. Here, with the joystick, you move the cursor over a fixed image, and the stage is only moved from frame to frame. Outlining a boundary in this manner is shown in Fig. 4-13. Using this system, it is easy to move a cursor rapidly along an outline and let the computer capture points one after another at a fixed spatial interval to map the outlines.

Three different point types are available for doing this outlining. A boundary is traced by entering one "outline start," a series of "outline continuation points," and finishing the boundary by entering an "outline end." The boundary may be either open or closed, i.e., ending at the original starting point. Somas may also be quickly outlined with

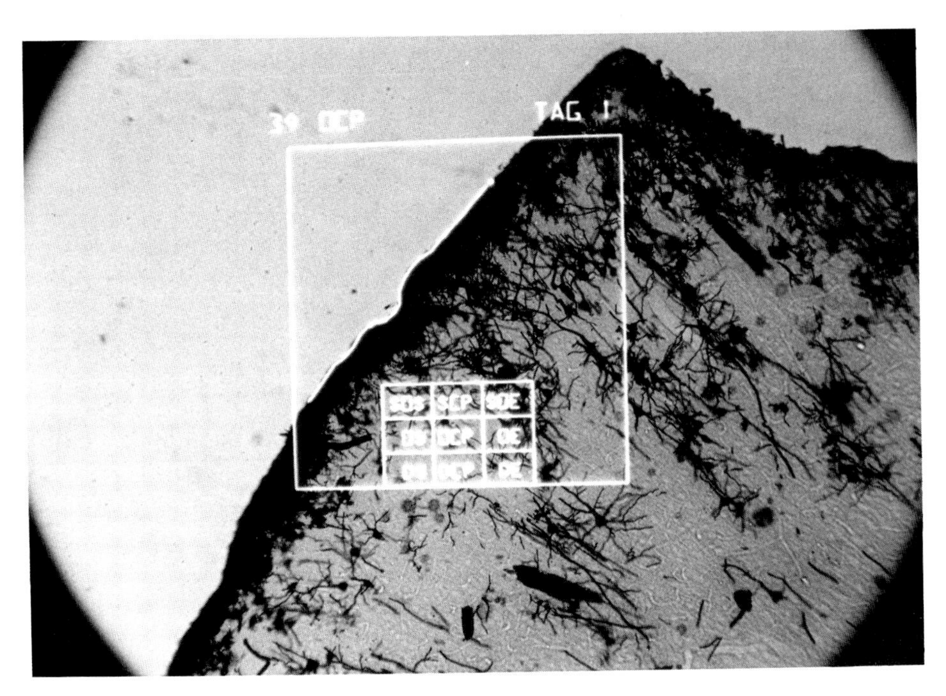

FIGURE 4-13. The view in the microscope while outlining a tissue boundary with the mapping function of the UNC neuron-tracing system. With the joystick, the user moves the cursor over the stationary image, and at fixed spatial intervals, points are recorded automatically. The computer generates an overlay of the just-recorded points so that he may see that he has entered the boundary correctly. The grid at the bottom of the box is a map of the computer keypad, and the letters inside the grid indicate the types of points that may be recorded. An outline starts with an outline start (OS), continues with a series of outline continuation points (OCP), and ends with an outline end (OE). The tag value shown in the upper right is an arbitrary numeric value that may be recorded with each point to distinguish among types of outlines. This tracing is three-dimensional; i.e., the focus axis is enabled during the tracing, and X, Y, and Z values are recorded with each point.

this program by entering a ‘‘soma outline start’’ (SOS), a series of ‘‘soma continuation points’’ (SCP), and a ‘‘soma outline end’’ (SOE).

A slightly different task is to locate items in the tissue. This involves no tracing; rather, the X,Y,Z coordinates of each point are located for later display, plotting, and/or analysis. Figure 4-14 shows the microscope view during this locating procedure. A ‘‘dot cluster’’ is defined as a group of dots and consists of three point types: ‘‘dot start,’’ ‘‘dot continuation point,’’ and ‘‘dot end.’’ These points are used to enter a dot for each structure. The cluster of dots is logical, not spatial; i.e., a name can be given to each cluster. Then, the entire set of dots, though they may be spread over wide areas of tissue, may be displayed, plotted, and analyzed by selecting them from the data base by name. Each point within the dot cluster may, of course, be identified with a distinguishing tag value.

4.3.7. The Storage of Traced Data

As tracing continues, a data base is generated in the computer memory. The data base forms a mathematical model of the neuron from which summaries, displays, and

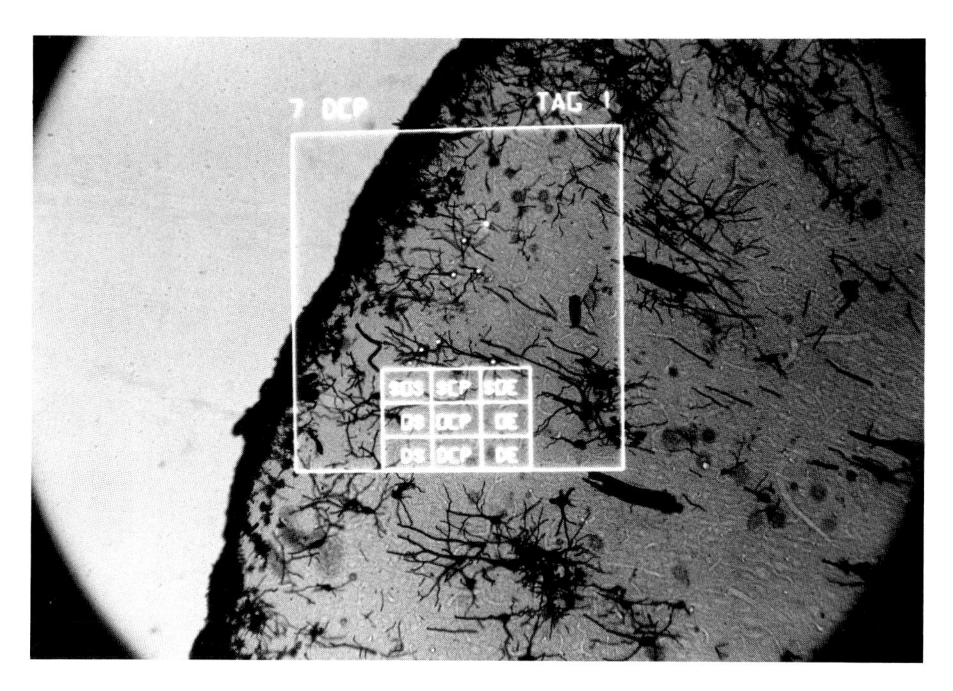

FIGURE 4-14. The view in the microscope while locating cell bodies with the mapping function of the UNC neuron-tracing system. The user moves the joystick to position the cursor at each soma and presses a button on the keypad to record that soma's location in X, Y, and Z. A dot is generated at each just-recorded point so that the user is prompted not to miss any or to record any twice. The keypad grid's bottom row indicates the three types of points that may be recorded: a dot start (DS), a dot continuation point (DCP), and a dot end (DE).

plots can be made. Figure 4-15 is a listing of part of a typical data base. You see a point number, a point type, a tag, X,Y,Z coordinates in micrometers, and a thickness value given for each point. In the 1988 version of the UNC neuron-tracing system, a maximum of 500 trees and 20,000 points can be stored in memory at one time. Storage on disk is also possible at a rate of 10 bytes per coordinate. A complex tree, such as the one shown in Fig. 1-14, consists of about 8000 points. Later chapters will describe what can be done with all these data.

4.3.8. Advantages of Vector Graphics in Neuron Tracing

The screen that presents the image to the camera lucida in the UNC system is a vector computer graphics screen. A vector graphics screen is more useful for tracing than a raster display system for two reasons. First, in order to draw the variable-diameter cursor to the fine resolution needed for measuring dendritic diameter, the high resolution of a vector system is required. A raster system is not able to generate a circular cursor with sufficient resolution. Secondly, a vector system presents a brighter image at any particular point on the screen than a raster system does. Because it is necessary to use a high light level to illuminate the tissue, the computer-generated image that is overlaid on it must also be bright. So for these reasons, a vector graphics system is recommended for tracing.

Point	Type	Tag	X	Y	Z	Thick	Name	Attachment
1	Soma outl start	0	-7.5	-3.5	.0	.5	soma01	
2	Soma continu pt	0	-9.5	2.0	.0	.5		
3	Soma continu pt	0	-4.0	5.5	.0	.5		
4	Soma continu pt	0	4.5	7.0	.0	.5		
5	Soma continu pt	0	10.0	6.0	.0	.5		
6	Soma continu pt	0	13.0	-3.5	.0	.5		
7	Soma continu pt	0	5.5	-9.5	.0	.5		
8	Soma continu pt	0	-1.5	-8.5	.0	.5		
9	Soma outlin end	0	-7.5	-3.5	.0	.5		
10	Middle tree org	0	-8.0	-2.5	2.0	2.6	tree01	------ 0
11	Continuation pt	0	-13.0	-4.5	2.0	2.4		
12	Branch point	0	-16.5	-5.0	2.0	2.4		
13	Continuation pt	0	-21.0	-4.5	5.0	2.0		
14	Bottom art end	0	-25.5	-4.5	7.5	1.9		
15	Continuation pt	2	-21.0	-8.5	1.5	2.1		
16	Natural end	2	-25.5	-12.0	2.0	1.9		
17	Top tree origin	2	-24.5	9.0	.5	1.6	tree02	tree01 14
18	Continuation pt	2	-29.5	13.5	.5	1.6		
19	Branch point	0	-29.5	19.0	.5	1.5		
20	Fiber swelling	0	-32.0	24.0	.5	1.5		
21	Top artific end	0	-30.5	27.5	-2.5	1.5		
22	Continuation pt	0	-36.0	21.0	1.0	1.5		
23	Spine base	0	-40.0	19.0	1.5	1.5		
24	Natural end	0	-42.5	22.5	1.5	1.5		
25	Outline start	4	-16.5	31.5	2.5	.1	outl01	
26	Outline cont pt	4	-32.5	33.5	2.5	.1		
27	Outline cont pt	4	-45.5	27.5	2.0	.1		
28	Outline end	4	-50.5	16.0	2.0	.1		
29	Dot start	3	-34.5	5.0	-1.0	.1	dots01	
30	Dot cont point	4	-37.0	8.0	-1.0	.1		
31	Dot end	4	-40.5	3.5	-2.0	.1		

FIGURE 4-15. A listing of points from the data base of the UNC neuron-tracing system. The type, tag, coordinates, and thickness of each point are shown. Trees (as opposed to somas, outlines, and dots) may also have attachments to other points. Further details are given in the text.

Chapter 8 presents the advantages and disadvantages of using vector graphics systems for displaying the three-dimensional image of completely reconstructed neurons.

4.4. OTHER NEURON-TRACING TECHNIQUES

Tracing with the UNC neuron-tracing system has just been described. Other techniques for neuron tracing have been devised during the last quarter century. Some of these techniques are now reviewed, listing their advantages and disadvantages relative to those used in the UNC system.

4.4.1. Alternatives to Motorizing the Stage

Using motors to drive a stage and focus axis to submicrometer resolution is quite expensive. In 1988, a motorized stage, necessary software, and a controller box to allow it to be driven by the computer cost between $8000 and $20,000 and are available from several vendors. A much less expensive alternative is not to motorize the stage and focus but rather to let them remain passive, putting sensors on them. The computer could read the sensors, though sacrificing the ability to drive the stage and focus, but would at least always know the current stage coordinates. The sensors can take the form of rotary potentiometers on the fine positioning knobs of a mechanical stage and a rotary potentiometer on the focus axis knob. Rotary shaft encoders could be used instead of potentiometers, yielding higher resolution, though at a higher cost. Alternatively, linear potentiometers could be used to monitor the translational movement of the X,Y stage. Systems using these concepts are not commercially available, so they are usually constructed in university research departments.

The advantage of the passive stage and focus concept is its low cost compared to motorized systems. It has the disadvantage that the computer cannot drive the microscope stage to individual points, for example, to prompt you not to miss a tree by moving you back to a branch point. It has the further disadvantage of not allowing the computer to scan a region of tissue systematically, looking for individual trees. A final disadvantage of the passive systems is that they usually do not have submicrometer resolution, although with good-quality optical shaft encoders, it may be achieved. However, the cost may rise to several thousand dollars.

4.4.2. Alternatives to the Computer-Generated Overlay

Another alternative has to do with the computer-generated overlay. The computer-generated overlay requires a computer graphics system and a camera lucida. Two of its primary functions are simply to provide a variable-diameter cursor and to provide instantaneous feedback that the tracing is proceeding correctly. If you are willing to forego a cursor of variable diameter and are willing to forego any instantaneous feedback that the tracing is going well, then there is no need to have a camera lucida looking at a computer CRT. A cursor can be established simply by placing a cross-hairs reticulum inside one of the microscope eyepieces. Dendritic thicknesses can always be measured by positioning the cross-hairs cursor to each side of the dendrite and letting the computer calculate the diameter of the dendrite from these two points. This thickness measurement will not be as accurate but may suffice for lower-resolution plots and statistical summaries.

4.4.3. Alternative Stage Control Methods

Another option for tracing neurons has to do with control of the microscope stage and push-button selections. Of the interactive devices reviewed in Chapter 2, only the three-dimensional joystick is comfortable for controlling a continuous, long-lasting, fine-resolution three-dimensional tracking task. But a quality joystick with its associated analog-to-digital converters is a moderately expensive (perhaps $1000) and uncommon device in the laboratory. Data tablets and mice are readily available, so researchers tend toward them. A data tablet is an excellent device for controlling motion in two dimensions, so it is reasonable to use it for controlling a microscope stage. The stage tracks the position of the data tablet cursor as you move it over the surface of the tablet. You must, however, use your other hand to control the focus, perhaps by rotating a knob attached to a potentiometer.

This stage-controlling scheme has the disadvantage that you now have no free hand, so when you want to make a push-button selection, you must interrupt tracing. A frequently used idea illustrated in Fig. 4-16 solves this problem with a perimeter menu presented in the eyepieces. You can move the cursor to an item on this menu to select point types, and then resume tracing. This way, the X,Y cursor control and push-button selection function are done with the right hand, and the focus-axis positioning is done with the left hand.

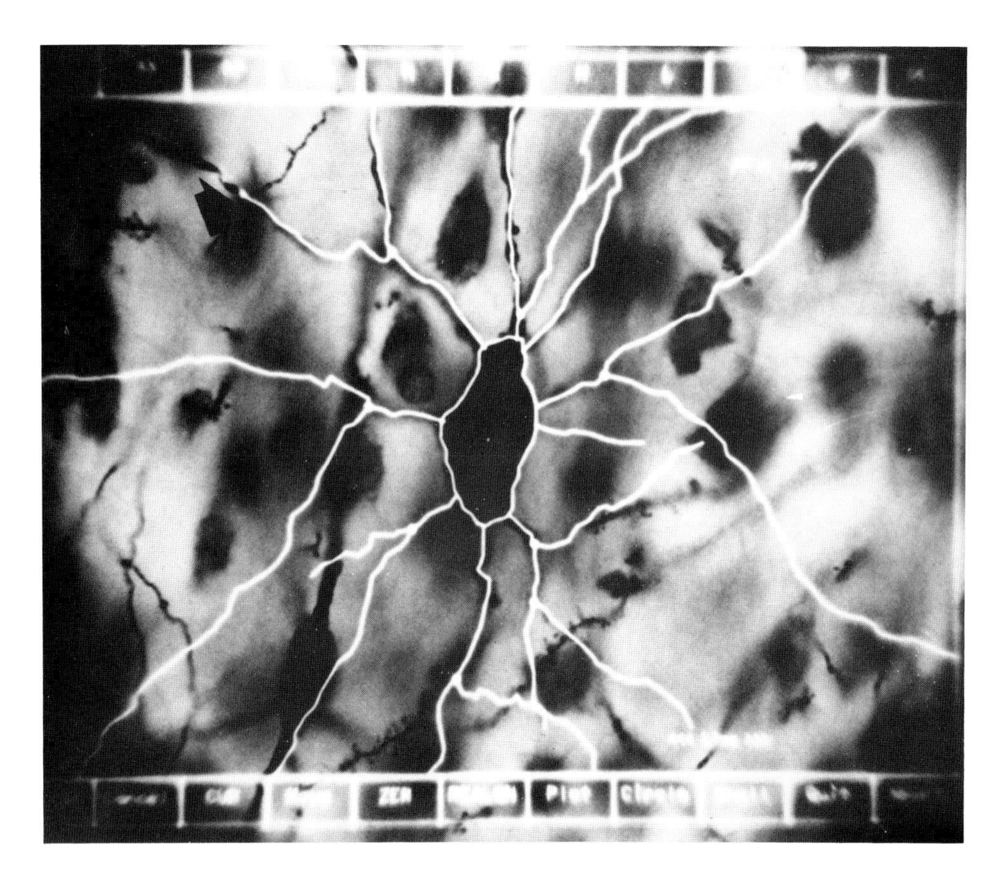

FIGURE 4-16. The microscope image during semiautomatic neuron tracing with the Glaser et al. (1983) system. The computer representation of the already-traced dendrites is overlaid on the actual neuron. Tracing is proceeding from the point marked by the black arrow. Menus at the top and bottom of the image are used to select special functions, for example, to specify a branch point. The menus are out of focus because of technical difficulty in taking the photograph. Reproduced from Glaser et al. (1983) with permission.

Two buttons on the mouse or on the data tablet hand cursor may be dedicated for focusing—one for focusing up and through the tissue, and the other for focusing down. This allows the cursor movement in all three axes to be controlled with one hand; however, the continual focus corrections required during three-dimensional tracking are difficult to perform accurately.

Much simpler neuron-tracing software can be generated. It is possible to move a motorized stage and focus by simply using cursor keys on the computer keyboard. Every time you hit a key you step a little bit in one direction. This saves money in hardware but makes it more difficult to trace.

Finally, the whole system could be completely passive. As you trace, you could simply read the scale lines on the X,Y stage and focus axis knob of the microscope and write the results down on a sheet of paper. These results, a series of coordinates that form a mathematical model of the dendrite, could be manually entered into the computer for further display and summarization. This inexpensive and crude system will indeed provide reasonable reconstructions, although it is labor intensive.

4.5. FOR FURTHER READING

The first description of a semiautomatic neuron-tracing system in which the researcher looks directly into the microscope and the computer records point coordinates was reported by Glaser and van der Loos (1965). The next report of such a system was by Wann et al. (1973). Since that time, numerous other similar systems, each with new improvements, have been published. These include reports by Capowski (1977, 1979), Glaser et al. (1977), Overdijk et al. (1978), DeVoogd et al. (1981), Capowski and Réthelyi (1982), Amthor (1985), Moyer et al. (1985), Freire (1986), Somogyi et al. (1987), and Zsuppán et al. (1987).

The use of systems in neuronal morphology research have also been well documented. Examples in which special attention was given to the neuron-tracing system include reports by Réthelyi and Capowski (1977), Capowski and Réthelyi (1978), Glaser et al. (1978), Light and Perl (1979), Réthelyi (1981), Tömböl et al. (1983), Blackstad et al. (1984), Leuba and Garey (1984a,b), Simons and Woolsey (1984), and Geröcs et al. (1986).

A significant improvement was the mixing of the computer graphics image with the microscope image so that the user could see both the original microscope image and computer-provided instructions and feedback simultaneously. The first report of such a system was by Glaser et al. (1979a). Then numerous other reports followed, including those by Glaser and van der Loos (1980), Capowski and Sedivec (1981), Glaser et al. (1983), Capowski (1985, 1987), Freiherr (1987), Kusinitz (1987), and Lieth (1987).

Such neuron-tracing systems with visual feedback are also receiving extensive use in the neuroscience community. A sampling of recent reports include those by Bregman and Cruce (1980), Sedivec et al. (1982, 1986), McMullen et al. (1984), Rosenthal and Cruce (1984, 1985), Ritz and Greenspan (1985), Sugiura et al. (1985, 1986), Arkin and Miller (1986, 1988), Capowski et al. (1986), Claiborne et al. (1986), Hitchcock and Easter (1986), Miletic and Tan (1987), and Light and Kavookjian (1988).

An interesting technique to increase the focus-axis working distance to enable a researcher to trace dendrites to a greater depth into the tissue was described by Glaser and van der Loos (1981).

Several alternative methods for configuring hardware into a neuron-tracing system have been described by Capowski and Cruce (1979).

Review articles on the state of the art of neuron tracing have been published by Marx (1976), Woolsey and Dierker (1978, 1982), and Sobel et al. (1980).

Many of the reported software packages for neuron tracing have been reviewed by Huijsmans et al. (1986).

5

Input from Serial Sections

JOSEPH J. CAPOWSKI and ELLEN M. JOHNSON

5.1. INTRODUCTION

Serial section reconstruction has a wide range of applications. The technique of serial section reconstruction is used to build models of all kinds of structures, both biological and mechanical. Indeed, examples of serial section reconstruction can be found in boat building, dentistry, and the manufacturing of machine parts. We limit this chapter to the use of serial section reconstruction in neuroanatomy.

The concept of serial section reconstruction is illustrated in Fig. 5-1. Imagine that you bake an object inside a loaf of bread; obviously, you can no longer see it from the outside. Next, you slice the loaf of bread with a bread slicer. The resulting 20 to 30 slices allow you to see slices of the object, but, unfortunately, you have destroyed its three-dimensional nature. The three-dimensional properties are now changed to a series of two-dimensional slices, called serial sections because they come from sequential slices. The process of serial section reconstruction allows you to "reassemble" the two-dimensional slices into a three-dimensional structure.

After you do the serial section reconstruction, your options for viewing the result are greatly enriched. You may look at the original two-dimensional slices, you may look at the three-dimensional surface viewed from any perspective, and you may look at the three-dimensional internal structures viewed from any angle and located within the environment of the external structure.

As for the neuron reconstruction described in Chapter 4, there are two major reasons for performing serial section reconstruction. The first, here most important, is to see the otherwise hard-to-view properties of the structure. The second, though somewhat less important, is to perform some mathematical summarizations of the data.

This chapter describes the techniques used for entering neuroanatomical data from serial sections into a computer. Since many of the same computer graphics techniques that are used to generate displays of serial section reconstructions are also used to generate displays of reconstructed neurons, the display techniques for both types of displays are presented together in Chapter 8. This makes further sense, because frequently you wish to

ELLEN M. JOHNSON • Department of Physiology, University of North Carolina, Chapel Hill, North Carolina 27599.

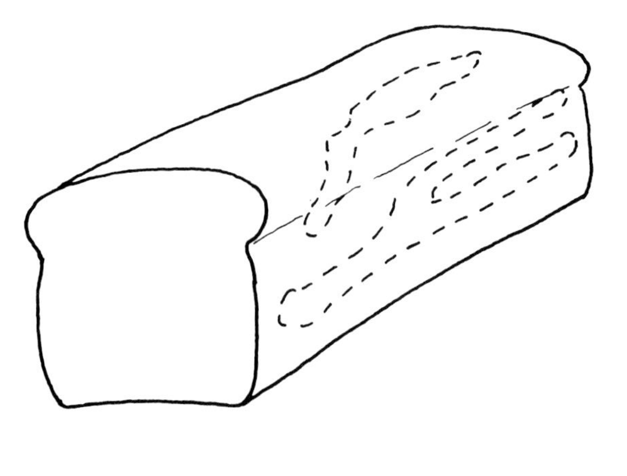

FIGURE 5-1. Diagram of the serial section reconstruction dilemma. Top: A three-dimensional structure is baked inside a loaf of bread; it is intact but not visible. Bottom: After the loaf is sliced, the internal structure is visible but is no longer intact and must be reassembled or "reconstructed."

combine the two reconstruction methods, for example, to build a display of a group of neurons in their anatomic environment.

5.2. HISTORY

Anatomists have been doing serial section reconstruction for many years. A German anatomist named His (1880) used a graphics technique to project points from the outlines of features in sections onto a series of straight lines. The lines were separated by the distance proportional to the thickness of the sections. His's results were a drawing of the original structure viewed perpendicular to the direction of sectioning.

Kastschenko (1886) and then Odhner (1911) used the technique of tracing individual tissue sections onto transparent sheets and then stacking the transparent sheets to build a three-dimensional reconstruction. This method is still used in a number of laboratories (for example, Stevens et al., 1980).

Solid models built of clay or wax have been made for many years. Born, a pioneer in the building of solid models from waxed plates, provided a classic reference (Born, 1883) on his work. A model by Spacek and Lieberman (1974), shown in Fig. 5-2, is a beautiful example of some recent work in solid modeling.

There are two excellent documents describing various physical aspects of serial section reconstruction: an article by Ware and LoPresti (1975) and a book by Gaunt and Gaunt (1978). Both of these sources append large bibliographies. Gaunt and Gaunt also

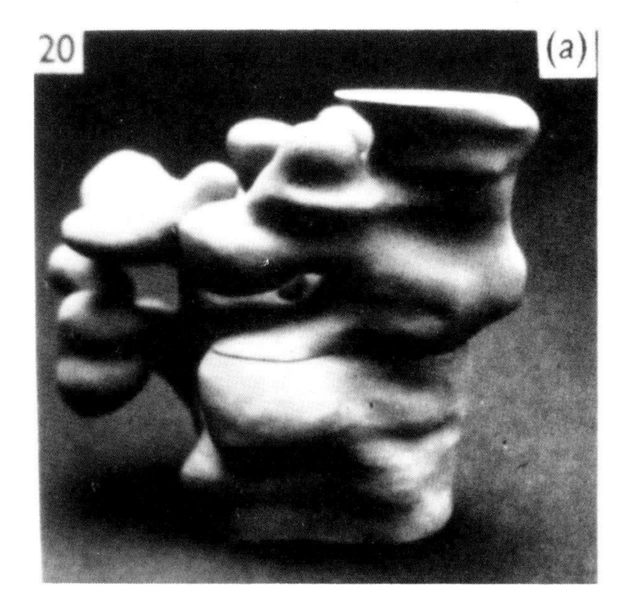

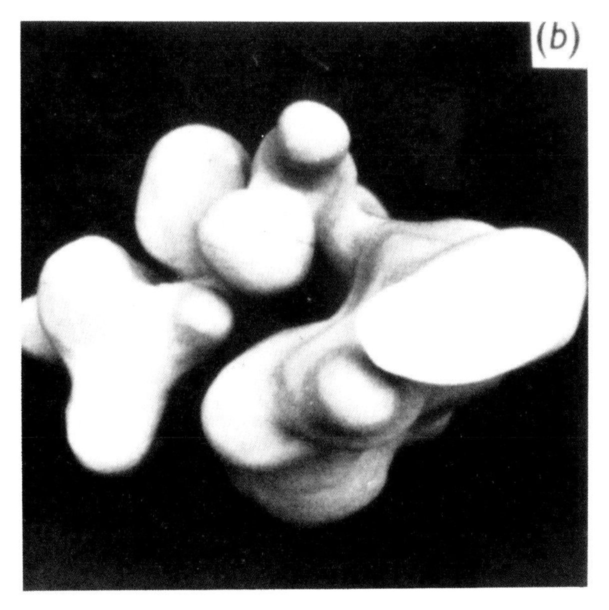

FIGURE 5-2. Two views of a wax reconstruction of a rat synaptic glomerulus. Reproduced from Spacek and Lieberman (1974) with permission.

present stereomicrophotography and holography as techniques for recording and displaying three-dimensional data used in serial section reconstruction.

Many authors have displayed three-dimensional serial section reconstructions in "shear" diagrams as illustrated in Fig. 5-3. Here, drawings of each tissue section are displayed with those deeper into the structure slightly above and possibly to the right. These shear diagrams have the advantage of letting you see some three-dimensional depth, but you cannot rotate it to view it from any orientation.

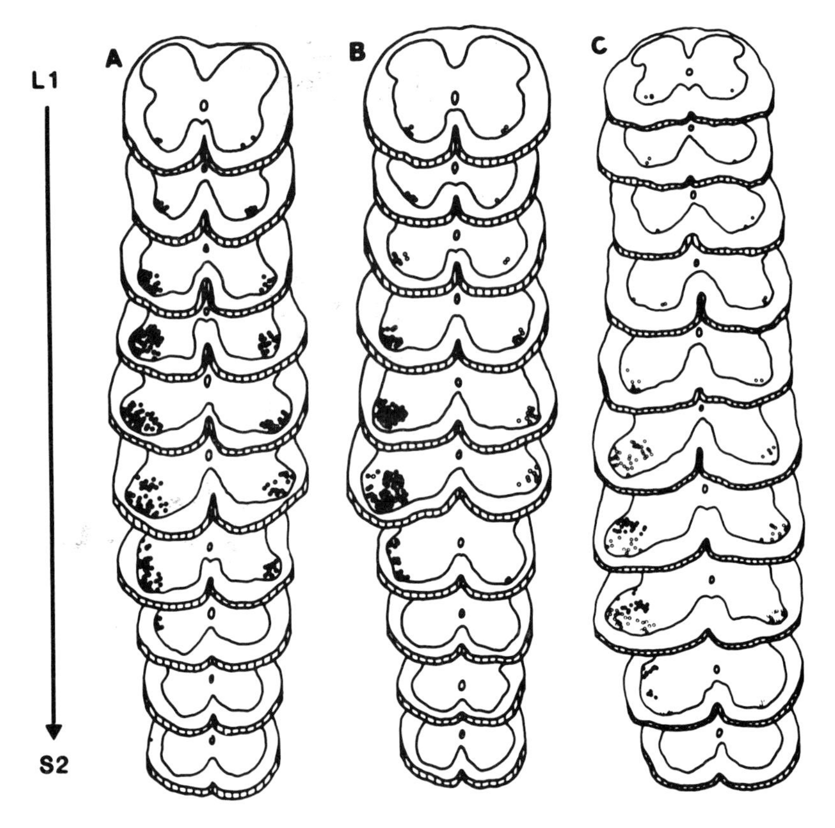

FIGURE 5-3. An anatomic shear diagram. Three series of transverse spinal cord sections with neuron locations marked with circles are presented. Each section is offset from the next one in order to illustrate the three-dimensional structure. Reproduced from Comans et al. (1988) with permission.

5.3. PURPOSE OF SERIAL SECTION RECONSTRUCTION

Why would one want to use serial section reconstruction in neuroscience? As an example, neuroanatomists wish to study what is called the ultrastructure. These are the very fine parts of neurons that are typically studied in an electron microscope (EM) using tissue sections approximately 1 μm thick. By taking EM photographs of these sections and stacking a series of these photographs, we can build a three-dimensional model of that part of the neuron. Scientists use this technique to look at boutons, swellings near the ends of the fibers; the shapes of somas, the spherical or football-shaped centers of neurons; or to generate three-dimensional maps of any piece of tissue that would contain neurons.

When a thick slab of tissue is cut into serial sections for entry into a computer reconstruction system, assumptions are usually made. The *Z* axis is defined to be the depth-of-focus axis, the one perpendicular to the plane of sectioning. Each section is considered to have zero thickness; i.e., no attempt is made to distinguish among the depths of various structures within a section. As an analogy, if you pull a sheet of paper off the top of a stack of papers, the sheet is so thin with respect to the stack that it is considered to have no thickness; all minute elements within the sheet of paper are considered to have the same *Z* value. Each sheet of paper or section of tissue is assigned a *Z* value, however, representing its depth in the stack. This is the physical sectioning process that assigns the third dimension to a series of two-dimensional slices.

This thin-section reconstruction is different from the neuron reconstruction in Chapter 4. There, each tissue section is thick, and microscope optics are selected to focus on only a small range of depth within the section. This technique is called "optical sectioning" and may be applied to looking at any kind of structure, including neurons. Optical sectioning has the advantage over physical sectioning that it is not necessary to cut the tissue physically. However, the depth into which one can optically section is restricted by the microscope optics and by the amount of light that can be passed through the tissue.

Another significant difference between the tracing of neurons and serial section reconstruction is that serial section reconstruction is not used to follow repeatedly branching dendrites as they meander through the tissue over a long distance. Often a small portion of a dendrite, for example, a cluster of synaptic boutons, is reconstructed from serial sections. Outlines are entered from each section, and these outlines, when stacked, form the mathematical model of the dendritic portion.

5.4. ENTERING SERIAL SECTIONS INTO THE COMPUTER

Computer input techniques for serial sectioning data are typically semiautomatic; that is, the user does the pattern recognition, and the computer does the bookkeeping. It is possible to a degree, however, to do automated input; this is covered in Chapter 15.

The basic technique for semiautomatic input of serial sections is to pass a cursor over each section; the user controls the cursor, and the computer records where the cursor is located. It does this by taking a series of *X,Y* coordinates, which are the positions of the cursor at a series of points along the structure's outline, and appending a *Z* coordinate, a constant that is the same for each tissue section. Thus, a series of *X,Y,Z* coordinates defining the outline are recorded into the computer memory.

5.4.1. Using a Data Tablet

A data tablet is the most often employed tool for entering serial sections into a computer. In brief, the researcher places an image of each tissue section on the data tablet and then traces it.

5.4.2. Types of Images

There are five ways images are placed on data tablets. As shown in Fig. 5-4, a photograph of each tissue section can be placed on the data tablet, and then the researcher traces over the outlines in the photograph with a cursor. Another option is to use drawings made on a sheet of paper with pencil. Another way, as shown in Fig. 5-5, is to project an image of the tissue onto the data tablet with a projection microscope and then pass a cursor over the tissue image. A fourth way is to project a 35-mm slide onto the data tablet. A fifth way is to use a camera lucida or drawing tube. This allows the researcher to look in the microscope and see the cursor, as shown in Fig. 5-6, overlaid on top of the actual tissue.

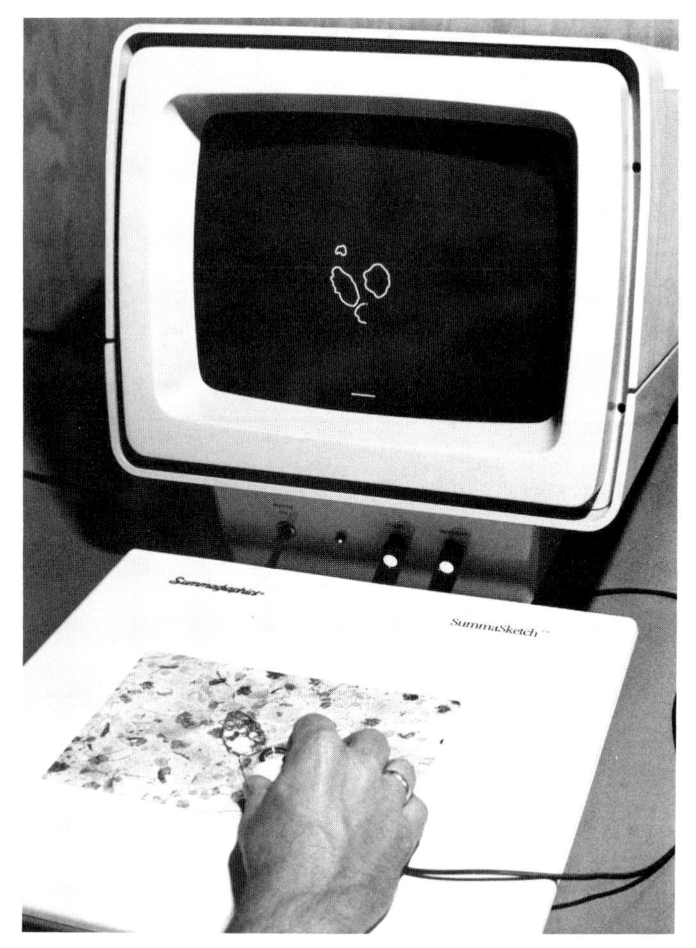

FIGURE 5-4. Entering data from an electron micrograph into the computer. The photograph is placed on the data tablet, and the user moves the hand cursor's cross hairs around the outlines of the individual structures. The CRT presents a view of the structures being traced so that he can see that the data are being accepted by the computer correctly.

5.4.3. Calibrating the Data Tablet

The first function that must be performed is to calibrate the data tablet. An easy way to understand what this means is to visualize a map of the United States on the data tablet. Consider, for example, the approximately 2500-mile distance from Los Angeles to Boston. You must tell the computer that the distance on the data tablet from Los Angeles to Boston is 2500 miles, since the perhaps foot-wide tablet is obviously not really 2500 miles wide. The usual way to do this is to place the cursor at one point on the tablet, say Los Angeles, and then at a second point, Boston. Next, instruct the computer to calculate the distance in data tablet units between the two points; then enter the distance in real units, that is, miles, between the two points. Now the distance on the data tablet, which

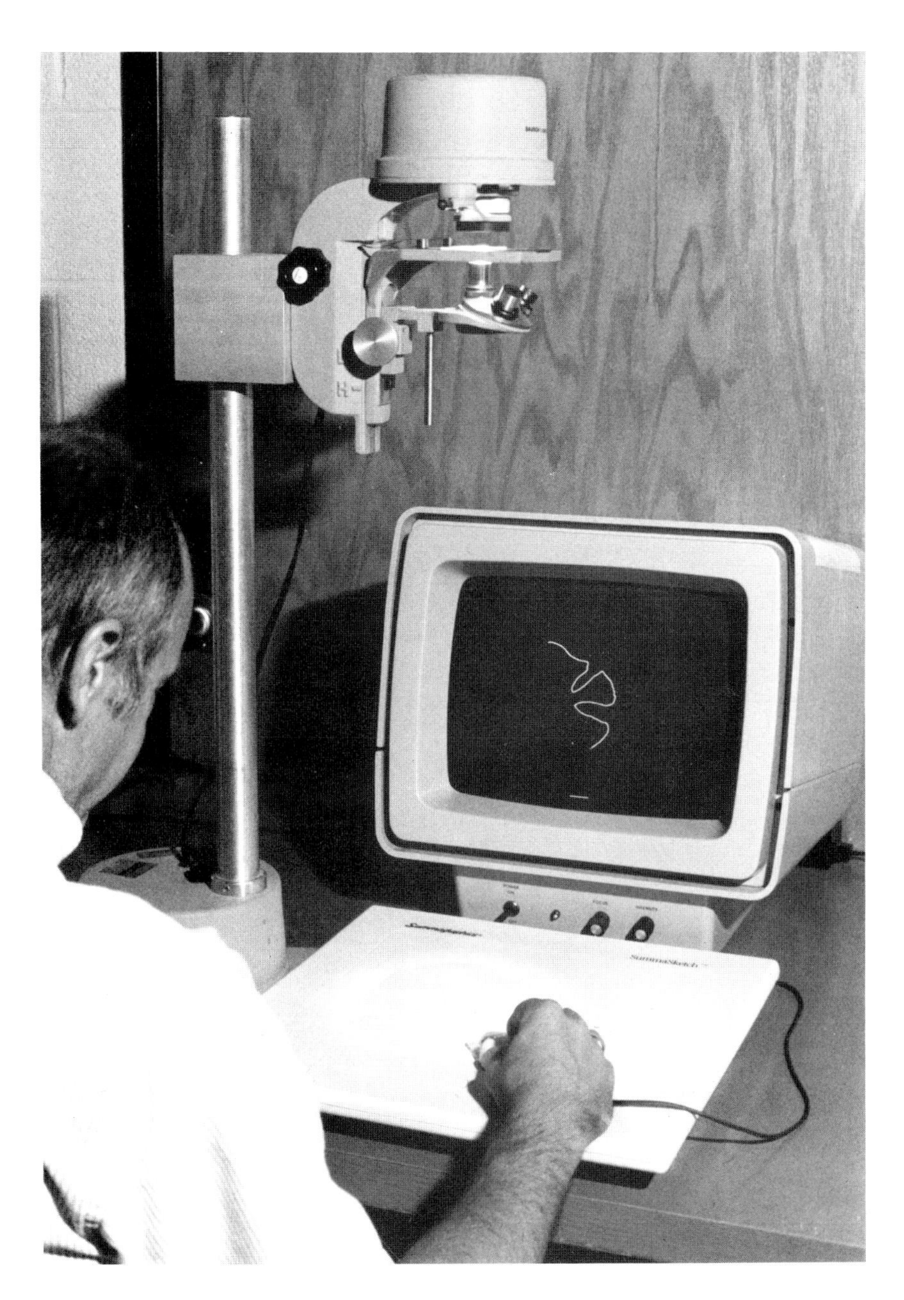

FIGURE 5-5. Using a projecting microscope to trace a structure. The image is projected onto the data tablet, and the user outlines the structure with the cursor. The CRT presents a view of the structure being traced so that he can see that the data are being entering into the computer correctly.

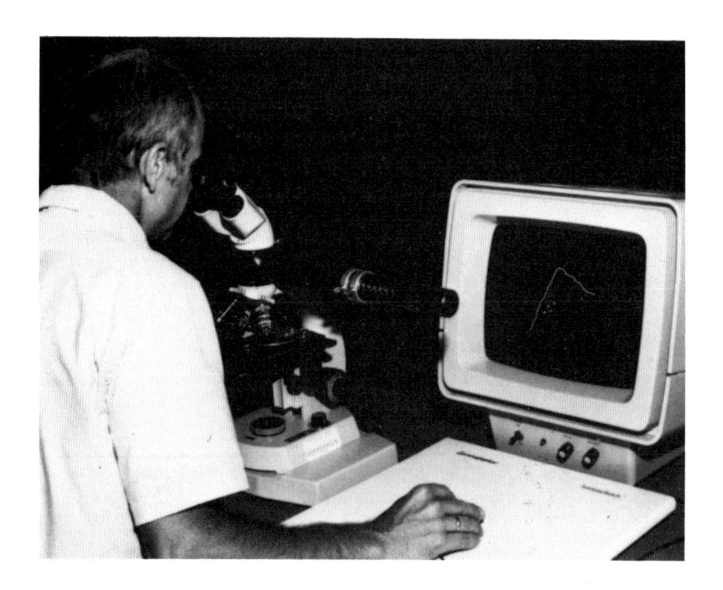

FIGURE 5-6. Using a microscope with drawing tube to trace a tissue section. Top: While looking at the tissue in the microscope, the researcher moves the tablet's hand cursor over the tablet surface. The drawing tube overlays the image of the cursor on the image of the tissue. Bottom: The microscope view during the tracing. Bottom photograph reproduced from Capowski (1988b) with permission.

might be 10 inches, is equal to the distance in real units, 2500 miles. Once this calibration process is finished, the computer can always associate a real distance or real coordinate with each tablet distance or each tablet coordinate. Then all data entered may be stored in real units, such as micrometers, millimeters, or miles.

5.4.4. Entering the Data

The tracing process is typically done in the following manner. Suppose that you want to follow an outline of tissue as shown in Fig. 5-4. You move the cursor to a beginning point on the outline, depress and hold down the button on the cursor, then move the cursor to follow the outline in a smooth, continuous motion. At frequent intervals (e.g., 10 times per second), the computer software samples the X,Y coordinates of the data tablet cursor. Whenever the data tablet has moved a certain distance, called a spatial filter, it captures the X,Y coordinate and scales it into user units. You then trace the entire outline, and the computer stores a series of X,Y coordinates. The Z value, which you have previously entered for the entire tissue section, is appended to the X,Y coordinates that are entered.

Tracing software may have an optional automatic closing feature so that when you release the button, a copy of the starting point is added to the outline. Thus, a straight line is drawn from where you end the outline back to the start of the outline.

The outline stored in the computer is a polygon consisting of a starting point, a series of points along the outline, and perhaps an ending point. If the polygon is closed, the ending point has the same coordinate as the starting point.

There are variations on this type of tracing. Instead of moving the cursor while holding down a button, you can move to each point along the outline and press the button once. With this technique, you move the cursor, press a button, move the cursor, press a button, etc., and the computer stores a coordinate each time you press the button.

Another variation on this tracing theme is to define indicators for the start of a polygon, for points along the polygon, and for the end of the polygon. Indeed, we could define three kinds of point types: an outline start, a point along the outline, and the end of an outline. You may indicate these to the computer by three separate cursor pushbuttons, or the computer program could generate them when you press a single button to start the outline, hold it down to continue the outline, and release it to finish the outline.

Yet another variation on this entering scheme has to do with the pixel-versus-vector representation of an image. In a vector representation, an image is represented by a series of X,Y,Z coordinates, and straight lines are drawn between them; in a pixel representation, all pixels are stored, and each pixel has a gray value. If there are only two shades of gray, then each pixel may be on or off. In the pixel representation, as you move the cursor over the data tablet, all pixels along the outline you are tracing are turned on. So it is possible to store an image and, as you trace it with the data tablet, turn on a series of pixels defining the image. This is the type of tracing that is used in the common "MacPaint" software packages on the Apple Macintosh computer. An example of this kind of tracing is shown in Fig. 5-7.

For storage of the data and for some statistical summarizations, it is frequently more efficient to convert the pixel representation of structure outlines to a vector representation. When this is done during tracing, a coordinate is stored only if it is the spatial filter distance from the previous point. Alternatively, this may be done after tracing by applying a spatial filtering program to the pixel representation. Such a program records only the coordinates of outline pixels that are a minimum distance from each other.

You may not always want to outline regions of neurons; sometimes you may want to locate items with dots. To do this, simply move the cursor to the item you want to locate and press a button. The computer records a point, that is, the X,Y,Z coordinate of the cursor at that point, and some indicator, called perhaps a dot, to indicate the location of the structure. No line is drawn to it, and no line will be drawn from it, but only a dot is drawn. It may be desirable to filter the dots spatially so that they cannot be drawn too closely together.

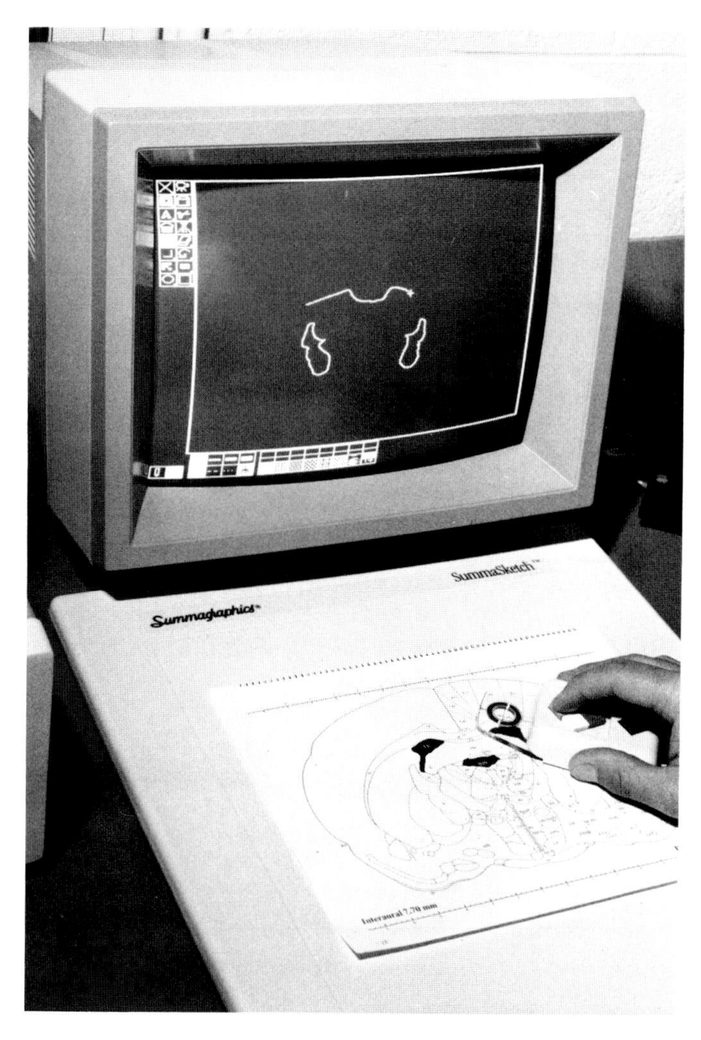

FIGURE 5-7. Tracing a structure using a graphics editor. A drawing of the structure is placed on the data tablet, and the user follows its outlines with the hand cursor. A representation of the just-traced structure is presented on the screen. Various nontextual menu options are presented on the computer screen. Contrast the pixel presentation of the structure to the vector presentation as shown in Fig. 5-4.

Another option for entering serial section data into the computer is to work directly from the microscope image rather than from an image projected or drawn onto the data tablet. This is similar to the neuron-tracing problem described in Chapter 4 except that you need not be concerned with the focus axis or with repeated treelike branching of the structure. Two techniques are used when working directly from the microscope image. With one you could use a passive microscope with potentiometers mounted on the stage and move the part to be traced underneath a cross hairs inside the eyepiece. With this technique the computer continually monitors the potentiometer output and thus keeps track of the current coordinate of the cursor. With the other you could use a motor-driven microscope stage with some interactive device, such as a data tablet or a mouse, to drive the stage to focus points along the outline under the cross hairs. In either case, when a button is pressed, a coordinate is recorded.

When you trace data into the computer, you may want to name each structure to differentiate one from another. So typically, you would name various structures and provide either icons, as shown in Fig. 5-7, or various colors while tracing to distinguish one structure from another, one dot from another, or points within an outline one from another.

5.4.5. Aligning Traced Sections

A classic problem in serial section reconstruction is how to align the sections as they are entered into the computer. Anatomic material usually has an irregular shape, and frequently there are no multisection structures that are perpendicular to the plane of sectioning. Therefore, it can be difficult to align one section with another. It makes little difference if you enter the second section with a small misalignment in rotation or translation relative to the first section. But if the third section has a similar misalignment relative the second one, you can imagine the magnitude of the result if a similar error occurs through every section, particularly because a tissue slab may be sectioned 25 times. Then cumulative misalignment errors may seriously distort the reconstruction. Some way to align tissue sections is needed so that there is a tolerable amount of alignment error from one section to another, with no cumulative error.

Alignment of any pair of adjacent sections is done by image superposition. The already-traced section is displayed, and the next adjacent section is overlaid on it; then you rotate or translate this next section until it aligns with the displayed section. This technique minimizes sections-to-section error, but a strategy is needed for minimizing cumulative error. For instance, if there are 25 tissue sections to be done in a reconstruction, you could start with the 13th, or middle, section and enter that one first. Then, it is useful to align all tissue sections with respect to the middle section rather than aligning each section with its adjacent one.

How does one do this alignment overlay? A crude method is shown in Fig. 5-8, with tracing paper and transparent tape. After you place the 13th tissue section on the data tablet, you simply outline an area or make other marks on the tracing paper that show the position of the section. Then, you fold back the tracing paper and put the next section, say the 14th section, in place. Then you put the tracing paper back over the 14th and align the 14th section with marks on the tracing paper. Then fold back the tracing paper and trace the 14th section, etc. In this crude mechanical way, all sections can be referenced to the 13th section, and alignment errors are minimized.

The overlay can be done in a much more sophisticated fashion than using tracing paper. If you are projecting slides, you can simply project two slides simultaneously and optically align them, or you can use computer graphics capabilities. If, as shown in Fig. 5-9, you have already traced the 13th section, you can move the cursor on the CRT to a point on the 13th section's display and slide the paper of the 14th section underneath the tablet cursor. You may therefore align various points on one section with corresponding points on an already-traced section.

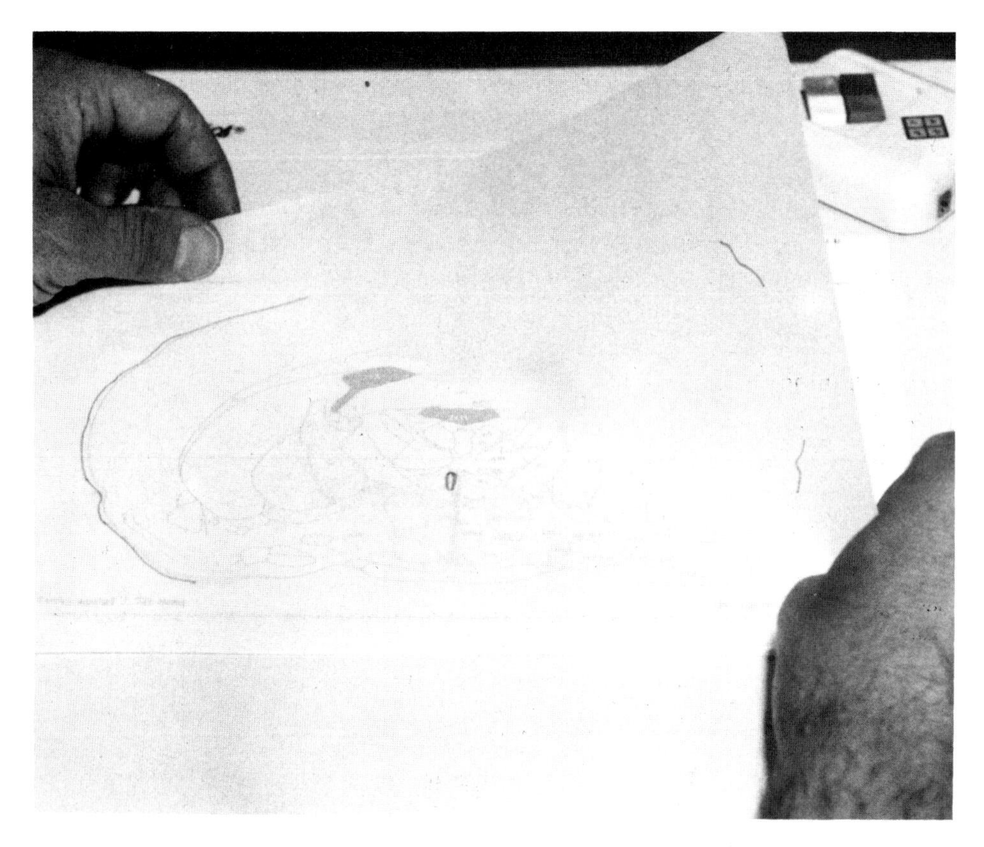

FIGURE 5-8. Aligning sections on a data tablet with transparent tape and tracing paper. A photograph of a tissue section is placed on the data tablet, and a piece of tracing paper is taped to the tablet near its left edge. A few tissue regions are marked on the tracing paper, and then the tissue is outlined into the computer. When the second section is placed on the tablet, it is aligned with the marks on the tracing paper.

Because anatomic structures are irregular, it may be helpful to manufacture artificial or ''fiducial'' marks within the tissue to help with alignment. If, before sectioning, one edge of the tissue slab is cut square, then after sectioning, it is possible to align each section by its straight edge. Holes in the tissue slab may be made with a pin or a fine drill bit perpendicular to the planes of sectioning. Then, after slicing, the sections may be aligned according to the pinholes.

Automatic alignment algorithms have been written. With these, already-traced data from two sections are oriented in many positions, and the computer calculates an error term proportional to the distance between known corresponding landmarks in the two sections. The orientation that generates the smallest error is thus selected. The alignment task is easy for the human and difficult for the computer, so it is debatable whether the programming effort required to write these algorithms is justified. More details about automatic section alignment are given in Chapter 15.

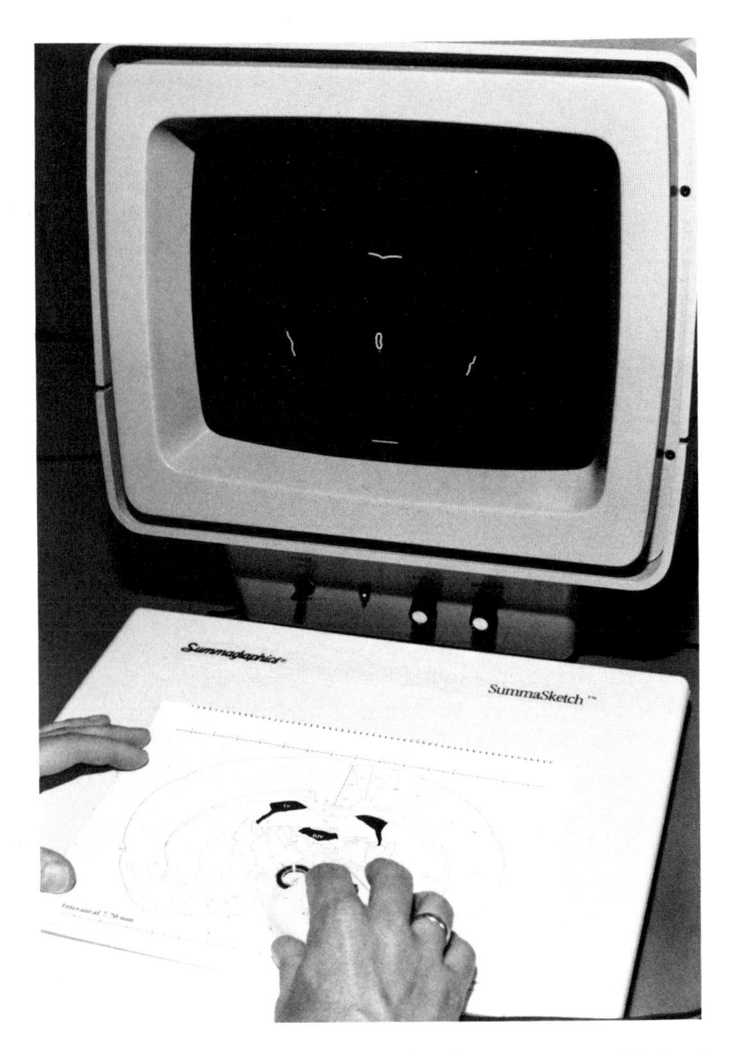

FIGURE 5-9. Aligning a section with a previous one using the screen cursor and the hand cursor. Several previously traced outlines are displayed on the CRT, and the user moves the cursor (a small dot below the small outline in the center of the screen) to a spot on the CRT. He then aligns the photograph to place the corresponding point under the hand cursor.

5.5. THE STORAGE OF DATA

As the data are entered into the computer memory, they form a mathematical model of the structure. Their storage is in a form tailored to serial section reconstruction. In this section we discuss specific details of this form.

To understand how the computer stores data, it is important to understand what is contained in each coordinate and what assumptions are made about how the coordinates are linked together. In the UNC serial section reconstruction system as reported by Johnson and Capowski (1983, 1985), each stored point consists of X,Y,Z coordinates, a point type, and a name. The coordinates are stored in user units, i.e., after calibration. Four types of points are allowed: a "move" indicates the start of a polygon, a "draw" indicates a subsequent point along it, a "close" indicates the end of a closed polygon, and a "dot" indicates a point marking some structure but not part of a polygon. Each point may also have a six-character name to distinguish among structures for display and statistical summarization. This storage arrangement assumes that the structure is formed by a straight line drawn from coordinate to coordinate, i.e., a vector representation of the data.

A pixel-based representation of an outline may also be stored in a frame buffer of size $512 \times 512 \times 1$ bit by giving the value 1 to those pixels that are along the polygons and the value 0 to those pixels that form the background. Pixels whose value is 1 are called "on" pixels, and those whose value is 0 are called "off" pixels. Since the number of on pixels is usually much smaller than the number of off pixels, storing outlines in this way is quite wasteful. So the techniques described below have been designed to compress the information.

5.5.1. Compressing the Data

Two classic computer science schemes for compressing the storage of a pixel-based representation of an outline have been devised: "run length encoding" and "pixel chaining." Run length encoding is illustrated in Fig. 5-10. A scanning pattern of the frame buffer memory is defined (usually from left to right along each line, and lines from top to bottom), and the number of pixels with the same value is counted. Each sequence of constant-value pixels is called a "run." The length of each run and its gray value is stored. This is a general-purpose storage compression technique that is frequently applied to the storage of any image but is quite useful in the storage of outlines.

The other storage compression technique is called "pixel chaining" and is illustrated in Fig. 5-11. Instead of storing a gray value for every pixel in the entire image (outline and

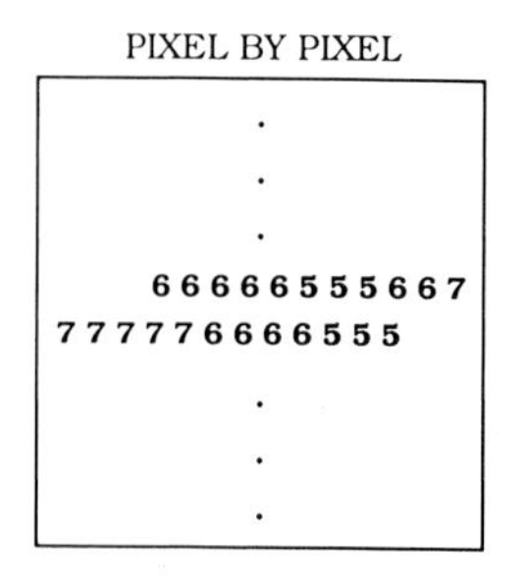

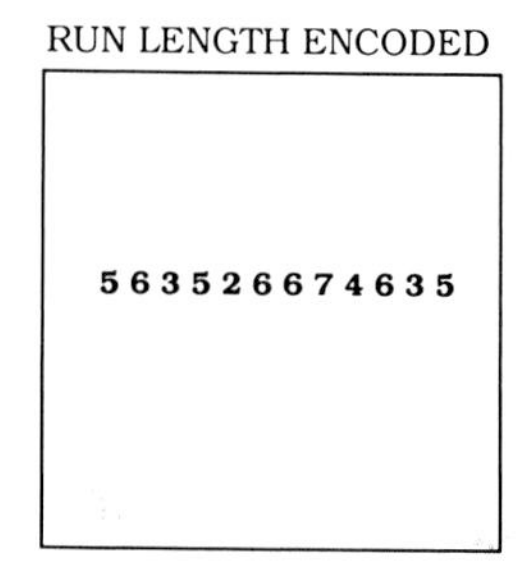

FIGURE 5-10. Run length encoding of an image in order to save memory. Left: Each pixel value requires a byte of memory, and each byte is stored in a memory location. In this example, 23 bytes are required to store the pixel values shown. Note that many pixels are repetitive; i.e., there are "runs" of identical pixels. Right: Each run is stored only as a count of the number of identical pixels followed by the value of each pixel in the run. Thus, each run requires two bytes of storage, or in this example, the same pixels are now stored in 12 bytes. The storage savings is data dependent, but a 512×512 byte image might be reduced from 256,000 bytes to 100,000 bytes.

PIXEL CHAINING

1 1 1	0 0 0	0 0 1
1 1 0	•	0 1 0
1 0 1	1 0 0	0 1 1

FIGURE 5-11. Pixel chaining to save memory of a stored outline. From any pixel in the outline, it is necessary to move in one of eight possible directions to find the adjacent pixel in the outline. The direction may be stored efficiently as a three bit code. Thus, one pixel is "chained" to the next one, and the outline storage requires little memory.

background), a 3-bit code is stored for every pixel in the outline. The code indicates which of the eight possible directions to go in order to reach the next pixel in the outline.

5.5.2. Organizing the Data

Another part of the data storage problem is how to organize all the data in a manner suitable for serial section reconstruction. Figure 5-12 and the following discussion illustrate the problem.

When you enter serial sections, you need the ability to enter data from one section without affecting the data that are stored for other sections. You also need to edit the data in a section, making insertions and deletions, without changing the data in other sections. Though the data may be stored temporarily in computer memory as it is entered, it must eventually be written to a disk for permanent storage. Most computers offer two kinds of disk data files: serial data files and random access data files. In a serial data file, coordinates are written one after another. In a vector representation of serial section data, you

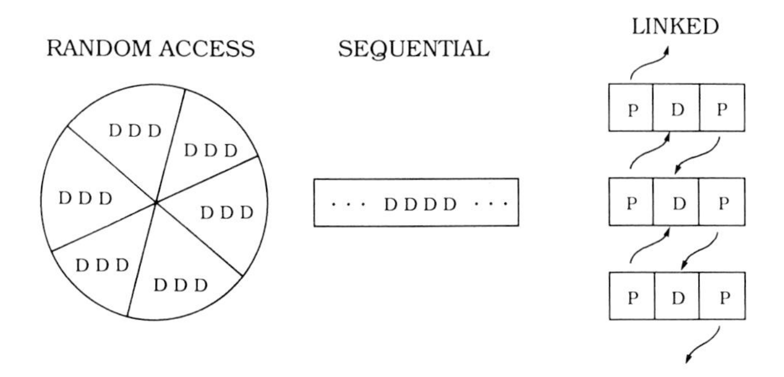

FIGURE 5-12. Three types of data storage used in serial section reconstruction. Left: In random-access storage, the memory is divided like a pie into equal-size slices, and data points (D) from each section are placed into one of the pieces. Data from each serial section usually occupies one pie slice. Editing the data is reasonable, but storage is inefficient because the same space is required to store complex sections and simple ones. Center: In sequential storage, data are stored as coordinates one after another. Storage is efficient, but editing may be slow because inserting a data point requires moving all the data above it. Right: In linked data storage, each data point is linked by pointers (P) to the previous and next data points. The data may be located anywhere in memory, and editing is reasonable because the chain may easily be broken to insert or delete data points.

can write a series of data points to a disk file and, when you finish entering the first section, continue writing the data points in the same file for the second section. Thus, the coordinates are stored on disk, one after another, with no space between the data from the first tissue section and the data from the second tissue section. This is an efficient way to store data but has the problem that all of the coordinates are stored serially, one right after another. Consequently, if you want to delete a 100-point polygon in the third section and replace it with a 150-point polygon, you would have to rewrite the entire data file for all of the data that are above the third section. This is awkward and inefficient, to say the least.

One alternative is to use what is called a "random access" data file. The data file is like a pie—you cut the pie into equal-size pieces, and data for each tissue section are written into each piece of the pie. With random access file storage, the advantage is that each tissue section is treated independently; the disadvantage is that a fixed amount of storage must be allocated for each section, even though in practice you might have some sections with very little data and some sections with a lot of data. So indeed, all pieces of the pie have to be cut so that they may contain the largest tissue section. This too can be inefficient.

A compromise between serial data files and random access files is to use "linked-list" storage; with linked-list storage, data may be stored in any order with pointers that link one piece of data to another (Fig. 5-12). This makes insertions and deletions easier, but each data point requires more storage, for the pointers must also be stored. Unfortunately, linked-list storage is most usually implemented in computer memory, not on a disk file. So it is necessary to convert linked-list data into serial or random access format for storage on a disk.

Another compromise is to trace data, section by section, into a disk-stored random access file and then, after editing, convert the data into a serial file. This eliminates the gaps between the unused parts of the random access file, thereby compacting the data. The serialized data may now be read repeatedly and efficiently from the disk for generating displays and statistical summaries.

Serial section data are best stored in some format that is compatible with data entered in a neuron-tracing system (Chapter 4) and captured by a television-based system (Chapter 6). Because of the many incompatibilities among the data types, and because these systems have been designed in different laboratories and companies, no standards have yet been established to ensure this compatibility.

5.6. EDITING OF DATA

Entering data for serial section reconstruction, like entering data for neuron reconstruction, is an error-prone process, and a capability for editing errors must be provided. This subject is covered in some detail in Chapters 7 and 12, but for continuity, here are some primary guidelines.

If an error in outlining is discovered as it is made, it must be easy to delete back around an outline, discarding erroneous points until the error is deleted, and then resume tracing as if nothing had happened. Fixing errors that are not found until later is more difficult. When this is the case, it is necessary to identify a part of the already-entered structure and have the ability to delete the identified part, to change any parameter of the identified part, or to insert a new structure into the "hole" left by the deletion. A method for moving a structure around interactively with respect to other structures is also desirable. In this technique, commonly called "dragging," several structures are displayed on a screen, and one of the structures is somehow identified. Using a mouse or data tablet cursor, you may drag the identified structure into its proper relative location and leave it there. The computer does the necessary manipulation of the coordinates of the moved structure to place it correctly in the data base.

Finally, filtering should be provided. Filtering is used to smooth a structure to eliminate small digitizing errors and may also be used to reduce the number of data points without changing the structure appreciably.

5.7. DISPLAYS AND PLOTS OF SERIAL SECTION RECONSTRUCTIONS

Once the data are entered, section-by-section, into the computer, computer graphics techniques are used to display the stored mathematical model. Realistic views of 3-dimensional structures viewed from any angle may be seen. Artificially enriched views, say of an inner structure within a transparent surface may also be presented as shown in Fig. 8-1. This presentation may be generated on a computer graphics screen or drawn on a sheet of paper by a computer plotter. Chapter 8 covers the details of these presentations.

5.8. STATISTICAL SUMMARIZATIONS OF SERIAL SECTION RECONSTRUCTIONS

A modest number of summarization are commonly used to describe and compare the structures that have been reconstructed. These summaries include counts, lengths measured in three dimensions, cross-sectional areas, surface areas of irregular patches, and form factors (a measure of roundness of an object). Chapter 9 covers the calculation of these parameters.

5.9. FOR FURTHER READING

The classic plotting light microscope or "pantograph" in which the microscope stage position is reproduced by the pen of an X,Y plotter has been described by Boivie et al. (1968), Patterson et al. (1976), Eidelberg and Davis (1977), Reed et al. (1980), and Davis (1985).

The pantograph may be computerized to include the function of recording the coordinates into computer memory as the structure is traced. Such computerized pantographs have been reported by Forbes and Petry (1979), Williams and Elde (1982), Mize (1983a, 1985b), and Prothero et al. (1985).

The plotting microscope may be used to provide input to a serial section reconstruction system. Such systems have been described by Dykes and Clement (1980), Foote et al. (1980), Dykes and Afshar (1982), Zsuppán (1985), and Upfold et al. (1987). Serial section systems using the plotting microscope for input have been used in biological research. Reports of research using this system are, as examples, Afshar and Dykes (1982, 1984), Sivapragasam et al. (1982), and Afshar et al. (1983).

A data tablet may be connected to a computer so that when a photo or drawing of a structure is placed on the tablet, measurements of the structure (lengths, areas, etc.) may be made by the computer. Such tablet-measuring systems, which do not include analyzing the branching patterns of arborizations, have been described by Cowan and Wann (1973), Dunn et al. (1975, 1977), Curcio and Sloan (1981), and Mize (1983b). These systems have been used in biological research. Some reports of their use are provided by DuVarney and DuVarney (1985), McKanna (1985), McKanna and Casagrande (1985), and Ramm and Kulick (1985).

The tablet-based measuring systems may also include computer programs for analyzing the branching patterns of dendritic trees. These systems have been described by DeVoogd et al. (1981), Pullen (1982), and Usson et al. (1987). Their use in research is described, for example, by Becker et al. (1986), Brown et al. (1979), Tamamaki et al. (1988), and Vaughn et al. (1988).

An unobvious measuring error that arises when using a data tablet to measure length is described by Cornelisse and van den Berg (1984).

Rakic et al. (1974) described a serial section reconstruction system in which the coordinate data were read from the digitizer and punched into computer cards. These cards were then read by the computer for input into the reconstruction program.

In a similar way, Tolivia et al. (1986) described a serial section system in which the coordinates were entered from the keyboard into the program.

Many data-tablet-based serial section reconstruction systems have been described in the literature by Levinthal and Ware (1972), Capowski (1973), Prothero et al. (1973), Willey et al. (1973), Levinthal et al. (1974), Cahan and Trombka (1975), Ware and LoPresti (1975), Macagno et al. (1976), Veen and Peachey (1977), Macagno (1978), Shantz and McCann (1978), Marino et al. (1980), Stevens et al. (1980), Moens and Moens (1981), Briarty et al. (1982), Chawla et al. (1982), Falen and Packard (1982), Perkins and Green (1982), Prothero and Prothero (1982), Freeman and Meltzer (1983), Gras and Killman (1983), Hengstenberg et al. (1983), Johnson and Capowski (1983, 1985), Speck and Strausfeld (1983), Street and Mize (1983), Sundsten and Prothero (1983), Stevens and Trogadis (1984), Fram (1985), Young et al. (1985, 1987), Braverman and Braverman (1986), Sinha et al. (1987), Smith (1987), and Yaegashi et al. (1987).

The tablet-based serial section reconstruction systems have been widely used in anatomic research. Some reports of their use include those by Lopresti et al. (1973), Macagno et al. (1973), Levinthal et al. (1976), Kimura et al. (1977), Ellias and Stevens (1980), Stevens et al. (1988), Chawla et al. (1981), German et al. (1983), Sasak et al. (1983), Nierzwicki-Bauer et al. (1983), Spacek and Hartmann (1983), Thompson et al. (1983), Stevens et al. (1984), Slepecky et al. (1984), Gambino et al. (1985), Greenberg et al. (1985), Antal et al. (1986), Coombs et al. (1986), Henson and Henson (1986), Jacobs and Stevens (1986), Mercer et al. (1987), Mercer and Crapo (1987), Villa et al. (1987), Harris and Stevens (1988), Kinnamon et al. (1988), Royer and Kinnamon (1988), Stevens et al. (1988), and Wind et al. (1988).

The problem of aligning consecutive serial sections has been addressed. Interactive or semiautomatic solutions to the problem have been reported by Capowski (1977),

Capowski and Sedivec (1981), Zsuppán (1984), Zsuppán and Réthelyi (1985), and Prothero and Prothero (1986). More fully automated solutions to the serial section alignment problem have been described by Dierker (1976a) and by Gentile and Harth (1978). The use of a double microscope for serial section alignment has been described by Fahle (1988).

CADCAM (computer-aided design and computer-aided manufacturing) software has been used for serial section reconstruction work. A description of such a system has been presented by West (1985). Use of this type of software in modeling anatomic structures has been reported by Ameil et al. (1984) and by Wind et al. (1986).

The state of the serial section reconstruction art has been reviewed by Woolsey and Dierker (1978, 1982) and Sobel et al. (1980). The many software packages available for serial section reconstruction have been reviewed by Huijsmans et al. (1986).

6

Video Input Techniques

6.1. INTRODUCTION

In Chapter 4, you entered anatomic data from tissue into the computer by looking directly at a microscope image and passing a cursor over the image. The cursor was mixed with the microscope image by an optical device such as a camera lucida. In Chapter 5, you placed a photograph or drawing of the tissue on a data tablet and passed a cursor, the data tablet cross hairs, over the image. In this case also, the cursor was mixed optically with the image. In both instances, the data that were stored in the computer were a series of coordinate locations of the cursor as it moved about the image.

In this chapter a third method of input, video or television, is introduced. Here you will look at a televised image of your tissue and control a cursor. Television hardware will mix the cursor with the image, and the computer will record a series of cursor coordinates as the cursor moves about the image.

In all three of these cases, the computer stores a vector representation of the data as you trace it. This is a series of stored coordinates with the assumption that the coordinates should be connected by straight lines.

Using a television system offers functions not available in the systems described in Chapters 4 and 5. The television image may be digitized, i.e., disassembled into an array of picture elements or "pixels," each containing a number that represents the brightness or gray value of the image at a location. These pixels are stored in a memory called a "frame buffer," which the computer may access. The computer may read the pixels to measure the gray value of locations in the image and it may write values into the frame buffer, thus modifying the image. The frame buffer's contents are presented on a TV monitor so that you may view the stored image. This image-digitizing hardware, when used with appropriate software, offers two new capabilities: image enhancement and optical density measurements. Image enhancement is the use of a computer program to manipulate the gray values of pixels so that specific features of the image become more outstanding. The usual techniques for enhancing a biological image are described in Chapters 13 and 14. Once the image has been enhanced, you may view it to appreciate its now enriched features, or you may trace it, extracting coordinate information such as contours. Optical density measurements are those that read and summarize the gray levels of the pixels in the image to extract biological information from it. This chapter and Chapter 14 discuss this feature.

The added capabilities have a cost, however. Figure 6-1 shows a televised image of a Golgi-stained neuron. Capturing the image with a TV camera, storing it in a frame buffer, and regenerating the image on a TV monitor all reduce its quality somewhat, so the measurement techniques that are described below do not work on the highest-fidelity original image. This tradeoff is tolerated to enjoy the benefits of video imaging.

6.2. A VIDEO-BASED ANATOMIC DATA-COLLECTING SYSTEM

In this section, the hardware that comprises a typical video-based anatomic data-collecting system is presented. With the appropriate software, you may use the system to extract both vector information (Section 6.3) and gray-level information (Section 6.4) from tissue viewed by a television camera.

Figures 6-2 and 6-3 show the hardware and a block diagram of the data-collecting system. Two television cameras are shown in Fig. 6-2, one mounted on the microscope and one mounted on a copy stand. A copy stand, shown to the right of the microscope, consists of a flat table on which a photograph, a negative, or a drawing of the specimen is placed. Spotlights may illuminate a photograph from above, or a light box may illuminate a negative from below. The camera is mounted on the vertical support, which rises from the rear of the light box, so that it looks down at the specimen.

The computer (in this case, an IBM AT) contains a video digitizing and display board or "frame grabber," which may capture the video signal from the TV camera into a frame buffer memory with a resolution of 512 × 512 pixels, each with 8 bits of resolution. Each pixel thus stores a gray value of range 0 (for black) through 255 (for white). A television monitor, shown to the right of the computer, presents the image that is stored in the frame buffer. A dot-matrix printer may be used to make a hard copy of the computer-stored image. The computer may read the gray values of individual pixels of the image and may also write data into the pixels to modify it.

A mouse and a data tablet are also seen in Fig. 6-2. Either device may be used for controlling the data collection software, both for menu selection and for moving a cursor over the image. The systems described in this chapter are controlled by the mouse; however, with the appropriate software, you may use the tablet with equal ease.

Finally, a pair of stepping motors that drive the microscope stage are shown in Fig. 6-2; a third stepping motor drives the fine focus knob of the microscope. All three of these stepping motors are driven by a motor controller box, which is in turn controlled by the computer. This system of motors provides the computer software the capability to move the tissue specimen on the stage so that you may examine any portion of it.

6.3. EXTRACTING VECTOR INFORMATION FROM A TELEVISED IMAGE

The television image of a microscope or copy-stand specimen may be captured into a frame buffer, where it is stored as a matrix of pixels, each pixel with a gray value. When you are presented with the image on a TV monitor (Fig. 6-1), it is very easy for you to see the shape of the structure contained in the specimen. The structure is simply part of the picture that is stored in the frame buffer, pixel by pixel, and the computer has no a priori

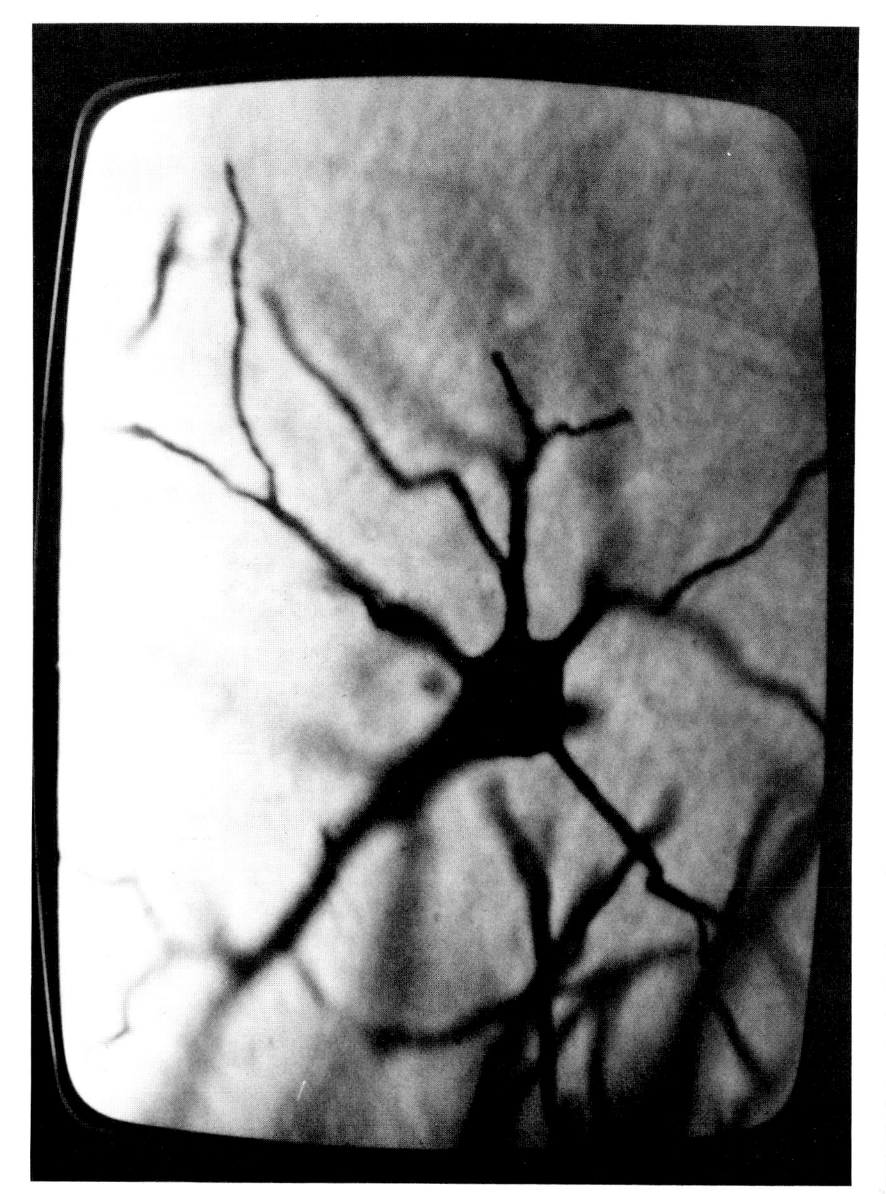

FIGURE 6-1. A televised image of a neuron. Its image quality is somewhat inferior to the original microscope image, shown in Fig. 4-1.

FIGURE 6-2. A general laboratory video data collection system. From left to right, TV cameras may view specimens from either a microscope or from a copy stand. An IBM AT computer containing a video digitizing and display board captures, stores, and displays images on two TV monitors. A mouse and data tablet are used to control the computer programs that perform the video functions. A large-screen vector CRT with its display processor presents three-dimensional reconstructions. A printer and plotter also present text and pictoral output to the researcher.

way of knowing which pixels in the image are part of the structure and which are part of the background.

To extract the structure from the image is to change the representation of the structure from pixel gray values to a series of X, Y, and possibly Z coordinates that define the outline of the structure or parts of it. This is a pattern recognition task that you may do by tracing the structure with a cursor or that the computer may do, to a limited degree, by automatically finding the edges of the structure in the image. Automatic tracing with all its problems is covered in Chapters 13 and 15; this section covers semiautomatic tracing, using your superior pattern recognition capacity. Several examples are given to illustrate the technique.

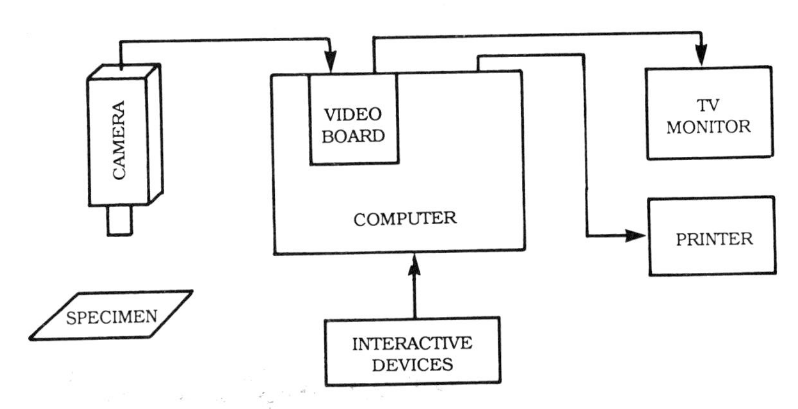

FIGURE 6-3. A general block diagram of a video-based data-collecting system. The camera views the specimen and delivers its video signal to the video digitizing board inside the computer. The computer may access the pixels in the stored image to extract information from it and may also write information into the memory, either to enhance or to annotate the image. A television monitor displays the image stored in the video board, and some type of printer may make a paper or "hard" copy of the image. Some interactive device (keyboard, mouse, tablet) is used to control the computer program.

6.3.1. Tracing a Dendrite

Figure 6-4 shows how you may trace the dendrites of a neuron semiautomatically using a video data collection system such as that pictured in Fig. 6-2. An image of a neuron, containing a soma and several dendrites, has been digitized into the computer frame buffer and is displayed on the TV monitor. A white rectangle encloses the image area, about 90% of the screen. At the left and bottom edges of the screen is a perimeter menu, from which you may select items; their use is described below. By moving the mouse with your hand, you may move a cursor over the televised image. The cursor can be seen as a circle in the lower left center of the figure. As you move the mouse and thus the cursor, the computer continually keeps track of the X,Y coordinates of the cursor. You may pass the cursor along the dendrites, and as long as you hold down the left button on

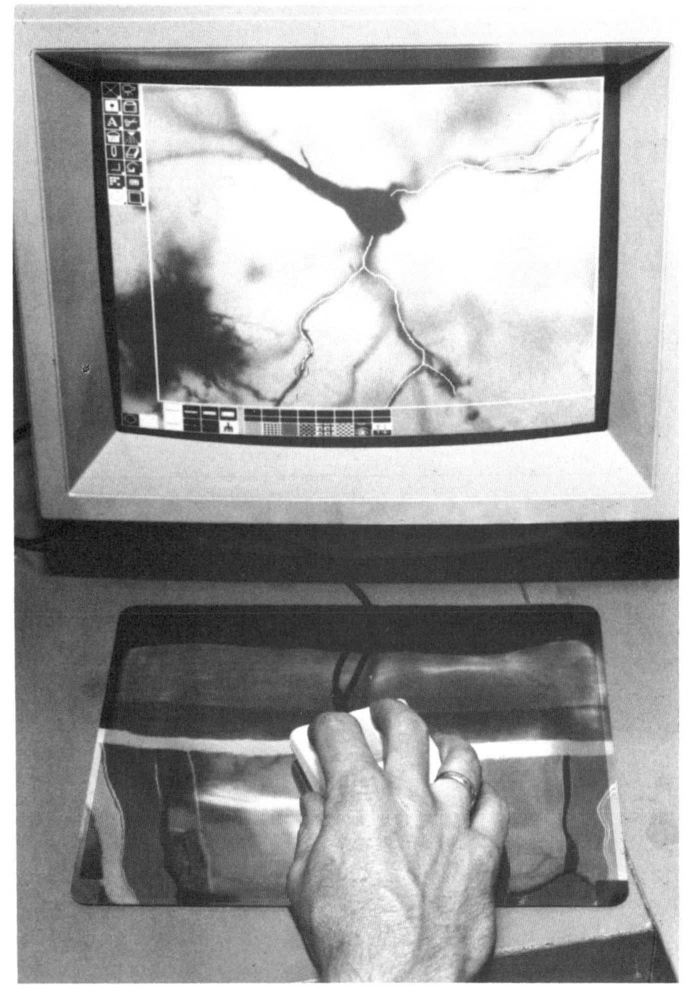

FIGURE 6-4. Using a video system to trace dendrites. After the image is digitized, the user moves the mouse to follow each dendrite, and the computer records a series of coordinates along the dendrite.

the mouse, the computer records in its memory a series of the cursor coordinates. Thus, the computer records the vector representation of the dendrites.

As you trace along a dendrite, the computer writes white pixels into the frame buffer at each location of a recorded point. When these are displayed on top of the dendrite, they provide you feedback that your tracing is proceeding well.

Because a vector representation of a structure consists of a series of coordinates and the assumption that straight lines are drawn from coordinate to coordinate, there is no need for the computer to store the coordinates of each pixel along the dendrite. Indeed, to do so would demand a lot of memory. So spatial filtering is done at some point in the data-recording process; it may be done during the actual tracing. As you move the cursor, the computer continually calculates the distance between the just-recorded point and the current cursor coordinate. When this distance exceeds a certain amount, called a "spatial filter," the current coordinate is recorded in computer memory. The software still writes a white value into each pixel along each line, so that you do not see gaps in the computer-stored data, but individual pixels are not stored in the data base that describes the structure.

Alternatively, the spatial filtering may be done after tracing. Every pixel that the cursor passes over is turned white, and its coordinates are stored in the computer memory. Then the software selects pixels that are separated by the spatial filter distance for storage in the data base. The two spatial filtering techniques yield the same vector representation of the dendrites.

As you trace along the dendrite, you need to signify to the computer that some points (e.g., branch points, spines, endings) have special significance. Since there are only three buttons on a mouse and there may be more than a dozen types of points, you cannot simply press a unique button for each point type. A standard manner for handling this problem has been devised. When you arrive, for example, at a branch point, you remove your finger from the left button to indicate to the computer that you have temporarily interrupted the tracing. Then you move the cursor to the item in the perimeter menu that signifies a branch point and press the left button. The software senses that you are not tracing because the cursor is in the menu area of the screen rather than in the image area, and it notes that the next point that you will enter should be a branch point. Then you move the cursor to the point in the image and repress the left button. The computer records this point as a branch point. Depending on the sophistication of the perimeter menu, all the various point types that define a neuron may be entered using a mouse.

Tracing in this manner consists of your moving a cursor over a stored and now stationary image; the distance over which you may trace is thus limited by the magnification of the image. While you trace, when you move the cursor to the edge of a frame, the computer will temporarily suspend the tracing process. It will move the microscope stage to position the most recent cursor location in the specimen in the center of the field and will capture a new image into the frame buffer. It will then rewrite the white pixels of the already-traced dendritic portions into the frame buffer in their new locations and will place the cursor in the center of the field. You may then resume tracing as if nothing happened.

In a manner similar to that described in Chapter 4, when you reach the end of a dendrite and record an endpoint of some type, the computer moves the cursor and, if necessary, the stage back to the most recent branch point from which you have not yet traced both exiting paths. This will prompt you not to miss any branches of the tree.

The dendrite-tracing problem is three-dimensional, but the television image is captured from a two-dimensional plane of the tissue. Therefore, you must be able to control the focus continually as you trace the structure. A response to this problem is to define the center and right mouse pushbuttons as the focus control. Whenever you depress the center pushbutton, the software focuses up through the specimen; whenever you depress the right button, it focuses down. Unfortunately, however, you see nothing on the TV monitor until an image is captured into the frame buffer and regenerated on the monitor. Most image-digitizing boards may be placed into ''continuous capture'' mode, which means that the board captures and displays a frame every $\frac{1}{30}$ sec. This allows you to focus the specimen interactively using the mouse but has the minor disadvantage that during focusing, no computer-stored data, cursor, or menu may be displayed. During focusing, the computer software keeps track of the current focus coordinate; during tracing, the focus coordinate forms the Z value that is recorded in computer memory with each point.

With the appropriate software, you may use the perimeter menu for numerous functions. For example, a stage movement menu item is valuable. When you select this menu item, you may then move the mouse north, south, east, or west (or combinations of these directions), and the X,Y stage will track your motion. Focus can be performed simultaneously using two mouse buttons so that the stage can be controlled in three dimensions.

You may enter the thickness of any point along the dendrite by selecting the proper perimeter menu item. Then you move the cursor to one edge of the dendrite and press a button. You then move to the other edge and press the button again. The computer calculates the distance between the two points (i.e., the thickness of the dendrite), and the thickness is recorded along with the point's coordinates. Thickness measurements done in this way tend not to be accurate, because you may only position the cursor exactly at a given pixel, and the spatial resolution of the pixels is modest. A one-pixel error in positioning the cursor may yield a 25% error in thickness measurement.

By using perimeter menu items you may delete points, branches, or trees, set the origin of the tissue, or give names to individual trees and neurons; finally, you may set the spatial filter value to control how fine or coarse you wish the resulting data base to be.

6.3.2. Outlining a Structure

Another function that is typically performed with a video-based measurement system is the outlining of somas. As Fig. 6-5 illustrates, you may focus on a soma and grab an image. Then you move the cursor around the perimeter of the soma while holding down the left button on the mouse. The computer records coordinates along the soma outline into the structure's data base at a fixed spatial interval.

You may also use the system to outline structures other than neurons, as Fig. 6-6 shows. Tissue boundaries, lamina borders, and blood vessels may be traced and recorded into the computer data bases.

A final tracing process is to locate structures. As Fig. 6-7 illustrates, you may be less interested in the shape or size of the structures than you are in their location in the tissue. Because you frequently wish to distinguish among different kinds of structures, you may use the perimeter menu to place different icons (e.g., triangle, square, happy face) at each type of structure. To do this, you select an icon from the menu, and the cursor becomes

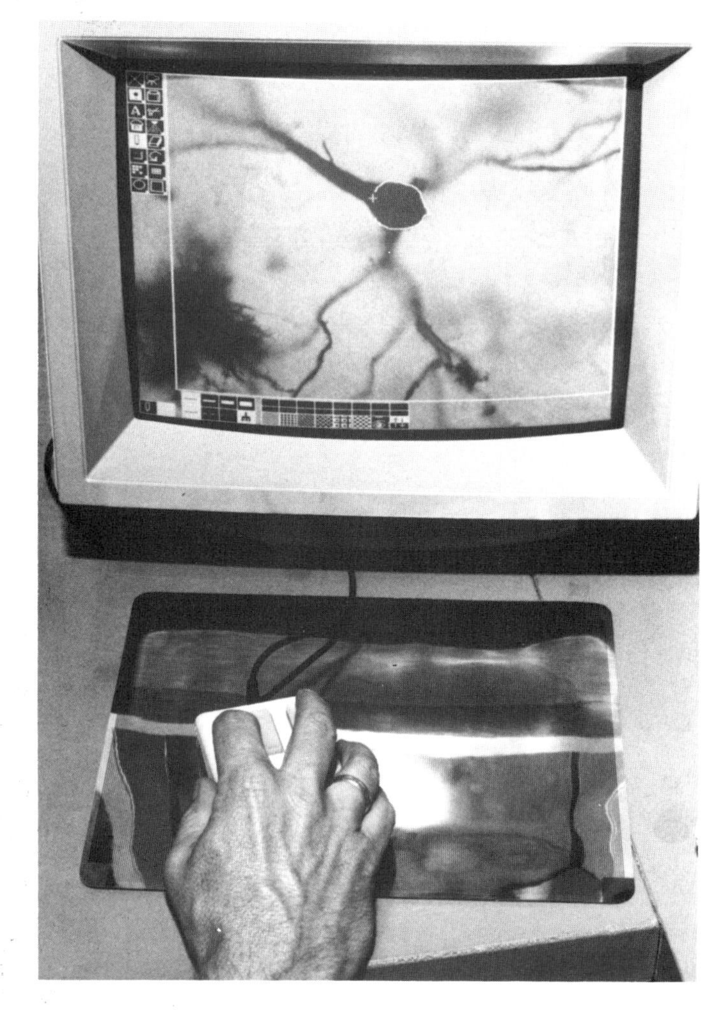

FIGURE 6-5. Using a video system to outline a soma. After the image is digitized, the user moves the mouse to outline the soma, and the computer records a series of coordinates representing the outline.

the shape of the icon. You then move the cursor to the structure and press the left button on the mouse. The computer records the X,Y,Z coordinates of the structure and a code number specifying its type. Finally, the software deposits a copy of the icon in the frame buffer memory so that it is presented on the TV monitor. You then can see that you have already located this structure and need not do it again.

6.4. EXTRACTING OPTICAL-DENSITY INFORMATION

You may recall from this chapter's introduction that a video-based data collection system offers two advantages over a directly viewed system: image enhancement and

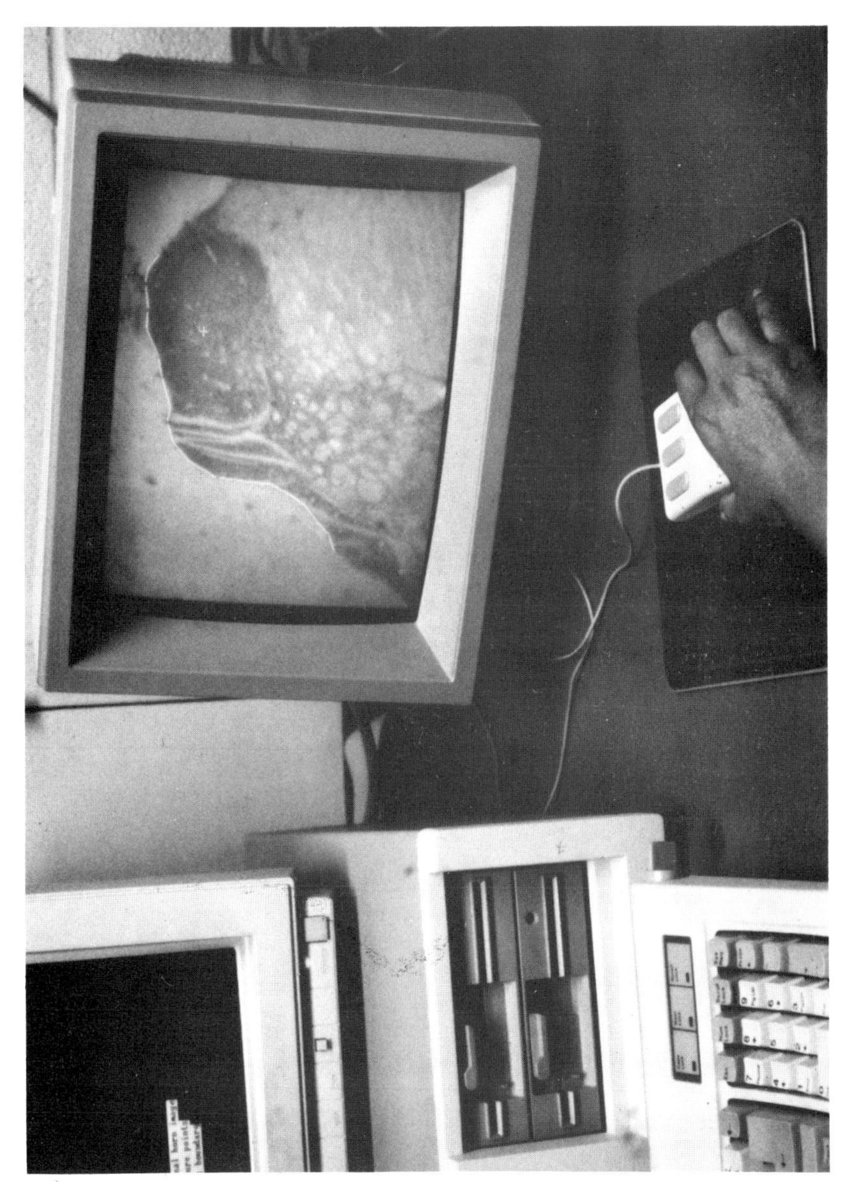

FIGURE 6-6. Using a video system to outline the dorsal horn of the spinal cord. The specimen is placed in the microscope (not shown), and a camera passes the image to a video digitizer board. The mouse is used to outline the region on the TV screen. The cursor may be seen as a cross in the top center of the picture. Elementary statistics such as the area and perimeter of the outlined region may easily be calculated. Reproduced from Capowski (1988b) with permission.

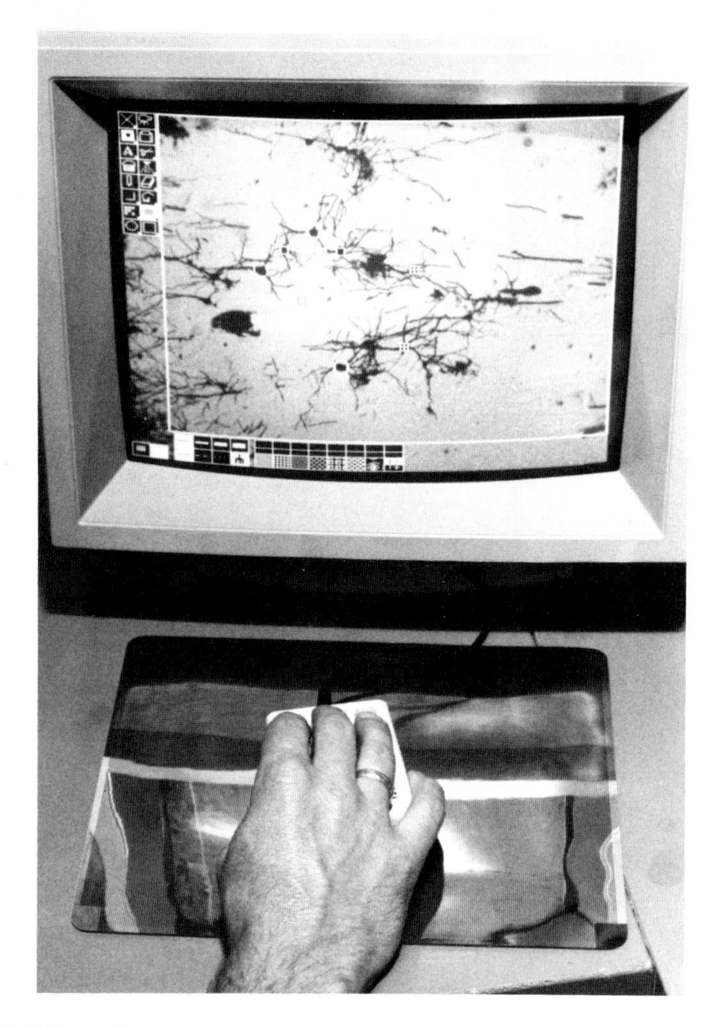

FIGURE 6-7. Using a video system to locate structures. After an image is digitized, the user moves the mouse to place an icon (square, circle, grid) at the location of each soma.

optical-density measurements. Image enhancement is the use of a computer program to manipulate the gray values of an image's pixels to enrich the image in some way; this is covered in Chapters 13 and 14. Optical-density measurements are those that read and summarize the gray levels of the pixels to extract numeric information from the image. These measurements are discussed in the following paragraphs.

Generally, you may use a video-based data collection system to capture an image, mark a region of interest in the image, and then summarize the gray levels within the region of interest. Figure 6-8 shows the simplest form of this technique, where the region of interest is a point. With the mouse, you move a cross-shaped cursor over the image, positioning it at any pixel. The computer software continually reports the coordinate of the

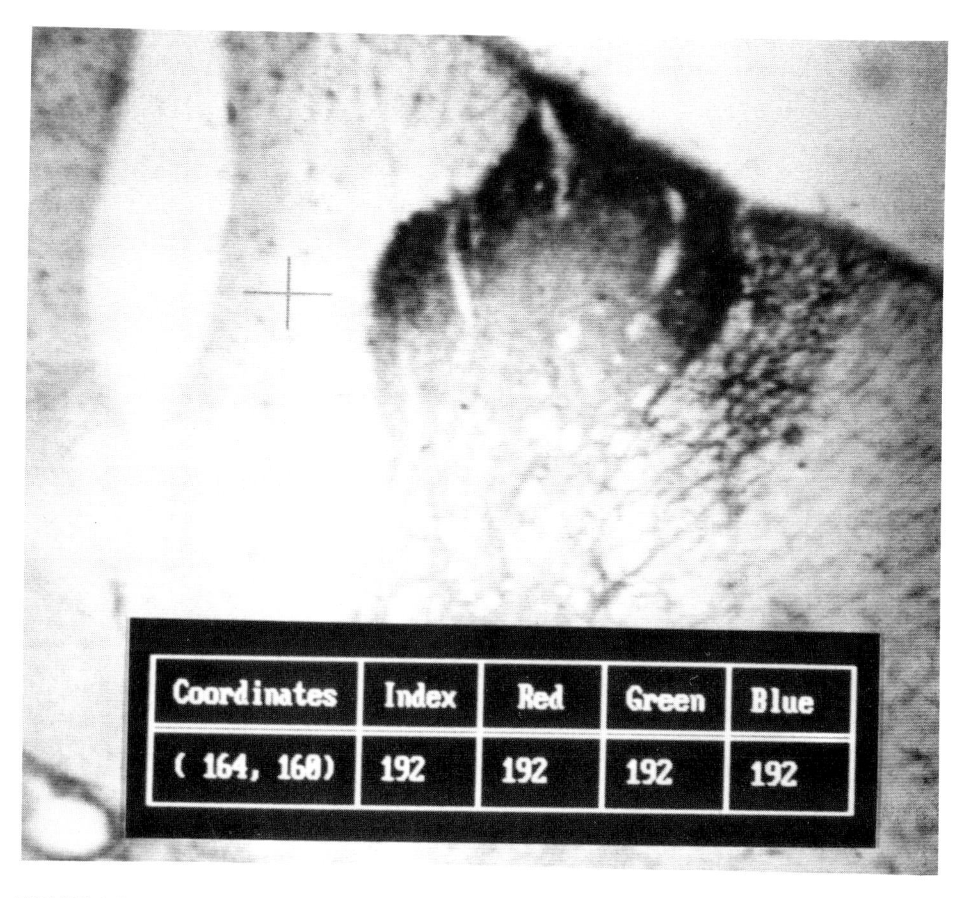

FIGURE 6-8. Using a video system to measure the gray level of a point. Once the image has been digitized, the user may move the mouse to position the cursor (the cross) on the image. The computer reports the coordinates and the gray level (here called the "index") of the point. If a color camera had been used, separate levels for red, green, and blue would have been available.

cursor and the gray value of the pixel at that location. This gray value may indicate, for example, staining intensity at any point.

A somewhat more complex example is shown in Fig. 6-9. You may use the mouse to position the endpoints of a line anywhere on the image. The computer software draws the line on the image and plots the gray values of the pixels along the line.

In the two examples just shown, the region of interest was a single pixel and a line, respectively. It may also be a rectangle or even an irregularly shaped polygon. You may easily use the mouse to define a rectangular region. A small square is drawn on the image, and you move it around the screen by moving the mouse. When you press a button on the mouse, the upper left corner of the square becomes fixed, and with the mouse, you move only the lower right corner of the square. The square stretches as you move the mouse, possibly deforming into a rectangle. When you are happy with its size, shape, and location, you may remove your hand from the mouse button.

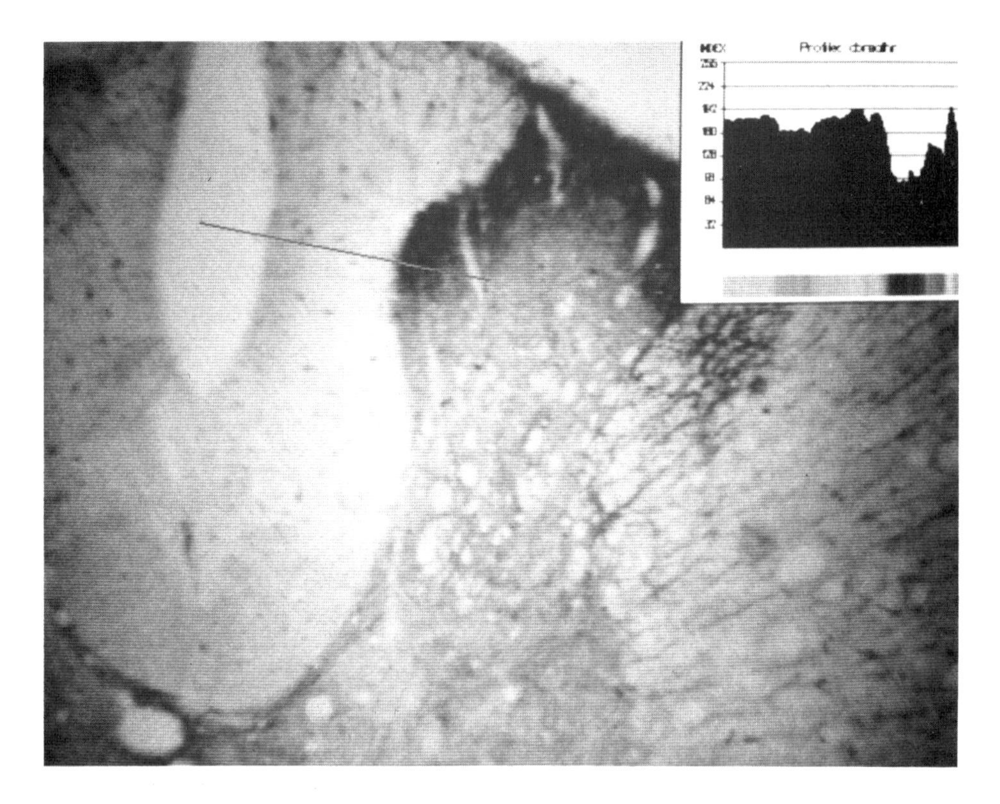

FIGURE 6-9. The gray-level profile of a line. After an image is digitized, the user stretches a line across a region of the image. The computer plots the gray level for each pixel along the line in the insert in the upper right.

You may also define an irregularly shaped region of interest. Figure 6-10 shows several areas of a spinal cord cross section that you may outline using the mouse. After you have outlined them, the computer software reads the gray value of each pixel within each region. The maximum gray value of 255 occurs when a pixel is the whitest, so each pixel's value is subtracted from this maximum to change it to a measure of darkness or "optical density." The computer sums the density values from all the pixels in the area to obtain the integrated optical density (IOD) of the area. Next, the software calculates the mean optical density for each area by dividing the IOD of the area by its number of pixels and then writes this mean gray value into every pixel in the area. All the pixels in each region are thus displayed at their mean optical density. Each pixel in the image has the same area, so the number of pixels in a region is proportional to the region's area. Consequently, the mean optical density represents the mean gray level per unit area. Figure 6-10 shows the mean gray level per unit area painted consistently across all the pixels in a region and yields a striking presentation of the biological parameter that is manifested by gray level. This figure, for example, shows the number of labeled cells per unit area.

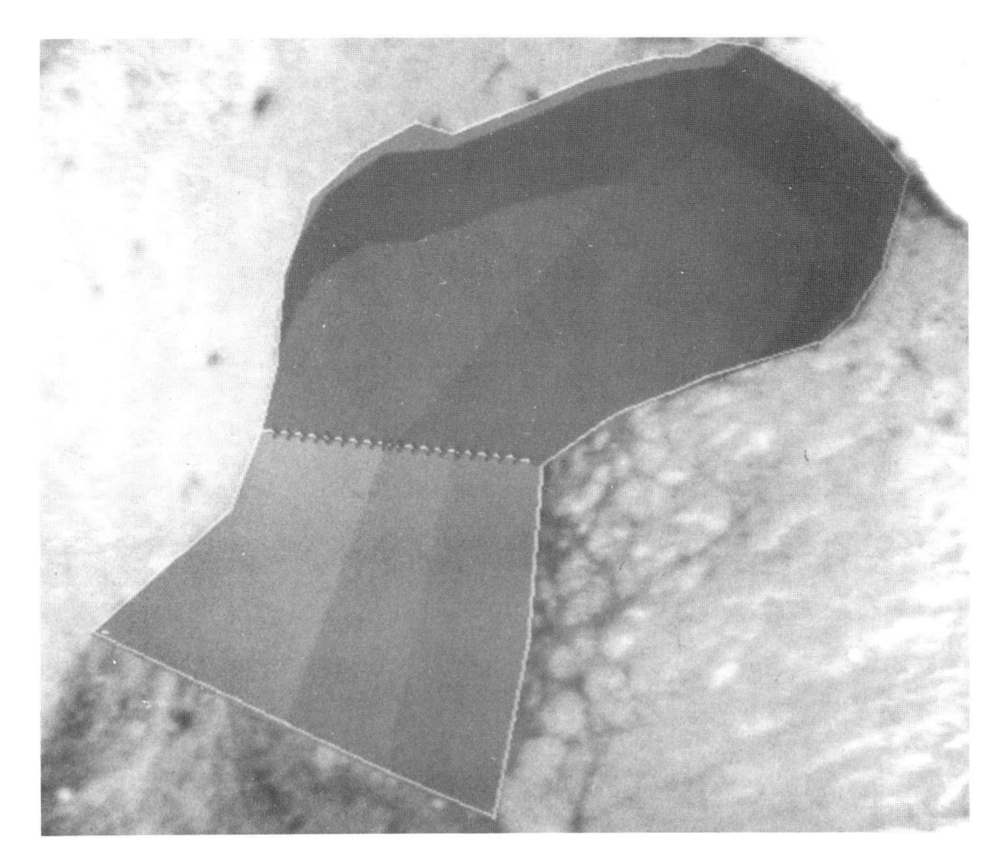

FIGURE 6-10. Calculation of the integrated optical density. After tracing several irregular outlines of regions of the spinal cord dorsal horn, the computer calculates the optical density of each region and fills that region with a gray value proportional to its optical density. Reproduced from Capowski (1988b) with permission.

6.5. MASKING OF IMAGES

Image masking is an interesting benefit of video-based data collection that is not easily done with directly viewed systems. Frequently you may be interested in comparing the gray levels of a small and irregularly shaped portion of numerous tissue samples. Figure 6-11 illustrates the technique. You may use the mouse to outline an irregular region of tissue. Computer software then generates a ''mask'' by filling the region you have traced with white pixels and setting the remaining pixels to black (Fig. 6-11, left inset). Then an image is digitized (Fig. 6-11 background), and the mask is applied. Here, each pixel in the image is transformed according to the value of the corresponding pixel in the mask. If the mask's pixel is white, the image's pixel is unchanged; if the mask's pixel is black, the image's pixel is set to black. The result is that only the pixels of the original image that are inside the mask retain their gray values, and those outside the mask are set to black. Thus, only the data within the relevant region remain, and the optical density

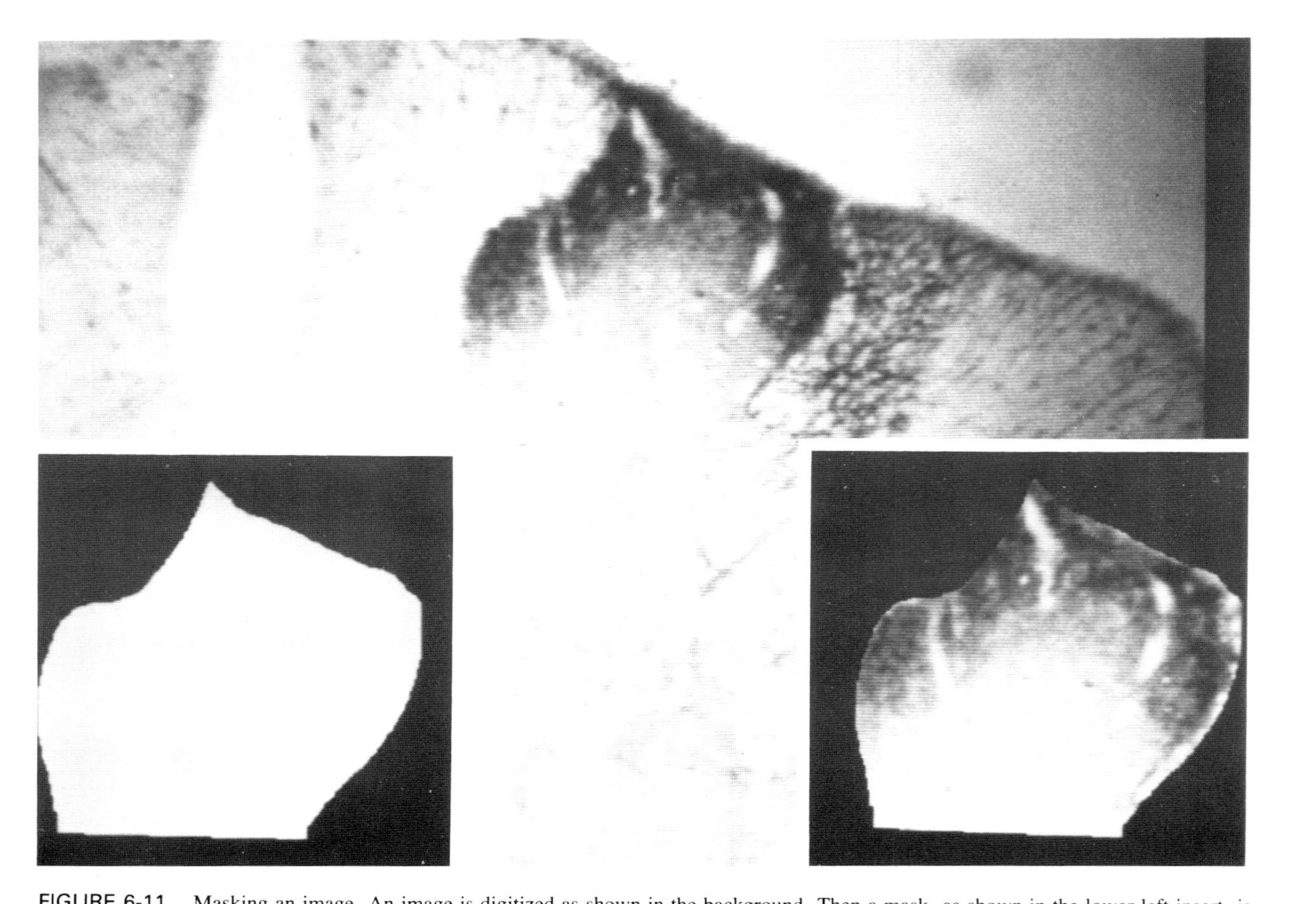

FIGURE 6-11. Masking an image. An image is digitized as shown in the background. Then a mask, as shown in the lower left insert, is generated by tracing a region of the image with the mouse. When the mask is ''anded'' with the original image, only that portion of the image remains, as shown in the lower-right insert. This technique is valuable when the same region of several images is of interest.

measurements detailed above may be performed repeatedly on different samples, ignoring the data outside the mask.

6.6. PARTICLE COUNTING

Particle counting is a video-based summarization technique that may be fully automated, provided that the particles have a regular shape and that they appear in the image with high contrast relative to their background. Two classic applications, silver grain counting and cell counting, have been widely automated. For silver grain counting, the grains are illuminated using "darkfield" illumination; i.e., light strikes them obliquely from below the microscope stage, and only light reflected from them passes through the microscope objective lens and on to the TV camera. The resulting digitized image appears as stars in a cloudless nighttime sky. Software may easily count the grains (stars) by looking for individual white pixels or clusters of them.

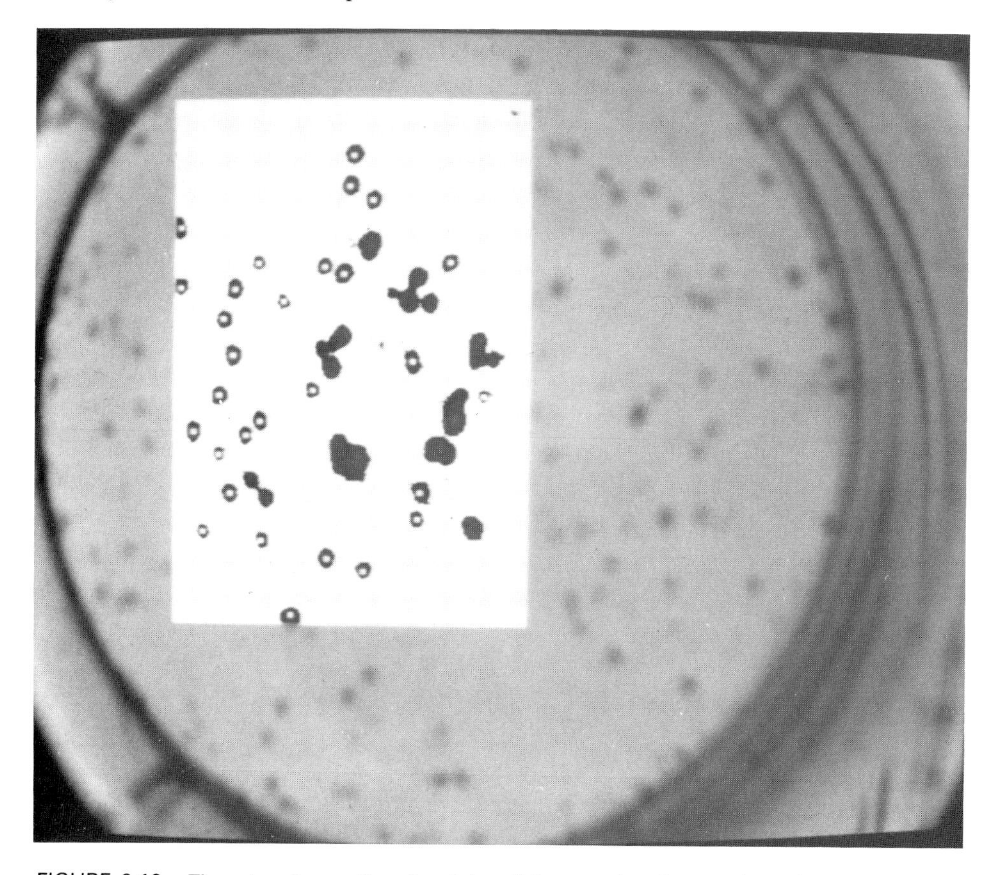

FIGURE 6-12. The automatic counting of particles. Cells grown in culture are imaged, and a rectangular region of the image is enhanced. Then a particle-counting algorithm is applied to the enhanced region. Counted cells are indicated by a large dot placed in their center. Clumps of cells are indicated by making them darker and not placing a dot within the clump. Specimen courtesy of J. Cidlowski, UNC Department of Physiology. Figure reproduced from Capowski (1988b) with permission.

You may also count cells automatically, as Fig. 6-12 illustrates. This is a more difficult problem than counting silver grains because the cells have less contrast from their background than do the stars in a nighttime sky. Consequently, the image is digitized and then enhanced to increase this contrast. You then select a region of interest using a mouse; in Fig. 6-12 the region is rectangular. A computer program then counts the cells. For a cell to be counted, it must fall within two windows. All its pixels must be between two gray levels, and the number of continuous pixels (i.e., the cell's area) must fall between two limits.

You may use the "gray-level profile" shown in Fig. 6-12 to help you set the gray-level window for the particle counting. You place the line across several cells and note the gray values inside and outside the cells. You then select the gray limits to include only the gray levels of the cell. You may determine the limits on cell area in two ways. One way is to trace several cells with the mouse and instruct the computer software to calculate their areas. You then select area limits to bound these areas and to reject those that are much larger or much smaller. Alternatively, you may use trial and error: guess at the average number of pixels in a cell and set bounds that include this guess; then instruct the computer to count the cells. Because the computer provides feedback as to which cells it has included in its count, you may note whether too few or too many have been counted, adjust the area limits, and instruct the computer to count again.

Because biological images are "noisy," i.e., contain artifacts, any particle-counting algorithm will make errors. So the algorithm provides you feedback about which particles it has counted, usually by loading a brightly colored pixel in the middle of each counted particle. You should visually inspect the counted image to see where the software has erred and then use an interactive editor to correct the machine's mistakes.

6.7. AUTOMATIC FOCUSING

An advantage of using video input techniques is that focusing on objects in a three-dimensional specimen may be automated. Mendelsohn and Mayhall (1972) showed that an object is in sharpest focus when its integrated optical density over a threshold is maximum. This is illustrated in Fig. 6-13. The computer algorithm to focus on one of the cells of Fig. 6-12, for example, could capture an image at a given focal plane. It then

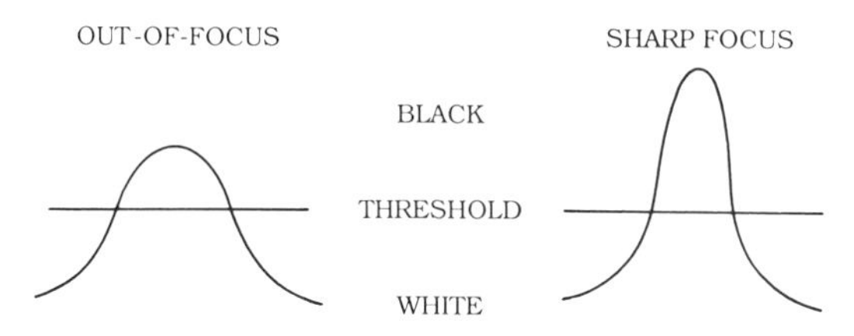

FIGURE 6-13. The autofocus scheme of Mendelsohn and Mayall (1972). An image of a small dark object on a light background is digitized at two different focal planes, and a scan line is drawn across the object in each image (not shown). The response lines shown give the gray values for pixels along the lines. Left: The response from an out-of-focus object. Right: The response from a sharply focused one. An arbitrary Right: The response from a sharply focused one. An arbitrary but constant gray-level threshold is established for both responses. The best focus is achieved when the integrated optical density above this threshold is the highest.

draws a scan line across the cell and reads the gray level of each pixel along the line. It repeats this process at numerous focal planes. For all scan lines, the gray-level response is a mound-shaped curve as shown in Fig. 6-13, peaking toward its blackest value where the line crosses the cell. For the best-focused scan, however, the peak is the sharpest. The Mendelsohn and Mayhall algorithm quantifies the sharpness of the peak by establishing an arbitrary threshold and summing the gray values of all the pixels that are darker than the threshold. The scan-line response for which this sum is maximum is the one in best focus. The computer then drives the focus axis to this focal plane, digitizes the image again, and uses this image for any subsequent calculations.

6.8. VIDEO INPUT COMPARED TO DIRECTLY VIEWED INPUT

For the rest of this chapter, directly viewed anatomic data collection systems, those in which you look at the original microscope image, are compared to television-based systems, where both you and the computer have access to only a televised image. Each system has advantages and disadvantages.

6.8.1. Advantages of Television-Based Data Collection

The first advantage of the TV-based system is that cameras are more accurate and consistent in measuring gray values (i.e., light intensities) than people are. The human visual system is quite poor at measuring absolute light intensities, although excellent at noting differences in intensity. Thus, any biological process that presents a gray-level result must be measured by a television-based system.

The TV system offers all the possible image enhancement techniques that are described in Chapter 13. For example, by averaging many images, the system may gather images for quantization that are too dim for our naked eyes.

Video-based data collection systems use readily available equipment. Cameras and monitors of all qualities are available in black and white and color from many sources. Microscope manufacturers routinely provide the necessary adapters for mounting cameras on their microscopes. Indeed, most of the microscope companies even sell TV cameras. Video digitizing boards are easily found for standard laboratory computers, and software to perform the algorithms described above and in Chapter 13 is also commercially available.

Finally, video input techniques offer the potential to fully automate a limited amount of feature extraction. You may, for example, instruct the computer to count particles more consistently and objectively than a human can, relieving you of a very tedious task. A directly viewed data collection system with which only you and not the computer can see the image has no capability to count particles automatically.

When you use a mouse or a data tablet to move a cursor over a televised image of your specimen, you are preserving visual continuity because you look only at the TV monitor, not also at your hand. You preserve tactile continuity also because you do not remove your hand from the mouse. Because the computer may write information into the frame buffer, it may provide you menu options and feedback right before your eyes so that you can easily control and monitor the data collection process.

6.8.2. Disadvantages of Television-Based Systems

The use of video for data collection does have some disadvantages, however, when compared to a directly viewed data collection system. The quantization of the image, first into the TV camera's raster scan lines and then into the pixel-by-pixel frame buffer, reduces the fidelity of the image. The highest-fidelity image is presented using only microscope optics, not television. Some of this fidelity loss may be recovered using the image enhancement techniques described in Chapter 13, although fine details such as dendritic spines will never be seen as clearly as in the original image.

Measuring the fine details is difficult because of this loss of fidelity. The variable-diameter cursor (Chapter 4) used to measure dendritic thickness is generated on a vector graphics system and mixed optically with the anatomic image. Such a cursor could not be written into the frame buffer and overlaid on the displayed image because the spatial resolution of the frame buffer is not sufficient. When you attempt to measure the thickness of fine processes, the cursor must be small, and only a few pixels are used to present it. The circle is then crudely represented, and measurements made from it become inaccurate.

You lose color information when you choose a television-based system. The spectral sensitivity of a black-and-white camera depends on the chemical compounds coated on its tube face. The camera may be most sensitive to low-frequency red light, for example, while ignoring the blue range. And many cameras respond as much to infrared radiation as to visible light. In any case, the original color image is converted to only gray levels, discarding the contrast between the colors of the original image. Filters may be placed in the light path to block the infrared and to enhance specific contrasts, though. An alternative is a color camera and true color digitizer, consisting of three frame buffers, one each for the red, green, and blue outputs from the camera. But a high-resolution color camera and the three frame buffers form a very expensive system.

The television-based system is more expensive than a directly viewed system. In both cases, identical microscope equipment must be purchased, and in the television system video equipment must be added.

A final disadvantage of video-based data collection systems is that many researchers have spent their professional lives looking at the original image in the microscope eyepieces. They are accustomed to the microscope and the images it provides, and a video system is a deviation from this experience.

6.9. FOR FURTHER READING

Semiautomatic neuron-tracing systems using video input have been described in the literature by Hillman (1976), Lindsay (1977a,b), Paldino and Harth (1977a), Paldino (1979), and Yelnik et al. (1981). The use of these systems for mapping neuroanatomical structure has been reported, for example, by Lindsay and Scheibel (1976), Paldino and Harth (1977b), and Calvet et al. (1985).

Semiautomatic serial section reconstruction systems using video input have been reported by Huijsmans (1983), Hibbard and Hawkins (1984), Huijsmans et al. (1984), Augustine et al. (1985), Curcio and Sloan (1986), Rogers (1986), and Toga and Arnicar-Sulze (1987). The use of these systems for mapping anatomic material has been reported by Hibbard et al. (1987) and by Schwaber et al. (1987).

Some full automation of the data input procedure using edge-detection techniques has been achieved. Video-input serial section reconstruction systems that automate the input process to some degree are reported by Reddy et al. (1973), Llinas and Hillman (1975), Hillman (1976), Hillman et al. (1977), Hibbard and Hawkins (1984), Kropf et al. (1985), Jimenez et al. (1986), and Winslow et al. (1987). The use of fully automated systems in anatomic reconstruction has been reported, for instance, by Hibbard et al. (1987) and Radermacher et al. (1987a,b). The automation of the alignment of serial sections has been reported by Hibbard and Hawkins (1988).

Automatic edge-detection algorithms have been specifically described by Hibbard and Hawkins (1984) and by Augustine et al. (1985). Successful use of these algorithms in serial section reconstruction has been reported, for instance, by Hibbard et al. (1987).

Imaging systems that have been constructed to improve the view of neurons may be read about in Shantz (1976), Kater et al. (1986), and Webb (1986).

A description of how to use a video-based imaging system to generate stereo pairs of an anatomic structure is available from Tieman and Murphy (1985) and Tieman et al. (1986) and in Chapter 13 of this book.

Automated particle counting, for example, of autoradiographic silver grains, have been the subject of several computer systems, described, for example, by Boyle and Whitlock (1974, 1977), Wann (1976), and Wann et al. (1974).

Algorithms for the automatic focusing of a specimen have been described by Mendelsohn and Mayall (1972), Kujoory et al. (1973), Wann (1976), and Harms and Aus (1984).

7

Intermediate Computations of Reconstructed Data

7.1. INTRODUCTION

The three preceding chapters have covered the entering of anatomic data into the memory of the computer. This computer-stored data may not model the original structure in the tissue as closely as you want because of histological, optical, and operator-related reasons. The tissue must be cut into sections thin enough to be viewed, one at a time, through a microscope. Consequently, some sort of reassembly process must be performed to align and reconnect the pieces entered from individual sections. Researchers experience numerous hurdles in the reconstruction process. The tissue sections may shrink and wrinkle, so the data recorded may not faithfully represent the original structures as they exist in vivo. Readings from the focus axis of a microscope are usually foreshortened because of unwanted refraction in its optical system. Shrinkage, wrinkling, and foreshortening result in coordinate errors that may be corrected by computer algorithms. A human operator entering data will make mistakes, even when using a well-designed anatomic recording system, and he may not detect some of them until the data have all been entered. Thus, a facility for error detection and correction must be provided. This chapter provides techniques to correct these errors in the computer-stored data in order to generate a data base that faithfully represents the original structure. These techniques are intermediate, for they are performed after the data are collected but before they are used for display and statistical summarization.

7.2. FOCUS-AXIS PROBLEMS

Microscopes have been designed classically to look at two-dimensional images, and they feature optical systems that maximize the depth of field, thereby integrating as much depth information as possible into each image. In recording three-dimensional structures from a microscope, a large depth of field is undesirable, for you want to be able to focus sharply on particular points in a structure and accurately record coordinates X, Y, and especially Z. These focus-axis problems and the solutions that have been developed are covered in the following paragraphs.

A microtome slices tissue into three-dimensional slabs whose thicknesses are known at the time of slicing. Further histological processing, e.g., embedding and dehydration, may cause shrinkage of the tissue so that it is thinner and possibly also smaller than it was when it was cut.

Furthermore, when you look at a piece of tissue in a microscope, the focus-axis measurements that you record will be shorter than reality (Fig. 7-1). To understand the focus-axis problem, consider trying to spear a fish in a barrel. As you look down the spear, there is unwanted refraction at the air–water surface, making the spear appear to bend. Thus, it becomes very difficult to spear a fish. A similar problem occurs in the microscope, resulting in focus-axis readings that are always shorter than reality. The tissue shrinkage problem combined with the improper focus axis reading problem may generate focus axis measurements that are 60% of reality.

You must consider whether focus-axis problems are important for your structures; for an almost planar one, it may not matter. For instance, for a neuron that extends 1000 μm in X and in Y but only 50 μm in Z, correcting a 25% error in the Z readings is probably not important because the correction will have a negligible effect on most of the neuron's statistical summaries. The "ten-to-one" rule in analytic geometry (Fig. 7-2) may guide you here. In a right triangle, the length of the hypotenuse is approximately equal to the length of the longer side if the longer side is greater than 10 times the shorter side. To apply this to neurons, if you wish to calculate the dendritic length, for example, and the neuron's range in the X or Y direction is ten times its Z range, then correcting for Z-axis errors is not justified.

If you decide that the focus-axis problem is important, how can you solve it? One solution is to scale the Z coordinates of the recorded structure linearly so that they range over the thickness of the tissue cut by the microtome. Then the data are stored with at least the accuracy of the microtome.

A more complex problem is tissue wrinkling. This problem is illustrated in Fig. 7-3.

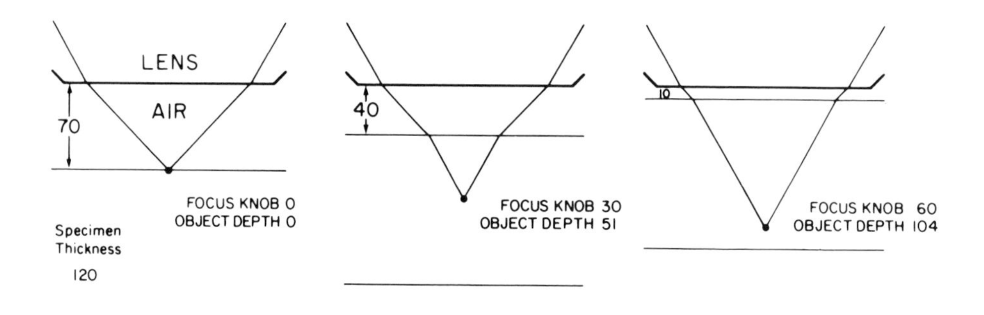

FIGURE 7-1. The unwanted refraction problem. Focus knob readings are significantly less than actual focal plane depth. In all three diagrams, a 120-μm-thick tissue section is shown beneath a dry objective lens. Left: The stage is set to focus on an object at the top of the tissue section (object depth equals 0), and the focus knob indicates 0. Center: The stage is raised 30 μm, but because of refraction at the tissue–air boundary, an object 51 μm deep is now in focus, even though the focus knob indicates 30 μm. Right: The stage is raised 30 μm further so the focus knob now indicates 60 μm. However, the object in focus is 104 μm deep in the tissue. Reproduced from Capowski (1988b) with permission.

FIGURE 7-2. The 10-to-1 rule of vectors. The lengths of three right triangles are shown. When the length of the smaller side is one tenth or less than the longer side, the hypotenuse is essentially equal to the longer side. Thus, when the range in Z of dendrites is less than one tenth of their X or Y range, accurate scaling of Z values is not important for the calculation of dendritic length.

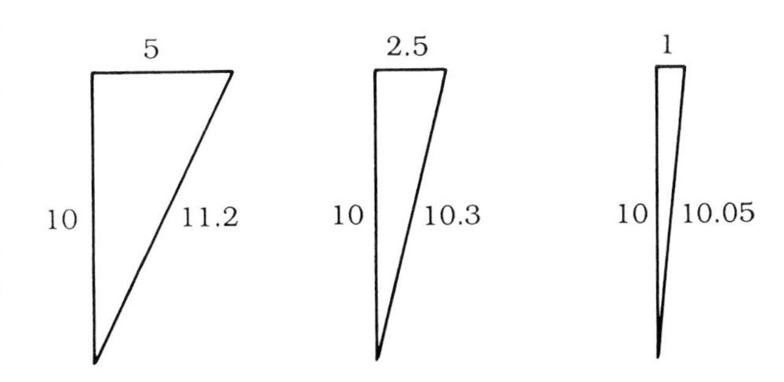

When you make Z-axis measurements from different X,Y locations of the tissue, you will record different Z values even though the values should, in theory, be constant. To correct for this wrinkling, it is necessary to deform the tissue back into its unwrinkled form, the flat slab it was at the time of sectioning.

One solution to the wrinkling problem is what I call the "volleyball net algorithm" (Fig. 7-4). Imagine a volleyball net strung between two posts. At each end of the net, there are two fixed points where the net is attached to the posts. The net is formed by individual strings that are tied together by an array of knots. If you remove the two fixed points at one end of the volleyball net and pull on them as shown in Fig. 8-4, lower left, you will distort the net. Each knot in the net will move a certain amount, depending on how far it is from the fixed points that you have pulled. Each knot will move until the forces acting on it achieve equilibrium, and there will be a smooth distortion of the net.

A neuron traced from its tissue into the computer is analogous to the volleyball net. As you trace the neuron, you record top artificial ends and top tree origins at the top tissue surface, and you record bottom artificial ends and bottom tree origins at the bottom tissue surface. These points are similar to the fixed points on the volleyball net, and the branch points on the neuron are similar to the knots of the net. So if you mathematically take the top and bottom surface points and move them to constant Z values and then mathematically let the branch points relax into places of equilibrium, you will distort the tissue section (including the neuron it contains) into a reasonable approximation of what it was before the wrinkling. After you change the location of each branch point, you then proportionally scale each point between the branch points. This algorithm works best if the tissue section has many top artificial ends and many bottom artificial ends.

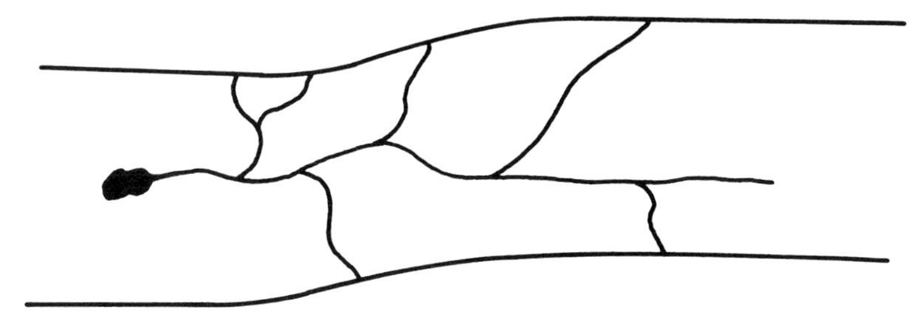

FIGURE 7-3. The tissue-wrinkling problem. As the tissue wrinkles in Z (the up–down direction), the neuron deforms, placing its surface cut ends at varying Z coordinates.

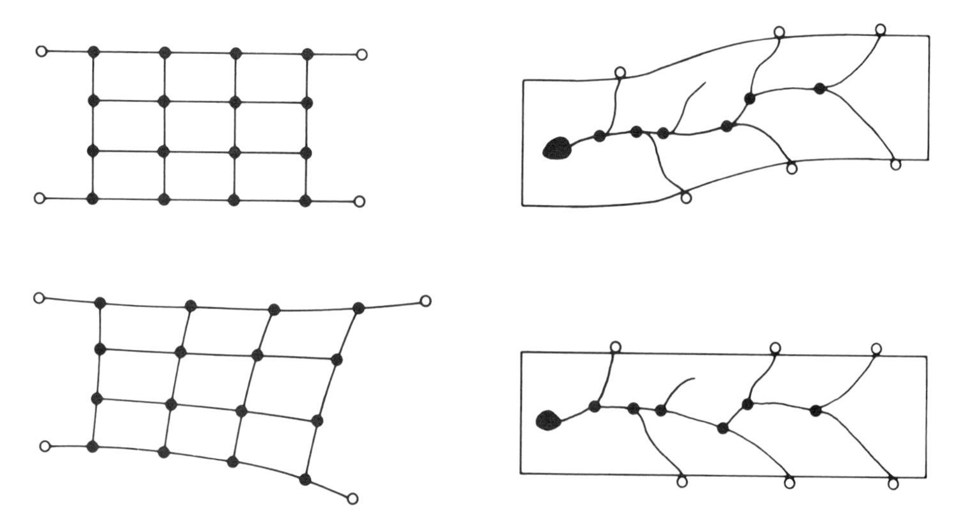

FIGURE 7-4. The "volleyball net" solution to the tissue-wrinkling problem. Upper right: The edge view (Z axis is up–down) of a wrinkled tissue section containing a neuron tree with cut ends indicated by open circles and branch points indicated by closed circles. Since the neuron contained in the section has been wrinkled, the Z values of its cut ends vary. Upper left: A volleyball net in which open circles indicate the attachment points, and the filled circles indicate knots that tie the net strings together. Lower left: The right attachment points of the volleyball net have been removed from their posts and have been stretched to the right and down. The net deforms smoothly, and its knots move, each to establish equilibrium of the forces acting on it. Individual points along the strings between the knots also move proportionately. Lower right: Similarly, the cut ends of the neuron are moved to constant Z values, and the branch-point Z coordinates are deformed, each to establish its equilibrium Z value. When individual points between the branch points are scaled proportionally, the neuron has been successfully dewrinkled. Reproduced from Capowski (1988b) with permission.

7.3. MERGING OF MULTIPLE-SECTION DENDRITES

When a microtome cuts a piece of tissue into sections, any dendrites that meander throughout the tissue may be severed into individual tree pieces. Figure 7-5 illustrates the problem. To perform an accurate reconstruction, you must reattach the tree pieces to their artificial stumps, which were generated by the microtome. At the University of North Carolina, we have developed two solutions to this problem: merging during tracing and merging after tracing.

The merging-during-tracing procedure is illustrated by Fig. 7-6. This figure shows a computer-generated overlay of all the pieces of trees from a previously traced section overlaid on top of the tissue currently in the microscope. You may see where the tree pieces in the current section should attach to the stumps of the previous section. As you begin to trace each neuron piece from the current section, the computer will ask you if it indeed should be attached to a particular identified stump. If you reply "yes," then the computer does the bookkeeping and attaches the tree that you traced to the tree from the previous tissue section.

The actual attachment of a new tree piece to the artificial end of a previously traced tree usually requires three bookkeeping steps. First, if the tissue section containing the

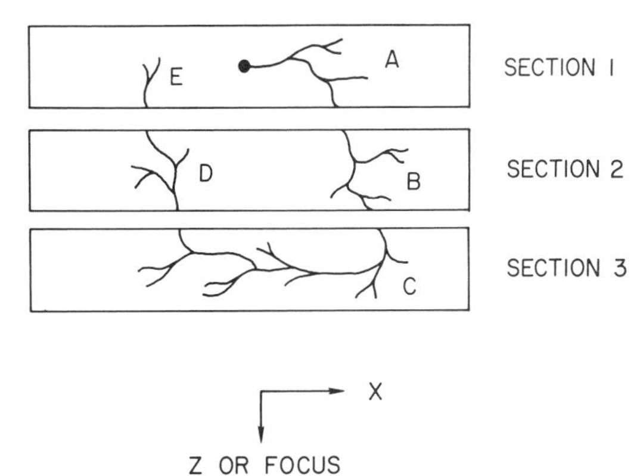

FIGURE 7-5. The merging problem. A contrived example of a dendritic tree whose tissue has been sliced into three sections. The tree pieces must be reassembled to form the intact tree. Reproduced from Capowski (1977) with permission.

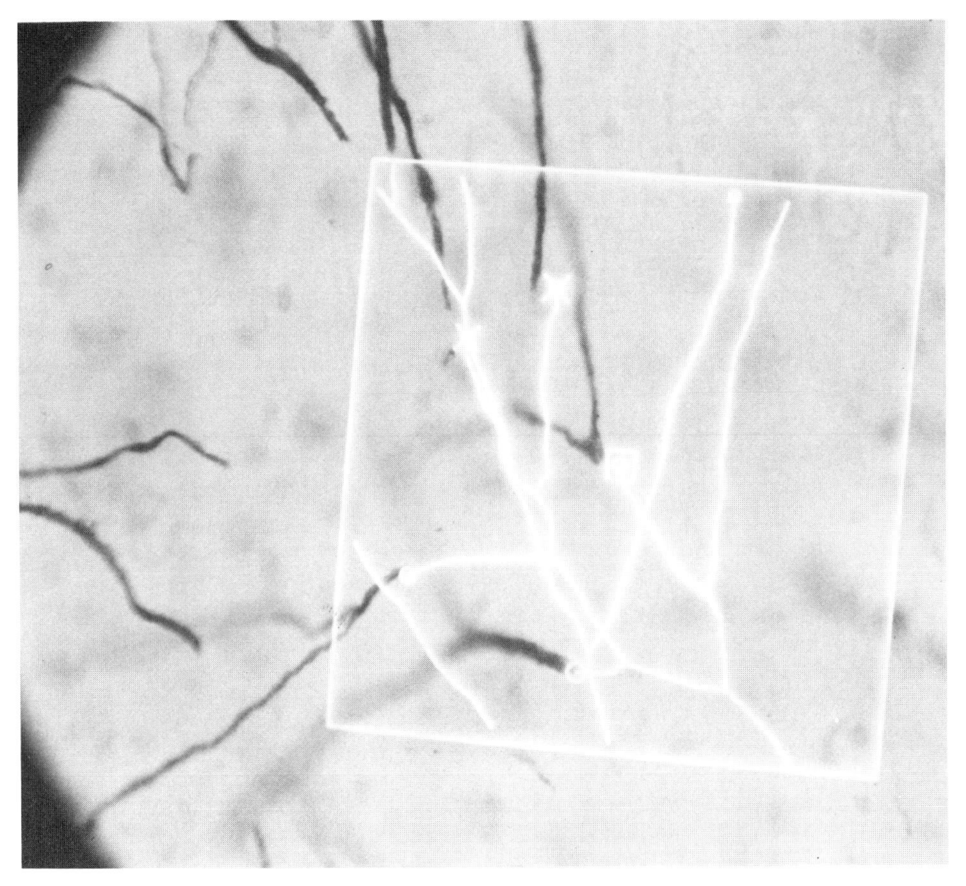

FIGURE 7-6. The view in the microscope when merging during tracing. A computer-generated overlay (white lines) presents the already-traced trees from the previous tissue section overlaid on the image of the section currently in the microscope. Circles indicate endpoints in the previous section; crosses indicate endpoints to which merging has already been done; the square indicates an ''identified'' endpoint. The user is about to trace the tree piece that starts at the identified endpoint, and the computer will attach it to this endpoint. Reproduced from Capowski (1988b) with permission.

new tree piece is placed in the microscope at a different rotational orientation about the focus axis of the microscope from that of the previous section, then the software must rotate the new tree piece to the previous section's orientation. The merging software must then translate the new tree piece so that the coordinates of its origin exactly match the coordinates of the stump. Finally, the software does the actual attachment. Here the software deletes both the artificial end from the previous tree and the origin of the new tree and replaces both of them with a continuation point. Now only one tree remains.

The alternative to merging during tracing is merging after tracing. With this solution, all tree pieces from each tissue section are traced without regard to their connections. Then, as illustrated in Fig. 7-7, all the trees from one section, called the top section, and all the trees from the adjacent section below, called the bottom section, are displayed on the CRT. A square is drawn around each candidate for merging, that is, around each top tree origin and top artificial end in the bottom section and around each bottom tree origin and bottom artificial end in the top section. With the joystick you may rotate both sections about their *Y* axis simultaneously. With these relative movements, you can match the squares and determine which tree pieces should be attached to which other tree pieces.

When you decide that a particular tree should attach to a particular stump, you interactively position the attaching tree's origin at the stump and move a circular cursor to this spot. The computer reports specific information (tree name, point type, thickness) about each point. You check that the attaching tree agrees with its stump in location, direction, and thickness. "Location" means that the two points are in the proper position relative to the rest of the structures. "Direction" refers to the direction of growth of the dendrite. Since dendrites make sharp directional changes infrequently, an attachment is improbable if the attaching tree grows in a different direction from the proximal stump. "Thickness" refers to the thickness of the stump compared to the thickness of the attaching tree. A very thin branch probably should not be attached to a much fatter proximal stump. When you review this information, if you believe that the merge should indeed by performed, you press a key, and the computer does the bookkeeping to merge the two trees into one.

Sedivec et al. (1986) showed a success rate of about 93% in merging the dendrites of very complex HRP-stained spinocervical tract neurons, though with some effort. These researchers spent approximately one fifth of their time doing merging. You may wonder whether taking the time to merge accurately is worth such an effort? If the analysis of the neurons that you want to do is simply to calculate the total dendritic length, then it is not necessary to merge. However, if you wish to analyze a parameter, e.g., number of spines, versus branch order, then merging is crucial. Also, if you want a final display of an intact three-dimensional neuron as it courses through many tissue sections, then merging its dendrites is important.

7.4. ALIGNING ONE TISSUE SECTION WITH ANOTHER

After the microtome cuts a block of tissue into individual sections, the structures within the tissue must be entered into the computerized microscope one section at a time. The sectioning process causes two problems in generating an accurate three-dimensional reconstruction. The first problem, applicable particularly to neuronal processes and covered in Section 7.3, it to merge the individual disjoint tree pieces. The second problem, covered here, has to do with correct rotational alignment of each section.

When a histologist mounts tissue sections on a microscope slide, he typically tries to put them all into constant rotational alignment by eye, but seldom are all the sections exactly aligned. It is thus necessary to rotate each individual section about the focus axis of the microscope either mechanically on the microscope stage before tracing or mathematically within the computer after tracing so that the sections are consistently oriented. If there exists an artificial cut edge of the tissue section, then you can simply align the cut edge of each tissue section with the X-axis motion of the microscope stage, thereby assuring the same rotational alignment of all tissue sections before the structures are traced.

An alternative to an artificial edge is any kind of natural or artificial multiple-section landmark (e.g., a straight blood vessel, a long axon, a pinhole that had been poked through the tissue before it was sectioned) that you may use for alignment. The section-to-section alignment is done semiautomatically; that is, you determine the proper relative rotation. To assist you in determining the rotation angle, an overlay is provided by the computer. The machine presents an overlay of structures traced from the previous section on top of the tissue currently in the microscope, similar to the display of Fig. 7-6. Then you may rotate the slide in the microscope specimen holder until the tissue is aligned with the computer-presented image.

Another option is to align the sections semiautomatically after you have traced the structures from both sections. Consider the technique shown in Fig. 7-7, where the structures from both sections are presented on the CRT. Using such a technique with a mouse or joystick, you may rotate one section with respect to the other so that they are aligned. When you have selected the best alignment, the computer will algebraically rotate the mathematical model of one section to bring it into the same orientation as the adjacent one.

Sophisticated computer algorithms have been written to automate this rotational alignment problem. These, covered in Chapter 15, have met with some limited success at a great programming effort.

7.5. EDITING

Once you have traced data into a computer, you may want to edit the data to correct tracing errors. The earlier you detect an error, the easier it is to correct, so that error correction is best done during tracing. If facilities to delete the most recent point, the most recent branch, the most recent tree, or the most recent structures are provided for this purpose, then you may simply retrace the piece of structure that you entered incorrectly. But some errors are not detected until later. For instance, during merging or during Z-axis scaling to correct for focus axis problems, you may detect incorrect point types. So you also need a data-base editor to correct these mistakes.

Editing procedures may be classified into "single-point editing" and "global editing." For single-point editing, you must identify a point in the data base and then change its individual parameters, i.e., its X, Y, and Z values, its point type, its tag, and its thickness. You must be able to delete a point or insert a point in some location. For more

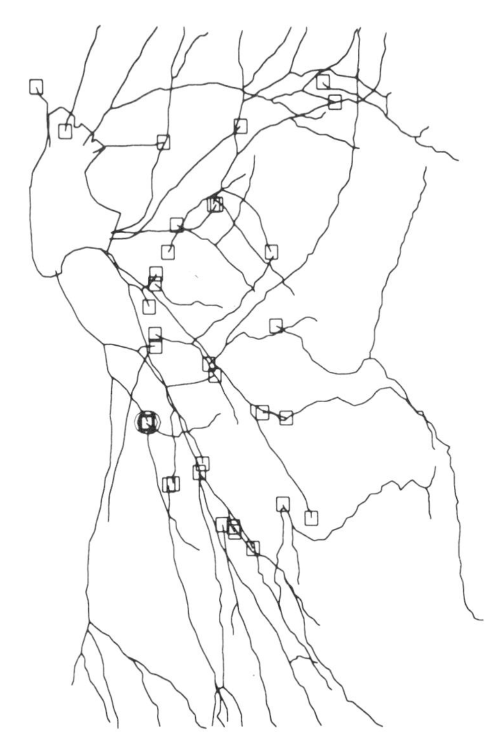

FIGURE 7-7. The CRT display during merging after tracing. Tree pieces from two sections, called the "top" and "bottom" sections are displayed on the CRT. Tree origins and endpoints at the common tissue surface are indicated by squares. The user may smoothly rotate and translate the top section with respect to the bottom section to match the squares. He then places a circle (lower center) around a particular pair of points and indicates to the computer via pushbutton to merge these two points. The computer performs the necessary bookkeeping in order to link these two tree pieces into one intact tree.

global editing, you must be able to specify a range of points within the data and delete the range or change the parameters of all the points within the range.

Two techniques allow you to identify a single point for editing. If you know it, you may simply enter the point number, or you may display the structure on the CRT, move a cursor near the point, and then instruct the computer to identify the point nearest the cursor.

You may identify a range of points in several ways. If the range of points defines a structure, you may identify it by simply entering the name of the structure that you specified when you traced it. Or you may build a display of the entire data base on the CRT and move a cursor to the structure you wish to identify. If the range of points forms a subset of a structure, you may enter the point numbers of the limits of the range, or you may build a display of the structure and move a cursor to the lower and upper limits of the range and instruct the computer to find the points nearest the cursor.

In any of the methods for identifying points within the structure, feedback is important so that you may be certain that the computer has identified the proper points. This feedback is best provided visually, as shown in Fig. 7-8, by highlighting the identified portion of the structure.

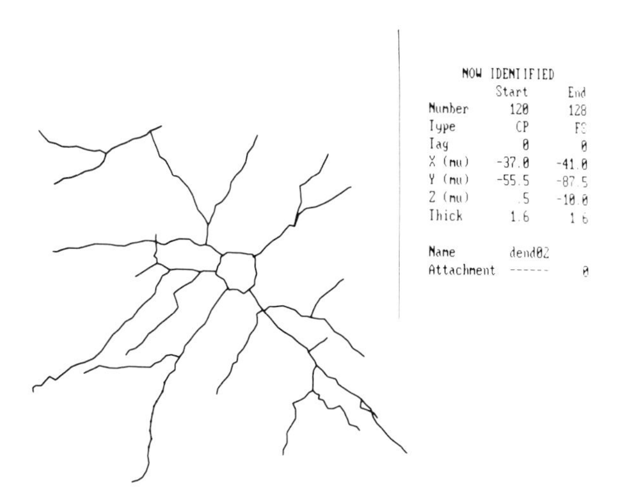

FIGURE 7-8. The CRT display during the editing of a structure. With the cursor (a small dot in the lower center of the screen) the user may identify a series of points along a dendrite. The points are highlighted with dots (in the lower center of the screen), and information about the points is also provided (insert in upper right).

A "detach, drag, and reattach" capability is also essential. Here you build a display of the data base on the CRT and, with a mouse or joystick-controlled cursor, grab one of the structures and drag it to a particular location. If the dragged structure should be attached to another structure, you should be able to instruct the computer to "attach this here." The computer does all of the bookkeeping.

A similar dragging capability without attaching is useful, for example, to move a dendrite to exactly the place on the traced soma from which it emanates. Using a cursor on the CRT, you may drag a dendrite around to this location and leave it there.

Finally, algebraic manipulation of the data base or a selected portion is needed. You may want to rotate the data base about any axis. You may want to multiply any of its coordinates or thicknesses by a constant, thereby scaling the structure. If you select a negative constant, then the structure is inverted. You may want to add a constant to each coordinate, thereby translating the data base. These capabilities are needed for practical reasons. One such example would be if the tissue section is placed upside down on the slide so that structures contained in it are traced with an inverted coordinate system.

7.6. MATHEMATICAL TESTING OF THE DATA BASE

Neuron trees are stored as binary trees, as illustrated in Fig. 7-9. They are so called because each branch point has two exiting paths, and this generates some interesting mathematical properties. A binary tree must have one origin, and it may have branch points. A complete binary tree must have one more ending than its number of branch points. It is useful to have a mathematical test that counts the number of origins, the number of branch points, and the number of endings in each tree in the data base to check that they are in the proper relation. If they are not, it will identify the erroneous tree.

Another mathematical test that you may perform is to check whether top tree origins

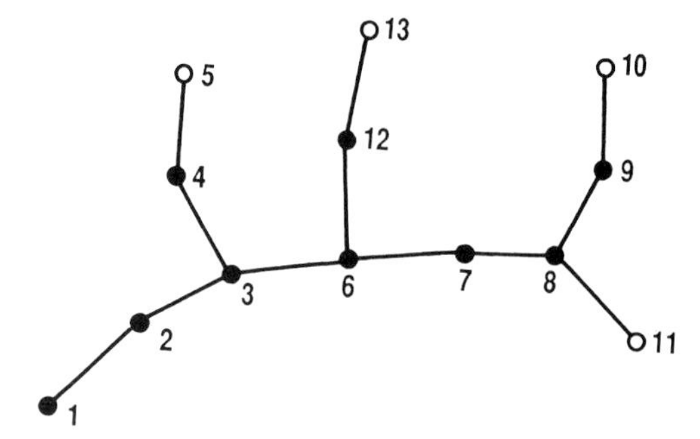

FIGURE 7-9. The storage of a point of a binary tree. Point numbers and coordinate information about them are stored sequentially. After a branch point (point 3), the next point stored is along either of the distal paths from the branch point. After an ending (point 5), the next point stored is the point along the other distal path from the branch point.

and top artificial ends are indeed near the top of the tissue section and that bottom tree origins and bottom artificial ends are indeed near the bottom of the tissue section. For each of these cases, two tests are performed by the computer. The first test asks whether, in the case of a top artificial end, for example, the end is in the top third of the tissue section. The second test asks whether the top artificial end is nearer the top surface than the branch segment that precedes it. This is, the computer calculates the mean Z value of the six points proximal to the top artificial end and compares their mean Z to the Z coordinate of the top artificial end; the Z coordinate of the ending should be closer to the top of the tissue section. This test is illustrated in Fig. 7-10.

7.7. SERIAL INSPECTION OF STRUCTURES

Occasionally, there may be an error in an X,Y,Z coordinate of a structure. Since humans are very good pattern recognizers, a quick method for finding such an error is to inspect each structure serially. The first structure in the data base is displayed on the CRT, and with the joystick you may rotate the structure smoothly to examine it quickly for flaws. Should you see a flaw, you may move a cursor over the display, and the computer will indicate which point is nearest the cursor. In this way you may determine which is the

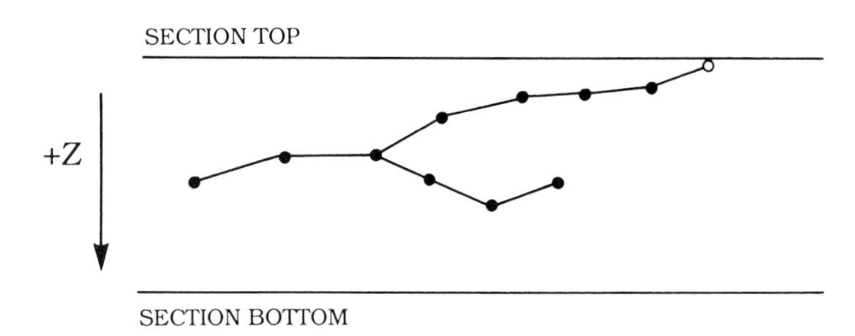

FIGURE 7-10. The test for top artificial ends near the tissue top surface. A top artificial end (open circle) should be near the top surface of the tissue. A two-part mathematical test may be run to confirm this. First, the Z coordinate of the endpoint must be in the upper 35% of the Z range of the tissue section. Second, the Z coordinate of the endpoint must be less than (nearer the top surface) the mean Z coordinate of several points proximal to it.

flawed point and, using the editing techniques of Section 7.5, correct the mistake. When you are happy that the structure is fine, you may press a button instructing the computer to display the next structure. You may inspect a large number of trees in this manner more quickly than a sophisticated computer algorithm can.

7.8. REORDERING OF TREES

Occasionally, when you trace a tree piece in a tissue section, you do not know which end of the tree is proximal and which end is distal. If you err here by starting to trace at a distal ending, you usually discover the error when you attempt to merge the tree to pieces from other tissue sections. You then find out which endpoint should have been the origin of the tree. The solution to this problem is to reorder or renumber the tree, as shown in Fig. 7-11. You specify the endpoint that should have been the tree origin, and a mathematical algorithm changes that endpoint to the origin and renumbers the rest of the points, keeping their coordinates, types, tags, and thicknesses unchanged.

7.9. FILTERING

A "filter" is a mathematical procedure that is applied to data to limit it in some fashion. There are two reasons to filter the coordinates of computer-stored anatomic data: (1) to smooth sampling errors in the digitized coordinates, possibly to make the structure more realistic or prettier, and (2) to decrease the number of points in the data base so that more trees can be stored and computation time can be reduced.

To illustrate the need to filter, consider the mechanical and optical limitations of the microscope as it is used for tracing data. It is easy to move a cursor accurately in X and Y on a microscopic image of a neuron and thus accurately record a series of X,Y coordinates. However, because of the depth of field of most microscope objective lenses, and because of backlash in the focus-axis gear system, it is not possible to record focus values (Z coordinates) as accurately. Therefore, when you trace a dendrite as it meanders in three dimensions, your Z-axis readings will not be as accurate as your X and Y readings. When

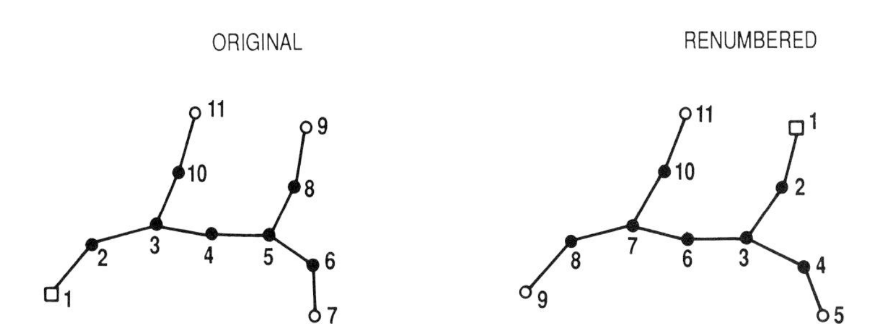

FIGURE 7-11. Renumbering a tree. Left: An original tree, with its origin indicated by a square, endings indicated by open circles, and point numbers shown. Right: Point 9 has been changed from an ending to the origin of the tree, and the remaining points have been renumbered.

the tracing is finished and you rotate the neuron 90° to see a different view, the dendrites will display a staircaselike effect, as shown in the left side of Fig. 7-12.

This problem is solved by low-pass filtering. Low-pass filtering allows lower frequencies or grosser structures to remain but smooths higher frequencies or minute changes in structure. It is done by averaging. To low-pass filter in Z is, for example, to replace each Z coordinate by the average of its Z value and the Z values of the previous and following points. This filter is particularly effective in removing the focus-axis recording errors that we just described; its result is shown in Fig. 7-12. It is also possible to low-pass filter the X and Y coordinates, resulting in a smoother neuron. This usually makes the display "prettier" but may or may not make it more realistic. Low-pass filtering does not remove any points from the data base.

Another kind of filtering is spatial filtering. Spatial filtering is used to remove any point that is closer than some distance (called the spatial filter distance) from the previous point. Thus, it will remove points from the data base that are closely spaced. This saving of points is permissible if you want to build a display at a low magnification, because at a low magnification, the closely spaced points would not be seen. Figure 7-13 shows typical

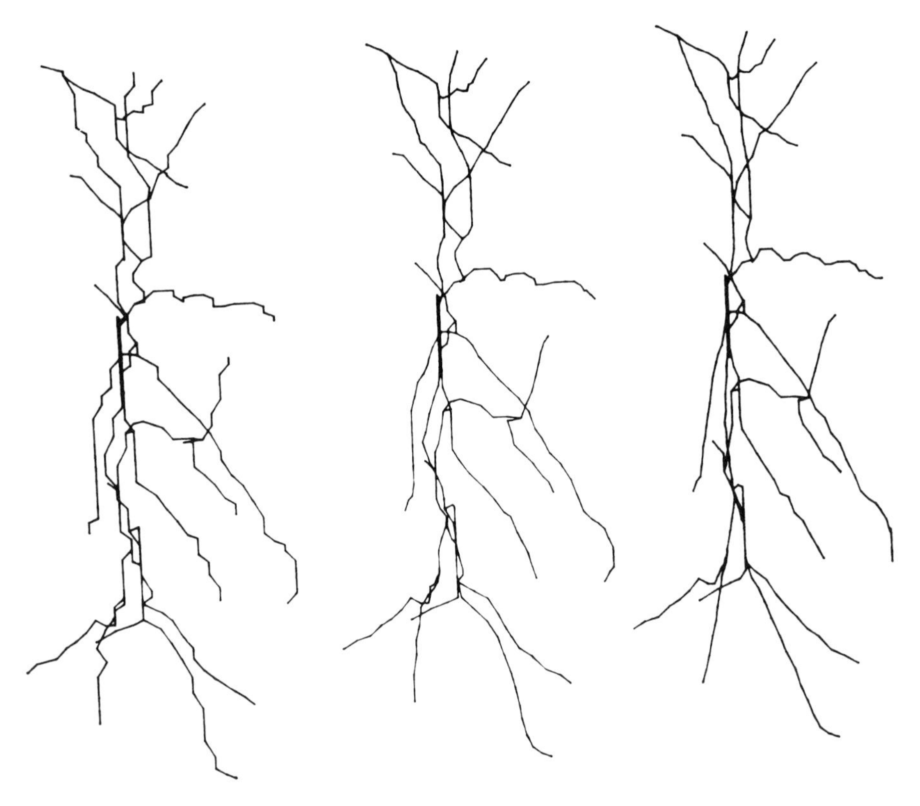

FIGURE 7-12. The effect of low-pass filtering. A neuron was traced in transverse section and rotated 90° about its Y axis to show a sagittal view. Its Z axis now extends from right to left. Left: No filtering has been done in Z, and the neuron appears jagged. Center: One pass of three-point low-pass filtering has been performed; the dendrites appear smoother. Right: Ten passes of filtering have been performed, and the result appears still smoother.

results of spatial filtering. To avoid destroying the binary tree structure, only continuation points are usually removed by spatial filtering. If the filter distance is set very large, the data base and display may be reduced to only origins, branch points, and endings, creating a "schematiclike" effect.

"Radius-of-curvature" filtering may also be used to reduce the number of points in the data base without changing its displays or statistical summaries appreciably. In this case, if there are three points that are approximately in a straight line, then the middle one of the three points will be deleted. This will reduce the number of points in the data base but will not change the display or analyses very much, because the vector data representation assumes a straight line from one point to the next. Radius-of-curvature filtering can be used to remove perhaps 30% to 40% of the points in the data base without changing its structures significantly.

Radius-of-curvature filtering derives from geometry. Consider that a small circle has a small radius and a larger circle has a larger radius. Similarly, an arc of a circle also has a radius, called its radius of curvature. A sharply bending arc has a small radius of curvature, and a gently bending arc has a larger one.

The filter technique calculates the radius of curvature of each set of three consecutive points in the structure as if they were on the perimeter of a circle (Fig. 7-14). If the three points have a larger radius of curvature than a threshold value, the middle point will be deleted. Continuation points are removed by this technique so that the binary structure of the tree is preserved. Typical effects of radius-of-curvature filtering are shown in Fig. 7-15.

7.10. STORAGE OF DATA ON DISK

When you trace a structure, data are stored in the memory of the computer. This memory is of limited capacity, is volatile (loses its data if there is a power failure), and may not be removed from the computer. Therefore, at some time during the tracing process, the data should also be stored on a disk. When and how should this be done? A variety of techniques for data storage to disk have evolved. The "when" portion is dealt with first.

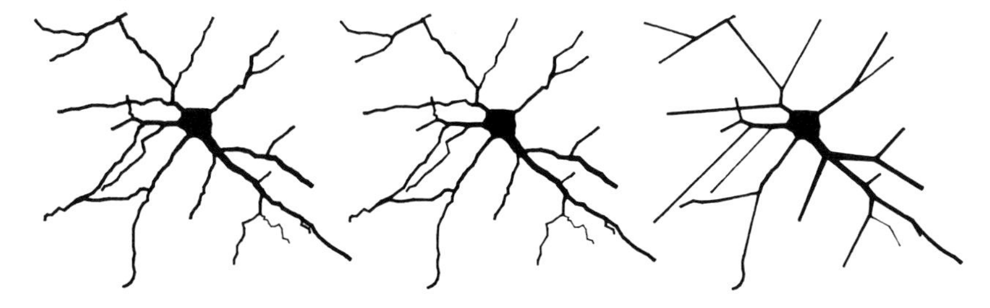

FIGURE 7-13. Three plots of a reconstructed neuron to show the effect of spatial filtering. Each cell is approximately 200 μm in diameter. Left: Unfiltered drawing contains 361 points. Center: Continuation points closer than 5 μm have been removed, resulting in a cell that now contains 271 points and whose appearance has changed little. Right: All continuation points have been spatially filtered, resulting in a schematiclike representation containing 86 points.

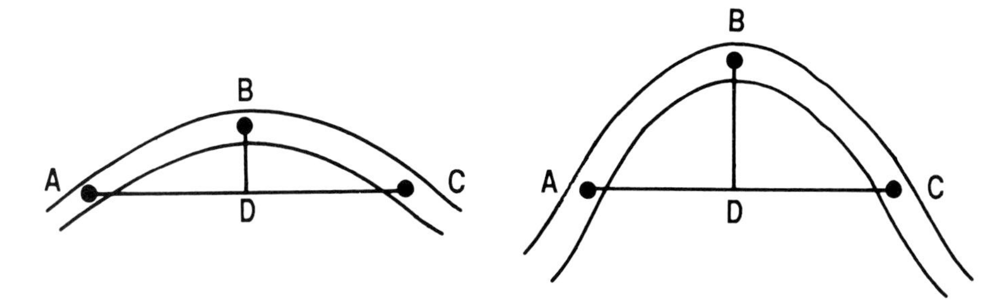

FIGURE 7-14. Radius-of-curvature filtering to remove points in relatively straight regions of a dendrite. Left: A slightly curving part of a dendrite with three points shown. Lines AC and BD are artificially generated. Since line BD is much shorter than line AC, point B may be removed from the neuron's structure without affecting its display or summarization significantly. Right: BD is not much shorter than AC, so point B should not be deleted. Generally, if the length of BD is less than a tenth of the length of AC, the point may be removed without effect.

The software may store data automatically every time you record a point or whenever you reach a milestone such as finishing a tree. The software may also store data at fixed time intervals or at a fixed number of point intervals. As examples, data could automatically be written from memory to disk every minute or every 100th point. Another technique is to mimic some common word processor programs, i.e., store the data only when instructed. With this method, you write data to the disk using any file name you choose. If you select a file name of an already existing file, the previous data in that file are destroyed, and the most recent data replace them; if you select an unused file name, a new file is made.

Now for the ''how'' portion. Data can be stored in memory as binary numbers. Binary files are those made by simply writing the data from memory to the disk without any transformation of the data. (The use of the work ''binary'' here refers to the base-2 number system used to store each data value; it is different from ''binary tree,'' meaning two exiting paths from each branch point.) Binary files are compact, but they cannot be read easily by any other computer program, and they cannot be transferred easily from one

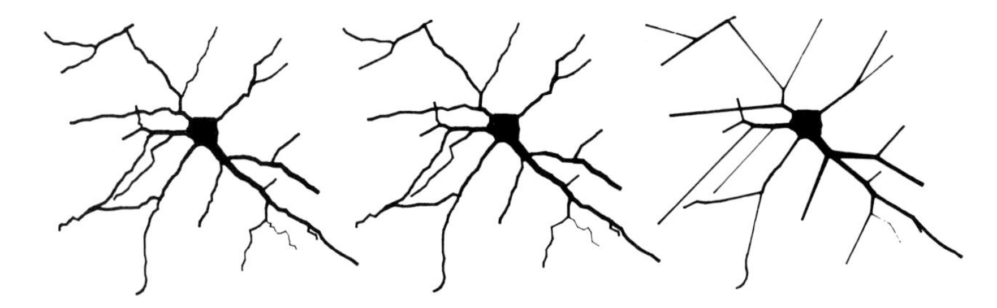

FIGURE 7-15. Three plots of a reconstructed neuron to show the effect of radius-of-curvature filtering. Each cell is approximately 200 μm in diameter. Left: Unfiltered, this drawing contains 361 points. Center: Continuation points in regions of low curvature have been removed, resulting in a cell that now contains 267 points and whose appearance has changed little. Right: With extreme radius-of-curvature filtering, all continuation points have been removed, resulting in a schematiclike representation containing 86 points.

manufacturer's computer to another. Furthermore, data stored in binary files can be neither printed nor edited with a word-processing program.

Files stored in ASCII format are composed of text characters; they look like the printout of the memory-stored data (Fig. 4-15). Such files can be easily sent from one computer to another; they can be typed so that you can read them; they can be edited with a text editor so that you can change their structure; and finally, they can be read by some commercial statistical packages. This last advantage of ASCII data storage means that structures can be summarized statistically with methods other than those provided with the tracing program. A limitation of ASCII storage is that files require about triple the storage space on a disk and take longer to read or write.

7.11. CONCLUSION

With the techniques described in this chapter, you may process the mathematical model of an anatomic structure in order to overcome the limitations of the devices on which it was entered. You may make the result more realistic or avoid some limitation in the computer hardware that displays it. Frequently these procedures are used to severely limit the amount of data in a structure so that many structures may be stored in memory at one time for display or for summarization of populations.

7.12. FOR FURTHER READING

Focus-axis measurements from a microscope are always shorter than reality because of tissue shrinkage and unwanted refraction. These problems have been described by Glaser (1982), Capowski (1985), and Uylings et al. (1986b).

Computer algorithms for renumbering dendritic trees in order to define a different point as an origin are described by Katz and Levinthal (1972) and by Capowski (1977).

The several ways to number the branching patterns of a dendritic tree are covered in depth in Chapter 10 of this book and may also be reviewed in Buskirk (1978), Glaser (1981), Triller and Korn (1986), and Uylings et al. (1986a).

8

Three-Dimensional Displays and Plots of Anatomic Structures

8.1. INTRODUCTION

This chapter presents methods for producing ("rendering" in computer graphics technology) displays and plots of anatomic structures that have already been entered into the computer. The goal is to generate views of the structure as they would have appeared naturally, before any tissue sectioning or other histological manipulations are performed. Since most of the structures are microscopic and a microscope imposes viewing constraints on its user, a display must actually present a view that is better than you could see if you looked into the microscope yourself.

Figure 8-1 shows an exquisite display of a part of a neuron. Although the computing resources required to generate this illustration are not readily available to the rank-and-file neuroanatomist, the illustration serves as a goal, an example of displays that are possible to generate.

Sutherland (1965) wrote a thesis entitled "SKETCHPAD" that marked the start of three-dimensional computer graphics. He could, for the first time, present drawings of a computer-stored three-dimensional structure on a cathode ray tube (CRT) and smoothly rotate and move them around the screen. Today, the CRT remains the most prominent computer display device, but the techniques that Sutherland pioneered to drive the CRT have been enhanced greatly. Using a vector graphics system, he was able to draw only "wire-frame" structures, i.e., to present only the lines that outline each structure. As you will see in this chapter, many more sophisticated displays are now possible, including the use of televisionlike raster graphics to shade the surfaces of a structure.

The CRT may be driven from the computer by either a raster or a vector computer graphics system. The CRT displays cannot be removed from the computer, so some system for making a copy on paper must also be provided; such a copy is called a "hard copy." You may, of course, obtain a hard copy by simply photographing the image on the CRT screen. You may also obtain a hard copy by using a felt-tip pen plotter, a laser printer, or a dot-matrix printer to draw the image on paper. This chapter discusses both CRT displays and paper plots.

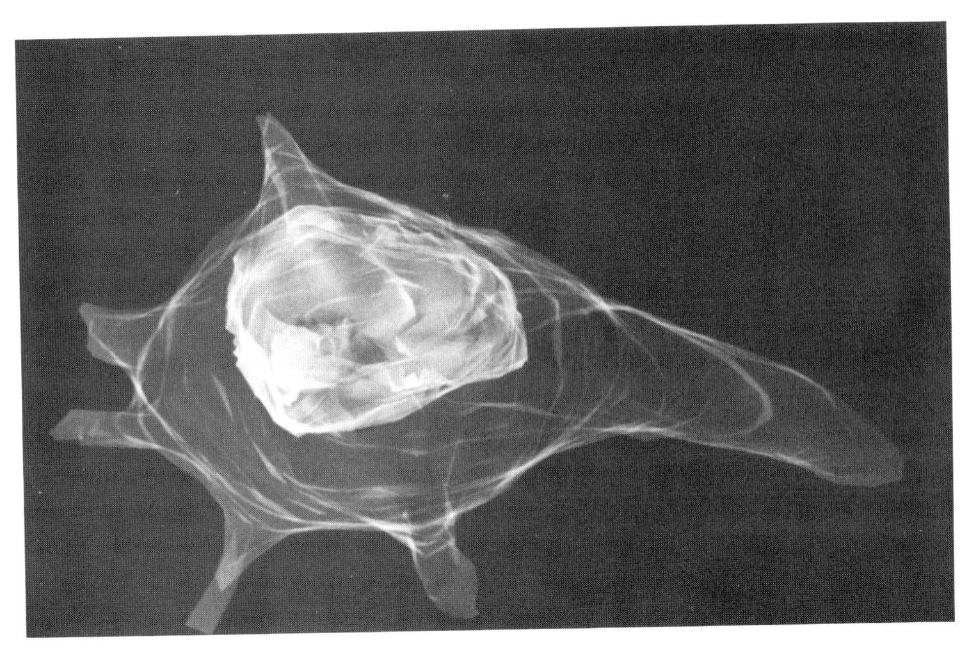

FIGURE 8-1. A three-dimensional reconstruction of the soma of a cortical pyramidal neuron from an Alzheimer's patient. The tissue containing this soma was cut into 100 sections, and the soma, nucleus, and nucleolus outlines were traced on a data tablet. Tiling triangles were generated to connect the sections, resulting in about 35,000 polygons in the complete image. It was rendered on a Cray supercomputer in about 1.5 min and presented on a Silicon Graphics workstation. Reproduced from Frenkel (1988) with permission. A color reproduction of this figure appears following p. xx.

An anatomic reconstruction system should allow you to build a display on a CRT screen of one or more structures in their proper anatomic relationship. You should be able to view them from any position and at any magnification. Furthermore, you should be able to see an entire structure on the screen at one time, or you should be able to zoom in to look at a highly magnified view of a small part of it.

Artificially enriched displays are also desirable. You should be able to highlight a portion of a structure so that it stands out from the rest of the structure. Either different colors or varying intensities may be used for this. Another example of an artificially enriched display is to paint an external structure transparent so that internal structures are visible through it. This technique, illustrated by Fig. 8-1, allows you to see the relationship between internal and external structures.

You may recall from Chapters 4, 5, and 6 that an anatomic structure may be entered into the computer in a vector-based format or in a pixel-based format. In vector-based format, the computer-stored model consists of a series of X,Y,Z coordinates and the assumption that the coordinates are connected by straight lines. In pixel-based format, the model consists of individual pixels, each with a gray value.

For the purposes of display, vector-based structures can be extended into polygon-based ones. This means that the line endpoints are assumed to be the vertices of planar polygons, and each polygon encloses a flat surface or "facet," so named because the many flat surfaces of an irregular structure resemble the facets of a cut diamond.

Displays may be "wire frame" or "surface-filled." Wire-frame displays (e.g., Fig. 8-6) are composed solely of lines, with no attempt to fill regions of the display. Surface-filled displays (e.g., Fig. 8-17) fill regions between the lines by painting pixels in these regions to various gray levels or colors.

Two types of display systems, vector and raster, are used to render these types of displays from the storage formats just described. To review from Chapter 2, a vector display draws lines on a CRT or on a plotter by moving the electron beam or plotter pen only from one point in the structure to the next. A raster display sweeps the electron beam of a CRT in a fixed pattern across the screen and draws individual pixels in different shades of gray or color.

Wire-frame displays may be drawn on either vector or raster systems. Surface-filled displays may be drawn only by raster systems. There are numerous tradeoffs between the types of display, complexity of the images that can be formed, speed, and cost. This chapter covers many of the standard features and explains some of the tradeoffs.

8.2. THREE-DIMENSIONAL DISPLAYS ON A TWO-DIMENSIONAL SCREEN

Although most neuroanatomical structures are three-dimensional, the face of the CRT on which the displays are generated has only two dimensions. This presents the question, "How do you create the third dimension?" Or rather, "How do you generate the illusion of depth?" Since Sutherland's thesis, the computer graphics industry has devised common techniques for generating the illusion of depth. These techniques, called "depth cues," are used to provide depth information about a three-dimensional structure when it is displayed on a two-dimensional screen. Depth cues that are relevant to neuroanatomical structures are described in the rest of Section 8.2.

8.2.1. Smooth Rotation and Kinesthesia

If I were to hand you a pineapple and ask you to examine it, you would probably hold it in front of you and rotate it smoothly in order to look at it from all sides. You would use your stereoscopic visual facilities coupled with the smooth rotation in order to comprehend its complex three-dimensional structure. You would receive additional input from your kinesthetic sense as you rotated the pineapple, and you would correlate this sense of movement to further support your three-dimensional appreciation of the pineapple's structure.

Unfortunately, though you want to appreciate its three-dimensional structure, you cannot examine a neuron in the same way that you can examine a pineapple. A reasonable attempt to reach this unrealistic goal would be to generate a two-dimensional display of the neuron on the CRT and to rotate it smoothly. As you watch it rotate, your visual system will provide you a three-dimensional illusion.

Your kinesthetic sense may also be employed here. If you control the rotation of the neuron on the screen with a joystick, and the neuron's position tracks the position of the joystick, then your three-dimensional illusion will be further enhanced. Even without a stereo display, it is possible to create an excellent illusion of depth in this manner.

Figure 8-2 formalizes the coordinate system normally used in three-dimensional computer graphics and demonstrates the rotations that may be employed. The most common rotation is about the *Y* axis, which you may easily control with the potentiometer on top of a three-dimensional joystick (Fig. 2-16). This rotation is analogous to putting your hand on top of the pineapple and rotating it about its *Y* axis. If you pull the joystick toward you, you will tip the neuron toward you about its *X* axis. Either of these rotations provides the illusion of depth. Rotation about the neuron's *Z* axis, however, provides no depth information.

How do you perform smooth movement or smooth rotation? In a motion picture film, individual still frames are generated with each frame slightly different from the previous one. When they are presented to you in rapid sequence, you interpret them as changing smoothly.

The same concept is used in computer graphics for presenting smoothly rotating objects. If you are presented a frame containing a pineapple rotated 34° about its *Y* axis and then, $\frac{1}{30}$ sec later, a frame containing the pineapple rotated 35°, and then, $\frac{1}{30}$ sec later, a frame containing the pineapple rotated 36°, etc., you will envision the fruit smoothly rotating.

Unfortunately, smooth movement is very costly. To see smooth motion, to prevent flicker (see Section 8.3.1), and to synchronize with standard television frame rates, about 30 frames must be produced every second. You may remember from trigonometry that a point in space is rotated about the three axes by multiplying its *X,Y,Z* coordinates by a 3 × 3 rotation matrix. This requires nine multiplications. If there are 1000 coordinates in a display and the display must be generated 30 times a second, then each matrix multiplication must be performed in about 30 μsec. This is too fast for normal laboratory microcomputers without special-purpose hardware called a "matrix multiplier." The matrix multiplier is usually included as an integral part of a dynamic computer graphics system, thus performing smooth rotation as well as other functions described below.

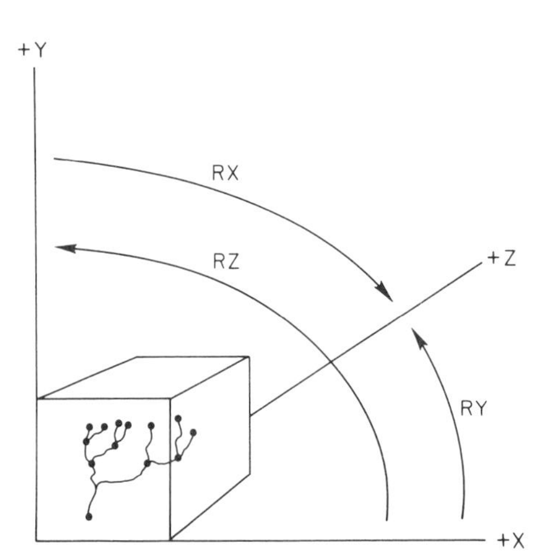

FIGURE 8-2. The usual designations for rotations of a display about an axis. When looking at the CRT screen, the positive *X* axis extends to your right, the *Y* axis extends up, and the *Z* axis extends away from you, into the screen. Positive rotation about the *X* axis is defined to be rotating the *Y* axis toward the *Z* axis, i.e., tipping the top of the structure away from you. Positive rotation about the *Y* axis is defined to be rotating the *X* axis toward the *Z* axis, i.e., pushing the right side of the structure away from you. Positive rotation about the *Z* axis is defined to be rotating the *X* axis toward the *Y* axis, i.e., rotating the structure counterclockwise in the plane of the CRT face.

8.2.2. Clipping

When you sit in your office and look out the window, you see a portion of the outdoors. there is a large environment outside, but the window frame and the walls of your office limit the amount of the environment that is visible to you. A similar situation exists in neuroanatomy. When you look in a microscope, you see a portion of the specimen, limited by the microscope's "window" or field of view. If you generate a CRT display of a structure with magnification great enough so that not all of it will fit on the screen at one time, you must somehow limit the display to a "window" the size of the CRT.

Let's assume that there is a long building outside your office window (Fig. 8-3) and that only the left end of the building can be seen through the window, while the right end of the building is hidden by the right side of the window frame and office wall. In computer graphics terminology, the building's right end is "clipped" by the right side of the window frame. There is a near-horizontal line that forms the top of the building's roof. Part of this line can be seen, from its left end up to the place where it intersects the right side of the window frame. This line is also clipped by the right side of the window frame.

In a vector representation of an object, coordinates are stored that define the endpoints of lines. So X,Y,Z coordinates are stored in computer memory for the left and right endpoints of the roof line. To display a portion of the building on a CRT, it is necessary to calculate the intersection of the roof line with the edge of the CRT screen and draw the line only from the line's left end to the calculated intersection. This calculation is called "algebraic clipping" and must be performed for every line that forms a partially visible structure.

For a dynamic picture, such as a smoothly rotating one, the calculations are very time-demanding, for as the picture moves, the intersection points of the lines in the structure with the edge of the screen change and must be continually recalculated. For-

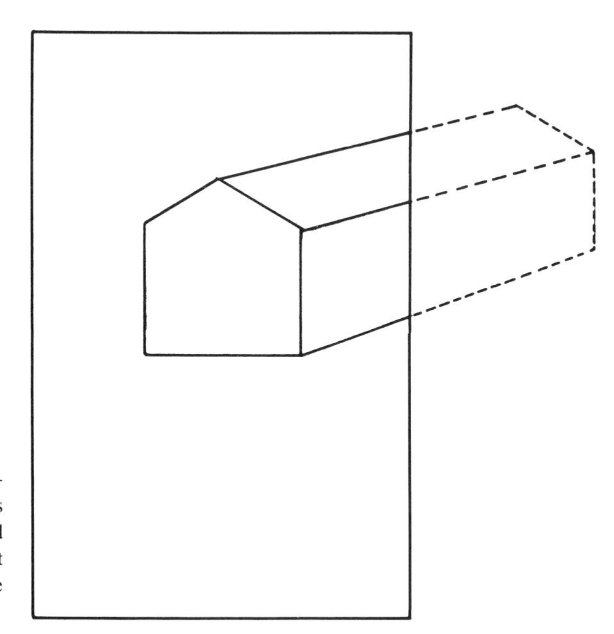

FIGURE 8-3. A simple building partially obscured by a window. The lines that form the building and that extend from the visible part to the hidden part are clipped at the right edge of the window.

tunately, however, as Fig. 8-4 illustrates, anatomic structures usually do not require this algebraic clipping.

As Capowski (1976) observed, anatomic structures are composed mainly of short straight lines; thus, if either end of a line is rotated or translated off the CRT screen, the entire line may be rejected without degrading the picture appreciably. The process of determining if a line endpoint is off screen is much easier than performing algebraic clipping, so a simpler computer graphics display processor may be used for presenting anatomic structures than would be necessary for presenting external environments.

Figure 8-3 presented the clipping problem as two-dimensional (i.e., finding where a partially visible line of a structure intersects with the lines that form a viewing window). The problem is really three-dimensional, however, for it is necessary to find where the lines that make up the structure intersect with six planes. The planes form the right, left, top, bottom, near, and far surfaces of a viewing volume.

The clipping process is usually used only to determine how much of a structure to display, but it may also be used for special effects, for example, to reveal parts of the inside of a structure. This is illustrated in Section 8.4.5.

8.2.3. Projection onto the Two-Dimensional Screen

The calculations described above are all performed in three dimensions, resulting in a modified three-dimensional structure. The computer must now project the three-dimensional structure onto the face of its two-dimensional screen. Two types of projection, "one-point perspective" and "orthographic" (Fig. 8-5), are commonly used.

Although you may not be familiar with its name, you are used to the concept of perspective projection. For instance, when you look down a railroad track, the two rails seems to come closer together with distance. This is a depth cue, for you expect things that are smaller to be further away. The computer graphics technique for perspective

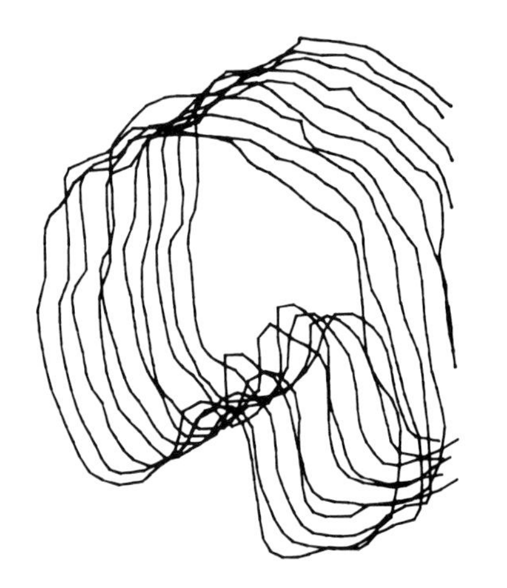

FIGURE 8-4. A comparison between simple line rejection and algebraic clipping. In both figures, a serial section reconstruction is clipped at its right edge. Left: Simple line rejection deletes entire line segments, resulting in a ragged margin. Right: Algebraic clipping computes the right border accurately and draws a piece of line segment exactly to it, resulting in a straight margin. The extra computation provides only modest improvement for structures composed of many short lines and plotted in one frame.

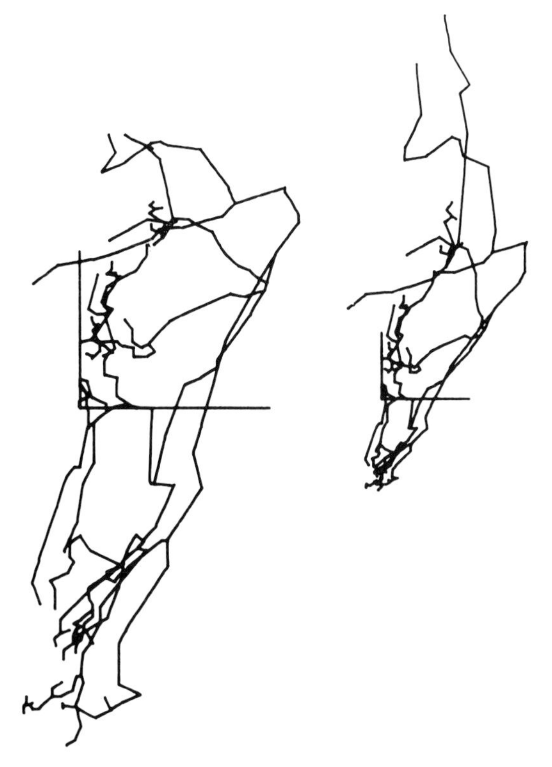

FIGURE 8-5. A comparison of the orthographic and perspective views of a neuron. An axon arborization that extends 200 μm in X, 800 μm in Y, and 80 μm in Z was entered into the computer. It has been tipped toward the viewer (rotated about the X or horizontal axis) 60° so that the bottom of the structure is further away from the viewer than the top of the structure. Left: An orthographic projection of the axon. Its presentation appears natural. Right: A perspective projection. Branches further away in Z now appear smaller, and the neuron seems distorted. Modified from Capowski (1976) with permission.

projection is called perspective division. In a simplified form, each projected X coordinate in two dimensions is calculated by dividing X by Z from its three-dimensional coordinates. Similarly, the projected Y coordinate is calculated by dividing Y by Z. Thus, as a coordinate is deeper into the picture (larger Z), it will be displayed nearer the center of the screen (smaller X/Z and smaller Y/Z).

Perspective division is expensive. Each division must be performed after rotation and after clipping, so that two divisions must be performed on each coordinate in the visible structure during each $\frac{1}{30}$ sec. This, like matrix multiplication, is beyond the power of most laboratory computers, so a "perspective divider" hardware capability may be included in a vector graphics display system.

Luckily, however, the perspective projection is not so important for looking at structures. If you are examining the pineapple in your hands or the neuron on the CRT screen, the perspective size reduction of the rear part of the structure is not much different from the size reduction of the front part of the structure, so you do not depend on it for your realization of the structure's depth. Contrast this to looking at environments. When driving a car down the highway, you receive crucial depth information from the perspective narrowing. So the design of a computer system for displaying environments (for example, in the field of flight simulation) demands a perspective division. The design of a computer graphics system for displaying structures as small as neurons may forgo the perspective projection.

A much simpler projection, orthographic projection, may be used for displaying structures. To generate an orthographic projection, the depth value is simply thrown away. So after the structure is rotated and possibly clipped, it is simply drawn in two dimensions on the CRT. When you see an orthographically projected structure rotating smoothly, you still see a striking illusion of depth, but it not possible to determine what is the front and what is the back of the structure. Restated, if the structure is rotating about the *Y* axis, you cannot tell if the structure is rotating clockwise or counterclockwise. Your visual system synchronizes to the rotation and assumes one direction.

8.2.4. Zooming and Magnification

You may want to look at a portion of a structure with high magnification, thereby ignoring the rest of structure. Such a change in magnification of a display has the computer graphics term "zooming." If there is a perspective divider in the graphic system, zooming may be done dynamically, creating the illusion of walking through a room, watching the structures in front of you grow larger and larger until they pass by you to your side. This zooming is done simply by changing the matrix-multiplier elements that affect the calculation of the displayed *Z* coordinates.

In a graphics system without perspective division, zooming cannot be done dynamically; magnification may be changed, though, by algebraically scaling the picture larger or smaller. With or without perspective division, clipping must be performed to eliminate the part of the picture that is scaled off screen.

You must specify four parameters to build a display of a structure at a given magnification. You must specify the size of the display, i.e., the viewing window in the units of the structure (e.g., micrometers). You must also specify the *X,Y,Z* coordinates of the part of the structure that is to be centered in the display. One option is for the computer software to calculate these four parameters automatically before it builds the display. In this case the computer finds the extremes of the structure and scales the display to fill the screen. An alternative is for you to specify these display parameters, giving you complete control of the amount and location of the structure that you wish to see.

In a vector display system, magnification may be changed to any degree by algebraically scaling the coordinates of the line endpoints. In a pixel-based or raster system, magnification is usually changed by multiples of two. If a picture is reduced by half, every other pixel along each axis is discarded, and the remaining pixels are condensed, yielding a half-magnification view of a larger area of the specimen. When a picture's magnification is doubled, each pixel is replicated into four pixels, magnifying a small region but providing no greater resolution, so the picture looks coarser. The coarseness may be eased with a time-consuming calculation. Here, instead of copying the gray value of each original pixel into three new pixels, the gray values of the three new pixels may be interpolated from the gray values of the adjacent pixels.

Sometimes you may want to distort the shape of the structure artificially by magnifying it to different degrees along the three axes. Most frequently, you will want to stretch or shrink the structure along its *Z* axis or, in other words, change its length-to-width magnification. This is often done during the display of serial section electron micrographs, where frequently the structure is extremely shallow in comparison to its width. Imagine stretching the structure along its *Z* axis, as you might elongate the shape of a tunafish can into that of a hot dog. This may render details of the structure more readily understandable.

8.2.5. Variation of Intensity

Another depth cue that you commonly use is intensity variation; i.e., things further away appear to be dimmer. If you build a CRT display with lines of varying brightness, then, in the absence of other depth cues, you will perceive dimmer lines to be deeper into the picture. This display technique is called ''intensity depth cuing.'' Like perspective division, it is a weak depth cue for structures whose depth is not extreme. But, since some computer graphics systems do provide this function, it is used in the display of anatomic structures.

8.2.6. Hidden-Line and Hidden-Surface Removal

An opaque surface of a structure blocks your view of the parts of the structure behind it. Therefore, you may eliminate ambiguity in depth if you delete parts of a display that are hidden by parts in front of them. Hidden-line removal techniques are usually applied to vector-based computer graphics displays. With these displays, the structure is made up of lines that are either drawn or are eliminated from the display depending on whether their view is blocked by any part of the structure. Figure 8-6 illustrates how hidden-line removal provides unambiguous depth information.

Hidden-line removal may also be used to provide an artificially enriched view of a structure. Figure 8-7 shows the use of hidden-line removal to provide an unambiguous view of an inner structure inside a view of an outer one.

Hidden-surface removal is usually applied to polygon-based structures displayed on raster graphics systems. The structure is composed of polygon vertices that define the perimeter of facets. Hidden-surface removal algorithms load pixels contained in a facet

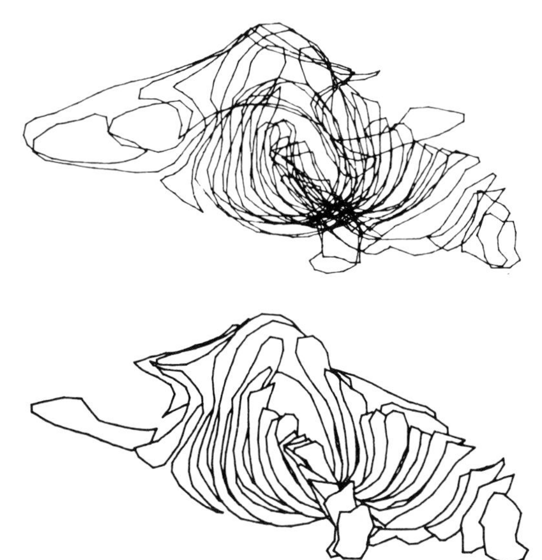

FIGURE 8-6. Wire-frame view of a serial section reconstruction of the larynx of a bat. Top: With all the lines shown, it is impossible to determine whether the front is at the left or the right. Bottom: With the hidden lines removed, the ambiguity is removed. Modified from Capowski and Johnson (1985) with permission

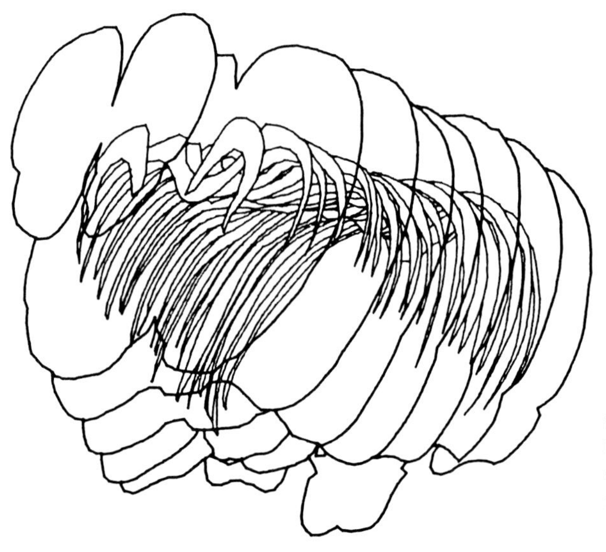

FIGURE 8-7. A view of a serial section reconstruction of the rat brain. A hidden-line-removed view of the corpus callosum is seen inside a hidden-line-removed view of the cortex. Further details are given in the text.

that are not hidden by other facets into the frame buffer. Section 8.4 and subsequent sections contain numerous hidden-surface-removed displays.

Hidden-line and hidden-surface removal algorithms are very time consuming and require significant resources to generate images. Line or surface removal must be done after all the above-described transformations are performed, so generating a dynamic hidden-line-removed image is beyond the means of laboratory computers unless they are equipped with special computer graphics hardware.

It is worthwhile to describe what is involved in the hidden-line-removal process, so that you may appreciate the difficulties. Two procedures, called the ''check-everything'' procedure and the ''mask'' procedure, are described in the next sections.

8.2.6a. ''Check-Everything'' Hidden-Line Removal. The check-everything method of hidden-line removal is a simple but slow method for removing hidden lines from a wire-frame structure consisting of planar polygons that form a series of stacked sections. There cannot be any lines going from section to section for this algorithm to function.

Here is a simplified description of the check-everything algorithm. The polygons are sorted into ascending order of their Z coordinates from nearest to farthest away. The front polygon, the one nearest the viewer, is drawn on the screen. Then each point in the second polygon is tested to see if it is surrounded by the first polygon. (The ''surrounder'' test is described in the next paragraph.) If the point is not enclosed by the first polygon, it is drawn on the screen; if it is enclosed, it is eliminated. At this time, all of the first polygon and possibly some of the second polygon are now on the screen. Then, each point in the third polygon is considered. Each point is drawn if it is not enclosed by either the first or second polygon. If either the first or second polygon encloses the point, the point is eliminated. Each point of each subsequent polygon is tested to see if it is enclosed by any previous polygon. Whenever a previous polygon encloses the point, the point is not added to the display.

The heart of the check-everything algorithm is called the ''surrounder test.'' This test determines if a point is enclosed by a polygon. The idea of the test is to extend a line upward (in the $+Y$ direction) from the point and count the number of intersections that the line makes with the polygon. If the number of intersections is odd, then the point is enclosed by the polygon. If the number of intersections is even, then the point is not inside the polygon.

The check-everything algorithm works well; Figs. 8-6 and 8-7 were generated using it. It is slow, however, because each point must be checked against all previous polygons. Figure 8-7 required about 3 min of calculation on an IBM AT computer.

8.2.6b. ''Mask'' Hidden-Line Removal. An alternative and more efficient method of hidden-line removal is sometimes called the ''mask'' method (Fig. 8-8). As with the check-everything method, the polygons are first sorted into front-to-back order, and the first polygon is drawn on the screen. The first polygon also forms a mask, the current portion of the display that blocks everything behind it. Each point of the second polygon is now considered. If it is surrounded by the mask, it is eliminated; if it is not hidden by the mask, it is added to the display. Then the mask is updated to include the outer boundaries of the first and second polygons. Each point in the third polygon is now checked to see if the mask surrounds it. If not, it is drawn; if so, it is not. Then again the mask is updated. The mask continually grows to be the union of all the polygons.

Mask methods of hidden-line removal operate easily and reliably on raster displays, for updating the mask means simply turning on pixels in the display. When operating on vector displays, however, mask methods also operate quickly but have occasional failing cases because the mask-updating process is subject to roundoff errors. When these errors occur, lines that should be hidden are drawn.

8.2.7. Stereo Pairs

The final depth cue to consider is called stereo pairs. Humans, with their two eyes several inches apart, have binocular vision. Thus, when you look at an object a few feet from your eyes, each eye sees the object from a slightly different angle. Your visual

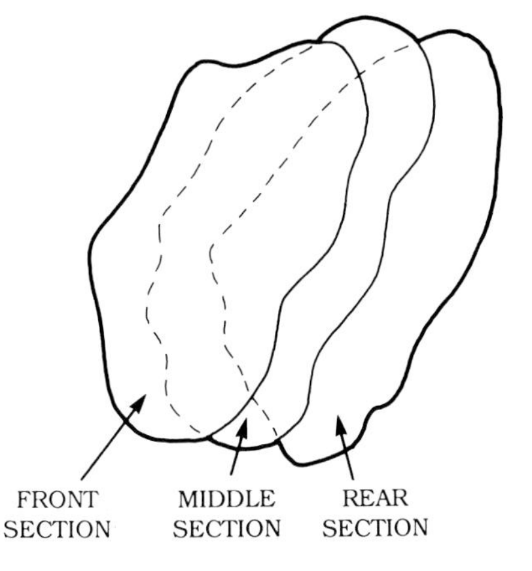

FIGURE 8-8. The mask method of hidden-line removal. The front section is drawn on the display. A mask is then defined to be the outline of the front section. Then each point in subsequent sections is checked to see if it is hidden by the mask, and if it is not, it is drawn on the display. Finally, the mask is updated to include any newly drawn points from the just-drawn section. The figures show the mask in boldface after three sections have been drawn.

system merges these two views into a three-dimensional representation of the object in your brain.

To take advantage of this human visual processing, you generate stereo pairs of a structure on a graphics screen. You generate two images, one with a *Y*-axis rotation about 6° different from the other. Then using some optical system, you direct only one image toward your left eye and the other image toward your right eye. You binocular visual system will create the three-dimensional illusion. The result can be very striking. A 6° rotation between the two displayed images is typically chosen because it is the angle between the two viewing paths of your eyes when you look at an object about 2 feet away, assuming that you are of average size. However, by varying the rotation of one view with respect to the other, you can artificially enhance the depth illusion. Figure 8-9 shows a stereo pair of a neuron.

Some of you will be able to see the stereo effect from Fig. 8-9 by simply relaxing your eye muscles, but I have never been able to do this. If you are like me, you must resort to an optical aid to direct only one image to each of your eyes. If the stereo pairs are produced side by side, a simple binocular viewer will allow you to focus on the two images simultaneously. If the two images are overlaid, then some filter system using colored or polarized lenses must be used to direct one image to each eye. An alternative is to time-share the images. If the image for the left eye is presented on the CRT for a few milliseconds as a shutter is opened in front of the left eye and then the right eye's image is presented synchronized with a right eye's shutter, the stereo effect may also be achieved. All of these optical aids have been manufactured during the last 20 years.

8.2.8. Line Generation

In a vector-based graphics system, the structural data are represented as a series of *X,Y,Z* coordinates of line endpoints, with the assumption that straight lines will be drawn from point to point. The transformations described in Sections 8.2.1 through 8.2.7 change the location of the endpoints, perhaps eliminating some of them. After these transformations, straight lines must be generated from point to point. A "line generator" is the special-purpose computer component that connects these lines.

Line generators use several common algorithms to construct ramps, either as a series of digital values or as analog voltages from the *X* coordinate of a line endpoint to the *X*

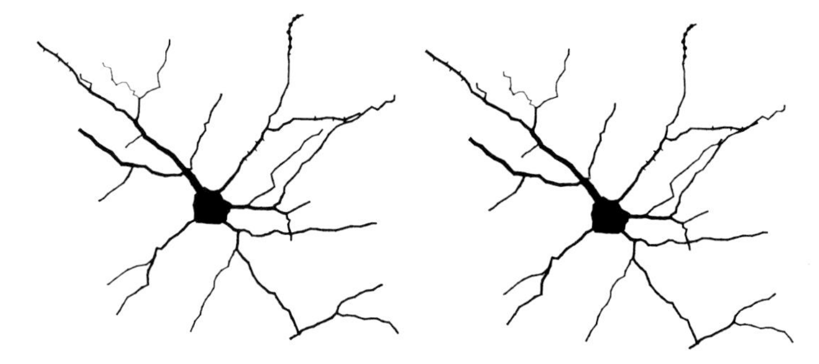

FIGURE 8-9. A plot of stereo pairs of a neuron. The two views are rotated 6° apart about the *Y* axis. Reproduced from Capowski (1988b) with permission.

coordinate of the next line endpoint as well as from one Y coordinate to the next. In a vector graphics system, these ramps are used to deflect the electron beam in the CRT to move the beam from one point to the next. In a raster graphics system, the digital ramp values are used as X,Y addresses of pixels in the frame buffer. Then a gray value that contrasts with the background is placed at each addressed pixel. This results in a series of pixels in the frame buffer that form a straight line. In either the digital or analog case, a line is drawn from one X,Y coordinate to the next, but since line generation must be done on the transformed coordinates, if the image is dynamic, new lines must be generated for all endpoints in $\frac{1}{30}$ sec. Therefore, to generate a dynamic image, the line generator must be special-purpose hardware built into the display processor. For static images, the line generation can be done in software on the laboratory computer.

8.3. VECTOR DISPLAY OF STRUCTURE

In this section, the characteristics of a vector graphics display used to present neuroanatomical images are discussed, building a contrast to the use of the raster display, which is discussed in Section 8.4.

You may recall from Section 8.1 that a vector-based image is stored as a series of X,Y,Z coordinates, with the assumption that a straight line will be generated between successive coordinates. Thus, the storage of the structure is moderately compact; a 1000-coordinate image might require 8 bytes per coordinate or 8000 bytes. Contrast this to a raster image of spatial resolution $512 \times 512 \times 8$ bits, or a total of 256,000 bytes. This great difference in the amount of data required to store an image in the two types of graphics systems forms the basis for many of the differing characteristics between the vector and raster systems.

8.3.1. Time Constraint to Avoid Flicker

When an electron beam strikes a point on the phospor-coated front surface of a CRT, it generates a display at that point that lasts for the persistence of the phosphor, typically a few milliseconds. Thus, when the electron beam is directed by the computer graphics hardware to draw an image, the image persists for only a short period. To fool you into thinking that you are seeing a continuous image, the display must be drawn repetitively or must be "refreshed." The number of times per second that the display must be drawn varies with the type of phosphor, display brightness, ambient light, and the characteristics of your visual system, but it is typically 35 times per second. If the display is not refreshed frequently enough, it will appear to blink off and on or to "flicker." A certain amount of flicker seems not to be too disturbing, so a goal for vector-refreshed graphics systems is to refresh the image 30 times per second. Thus, the maximum time that the hardware is allowed for drawing an image is $\frac{1}{30}$ of a second, or 33 msec.

This is a severe time constraint. It means that all the graphics processing that was described previously (matrix multiplication, clipping, projection, line generation) must be done in 33 msec. If there are 1000 points in a display, then all the processing on each point must be done in 33 μsec. This computing requirement far exceeds the capability of a small laboratory computer, so special-purpose computer graphics systems must be built to generate vector graphics displays.

If, for example, the structures to be displayed have 5000 points and the hardware has a speed capable of presenting 2000 points without flicker, how then can you generate a flicker-free display of the structures? Somehow, you must limit the data, perhaps by building a display of only certain structures or by building a higher-magnification view of a portion of all the structures. Another alternative is to filter the data base spatially as it is displayed. If you build a low-magnification view of a structure, some of the very fine detail will be lost. Therefore, you may eliminate very closely spaced points without degrading the display. My experience is that rejecting points that are closer together than $\frac{1}{200}$th of the size of the display will reduce the number of points displayed considerably and yet leave an image of about the same quality as the originally entered structure.

8.3.2. Only Wire-Frame Stick Figures without Thickness

With a vector system, it is possible to drawn only wire-frame figures as shown in Fig. 8-10. It is not possible to fill in regions between lines. This is probably the most severe limitation of a vector graphics system. The wire-frame display shows you the structure but not the texture of the structure's surfaces.

8.3.3. Smooth Rotation

The wire-frame model may, however, be smoothly rotated by updating elements of the rotation matrix between the drawing of successive frames of the image. As described above, this smooth rotation yields the best easy-to-calculate depth cue so that you may appreciate the three-dimensional structure. Unfortunately, it is not possible to show a smoothly rotating structure in this book.

8.3.4. Spatial Resolution and Aliasing

A vector graphics system usually enjoys a higher spatial resolution than its raster counterpart; most frequently coordinate data are stored to 12-bit accuracy, or to one part in

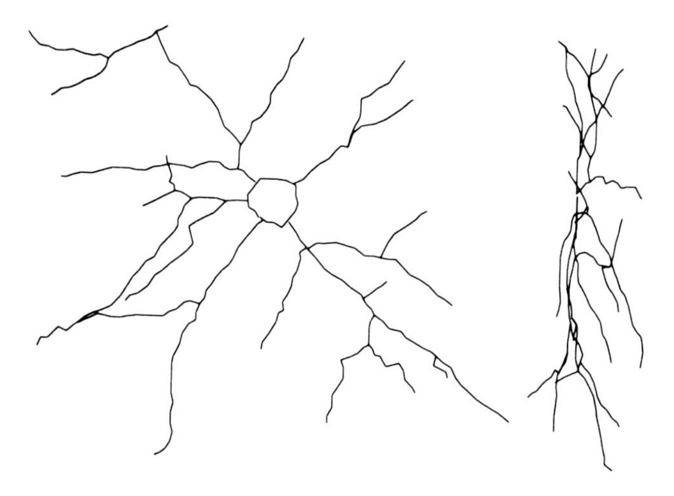

FIGURE 8-10. A vector CRT display of a traced neuron. Left: Transverse view. Right: Sagittal view. The three-dimensional structure of the cell can be seen.

4096. The graphics transformations described above are done to either 12- or 16-bit accuracy, and the line generation is done to 12-bit accuracy. The resolution is usually limited by the analog electronics in the CRT's deflection amplifiers, and depending on quality, the deflection amplifiers may position the electron beam to a resolution of one part in 1500.

You may recall from Chapter 2 that an aliasing error occurs, creating a staircase effect, when a line is digitized into pixels and drawn on a raster screen. Because this problem does not occur with vector graphics systems and because their spatial resolution is higher, the line quality of vector systems is superior to that of raster systems. Figure 8-11 compares the vector and raster graphics systems in this regard.

The superior line quality of a vector system has another benefit; details of small structures may easily be seen. You may thus display a number of small neurons in their proper relative environment and still see the structural details of the neurons. This is illustrated in Fig. 8-12.

8.3.5. Highlighting Vector Displays by Varying Brightness

You may want to indicate that a part of a displayed structure is somehow "special" with respect to the rest of the structure. In a monochrome vector display, highlighting of the special portion is easily done by increasing the brightness of the lines that comprise this special portion. You may vary the brightness of the lines in a vector graphics display in three ways. If the hardware permits, the actual intensity of the electron beam may be modulated, and thus more photons of light will be emitted where the beam strikes the phosphor. As an alternative, if the hardware permits, the speed of line generation may be slowed, so that the electron beam, still of the same intensity, spends more time at each location along its line. Or finally, some lines may be drawn more times than others, and the lines that are redrawn will appear brighter. This third technique is used (Fig. 8-13) to show a dendrite that has been highlighted to indicate that it is somehow different from the

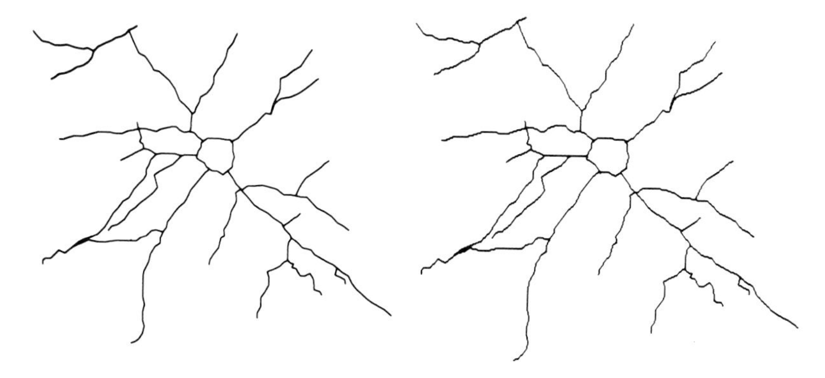

FIGURE 8-11. A comparison of vector and raster displays of a stick figure representation of a neuron. Left: Presented on a vector CRT where the limiting factor is the ability to deflect the electron beam to approximately one part in 1500. Right: Presented on a raster CRT with spatial resolution of approximately 500 by 500 pixels.

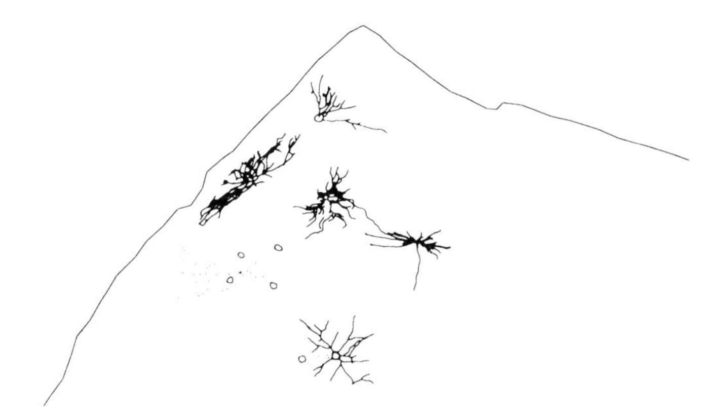

FIGURE 8-12. A vector CRT view of several neurons located in their environment. Small circles outline the somas of other cells, and dots represent still others. The tissue cortical outline is also presented.

rest of the neuron. Figure 8-14 illustrates the highlighting of certain points along a dendrite. The highlighted parts of the structure may be smoothly rotated with the rest of it.

8.3.6. Color Vector Graphics Systems

Two techniques are normally used to generate color vector graphics displays: RGB phosphors and beam penetration. The RGB (for red–green–blue) phosphor technique is the same technique that is used in normal color television. With this technique, three electron beams are generated, each aimed at phosphors that emit red, green, or blue light. Any color may be generated by varying the amount of these elemental colors. With the RGB technique, either spatial resolution is reduced or cost is increased. As of 1988, vector color monitors with a resolution of 1 part in 500 are available for about $1500; to achieve the high resolution of Fig. 8-13 but in color, a vector monitor costing about $10,000 is required.

An alternative method for creating color on a vector graphics screen is to use beam-

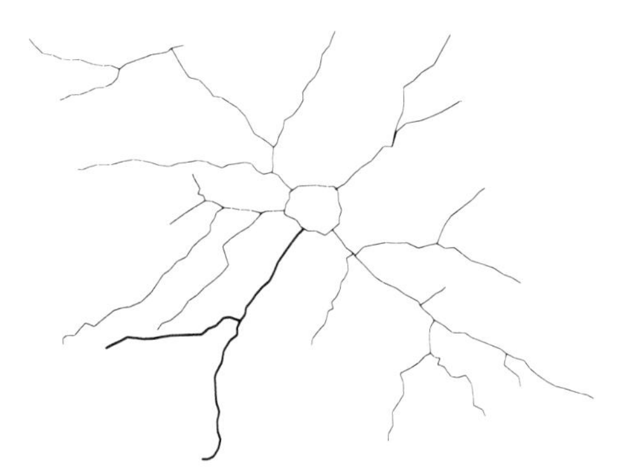

FIGURE 8-13. A vector CRT display of neuron with a highlighted dendrite.

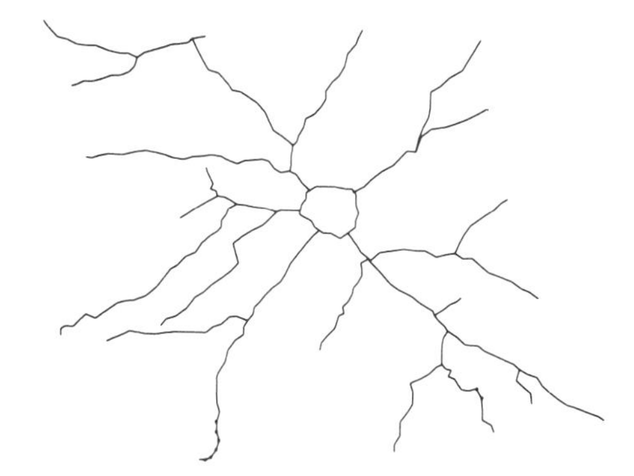

FIGURE 8-14. A vector CRT display of a neuron with dendritic spines near the bottom of the neuron highlighted with dots.

penetration phosphors. In such a system, several coats of phosphor are layered on the front surface of the CRT. Each layer of phosphor emits a different color when it is excited by the electron beam. When the intensity of the electron beam is modulated, the beam penetrates into the different layers of the phosphor, thereby presenting different colors of a display. Little spatial resolution is lost, so that a fine detailed color display is possible and the cost remains moderately low. A limitation of this technique is that the chemical compounds in the phosphor layer can generate only a few colors. Although their intensity is constant, it is not possible to combine them in any manner to produce the full rainbow of colors.

8.4. RASTER DISPLAY OF STRUCTURES

In a raster display system, the electron beam covers the entire image, illuminating every pixel to some degree. Thus, you may generate images in which every point on the screen has a gray level or color. This scheme provides superior capability for building sophisticated and realistic images. To illustrate, Fig. 8-15 shows a well-known computer-generated image that is realistic enough to be mistaken for a photograph of a panorama. Tradeoffs in memory, time, and cost in producing such realistic images are discussed in Section 8.4.7.

To help you understand how raster displays of anatomic structures are built, you should consider that a raster graphics system has, crudely speaking, two computing components: a "display processor" and a "refresher," both interacting with one frame buffer. The display processor fetches the picture data from the computer, possibly transforms the data in the manners described in Section 8.2, and deposits the transformed data into the frame buffer. Depending on the sophistication of the display processor and the complexity of the image, it may take from a tenth of a second to many hours to fill the buffer. Meanwhile, the "refresher" is fetching the picture data from the frame buffer, converting it into a video signal, and drawing the data onto the screen. This refreshing of the screen is done at a nominal rate of 30 times per second, fast enough to prevent flicker. Thus, in a raster system, even though the time required to update the image may be long, flicker will not be a problem.

An analogy can be made here to painting a house. A paint provider (the display

FIGURE 8-15. The "Road to Pt. Reyes," a famous computer-generated picture, to show the potential of raster graphics systems for generating lifelike images. Reproduced from SIGGRAPH (1983) with permission.

processor) fills a paint bucket (the frame buffer) with certain color paints, possibly selecting the colors by some sophisticated criteria. Meanwhile the painter (the refresher), in a machinelike fashion, dips his brush into the bucket and spreads the paint onto the wall of the house (the CRT screen). The painter works at a constant speed, independent of the speed of the provider and independent of the colors he provides. Similarly, the computer graphics refresh logic paints the CRT screen with the image at a fixed rate, independent of the rate which the computer may update or change the image and independent of the image's complexity.

8.4.1. The Aliasing Problem in Raster Displays

As we discussed in Chapter 2 and in Section 8.3.4, aliasing results in a staircase effect when drawing lines on a raster graphics system. Figure 8-16 illustrates the problem with a complex image of a serial section reconstruction.

Antialiasing techniques are available to minimize this problem somewhat, though at a significant computing cost. In the normal case with no aliasing correction, when a nearly horizontal line of a certain gray level is drawn across a screen, only one pixel is set to the line's gray level for each position along the X axis. Because the pixels occur in horizontal rows, the line's pixels form a little staircase. In reality, because the line is not exactly horizontal, for each position along the X axis, the line actually covers a percentage of the area of two vertically adjacent pixels. A simple antialiasing program would set each of the vertical pair of pixels at each X position according to the percentage of the line that intersects it. This results in a somewhat better representation of the line. More sophisticated algorithms are available, and better results are available if more computing effort is applied. Unfortunately, antialiasing techniques must be applied after all the transformations (e.g., rotation, projection) have been done, so that antialiasing techniques cannot be applied to dynamic images using only a general-purpose computer. Special-purpose antialiasing hardware is often included in the more expensive dynamic raster display systems.

8.4.2. Spatial Resolution

The spatial resolution of a raster display, typically $512 \times 512 \times 8$ bits, makes it difficult to display the very fine details typically seen in an anatomic structure. The cost of a raster display system of this resolution, excluding special-purpose hardware to perform transformations, may be quite low, because it can take advantage of the mass consumer market of United States television; 1024×1024 display systems are now available, but they are quite expensive because memory is quadrupled, electronics required to position the electron beam is difficult to build, and there is no mass consumer marketplace.

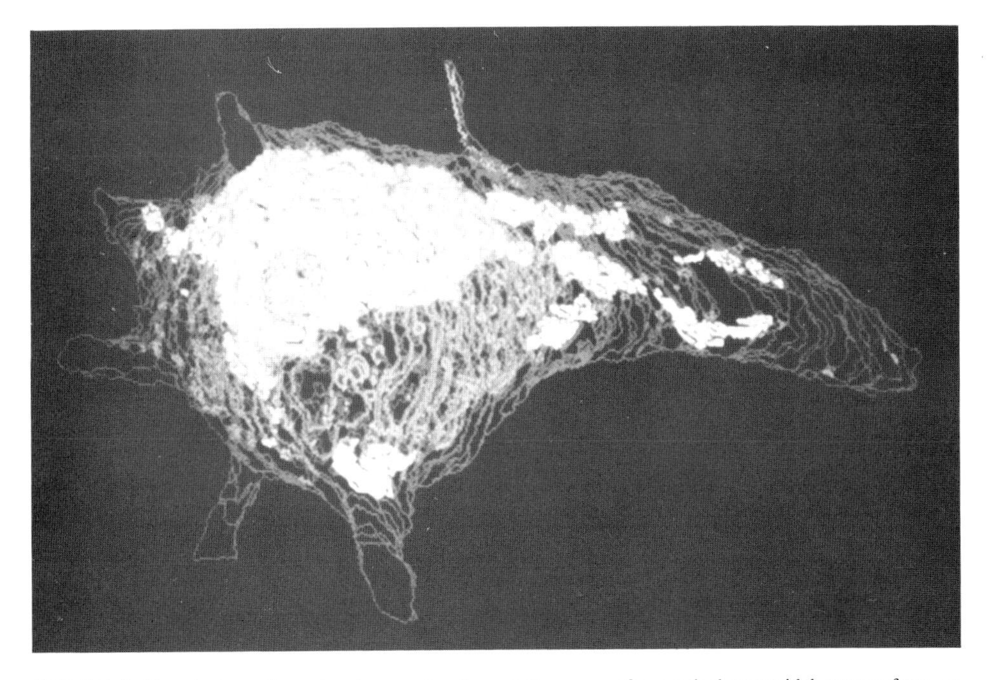

FIGURE 8-16. A three-dimensional reconstruction of the soma of a cortical pyramidal neuron from an Alzheimer's patient. Several internal structures are shown. The reconstruction is composed of about 3000 contours entered on a data tablet. It was rendered using software described by Young et al. (1987) in approximately 15 min using an IBM AT clone computer. This black-and-white reproduction unfortunately does not show how the use of color makes it easy to distinguish among the structures. Reproduced from Ellisman et al. (1987) with permission.

8.4.3. Surface-Filled Raster Displays

The primary advantage of raster displays is the ability to illuminate every pixel to generate a realistic image. Figure 8-17 shows such an image. In the sections that follow, techniques for generating raster display images are reviewed.

8.4.4. Current Sophisticated Image-Generation Techniques

The display of a reconstruction should look as close as possible to the original structure before the tissue was sectioned, or perhaps be enriched to emphasize certain parts of the original.

As you may recall from Chapter 5, serial sections are entered into the computer as a group of polygons, with each polygon comprised of a series of X,Y,Z coordinates that outline a planar region. Because the original tissue was sectioned into parallel slices and each slice was entered individually, there are no lines linking one section to the next. The three-dimensional surface of the structure, however, does span multiple sections, and some way must be found to connect the sections. The process is called "tiling," "triangulating," or "tesseling." To perform this procedure, a computer program generates a series of short lines from each polygon of a certain structure in one section to the

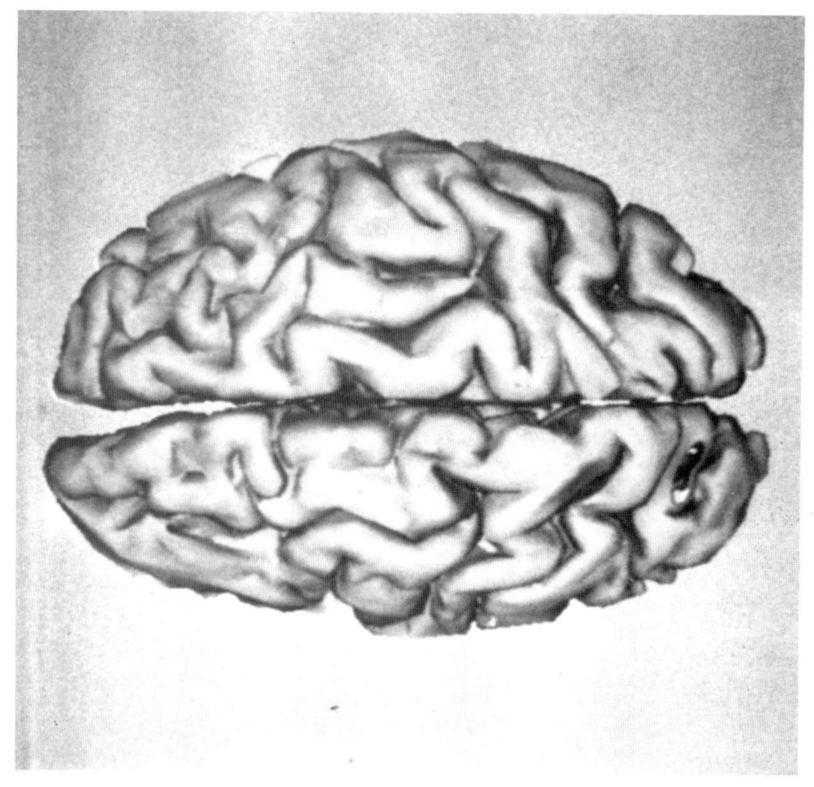

FIGURE 8-17. A 30,000-polygon reconstruction of the human cortex. This shows the realistic result of hidden-surface removal and shading. Reproduced from Shantz (1980) with permission.

corresponding polygon of the same structure in the next section. An illustration of this process is given in Fig. 8-18.

The result of tiling is a number of planar triangles that form the borders of facets. The collective group of triangles form a wire-frame representation of the irregular surface of the structure. Once all the tiles are generated, the image may be rotated to a desirable viewing position, and the hidden lines in the wire-frame display should be removed. The result of these procedures is shown in the bottom of Fig. 8-18.

Next, the facets should be filled. This is done by assigning each pixel inside a facet a gray or color value, depending on several parameters. The facet-filling algorithm takes into account the position of the light source illuminating the facet, the position of the

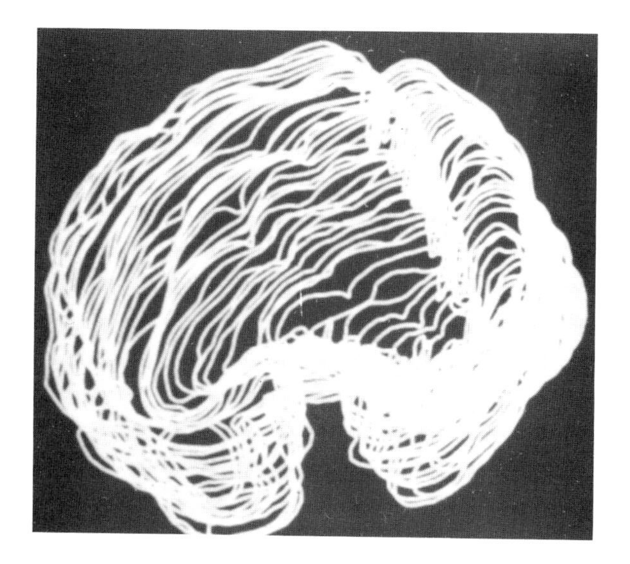

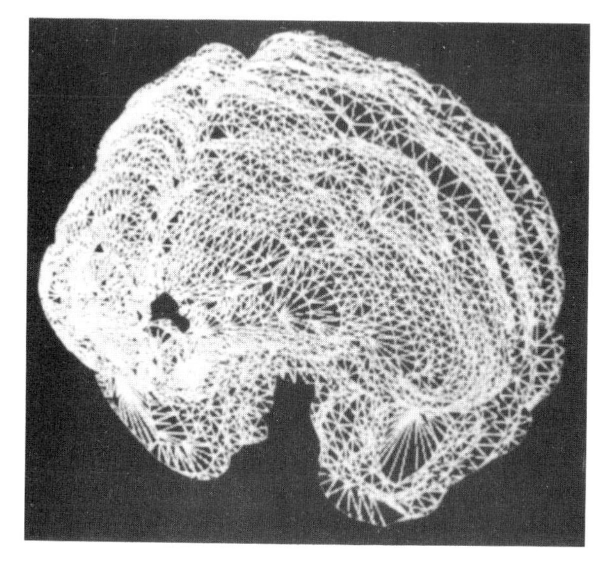

FIGURE 8-18. The effect of tiling. Top: A series of stacked contours of the cortical surface of a monkey brain. All lines are shown. Bottom: Small tiles are built by drawing triangles between the contours, using an algorithm described in Toga and Arnicar-Sulze (1987) and Toga and Arnicar (1985). Figure courtesy of A. W. Toga, UCLA.

viewer, whether there are any shadows cast on the facet by other parts of the structure, and the physical characteristics of the facet surface. The sophistication level of the filling program may be quite high. The top of Fig. 8-19 shows an image after its facets have been filled.

The surface-filled image in the top of Fig. 8-19 shows a disadvantage of filled images that must be corrected lest the image remain unrealistic. The borders between the facets are sharp, and, to make matters worse, your human visual systems strives to enhance the borders. A facet-to-facet junction that better mimics the curved surface of the actual structure should be generated. This is done by adjusting the gray levels of the pixels near

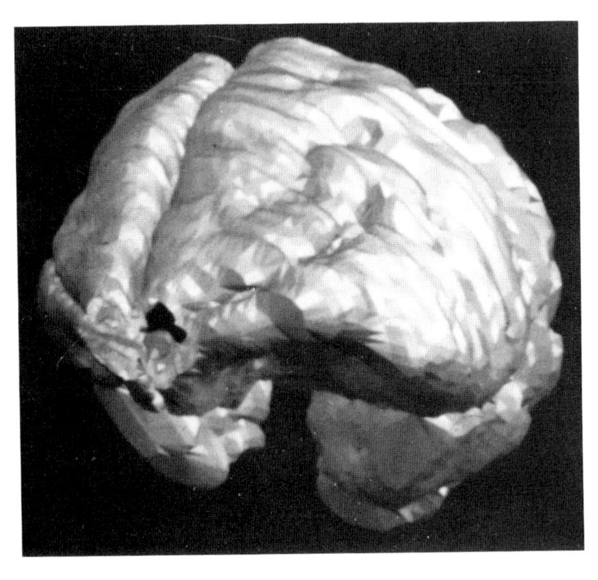

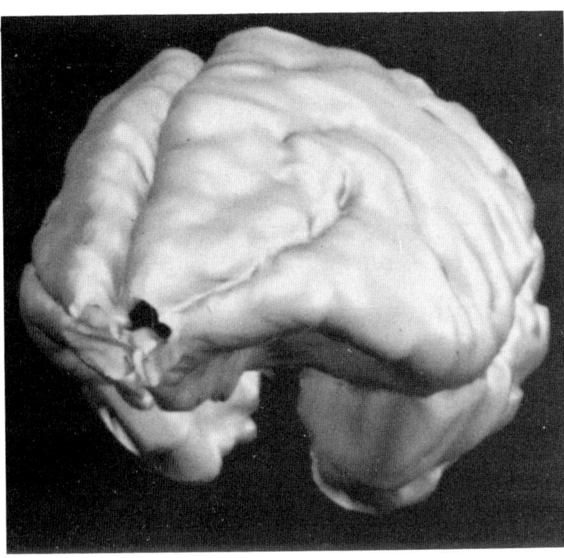

FIGURE 8-19. The effects of surfacing and smoothing. Top: The triangulated brain cortex of Fig. 8-18, but the triangular facets are filled to a gray level as a function of illumination angle and viewing angle. Note that the intersections between facets are still obvious. Bottom: A smoothing algorithm has been applied to the facet boundaries to blend each facet into its neighbors. The result is a realistic-appearing irregular surface. Figure courtesy of A. W. Toga, UCLA.

each border to blend each facet with the ones adjacent to it. The smoothed result is shown on the bottom of Fig. 8-19.

In the above description, the hiding of the rear parts of the structure was performed while the structure was represented in a vector format. That is, hidden lines were removed, not hidden surfaces. There is an alternative, hidden-surface removal, that applies well to raster graphics display hardware. Instead of determining the visibility of each point in the structure as is done in hidden-line removal, the visibility of each pixel is determined. For each pixel, surrounder tests are performed to determine what polygon, if any, of the structure surrounds that pixel. If more than one polygon surrounds the pixel, the one with the lowest Z coordinate (the polygon nearest the viewer) is used to determine the gray level that is assigned to the pixel. That gray value is written into the frame buffer at that pixel location. This is a time-consuming process, however, because of the number of pixels in an image.

8.4.5. Enriched Raster Displays

The raster display techniques described above may be slightly modified to generate better-than-life displays. For example, a transparent surface that surrounds an internal structure may be generated as shown in Figs. 8-1 and 8-20. A normal raster display is generated so that you may determine which structure is external, and you indicate this to the computer software. The software creates the transparent effect by setting the gray values of the pixels of the external structure to be the average of the pixels of the external and internal structures.

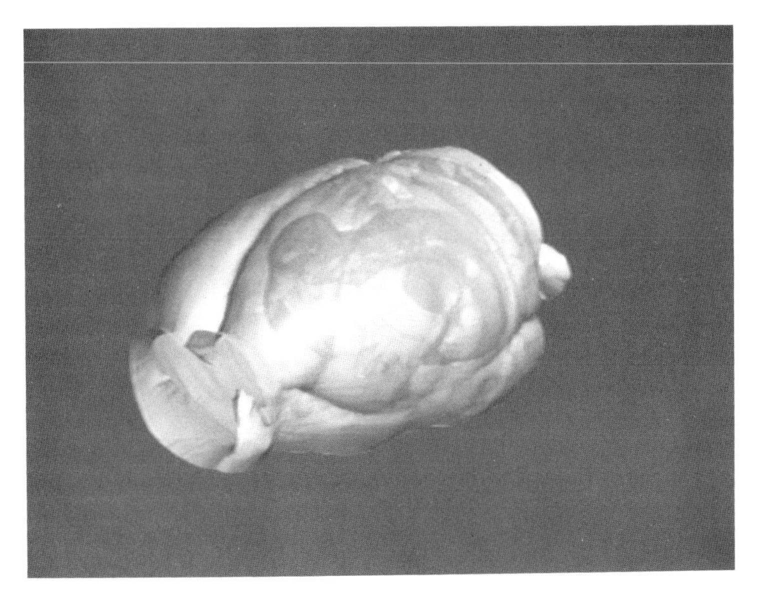

FIGURE 8-20. A view of a rat brain reconstructed from serial sections of an atlas. The hippocampus may be seen through the translucent cortical surface. Produced at the UCLA Laboratory of Neuro Imaging and provided courtesy of A. W. Toga. A color reproduction of this figure appears following p. xx.

A series of planes of pixels form a volume. The elements of the volume are called volume elements or "voxels" (i.e., a voxel is a three-dimensional pixel). If you combine clipping against a near plane as described in Section 8.2.2 with a three-dimensional voxel-based representation of the image, then a portion of the image is sliced away, thereby exposing voxels that would otherwise be hidden. Figure 8-21 shows the result of such an enrichment.

8.4.6. Time and Costs of Raster Displays of Anatomic Data

A raster image of resolution 512 × 512 bytes requires 256,000 bytes of memory. Handling this moderately large amount of information in a laboratory computer can be a time-consuming task. Reading an image from a hard disk and delivering it to the frame buffer with no additional graphics processing takes about 20 sec on an IBM AT. You easily become frustrated waiting for an image at this rate, especially if you need to examine a series of images quickly.

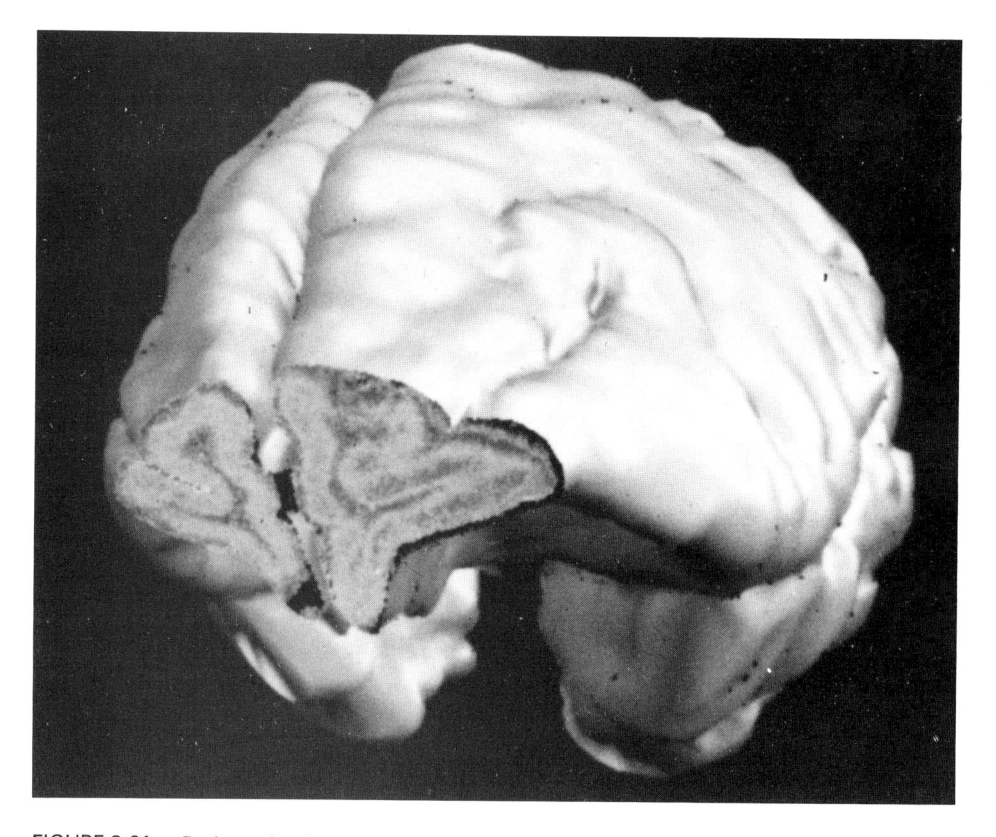

FIGURE 8-21. Cutting a plane into a serial section reconstruction. The cortical surface of Figs. 8-18 and 8-19 has been cut, exposing the densitometric data derived from the two-dimensional radiograms that were scanned when the data were entered. Although no new data result from this display, they are presented in their environment. Figure courtesy of A. W. Toga, UCLA.

Any processing to generate the values of pixels may require much more time. At the extreme, Fig. 8-1, were it generated on a laboratory computer, would take perhaps 2 days.

Smooth rotation of a raster image is not possible without the purchase of sophisticated and dynamic raster display hardware. To rotate a raster image of modest complexity smoothly is a demanding computational task and requires a display system costing from $30,000 to $70,000. A vector system still outperforms a raster system in this case, although it can only generate a wire-frame image.

If you need only a static image, it would be possible to provide a realistic rendering of an anatomic structure on an inexpensive raster graphics system (e.g., an IBM EGA card) without spending great sums of money, though a great deal of time may be required. Many computer graphics researchers are attempting to reduce this time consumption by building hardware and writing software to make rendering more efficient.

Unfortunately, however, to appreciate fully the three-dimensional structure of a complex image, smooth rotation is still needed. This requires either an expensive raster system or a less expensive, but wire-frame only, vector system.

8.5. THREE-DIMENSIONAL PLOTS OF NEUROANATOMICAL STRUCTURE

The majority of this chapter has discussed the generation of displays of anatomic structures on a CRT. You cannot, unfortunately, put these displays in your briefcase or mail them to a journal. So you need, in addition, a paper copy or "hard copy." The rest of this chapter discusses hard copies made on the two most common devices, a felt-tip pen plotter and a printer.

8.5.1. Felt-Tip Pen Plotter

A felt-tip pen plotter, as shown in Fig. 2-14, is most often used to draw the computer-stored anatomic structures on paper. Plotters are electromechanical devices and therefore function more slowly than computers, so typically, you would look at many displays on the CRT until you find one that you like before you plotting it on paper. Figure 8-22 shows a relatively simple plot of a neuron.

Sophisticated plotting programs can be written to draw more realistic plots of neu-

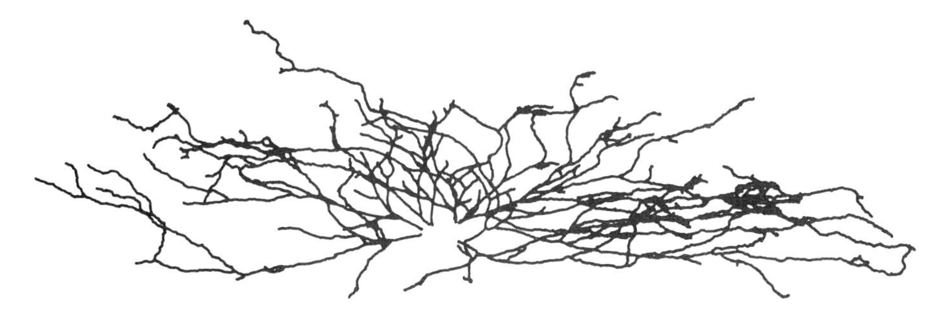

FIGURE 8-22. Plot of the sagittal view of a spinocervical tract neuron from cat spinal cord. This is a "stick figure" showing only the dendrites and not indicating their thickness.

ronal structure. Thickness of the dendrites may be shown, somas may be filled in, and special features (e.g., swellings, spines) may also be drawn. Figure 8-23 shows a more realistic plot of a neuron.

Plots using different colors, for example, to differentiate among cells are easy to make, because you may change the color of the plot simply by changing the pen. To facilitate this task, some plotters provide a carousel containing six or eight pens, and the computer software may select a new pen at any time, even during the middle of a plot. Therefore, you may control the color selection to a fine level of detail, for example, to plot different types of dendritic spines in distinguishing colors.

A laboratory plotter that costs about $1000 has a spatial resolution of about 1 part in 400. The spatial resolution may be increased almost infinitely by plotting many frames of a structure and then taping the frames together. Then the resulting montage may be photographically reduced. Although this process is time consuming, it results in finely detailed plots of anatomic structures. Such high spatial resolution allows the plotting of many structures in their proper environment, still with their details clearly shown. Figure 8-24 shows such a plot.

Finally, a plotter forms a general-purpose tool for the neuroanatomical laboratory. A text caption may be plotted on any structure. Graphs of statistical summaries as described in Chapters 9, 10, and 11 may be plotted for storage or publication.

8.5.2. Plots on a Laser Printer

The laser printer (Section 2.4.10b) may also generate plots of anatomic structures when appropriate software is provided. The printer presents its image as individual dots, spaced at several hundred per inch, which are either black or white. Either a vector or a raster display may be generated by a laser printer.

When used as a vector device, the laser printer functions as a felt-tip pen plotter, drawing lines from point to point. Line-generation software within the printer determines which dots to set to black in order to form individual lines.

When the laser printer is used as a raster hard-copy device, the individual dots are turned on or off as pixels in a frame buffer with high spatial resolution, but each pixel has only two possible values. Strictly speaking, therefore, it is not possible to generate gray

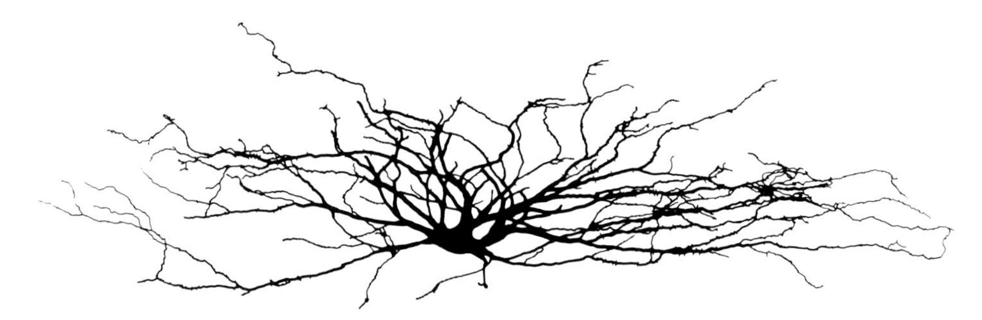

FIGURE 8-23. A plot of the neuron of Fig. 8-22, but the soma and the dendrites are filled, and the spines and swellings are shown. Modified from Capowski et al. (1986) with permission.

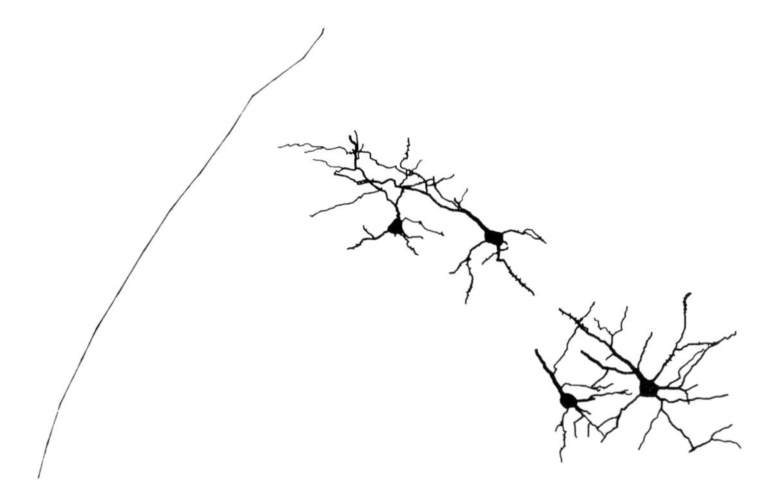

FIGURE 8-24. A plot of several neurons in their proper relative position in the tissue. The cortical surface is also shown.

values. But through the use of a technique called "dithering," a gray-level effect can be achieved. A small matrix of dots (e.g., a 4 × 4 array) is assumed to be a larger pixel, and a variable number of these dots is blackened, resulting in 16 shades of gray in the larger pixel. Because of the superb spatial resolution of the laser printer, the dithering technique may be used to create acceptable gray-level renditions of raster displays.

Laser printers are faster than felt-tip pen plotters, requiring only a few seconds to plot a page once all the calculations needed to form the plot are finished. The calculations demand a wide range of time, depending on the complexity of the structures, the transformations required, the software efficiency, and the speed of the computer. Currently, color is not available on laser printers.

8.6. FOR FURTHER READING

Most often, credit for developing the first interactive computer graphics system is given to Sutherland (1965). Capowski (1976) described the subset of computer graphics that are important for the display of neuroanatomical structure. Pearlstein and Sidman (1986) have presented a more contemporary general description.

The details of a special-purpose graphics display system tailored to neuroscience needs are available in Capowski (1976, 1978a,b, 1983b, 1988b) and McInroy and Capowski (1977).

Many computer graphics laboratories have developed hidden-line and hidden-surface removal algorithms that are applicable to neuroscientific work. A survey that has become popular in the computer graphics community and contrasts many algorithms was presented by Sutherland et al. (1974). Two types of hidden-line-removal algorithms have become popular in serial section reconstruction. The masking-type algorithm is described in detail in Gras (1984) and in Street and Mize (1985). The check-everything type is described in Capowski and Johnson (1985).

Several textbooks describe the standard techniques for generating three-dimensional displays, including those of biological structures. The best known are Newman and Sproull (1979), Foley and Van Dam (1982), Giloi (1978), and Artwick (1984).

Techniques for drawing surfaced three-dimensional images of anatomic structures on a raster graphics system may be found in Shantz and McCann (1978), Radermacher and Frank (1983), and Toga and Arnicar-Sulze (1987).

The varifocal mirror is a device to generate true three-dimensional displays of structures, anatomic or otherwise. Nelson (1986) illustrates its use in anatomy.

Dierker (1976b) presents a description of the standard mathematical manipulations that must be performed to generate a three-dimensional interactive display of an anatomic structure.

9

Mathematical Summarizations of Individual Neuron Structures

9.1. INTRODUCTION

From Chapter 1 you may recall that there are two major reasons for using a computer to model neuroanatomical structures: to present views of the tissue that are superior to those visible in the unassisted microscope and to generate mathematical summaries that describe individual structures and compare one population of them to another population. Figure 9-1 shows a scientist summarizing a botanical tree, making the measurements on the actual growing tree. You are not so lucky, for usually you cannot measure a growing neuron. However, with help from computer programs, you may summarize the data base that represents the neurons.

This chapter and Chapters 10 and 11 cover the techniques normally used in the mathematical summarization of neuronal structure. This chapter specifically covers the mathematical summarization of individual cells. ''Mathematical'' implies that the size of the neurons is as important as their shape. Chapter 10 describes topological summarizations of individual cells, looking more at the shape of the cells and how they grow rather than how large they become. Chapter 11 returns to the mathematical summarizations but considers populations of cells, not individual ones.

Mathematical summarizations of individual neurons may be subdivided into two groups: numeric summaries and graphing summaries. Numeric summaries are descriptions of cells or their parts that are presented as one number or a small group of numbers. These are usually presented in tabular form. Graphing summaries are descriptions that relate one numeric parameter to another. These are most often presented as two-dimensional graphs.

9.2. NUMERIC SUMMARIES OF A CELL

Figure 9-2 presents a numeric summary of the neuron that was shown in Fig. 8-10. In the sections following, each item in the summary is explained.

FIGURE 9-1. Measuring a tree. If only measuring a neuron were this simple.

As described in Chapter 4, when you traced the structure into the computer, you had the capability of "tagging" individual points or portions of a structure to indicate that the point or portion was somehow special. In all the numeric summaries (Section 9.2) and in the graphs (Section 9.3), you may limit the analysis to only the tagged points or portions. You may thus, for example, plot the distribution of synaptically active points along the rostrocaudal axis.

9.2.1. Counting Measurements

The most elementary summarizations are simply counts. In the example of Fig. 9-2, a total of 361 points were used to record a neuron that consists of one soma and five dendrites.

```
                    Numeric summary
Text:  Golgi-stained cell from cat neocortex.  Section 7, cell number 14.

Name:  ******
No. pts summarized (incl somas)       361
No. trees summarzd (incl somas)         6

 MTO TTO BTO    CP   FS   SB   BP   NE   ES  MAE  TAE  BAE SOS  SCP SOE
   5   0   0   275    8   14   16   13    1    1    4    2   1   20   1

Limits in X,Y,Z (mu)                -105.0   102.0 -105.0    70.5   -27.0    16.0
Total fiber length (mu)             1263.9
Membrane surface area (mu**2)       8046.5
Cell volume (mu**3)                 4829.6
Soma cross sectn area (mu**2)        325.4
Center of fiber length (mu)           -8.3   -13.4   -4.8
Center of surface area (mu)           -3.7   -13.5   -5.2
```

FIGURE 9-2. A numeric summary of a single neuron. Details of each of the parameters are covered in the text. Reproduced from Capowski (1987) with permission.

Somewhat more descriptive is the number of each type of point. These are the counts of the point types that were recorded by the computer when the neuron was traced as described in Chapter 4. The neuron's five middle tree origins indicate that all five dendrites emanate from the soma. Two hundred seventy-five continuation points were used to mark points along the dendrites that were of no special anatomic significance. Eight varicosities along the dendrites and 14 dendritic spines were observed and entered into the computer. Sixteen branch points occurred in the cell. Of the total of 21 endings entered, 13 naturally tapered to a point, one ended with a thermometer-bulb-like swelling, one could not be followed for some artificial reason while it was in the middle of the tissue, four were cut by the microtome at the top of the tissue section, and two were cut at the tissue section bottom. One soma was traced, requiring a total of 22 points.

9.2.2. Length-Based Measurements

Dendritic length is probably the most widely used summarization in neuroanatomy. The computer program calculates this by summing the length from point to point in the mathematical model along the dendrites. You may be interested in seeing whether the dendrites extend further when they are deafferented or starved for information. This calculation may test whether they grow in length (possibly compared to thickness, Section 9.2.3) much as a plant grows toward the sun.

Dendritic length is calculated along the dendrite as it meanders throughout the tissue, perhaps in a convoluted path. Another parameter that may be calculated is the dendritic "airline distance," i.e., the distance in a straight line from one point in a dendrite to another. A measure called "bending factor" or "tortuosity" combines this measure with length calculations. This parameter, whose calculation is described in Fig. 9-3, indicates how much the dendrite wiggles or deviates from a straight line. A value of 1.0 indicates a straight dendrite, and a larger value indicates a less direct path.

Dendritic length, airline distance, and bending factor may be calculated for any portion of a dendrite, but they are most often done for individual branch segments, that is, the portions from origin to the first branch point, from branch point to branch point, or from branch points to endings.

9.2.3. Area-Based Measurements

Several measurements of neurons are based on the calculation of a cross-sectional area. The outline of the area is represented in the computer by points along a closed irregular polygon. Figure 9-4 shows the procedure involved in calculating the area contained within the polygon. The cross-sectional area of a cell's soma is commonly calculated and presented as part of a numeric summary (Fig. 9-2). The cross-sectional area

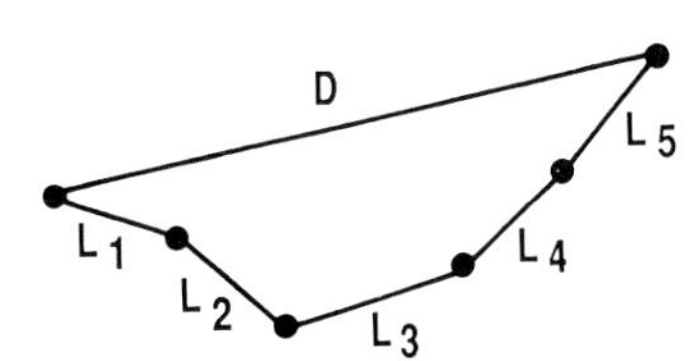

FIGURE 9-3. The calculation of the bending factor or tortuosity of a dendritic segment. A dendritic segment with six points is shown. The length of each element (L_1 through L_5) may be computed and summed to yield the segment's dendritic length. The bending factor is the dendritic length divided by the airline distance (D) between the start and the end of the segment. Thus, a larger number indicates greater bending or "tortuosity."

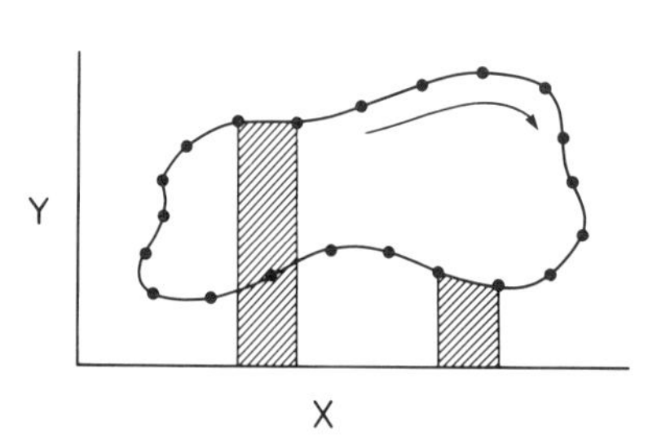

FIGURE 9-4. How to calculate the area of a closed polygon. Elements of area (shaded regions) are formed by defining an arbitrary X axis and extending lines from consecutive points on the polygon down to this X axis. The area of each element is calculated by multiplying the width of the area times its mean height. The area inside the polygon is found by summing the area elements in the clockwise direction as shown by the arrow. If the polygon is traversed in the counterclockwise direction, the calculated area will be negative and should be negated. Reproduced from Capowski (1988b) with permission.

calculation is also frequently applied to other outlines such as tissue borders. In addition, you may an outline a neuron's dendritic field (Section 9.2.5), and the computer can calculate the area of the field.

Three other mathematical parameters are often presented to describe a two-dimensional shape: perimeter, mean diameter, and form factor. Perimeter, the length along the outline, is calculated in the same manner as dendritic length. The mean diameter provides another measure of the size of the outline. It is computed from the cross sectional area using the following formula:

$$d = \sqrt{4A/\pi}$$

Form factor is a measure of roundness of a two-dimensional outline, not its size; Fig. 9-5 presents its calculation. The form factor has the value 1.0 for a perfect circle, decreasing toward zero as the outline becomes less round.

The areas described above are planar. A different kind of area, the surface area of the membrane that surrounds the neuron, may also be calculated. Figure 9-6 shows how to calculate the surface area of each element (i.e., the short piece of dendrite between two coordinates) of the neuron. The surface area of the neuron is the sum of the surface areas of all the elements. This parameter is important in summarizing neuronal structure since the membrane is the part of the neuron that accumulates charge from other neurons.

9.2.4. Volume-Based Measurements

The volume of a cell or parts of it may also be calculated. This refers to the actual volume of the neuron enclosed by its surface membrane, not the volume of tissue into which the cell sends its dendrites. Figure 9-6 shows how to calculate the volume of each element of the neuron, and the total volume is formed by summing the element volumes.

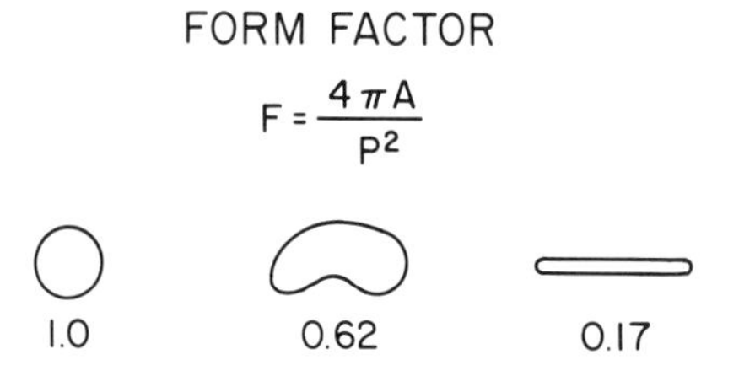

FIGURE 9-5. The form factor (F) is calculated from an outline's area (A) and perimeter (P) according to the equation shown. It is a measure of roundness whose value is 1 for a perfect circle and decreases as the outline becomes less circular. Form factor values are shown for three shapes. Reproduced from Capowski (1988b) with permission.

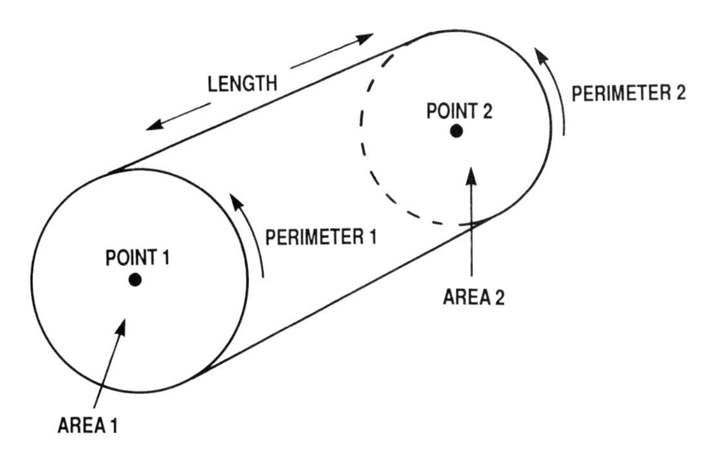

FIGURE 9-6. The calculation of surface area and volume of a dendritic element. Two adjacent points along a dendrite define an element. If the diameter of the dendrite at each point is known, the surface area of the element may be found by multiplying the length of the element times the mean of the perimeters at each point. The volume of the element may be found by multiplying the length of the element times the mean of the cross-sectional areas at each point.

9.2.5. The Region of Influence of the Neuron

The area and volume measurements described in Sections 9.2.3 and 9.2.4 are calculated for the actual neuron enclosed by its cell membrane. Similar measurements may be made for the region of influence of the neuron. In this case, the summaries describe the volume of tissue into which the dendrites grow, sometimes called the receptive field of the neuron.

The numeric summary of Fig. 9-2 shows the limits of the neuron in X, Y, and Z. These coordinate values describe the extremes of a rectangular solid tissue region that contains the neuron's dendrites.

The region of influence is rarely a rectangular solid, however. So, to calculate its area with some accuracy, it is better to outline the region with a polygon and compute the area of the polygon. Figure 9-7 shows how this is done. This value enumerates the cross-sectional area of the dendritic field. Since this area is computed from a closed polygon, it is possible to calculate its perimeter, mean diameter, and form factor. Thus, the area of influence may be characterized in the same manner as the soma may be.

The volume of the region of influence is also valuable, for it describes the three-dimensional space in which the neuron exerts influence. This may be calculated by dividing the rectangular slab of tissue specified by its coordinate limits into thin slices, perhaps 10 μm thick. The computer program then draws a polygon around each slice and calculates the cross-sectional area of the polygon. The area of the polygon times the thickness of the slice yields the volume of the dendritic field that is within the slice. The sum of all the volumes gives the volume of the entire neuron region of influence.

The surface area of the neuron's region of influence may be calculated in a similar manner. The tissue slab is divided into slices, and each slice is surrounded by a polygon. The perimeter of the polygon is computed and multiplied by the thickness of each slice to

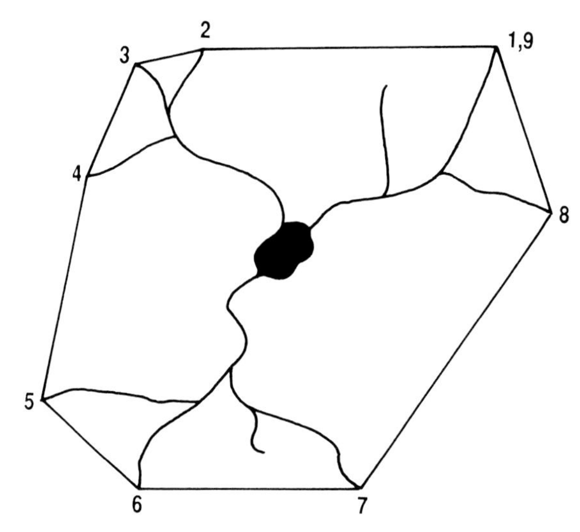

FIGURE 9-7. The calculation of the area of influence of a neuron. The point of the neuron farthest from the soma (point 1) is found. Then the headings (i.e., the compass direction measured clockwise from North) from that point to all the other points in the neuron are calculated and a line is drawn to the point with the greatest heading (point 2). Then headings are calculated for lines from point 2 to all other points and the next line segment is drawn to the point with the greatest heading. The perimeter of the neuron is thus traversed by always "turning left" until the starting point is reached. The result is a neuron circumscribed by a convex polygon, whose area can be calculated using the scheme shown in Fig. 9-4.

yield the surface area of the irregular band that outlines the slice. The sum of the surface areas of all slices forms the surface area of the neuron's region of influence.

9.2.6. Measurements of the Neuron's Center

You may want to quantify the symmetry of the branching of a neuron's dendrites. This is readily done by calculating the location of the neuron's center of gravity and comparing it to the location of the neuron's soma. The X coordinate of the center of gravity of a structure is computed by decomposing the structure into small elements. The product of the mass of each element times the distance along the X axis from the element to an arbitrary point is computed. The products of all the elements are accumulated, and their sum is divided by the total mass of the structure to yield the X coordinate of the structure's center of gravity. Similar calculations are done to calculate the Y and Z coordinates of the center of gravity. The center of gravity is the location the mass of the structure would have if it were all concentrated in a single point.

If the neuron is perfectly symmetrical, like a bicycle wheel with its soma at the hub and dendrites radiating outward like the spokes, then the center of gravity would be located at the center of the soma. On the other hand, if the dendrites grow primarily in one direction, the center of gravity will be shifted in the direction of maximum growth. The difference between the coordinates of the center of gravity and the coordinates of the center of the soma provide a measure of the magnitude and direction of the asymmetry.

Elements of mass have little meaning for a neuron, however, for it is not known what effect gravity has on it. Furthermore, it would be difficult to measure the mass of elements of a neuron; so elements of dendritic length between the coordinates of the computer-stored model are chosen as an alternative. Then the calculated coordinates of the neuron's center form the "center of dendritic length" instead of the center of gravity and are presented as in Fig. 9-2.

Since the surface area of the membrane has a biological significance in collecting charge, another calculation similar to the center of gravity is done. This uses elements of surface area to calculate the center of the neuron but gives more weight to fatter fibers. The coordinates of this "center of surface area" are also presented in Fig. 9-2.

9.2.7. Orientations of the Neuron

Another indicator of the directional growth of a neuron (or other nonneuronal structure) is the orientation of its major axis. As Fig. 9-8 illustrates, a principal axis line is drawn through the structure. The axis may be drawn in several ways. You may draw it, orienting it by eye, or the computer may draw it between the two points in the structure that are the most separated. As a final alternative, the computer may draw the line by finding the line that minimizes the error in Y for each X coordinate in the structure. Once the line is drawn, the computer calculates the angle between the line and the horizontal axis. This angle represents the structure's orientation.

9.2.8. Point-Type Densities

A density is the number of items per unit length, unit area, or unit volume. In neuroanatomy, densities are commonly computed to characterize neurons and to compare one population to another. The number of neurons per unit area of tissue may be easily calculated. In this situation, either you or computer software may count the cells; most frequently you would outline the relevant tissue area with a cursor, thus forming a surrounding polygon. The computer calculates the area enclosed within the polygon (Section 9.2.3) and divides this figure into the cell count, thus providing the density. Similar computations may be made for number of cells per unit volume, using the volume calculation described in Section 9.2.5.

A very common density is for dendritic spines. When you trace the neuron, if you record the location of each spine with the point type "spine base" (Section 4.3.5a), the computer may count them and divide the count by the total dendritic length of the neuron (Section 9.2.2). This provides the spine density, the number of spines per unit length of dendrite.

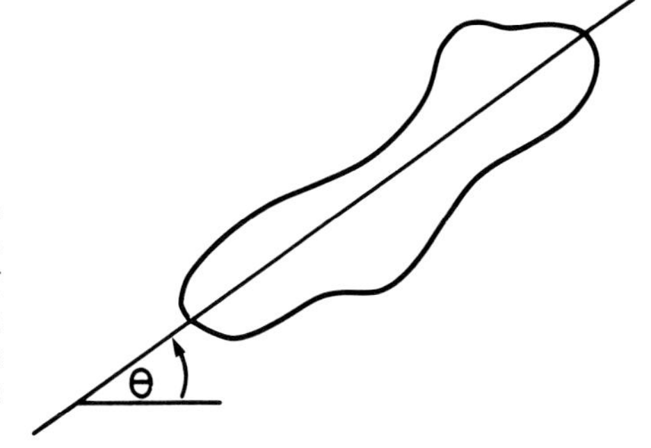

FIGURE 9-8. The major axis and orientation of a structure. A major axis line indicating the structure's orientation is drawn through the structure using one of the methods outlined in the text. The angle θ between the line and the horizontal axis is called the structure's orientation. Reproduced from Capowski (1988b) with permission.

9.2.9. Analysis of Dendritic Spines

More sophisticated analyses of dendritic spine growth are also possible. Figure 9-9 shows examples of three types of spines: regular, branched, and bulbed. When you trace the neuron, you may enter each of the three types into the computer, actually tracing each spine as if it were a little tree, indicating what type of spine it is, and measuring its thickness at points along the spine.

Various forms of analyses may then be performed on the computer-stored spines. Counts and densities of each type may be calculated. Spines may also be analyzed in a manner similar to regular dendritic trees, as discussed in Section 9.3; for example, the computer may plot graphs of the number of each type of spine versus any anatomic axis (Fig. 9-10), versus the thickness of the dendrites (Fig. 9-12), or versus the branch order of the dendrites.

9.3. GRAPHING SUMMARIES

In Section 9.2, numeric summaries of neuron trees and other structures were described, and the results are presented as tables of numbers. In this section, mathematical summaries that show relationships among various parameters of the cells are described. These are presented in visual form as graphs.

So that you may understand the summaries, three concepts used in them should first be described. The first concept is the "distribution axis." it is one of five axes that may form the abscissa of a graph. The distribution axis may be any of the standard three anatomic (e.g., dorsoventral) axes. You relate the anatomic axes to the *X*, *Y*, and *Z* axes of the microscope stage (Fig. 4-6) when you place the specimen in the microscope. The fourth distribution axis is the radial or "airline" distance from the soma. The fifth one is

FIGURE 9-9. Computer reconstructions of three types of dendritic spines from the cat lateral geniculate nucleus. Top: Guillery type I spines. Lower left: Guillery type II grapelike appendages. Lower right: Guillery type III branched spines. Courtesy of Chie-Fang Hsiao and Mark W. Dubin, unpublished.

the distance along the dendrite from the soma. Many parameters, e.g., the number of dendritic spines, may be plotted against this distribution axis.

The second concept, reviewed from Chapter 4, is branch order. Any point in a tree has a branch order, indicating the number of branch points between the soma and the point. Any point between the soma and the first branch point is defined to have branch order 1. Any point from the first branch point to the next distal branch point has order 2, and so on. Most parameters (e.g., the number of fiber swellings) may be plotted against branch order.

The third concept is a "branch segment." A branch segment is a portion of a tree between the soma and the most proximal branch point, between branch points, or between branch points and the dendritic terminations distal to them. Several statistical summaries are performed on branch segments, as you will see in Section 9.3.3.

9.3.1. Distributions of Point Types

As described in Chapter 4, you enter a neuron tree into the computer as a long series of points, and each may have one of 15 types. Some of these point types (e.g., dendritic spines) have special anatomic significance, and you may want to summarize their distribution. Figure 9-10 shows dendritic spines plotted against the rostrocaudal axis. This graph is an example of the more general case of plotting any type of point against a distribution axis.

Figure 9-11 also shows a distribution of a particular point type but, in this case, plotted against branch order. From this graph you may see whether the swellings occur proximally or distally.

You may also be interested in the distribution of a particular type of point versus

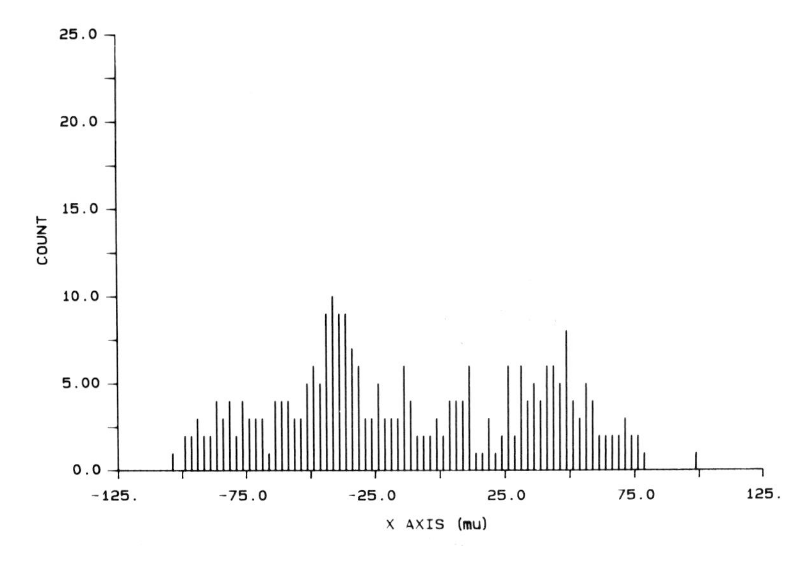

FIGURE 9-10. The number of dendritic spines along the rostrocaudal axis. Any point type, such as spines, may be plotted against any anatomic axis, thus presenting its distribution. Reproduced from Capowski (1987) with permission.

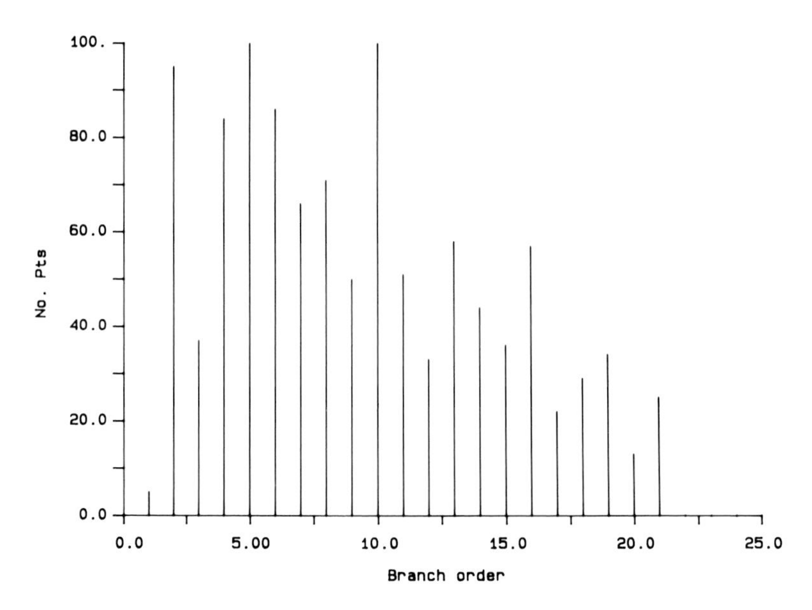

FIGURE 9-11. The number of fiber swellings versus branch order. With more branching (higher order) there are somewhat fewer swellings.

thickness. Figure 9-12 shows a plot of the dendritic spines versus the thickness of the fiber at the base of each spine. You may thus see whether the spines tend to occur at skinny or fat fibers.

9.3.2. Distributions of Dendritic Length

The dendritic length of a cell or a portion of it may also be analyzed according to distribution axis, branch order, and thickness. Figure 9-13 shows an example of dendritic length versus the distribution axis. By looking at this graph you may see, generally, where most of the dendrites grow.

The dendritic length may also be plotted against branch order, as shown in Fig. 9-14. In this example, the higher values on the left side of the graph indicate that most of the dendritic length occurs in the more proximal, lower-branch-order regions of this neuron.

Finally, as Fig. 9-15 illustrates, dendritic length may be plotted against thickness. This graph provides you the answer to the question ''For this cell, how much of the fiber is skinny and how much is fat?''

9.3.3. Analysis of Branch Segments

Several techniques are frequently used to analyze the individual branch segments of a neuron. Recall from Section 9.2.2 that the dendritic length of each segment may be calculated. Then, as illustrated in Fig. 9-16, the lengths of each segment may be accumulated into a histogram, so you may see the distribution of the lengths.

The relationship of the branch segment lengths to other parameters may also be

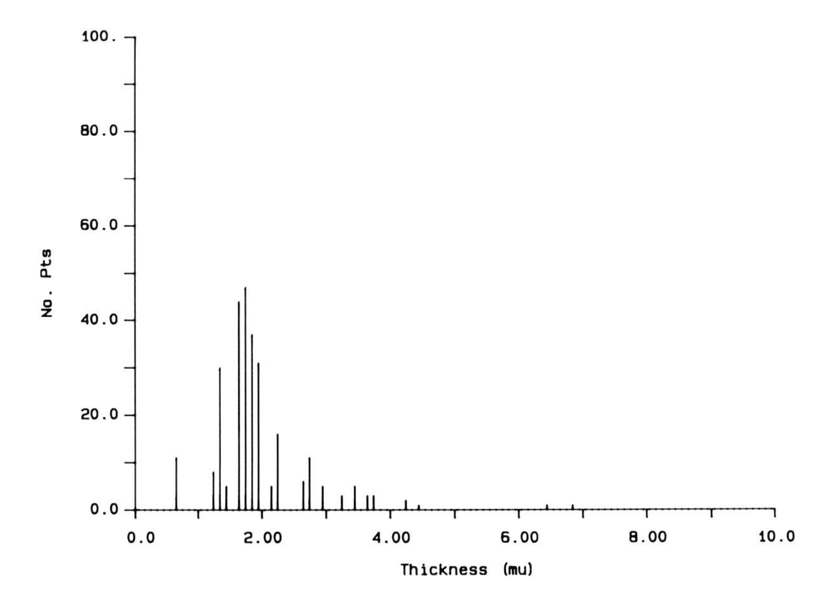

FIGURE 9-12. The number of dendritic spines versus dendrite diameter. For this cell, most spines occur at a 1.7-μm dendrite thickness.

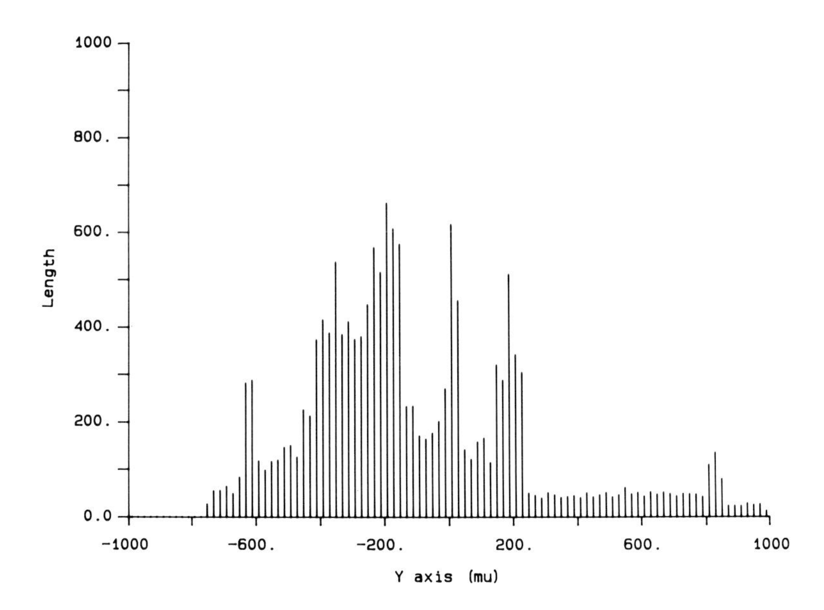

FIGURE 9-13. The dendritic length versus mediolateral axis. The total length of a dendrite may be plotted versus any anatomic axis, to show in what anatomic region most of the dendrite grows.

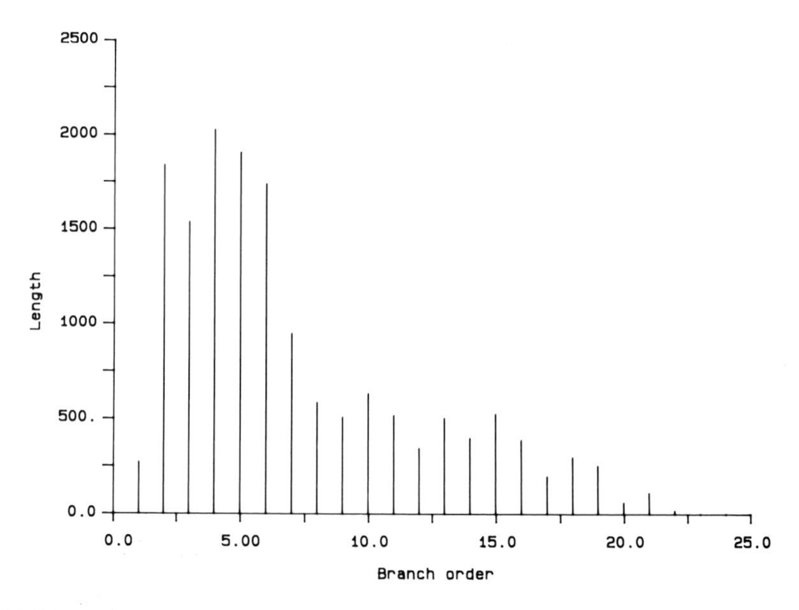

FIGURE 9-14. The dendritic length versus branch order. This plot shows the length of dendrite as a function of the amount of branching between it and the soma.

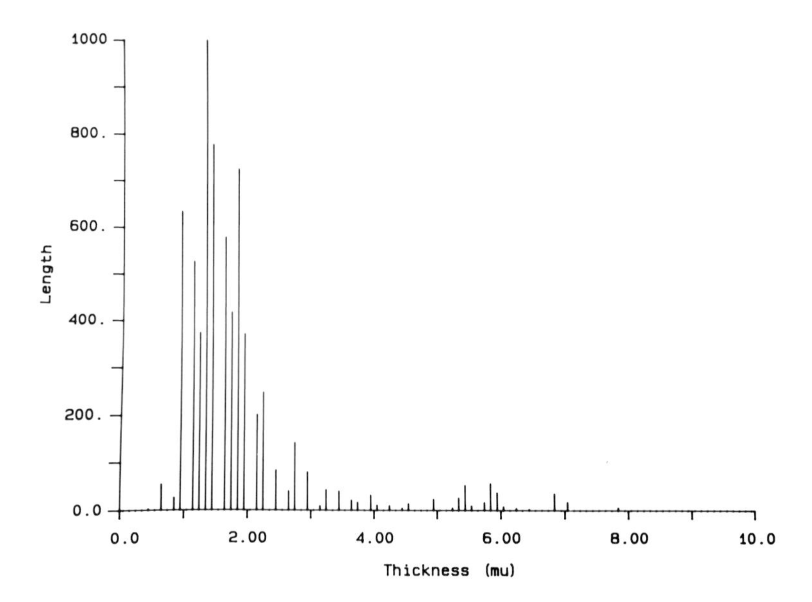

FIGURE 9-15. Dendritic length versus thickness. This plot shows how much fat and how much skinny fiber exist in a cell or population.

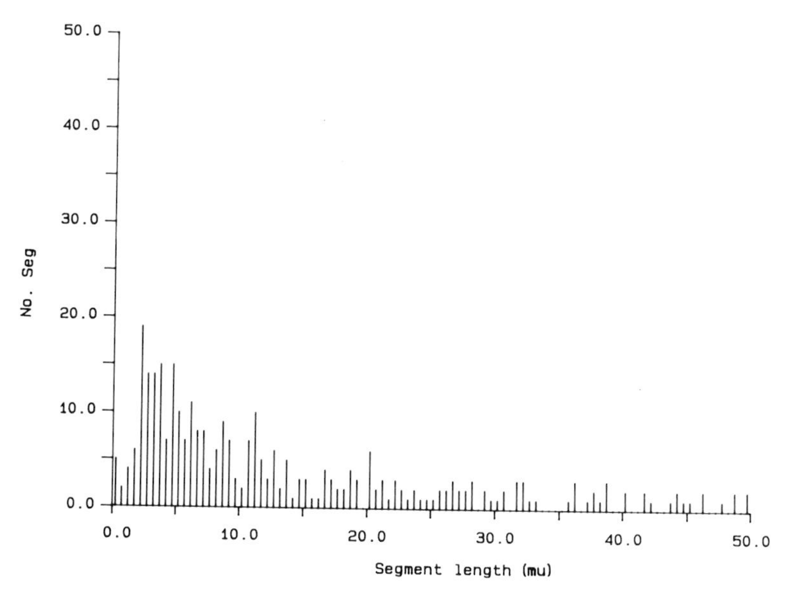

FIGURE 9-16. A branch segment length histogram. Here the length of each branch segment (soma to branch point, branch point to branch point, branch point to terminal) is calculated and binned into a histogram. This provides a distribution of the branch lengths.

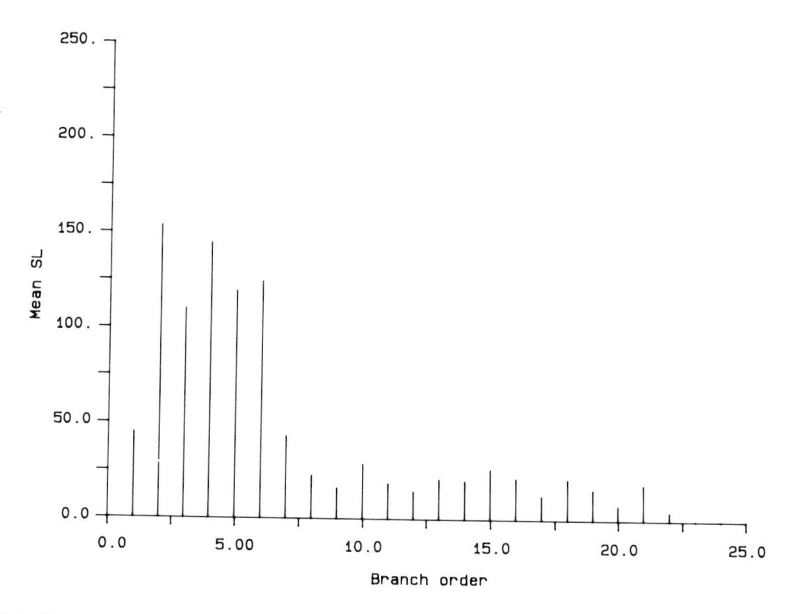

FIGURE 9-17. The mean branch segment length versus branch order. For each branch order, the mean of the dendritic lengths at that order is calculated. This results in a distribution of segment lengths as a function of amount of branching.

computed. For example, Fig. 9-17 shows a graph of the mean branch segment length versus branch order for a complex neuron. From this graph you may learn that the more proximal branch segments are longer than the more distal ones.

Similarly, you may wish to see the relationship between the thickness of the branch segments and their order. Figure 9-18 shows that for this population, the segment thickness seems to be independent of branch order.

You may be interested in the tapering of a dendrite or axon in the segment between branch points, for the movement of the electric charge in the dendrite is a function of the diameter of the dendrite. A convenient way to summarize this intrasegment tapering is to calculate a "tapering ratio" as shown in Fig. 9-19. This ratio demonstrates the amount of tapering in any one segment.

The tapering ratio may be plotted against a distribution axis, versus branch order, or versus thickness. In the example shown in Fig. 9-20, the tapering ratio is presented as a function of radial distance from the soma, one of the distribution axes. The graph shows that there is some tapering of the segments near the soma but nearly none at more distal regions. there is no directional information here, for the radial distance from the soma extends in all directions.

The tapering ratio may be plotted against branch order, as is shown in Fig. 9-21. In this graph you may see that the tapering ratio increases toward 1 (i.e., the tapering actually decreases) with higher branch orders in the more distal regions.

A final commonly used technique to help you understand the branching of a dendrite is to reduce the branching pattern to a sticklike or schematic form. Figure 9-22 shows such a graph, known variously as a "Sholl schematic diagram" (first described by Sholl, 1956), a "dendrogram," or a "tennis tournament" diagram.

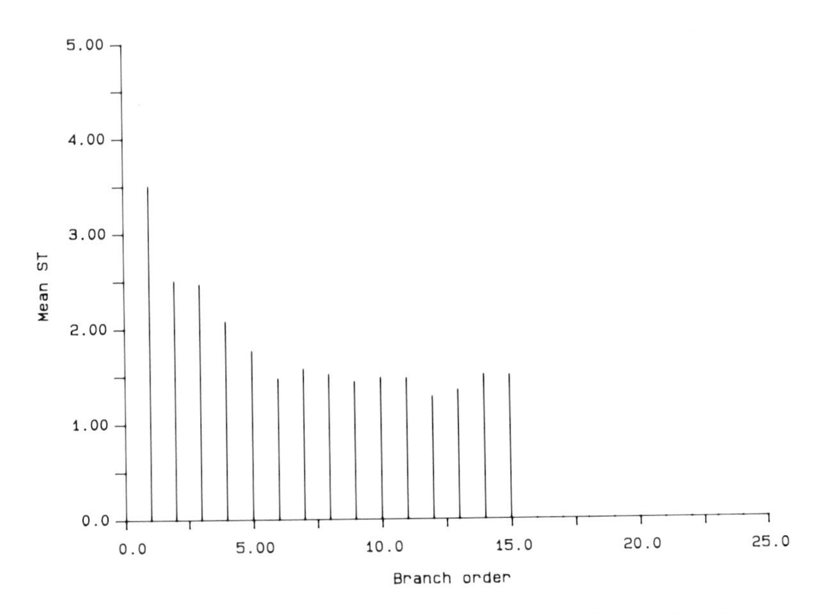

FIGURE 9-18. The mean branch segment thickness versus branch order. For each branch order, the mean thickness of all the dendritic segments is calculated. This results in a distribution of segment diameters as a function of the amount of branching.

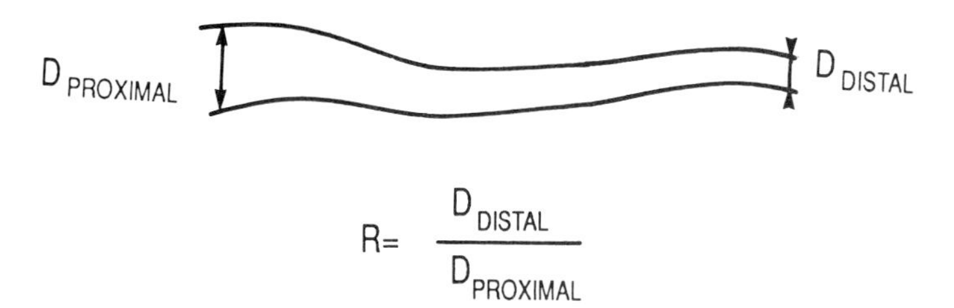

FIGURE 9-19. The calculation of a dendritic segment's tapering ratio. The tapering ratio is expressed as the diameter of a dendrite at the segment's distal end divided by the diameter at the proximal end. Thus, the ratio is 1.0 if there is no tapering, and the smaller the number, the greater is the degree of taper.

9.3.4. Analysis of Branch Points

Branch points themselves create several parameters. One is the angle between the two distal paths that exit from the branch point. Figure 9-23 shows a histogram of the angles for all the branch points of a particular neuron. With this graph, you may characterize a neuron according to the shape of its branch points.

Rall (1977) has presented a ratio that bears his name and describes the ideal relationship between the diameter of the proximal fiber entering a branch point and the two distal fibers that exit from it. The ratio is used particularly in modeling the electrical activity of neurons, by assuming that they form cables through which charges may flow. Figure 9-24 shows the calculation of Rall's ratio.

Rall's ratio, like the tapering ratio, may be plotted against the distribution axis, branch order, or dendritic thickness. Figure 9-25 shows the ratio of the branch points of a

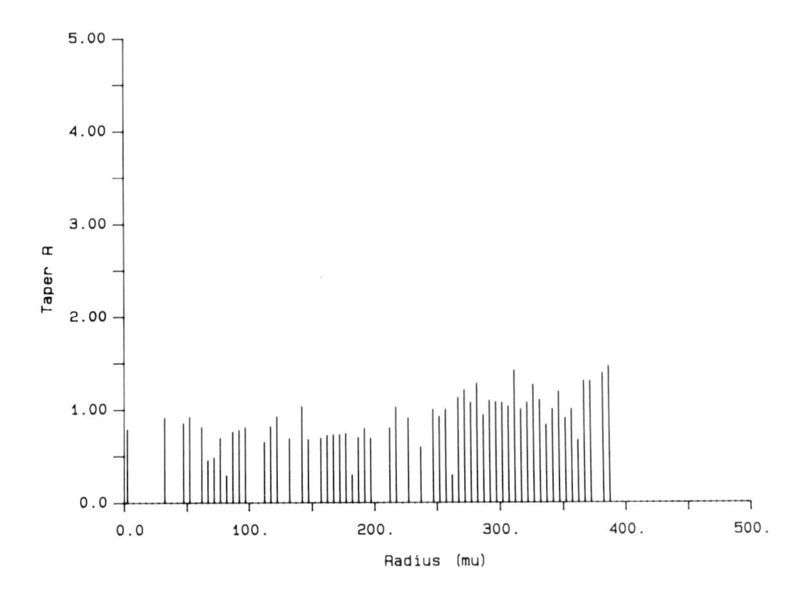

FIGURE 9-20. The distribution of the tapering ratio along the radial axis from the soma. This shows how dendrites taper as a function of airline distance from their cell's soma.

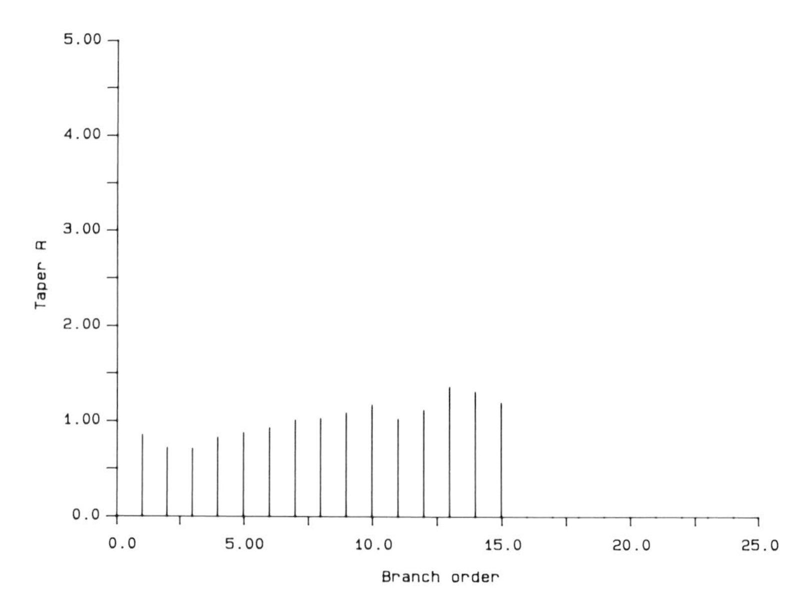

FIGURE 9-21. The distribution of tapering ratio versus branch order. This shows how dendrites taper as a function of their amount of branching.

neuron plotted against the dorsoventral axis. You may see that for many branch points, the ratio is nearly 2, indicating that the two exiting fibers have about the same diameter as the entering one.

When, as shown in Fig. 9-26, Rall's ratio is plotted against branch order for a neuron, you may see that at the higher branch orders (more distal), the ratio increases to nearly 2, indicating little difference between the diameters of the input and output fibers. Note that for this graph there are no data for branch order 1. This phenomenon occurs because of the definition of branch order. The branch order has the value 1 for the most proximal segment of a dendrite and increases to 2 at the first branch point.

9.3.5. Sholl Sphere Analysis

Sholl (1956) created another device that bears his name for summarizing a neuron, a "Sholl sphere diagram," introduced in Section 1.7 and illustrated in Figs. 1-12 and 9-27. He constructed a series of concentric spheres centered at the neuron's soma and counted the number of intersections that the dendrites made with each sphere. The Sholl sphere diagram displays the number of intersections versus the radius of each sphere. You may observe in this example that the number of branches of the neuron peaks at four-fifths of the way from the soma to the extreme of the dendritic field.

Because the Sholl diagram is generated by counting the number of intersections that each dendrite makes with each sphere, it gives equal weight to each dendrite regardless of its thickness. In a variant on this Sholl diagram, intersections are not counted; rather, the cross-sectional area of the intersection that the dendrite makes with each sphere is accumulated and plotted versus the sphere radius. This yields a graph similar to Fig. 9-27 except that more weight is given to fatter fibers.

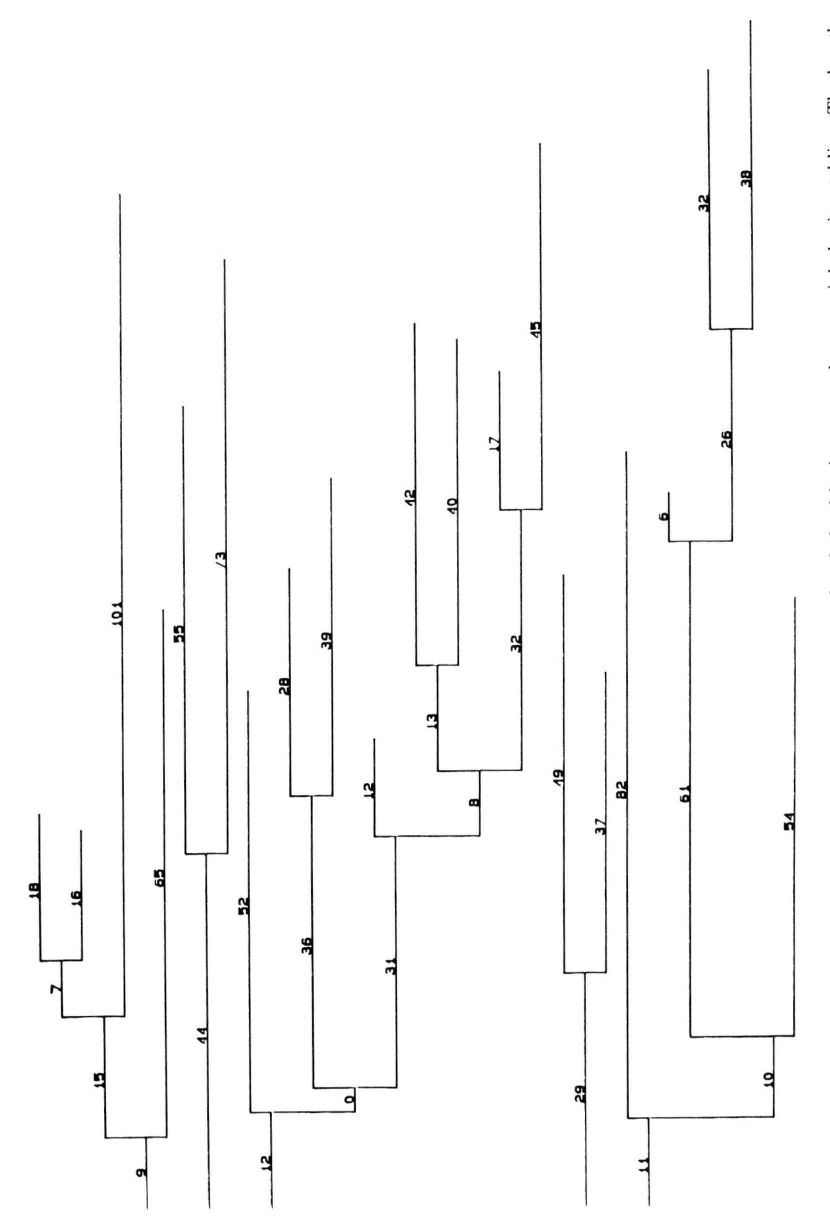

FIGURE 9-22. The schematic diagram of a neuron. Each segment of each dendrite is presented as a straight horizontal line. The length of each segment in micrometers is plotted above the line. Sometimes, though not shown here, the mean thickness of each segment is also plotted. This graph easily exhibits the branching pattern of simple dendrites. Reproduced from Capowski (1987) with permission.

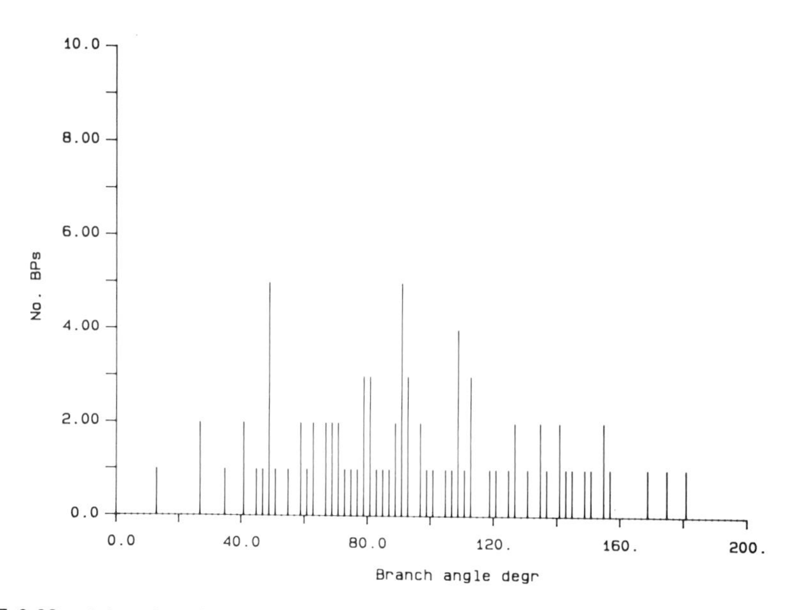

FIGURE 9-23. A branch-angle histogram. The exiting angle between the two daughter branches of each branch point is calculated and binned into a histogram.

9.3.6. Directional Analysis

Although the Sholl sphere diagram presents information about the amount of branching of a neuron's dendrites, it has no directional sensitivity. A dendrite growing northward will generate the same number of intersections as one growing southward. A convenient way to provide the directional information is a polar diagram, illustrated in Fig. 9-28.

A polar diagram is actually a histogram of dendritic length except that its bins are wedge-shaped and are arranged circularly. The diagram indicates the length of the dendrites in the various directions around the compass. A variant of this may also be constructed using dendritic surface area instead of dendritic length to give more credit to fatter fibers.

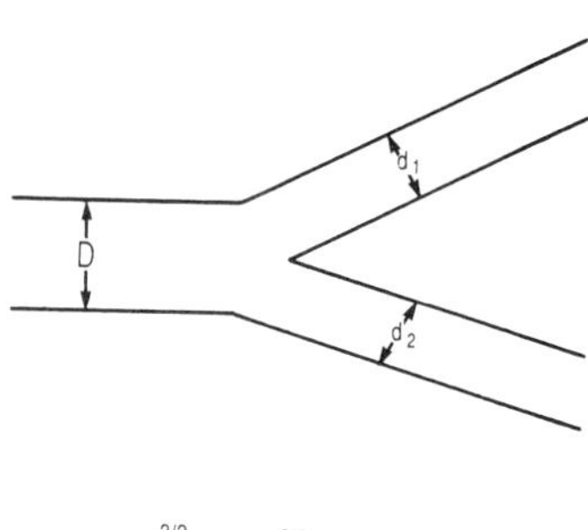

FIGURE 9-24. The calculation of Rall's ratio at a branch point. The diameters of the two daughter segments d_1 and d_2 exiting from a branch point are related to the mother segment's diameter D by the formula shown. Theoretically, the ratio should have the value 1.0.

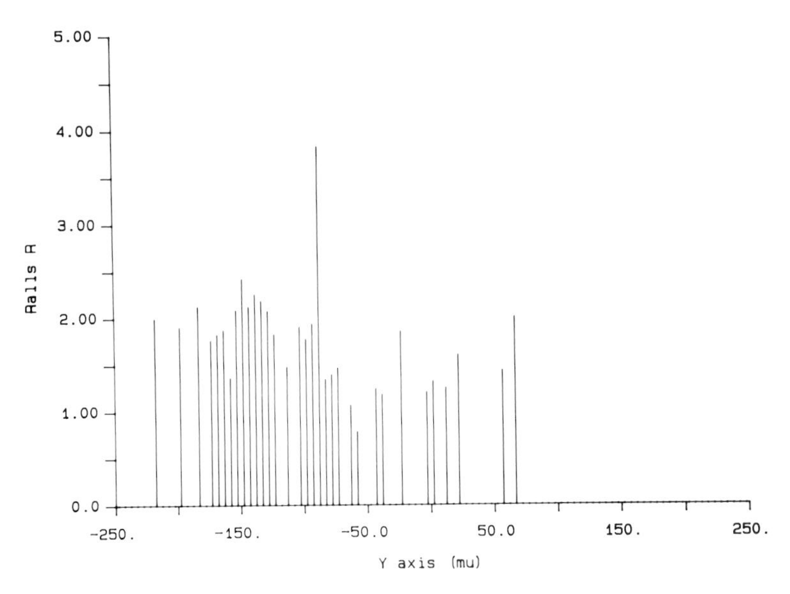

FIGURE 9-25. The distribution of Rall's ratio along the dorsoventral axis. This shows how well Rall's ratio performs as a function of anatomic region.

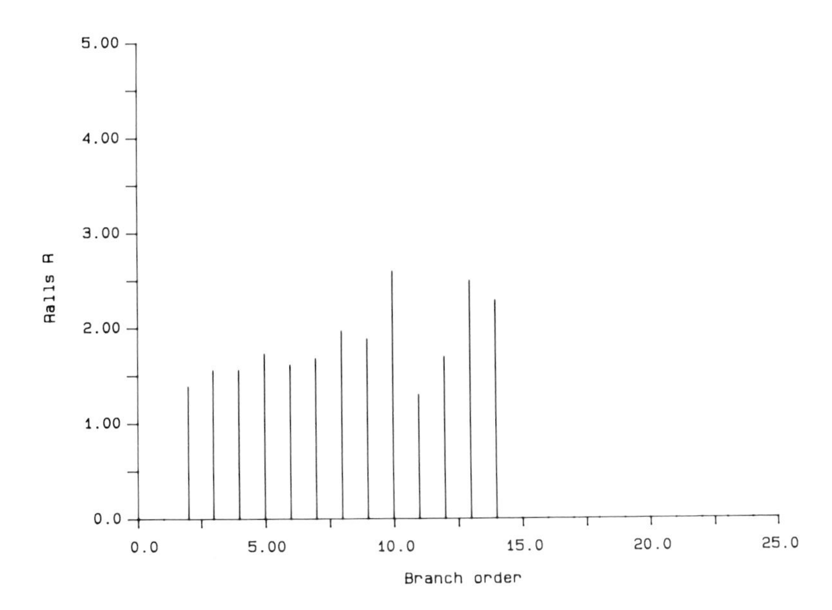

FIGURE 9-26. The distribution of Rall's ratio versus branch order. This shows how well Rall's ratio performs as a function of the amount of branching of a dendrite.

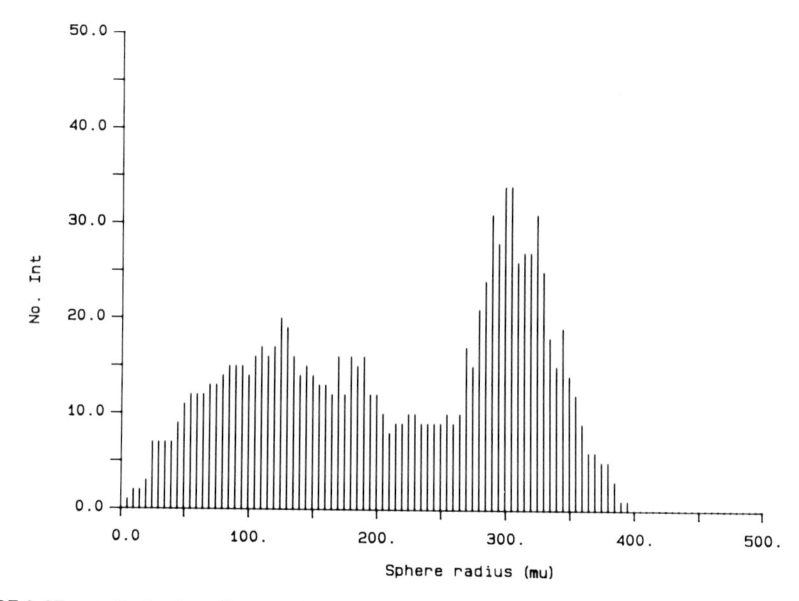

FIGURE 9-27. A Sholl sphere diagram. A series of concentric spheres centered at a cell's soma is constructed. The number of intersections that dendrites make with each sphere is counted and plotted as a function of sphere radius. This graph describes the amount of branching that a cell does as a function of airline distance from its soma.

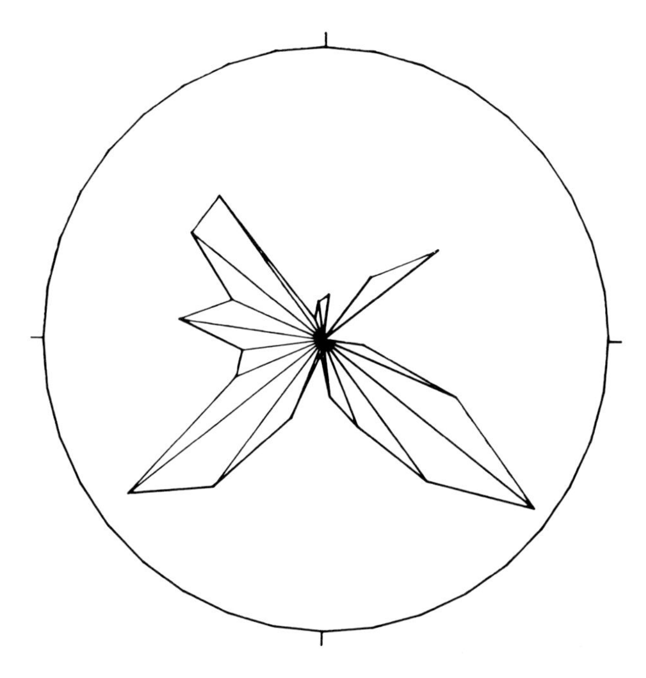

FIGURE 9-28. The polar diagram of a neuron. The dendritic length is accumulated into 15° wedge-shaped bins centered at the neuron's soma. A line whose length is proportional to the accumulated value in each bin is drawn radially to the extreme of each bin. Finally, an envelope is drawn around each bin. This graph shows the direction of growth of the dendrites. Radius is 150 μm. Reproduced from Capowski (1987) with permission.

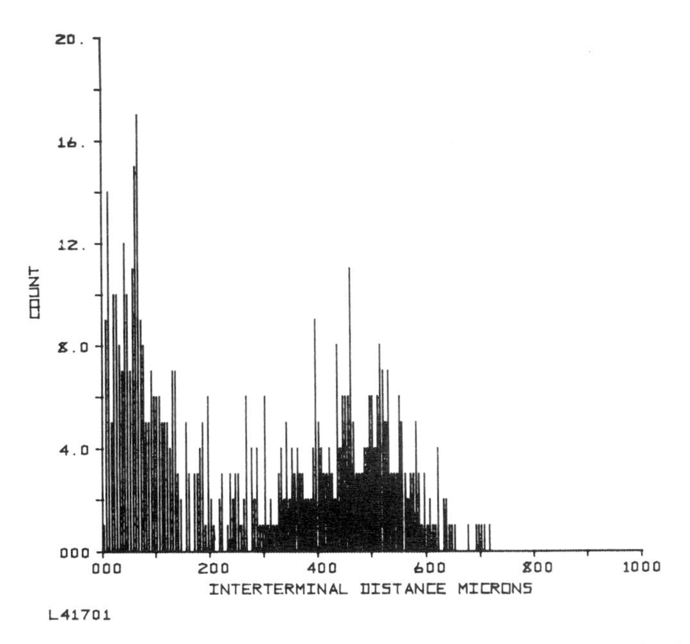

FIGURE 9-29. A histogram of interterminal distances. The airline distance between each pair of terminals (or any other kind of point) is calculated and placed into a histogram. The hump in the center of the graph indicates that there is a clustering of the terminals. Reproduced from Capowski and Réthelyi (1978) with permission.

9.3.7. Cluster Analysis

The final analysis presented here is used as a preliminary indication of the clustering of points. It is the interpoint distance histogram, illustrated by the example in Fig. 9-29. In this graph the airline distances between all pairs of points in a structure are calculated and binned into a histogram. If the points are evenly spaced, the histogram will have the highest values at the left (shortest distances) and will taper linearly to the right (longest distances). If the points are distributed randomly, as you might sow grass seed on a lawn, the distribution will be Poisson, decreasing continuously but nonlinearly from left to right. The distribution in Fig. 9-29 indicates clustering, however, for there are a great number of points near each other (causing the hump on the left) and a great number of points that are far from each other (causing the hump in the middle of the graph). This indicates two clusters of points.

9.4. FOR FURTHER READING

Further descriptions of statistical summarizations that have become routine for quantifying neuron trees are presented by Brown (1977), Lindsay (1977c,d), Yelnik et al. (1983), Glaser and McMullen (1984), and Rose and Rohrlich (1987). A particularly illustrative presentation of the use of these summarizations is given by Coleman et al. (1981).

Information on computer simulations of growth of trees and comparison of the results of the simulations to botanical trees may be found in Fisher and Honda (1977), Honda and Fisher (1978), and Verwer and van Pelt (1986).

A mathematical growth model for neuron trees, i.e., a description of the rules for the growth of the trees, is available in Ireland et al. (1985) and van Pelt et al. (1986).

A good source to learn about the cable properties of neurons and how to model cell operations may be found in Rall (1977).

10

Topological Analysis of Individual Neurons

HARRY B. M. UYLINGS, JAAP VAN PELT, and RONALD W. H. VERWER

10.1. INTRODUCTION

What is topology? Topology is the study of geometric surfaces with concern only for their mathematical form but not their physical size. A topologist thinks of structures as formed from rubber sheets that can be stretched at will, and as long as no holes are poked in them, the properties of the structure do not change. Carried to a humorous extreme as Fig. 10-1 indicates, a coffee cup and a donut are topologically equivalent structures, for they both have the same form, a surface with one hole.

Why should topological techniques be applied to neurons? Neurons are complex treelike branching structures with geometric forms. It is enlightening to examine their structural properties as well as, of course, their physical ones. In this chapter, we discuss some techniques for describing neurons topologically, concentrating on their mathematical form while ignoring their physical size. Then, the branching patterns are reduced to graphs and especially to rooted trees defined by (origin, branching, and terminal) points and by segments connecting these points in a particular way.

Neurons are connected with other neurons via their dendritic and axonal arborizations. A general feature of branching patterns in nature is that they are receptive and flow-conductive. These functions can be performed optimally by branching patterns since they show both a large interface between the structure itself and its surroundings and a shorter conductive pathway than unbranched structures would have. The shape of a neuronal arborization is usually characteristic for the neuronal cell type. The size of neuronal arborizations can reflect the maturational state. Diseases or experimental conditions frequently affect the size and may also affect the shape of the neuronal tree structures (see, e.g., Uylings et al., 1986a).

To study differences caused by disease or experimental conditions between groups of

HARRY B. M. UYLINGS, JAAP VAN PELT, and RONALD W. H. VERWER • Netherlands Institute for Brain Research, 1105-AZ Amsterdam, The Netherlands.

FIGURE 10-1. A topologist in the morning. He is confused because he cannot tell the difference between his coffee cup and his donut.

neurons, Sholl's method (Sholl, 1953) or the modified Sholl's method (Eayrs, 1955) is frequently used. Sholl's method determines the number of intersections of dendrites with a set of concentric spheres, and the modified Sholl's method determines the number of intersections of projected dendrites with a set of concentric circles (Fig. 10-2). Figure 10-2, however, illustrates what has already been shown by Ten Hoopen and Reuver (1970) and, e.g., Smit et al. (1972), that quite different neurons can have the same set of dendritic intersections. The reason for this is that the numbers of intersections are determined by both the metric and the topological characteristics of the neurons. In Fig. 10-2, you can see that the right neuron has fewer bifurcations and longer segments than the left one. The analysis of neuronal tree structures will, therefore, reveal more information when both the metric and the topological characteristics are studied separately. Metric

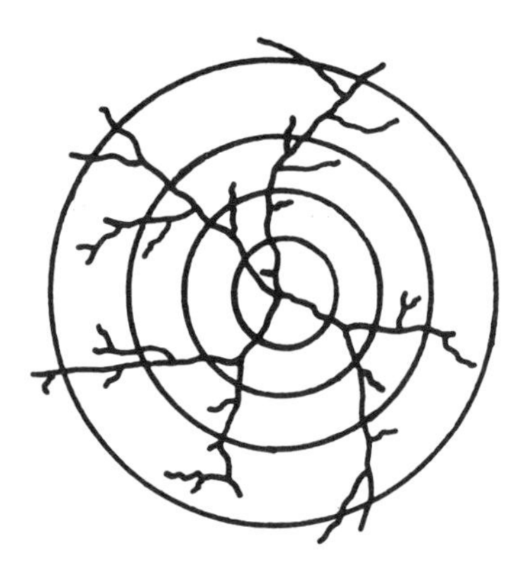

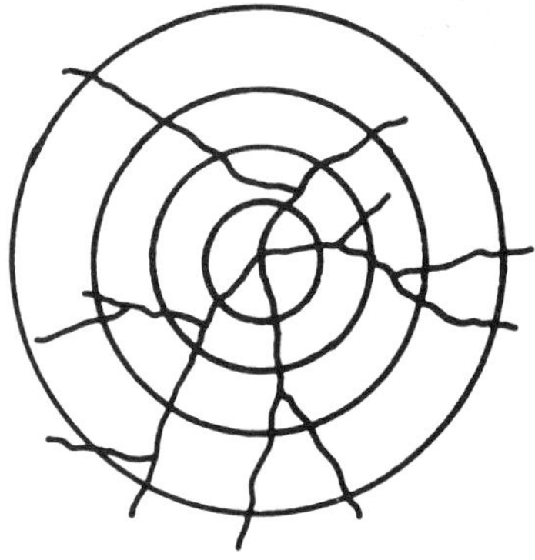

FIGURE 10-2. The numbers of dendritic intersections with the set of concentric circles are equal for the two very different neurons illustrated here. This figure shows that the topological and metric aspects are intermingled in the Sholl's and Eayrs' method. Reproduced from MacDonald (1983) with permission.

characteristics of trees, such as the dendritic length, the thickness or diameter of segments, and the extension in the three-dimensional space are considered in Chapters 9 and 11 of this book.

This chapter deals with the topological description of trees. In the topological description, a tree is reduced to a number of segments (topological size) that are connected to each other in a particular way (topological structure). Other elements of a topological tree are the branching points, the origin or root point, and the terminal points. For a given number of segments and branching points, only a finite number of different connection patterns form tree structures. These particular topological tree types can be distinguished and studied. A branching tree develops through a sequence of branching events. Many different ways of branching or growth modes can lead to the same particular tree type, and, in general, a particular growth mode can lead to many different topological tree types. A particular growth mode, however, induces a characteristic probability distribution of different topological tree types. When different groups of neuronal trees have been developed differently, these groups will also differ in the frequency distribution of topological tree types. Therefore, it is the frequency distribution of topological tree types that is of importance in the topological analysis of neuronal tree structures.

In practice, topological tree types are often unmanageable in statistical evaluation because of the large increase in number of topological tree types with increasing size, that is, number of segments (e.g., Van Pelt and Verwer, 1984a). Therefore, relatively simple topological variables are discussed in this chapter that (1) reflect special features of the topological structure of a tree, (2) still have discriminative power to distinguish different neuronal groups, and (3) are more easily manageable, even with increasing tree size. Topological analysis is also useful in testing growth models. A growth model defines rules for the branching events and is tested on the basis of a comparison of the measured and expected frequency distributions of topological tree types. We will also discuss the analysis of incomplete trees, which occur often (because of cutting, etc.). For the analysis of topological tree characteristics no expensive semiautomatic measurement systems are necessary. Computational facilities for the statistical evaluation, however, are often preferable and, in some types of statistical analysis, necessary (e.g., for a Monte Carlo test).

10.2. CLASSIFICATION OF TREE TYPES

A tree is topologically characterized by the way the branching points and terminal tips are connected with each other through segments. The topological size of a rooted tree is determined by its total number of segments, which is the sum of the terminal and intermediate segments and the root segment (Fig. 10-3). In trees with bifurcations only, i.e., binary trees, the number of terminal segments or tips, n, equals the number of bifurcations plus the root origin. The total number of segments, n_s, of a rooted tree is then

$$n_s = 2n - 1 \tag{10-1}$$

Thus, the number of terminal segments, n, called the degree of the tree (Harding, 1971) (Fig. 10-3C), is directly related to the topological size of a binary tree. With a definite number of terminal tips (i.e., a definite degree), a finite number of topologically

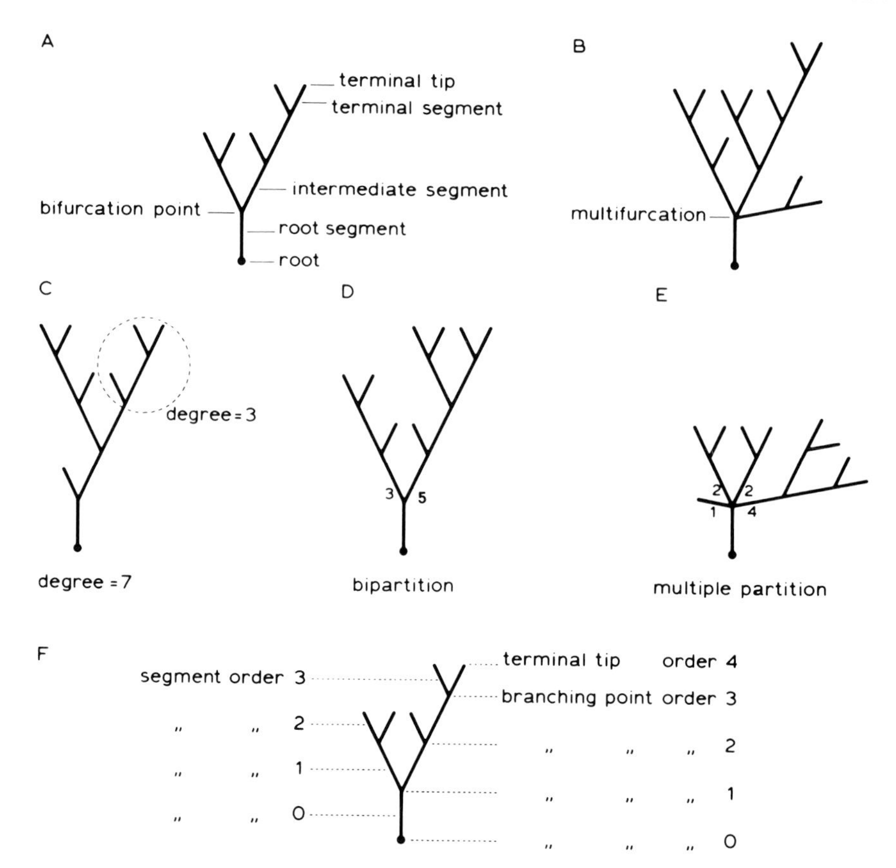

FIGURE 10-3. Elementary description of tree structures. A: Illustration of tree elements. B: More than two subtrees arise from a multifurcation. C: The degree of a tree or subtree is the number of terminal segments. D: A first-order partition of the degrees of two subtrees. E: Multiple partition, i.e., a partition of the degrees of the subtrees that emanate from a multifurcation. F: Centrifugal ordering of segments and centrifugal ordering of bifurcation points and terminal tips. The latter is equal to the topological path length to root. Reproduced from Verwer and Van Pelt (1986) with permission.

different connections can be constructed. Each construction represents a different type of binary tree. In Fig. 10-4 the different topological types for trees up to degree 8 are shown.

The topological tree types characterize (neuronal) trees embedded in the three-dimensional space, and (internal) rotations of subtree pairs do not change its topology (Smart, 1969).* In Fig. 10-4, the left–right arrangement of subtrees at bifurcation points has no meaning but is only chosen for the graphic representation, so that the smallest of the two subtrees is chosen as the "left" one. To obtain an unambiguous classification (Van Pelt and Verwer, 1983), the tree types are classified according to (1) their degree,

*The term "topological tree type" is used for river networks embedded in a two-dimensional space in which the two-dimensional "left–right" arrangement cannot be interchanged (Shreve, 1966). Smart (1969) has introduced the system of classification that allows (internal) rotation of left and right subtrees and called these types ambilateral types. These types characterize the topology of three-dimensional trees. We prefer to call the ambilateral tree type the (topological) tree type.

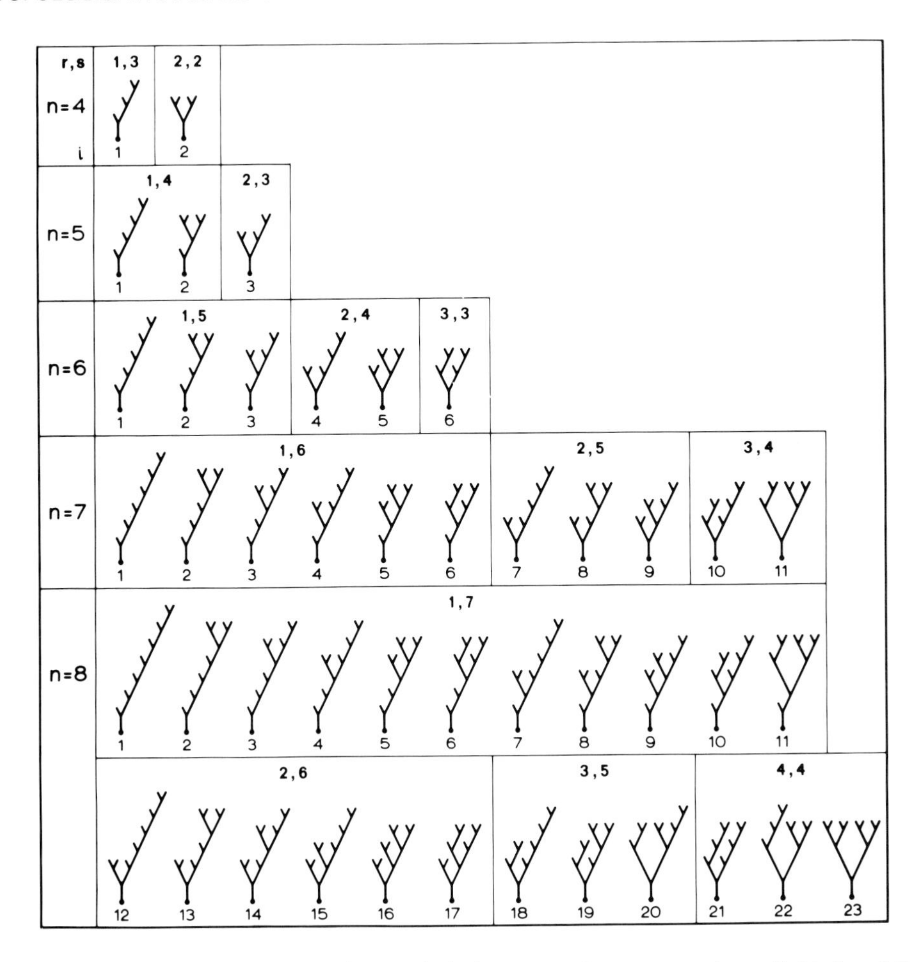

FIGURE 10-4. Schematic drawings of all topological tree types for trees up to degree 8: i is the relative number of tree type for a given degree n; r,s represents the first-order partition of degrees of the two subtrees. Reproduced from Van Pelt and Verwer (1983) with permission.

i.e., number of terminal tips, n; (2) the degree of their smallest first-order subtree, r, with $n(r,n - r)$ (Fig. 10-3D); (3) the class number of the smallest first-order subtree; and (4) the class number of the largest first-order subtree. This classification scheme is displayed in Fig. 10-4 and by the branching code in Table 10-1.

The classification scheme of topological tree types is iterative. Within the group of degree-8 trees the subclassification of the tree types with first-order subtree pair (1,7) is in conformity with the classification of degree-7 trees (see tree types 1–11 of degree-7 and degree-8 trees in Fig. 10-4). The subclassification of the degree-8 tree types with first-order subtree pair (2,6) in turn is in conformity with the classification of the degree-6 trees (see tree types 12–17 of the degree-8 trees and types 1–6 of the degree-6 trees in Fig. 10-4), and so on. As was indicated in the introduction, different growth models lead to characteristic probability distributions of topological tree types (e.g., Van Pelt et al., 1986). As an example the different expected probability distributions for the degree-8

TABLE 10-1. Classification Methods of Trees[a]

n	Rel. no.	Branching code tree type	$n(r,s)$	No. segm. c. o. 0,1,2,3,4,5,6,7	C	p	p_e	D
1	1	1	1	1	0	1	1	1
2	1	2	2(1,1)	1,2	2	5	4	2
3	1	3	3(1,2)	1,2,2	6	11	8	3
4	1	4(1 3)	4(1,3)	1,2,2,2	12	19	13	4
	2	4(2 2)	4(2,2)	1,2,4	10	17	12	3
5	1	5(1 4(1 3))	5(1,4)	1,2,2,2,2	20	29	19	5
	2	5(1 4(2 2))	5(1,4)	1,2,2,4	18	27	18	4
	3	5(2 3)	5(2,3)	1,2,4,2	16	25	17	4
6	1	6(1 5(1 4(1 3)))	6(1,5)	1,2,2,2,2,2	30	41	26	6
	2	6(1 5(1 4(2 2)))	6(1,5)	1,2,2,2,4	28	39	25	5
	3	6(1 5(2 3))	6(1,5)	1,2,2,4,2	26	37	24	5
	4	6(2 4(1 3))	6(2,4)	1,2,4,2,2	24	35	23	5
	5	6(2 4(2 2))	6(2,4)	1,2,4,4	22	33	22	4
	6	6(3 3)	6(3,3)	1,2,4,4	22	33	22	4
7	1	7(1 6(1 5(1 4(1 3))))	7(1,6)	1,2,2,2,2,2,2	42	55	34	7
	2	7(1 6(1 5(1 4(2 2))))	7(1,6)	1,2,2,2,2,4	40	53	33	6
	3	7(1 6(1 5(2 3)))	7(1,6)	1,2,2,2,4,2	38	51	32	6
	4	7(1 6(2 4(1 3)))	7(1,6)	1,2,2,4,2,2	36	49	31	6
	5	7(1 6(2 4(2 2)))	7(1,6)	1,2,2,4,4	34	47	30	5
	6	7(1 6(3 3))	7(1,6)	1,2,2,4,4	34	47	30	5
	7	7(2 5(1 4(1 3)))	7(2,5)	1,2,4,2,2,2	34	47	30	6
	8	7(2 5(1 4(2 2)))	7(2,5)	1,2,4,2,4	32	45	29	5
	9	7(2 5(2 3))	7(2,5)	1,2,4,4,2	30	43	28	5
	10	7(3 4(1 3))	7(3,4)	1,2,4,4,2	30	43	28	5
	11	7(3 4(2 2))	7(3,4)	1,2,4,6	28	41	27	4
8	1	8(1 7(1 6(1 5(1 4(1 3)))))	8(1,7)	1,2,2,2,2,2,2,2	56	71	43	8
	2	8(1 7(1 6(1 5(1 4(2 2)))))	8(1,7)	1,2,2,2,2,2,4	54	69	42	7
	3	8(1 7(1 6(1 5(2 3))))	8(1,7)	1,2,2,2,2,4,2	52	67	41	7
	4	8(1 7(1 6(2 4(1 3))))	8(1,7)	1,2,2,2,4,2,2	50	65	40	7
	5	8(1 7(1 6(2 4(2 2))))	8(1,7)	1,2,2,2,4,4	48	63	39	6
	6	8(1 7(1 6(3 3)))	8(1,7)	1,2,2,2,4,4	48	63	39	6
	7	8(1 7(2 5(1 4(1 3))))	8(1,7)	1,2,2,4,2,2,2	48	63	39	7
	8	8(1 7(2 5(1 4(2 2))))	8(1,7)	1,2,2,4,2,4	46	61	38	6
	9	8(1 7(2 5(2 3)))	8(1,7)	1,2,2,4,4,2	44	59	37	6
	10	8(1 7(3 4(1 3)))	8(1,7)	1,2,2,4,4,2	44	59	37	6
	11	8(1 7(3 4(2 2)))	8(1,7)	1,2,2,4,6	42	57	36	5
	12	8(2 6(1 5(1 4(1 3))))	8(2,6)	1,2,4,2,2,2,2	46	61	38	7
	13	8(2 6(1 5(1 4(2 2))))	8(2,6)	1,2,4,2,2,4	44	59	37	6
	14	8(2 6(1 5(2 3)))	8(2,6)	1,2,4,2,4,2	42	57	36	6
	15	8(2 6(2 4(1 3)))	8(2,6)	1,2,4,4,2,2	40	55	35	6
	16	8(2 6(2 4(2 2)))	8(2,6)	1,2,4,4,4	38	53	34	5
	17	8(2 6(3 3))	8(2,6)	1,2,4,4,4	38	53	34	5
	18	8(3 5(1 4(1 3)))	8(3,5)	1,2,4,4,2,2	40	55	35	6
	19	8(3 5(1 4(2 2)))	8(3,5)	1,2,4,4,4	38	53	34	5
	20	8(3 5(2 3))	8(3,5)	1,2,4,6,2	36	51	33	5
	21	8(4(1 3) 4(1 3))	8(4,4)	1,2,4,4,4	38	53	34	5
	22	8(4(1 3) 4(2 2))	8(4,4)	1,2,4,6,2	36	51	33	5
	23	8(4(2 2) 4(2 2))	8(4,4)	1,2,4,8	34	49	32	4

[a]Abbreviations used: n, degree = no. terminal segments; rel. no., relative number within degree; $n(r,s)$, first-order partition; No. segm. c. o., number of segments per centrifugal order; C, sum orders of all segments; p, sum path lengths of all bifurcations and end points; p_e, sum path lengths end points; D, diameter.

trees under the segmental-growth (i.e., random branching of all segments) and terminal-growth (i.e., random branching of terminal segments only) hypotheses, respectively, are given in Fig. 10-5. In addition, the expected probability distributions of degree-8 tree types 1–11 in Fig. 10-5 are equal to the respective probability distributions of the degree-7 trees, apart from a normalization constant.

The classification according to tree types is the only classification that uses the complete topological information of tree structures. The number of different tree types, however, increases very rapidly with the degree, n, as is shown in the second column of Table 10-2. A tree type is a discrete variable. You can compare several groups of tree types by using statistical tests for comparison of different discrete frequency distributions (e.g., chi-square test). The chi-square test requires that for each class the expected number of observations is at least one (Conover, 1980). Even medium-sized trees such as degree-8 trees have probability distributions that contain many classes (tree types). Therefore, several classes can be empty because of the finite size of the set of observed trees (cf. the probability distributions in Fig. 10-5).

In general, the number of classes has to be reduced. One way of reducing classes is to lump several tree types into a few classes such that a suitable number of classes is obtained. The topological interpretation of differences between groups of neurons will then fade. For topological analysis it is often preferable to use other topological parameters that have fewer classes but still have a topological meaning. These topological parameters necessarily contain less refined topological information than the topological tree types.

10.3. CLASSIFICATION BASED ON DIFFERENT TOPOLOGICAL FEATURES

In this section we consider different topological parameters that reflect special features of the tree topology and have fewer classes. A definite selection of a particular topological parameter that is optimal for use in topological analysis to distinguish differences between groups of neuronal tree structures cannot yet be given. For this reason, different topological parameters are discussed, and their (possible) importance is indicated.

10.3.1. Classification according to the Degree of Subtree Pairs

10.3.1a. Analysis of Partitioning of Terminal Segments in Subtree Pairs. A bifurcation gives rise to a pair of subtrees and, consequently, to the partitioning of terminal segments. The first-order partition is the pair of the degrees of the first-order subtrees [e.g., the pairs (1,6) and (3,5) in Figs. 10-3C and D, respectively; see also Table 10-1]. The partitions can be used to classify trees according to (1) the first-order partitions or (2) the set of the partitions of all-order bifurcations. Trees with seven terminal segments, for instance, may be classified according to their first-order partitions in classes (1,6), (2,5), and (3,4). The grouping of trees obtained is illustrated in Fig. 10-4 and Table 10-1. The first-order partitions form a distribution per degree. The third column in Table 10-2 shows that the number of classes (the size of the distribution) of the first-order partitions is

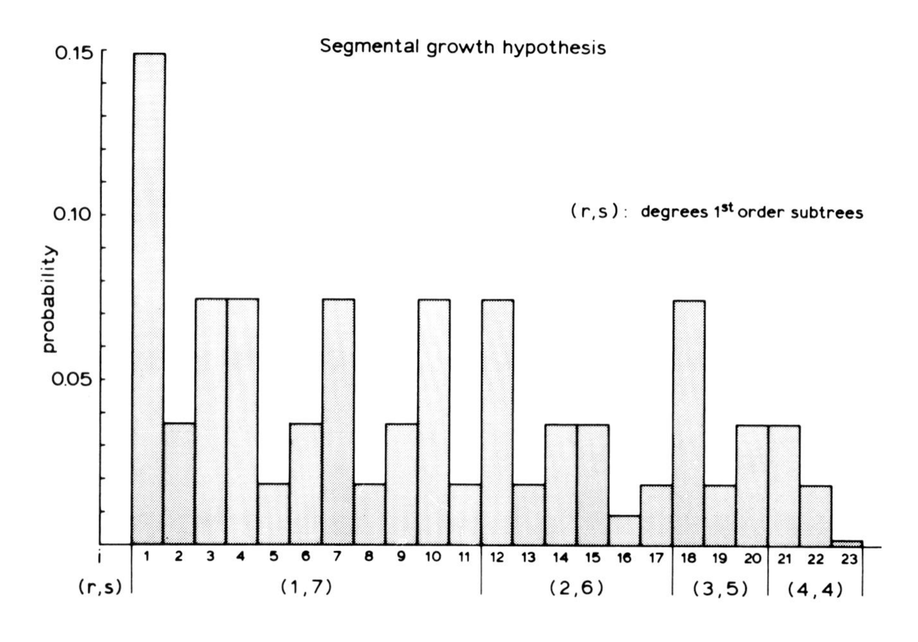

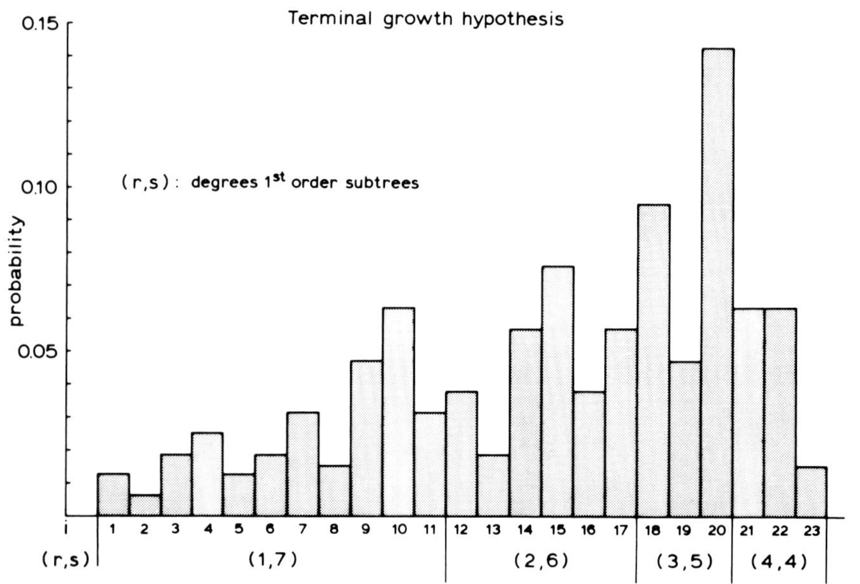

FIGURE 10-5. Probability distribution of the different tree types of degree 8 (eight terminal segments) under the segmental and terminal growth hypotheses, respectively: i is the relative number of tree type; r,s represents the first-order partition of terminal segments. Reproduced from Van Pelt and Verwer (1984a) with permission.

TABLE 10-2. Number of Classes for Several Classification Schemes for Binary Trees[a]

n	Tree type	First-order part.	All-order part.	$C,\bar{C},p,p_e$	D	V_r	$\overline{\text{PSAD}}$
4	2	2	2	2	2	2	2
5	3	2	4	3	2	2	3
6	6	3	7	5	3	3	5
7	11	3	10	8	4	3	10
8	23	4	14	12	5	4	18
9	46	4	18	16	5	4	40
10	98	5	23	21	6	5	78
11	207	5	28	27	7	5	154
12	451	6	34	34	8	6	298
13	983	6	40	42	9	6	690
14	2179	7	47	51	10	7	1106

[a]Abbreviations used: n, degree = no. terminal segments of a tree; part., partition; C, sum centrifugal order of all segments of a tree; $\bar{C}$, mean centrifugal order of all segments; p, total path length; p_e, total terminal path length; D, topological diameter; $\overline{\text{PSAD}}$, mean PSAD (see Section 10.4); V_r, vertex ratio; see Section 10.3.1c. Note that the number of classes of the all-order partition variable is the cumulative sum of the classes of the first-order partition given in the 4th column (see Section 10.3.1a).

considerably reduced in comparison with the number of classes of tree types. However, the loss of structural information is considerable as well. All degree-7 trees with a first-order partition (1,6), for example, are indistinguishable in this classification.

A tree can also be classified by a set of the partitions of all-order bifurcations, that is, its partition set. The 7(1 6(2 4(1 3))) tree, for example, gives rise to the partition set {(1,6),(2,4),(1,3)}. Each partition of the set is assigned to the partition distribution of the degree of that particular partition. The degree of a partition (r,s) is the sum of r and s. Each tree type contains partitions of different degrees, and these partitions are thus assigned to more than one partition distribution. For one tree, even for a large one, the partition distributions are sparsely filled. Usually many trees are required to have well-filled partition distributions. Alternatively, you may apply special lumping schemes to assign the partitions of different degrees to one distribution (see Verwer and Van Pelt, 1986, and Section 10.6). Note that partitions (1,1) and (1,2) are trivial ones, since there are no other possibilities to partition two- or three-terminal segments. These two partitions are therefore not considered in the partition analysis.

Column 4 of Table 10-2 shows that the number of classes of the all-order partition pair distribution is still much smaller than the number of tree types. The "all-order partition" decomposition is allowed if independence can be assumed for the partition at all bifurcations. In that case, two groups of branching patterns can be compared on the basis of the frequency distribution of their partition pairs rather than their tree types (Verwer and Van Pelt, 1986).

The partition pair distribution is also useful in testing growth models. Van Pelt and Verwer (1983, 1984a, 1985, 1986) and Verwer and Van Pelt (1983) have defined growth models on the basis of branching probabilities of intermediate and terminal segments.

These rules for branching result in probability distributions of tree types (e.g., Fig. 10-5). Additionally, they also calculated the expected probability distribution of partition pairs (e.g., Fig. 10-6). The partition probability distributions in Fig. 10-6 are calculated for the random terminal growth model (i.e., random branching of terminal segments only) and for the random segmental growth model (random branching of all segments). These distributions show striking differences such that in practice they allow statistical testing of observed partition frequency distributions (see, e.g., Van Pelt and Verwer, 1984a; Van Pelt et al., 1986; Verwer and Van Pelt, 1986). Furthermore, the existence of topological

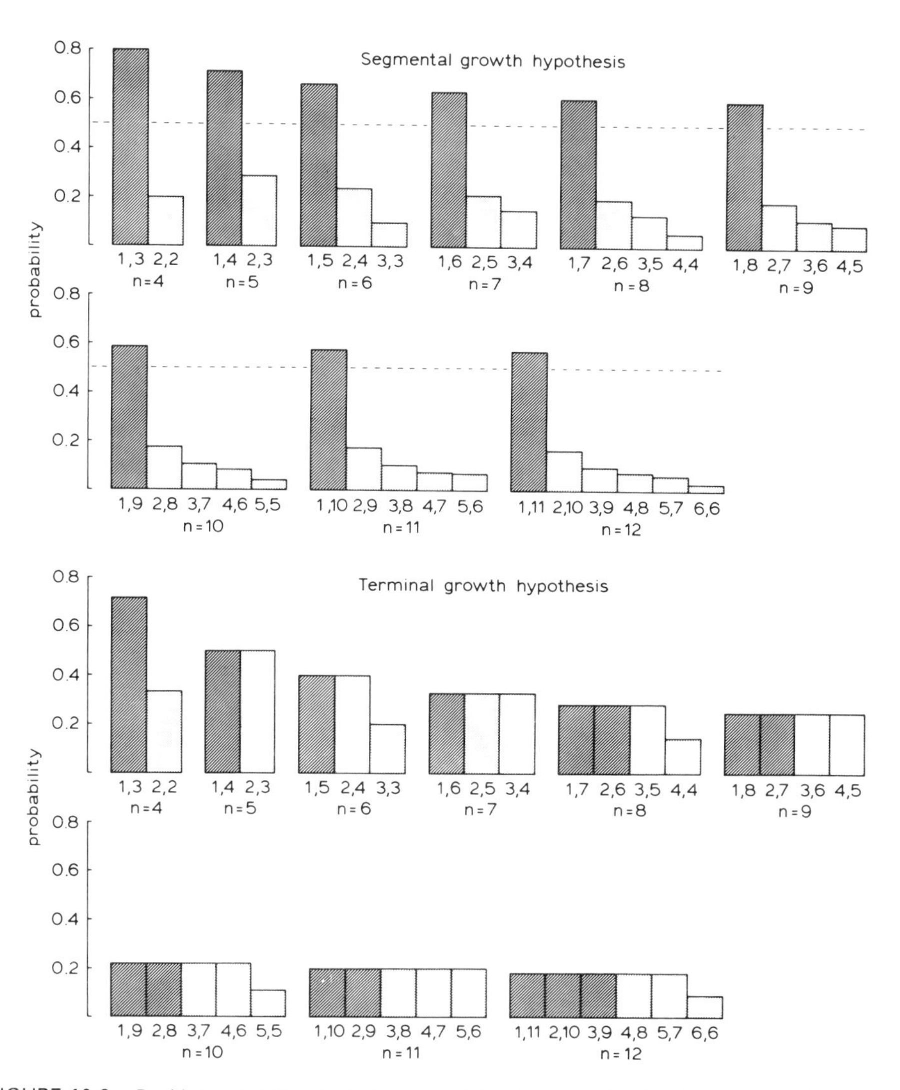

FIGURE 10-6. Partition probability distributions calculated for the random terminal growth and the random segmental growth models given for partitions up to degree 12 ($n = 4$–12). The distinction between shaded and nonshaded areas indicates a further aggregation of classes to obtain two-class distributions with equally probable classes for large values of n. Reproduced from Van Pelt and Verwer (1984a) with permission.

differences between neuronal trees of control and experimental groups can be detected by comparing the partition frequency distributions, using the chi-square test for differences in probabilities (Conover, 1980; Verwer and Van Pelt, 1986).

The classification according to partitions has been developed further to include multifurcating trees and essentially binary trees with some multifurcations by Verwer and Van Pelt (1987).

In addition, the best topological measure for asymmetry, i.e., the mean proportional sum of absolute deviations (PSAD) of degrees of subtrees (Section 10.3.1b) (Verwer and Van Pelt, 1986), for whole trees is based on the partitions of all bifurcations by taking the average of the asymmetry of partitioning at all bifurcations of a tree.

10.3.1b. Topological Asymmetry of a Tree Based on the Asymmetry in Partitioning. The asymmetry of a branching pattern can be determined directly by a measure for the asymmetry in the partitioning of the number of terminal tips (i.e., degrees of subtrees) at a branching point. The measure we have defined for the asymmetry of branching is the PSAD. For binary trees the PSAD is defined at a bifurcation with the partition of terminal tips $(r,n - r)$, in which r is the smallest degree number, as the absolute difference between the degrees of the two subtrees divided by the absolute value of the maximal possible difference (Van Pelt and Verwer, 1986).

$$\text{PSAD} = (n - 2r)/(n - 2) \qquad (10\text{-}2)$$

with $0 \leq \text{PSAD} \leq 1$.

The PSAD obtains the value 1 for the most asymmetric partition, $(1,n - 1)$, and the value 0 for the most symmetrical partition $(n/2,n/2)$ in case n is even. The topological measure for the asymmetry of a whole tree is defined as the mean of the PSAD values at all branching points leading to more than three terminal segments, $\overline{\text{PSAD}}$. An illustration of the calculation of the $\overline{\text{PSAD}}$ value is given in Fig. 10-7. In Fig. 10-7B the partitions are indicated by the abbreviated form n,r instead of $n(r,n - r)$, in which n is the number of terminal segments at a bifurcation, with $n > 3$, and r is the number of terminal segments in the smallest subtree.

In Fig. 10-7C the PSAD value at each partition is indicated from which the mean PSAD value ($\overline{\text{PSAD}}$) per dendrite is derived. The $\overline{\text{PSAD}}$ values for individual tree types are given in Table 10-3 up to degree 8. When we consider the $\overline{\text{PSAD}}$ values in Table 10-3 for $n = 8$, i.e., the degree-8 trees, we notice the irregular differences in size between the different $\overline{\text{PSAD}}$ values of the different tree types. When we make a qualitative comparison between these degree-8 $\overline{\text{PSAD}}$ values and the topological tree types with degree 8 as displayed in Fig. 10-4, these irregular differences appear to correspond quite well with the differences in the tree types. Tree types that have a small mutual difference in $\overline{\text{PSAD}}$ values appear to be more similar than those with a large mutual difference in $\overline{\text{PSAD}}$ values. A tree type has one $\overline{\text{PSAD}}$ number, and a set of tree types of the same degree shows a distribution of $\overline{\text{PSAD}}$ values that reflects the extent of asymmetry of these trees. With the increase of the degree n, the number of different $\overline{\text{PSAD}}$ class values increases generally much more than those of the "all-order partitions" (Section 10.3.1a); see Table 10-2). The $\overline{\text{PSAD}}$ is a sensitive measure of the topological structure and suitable for

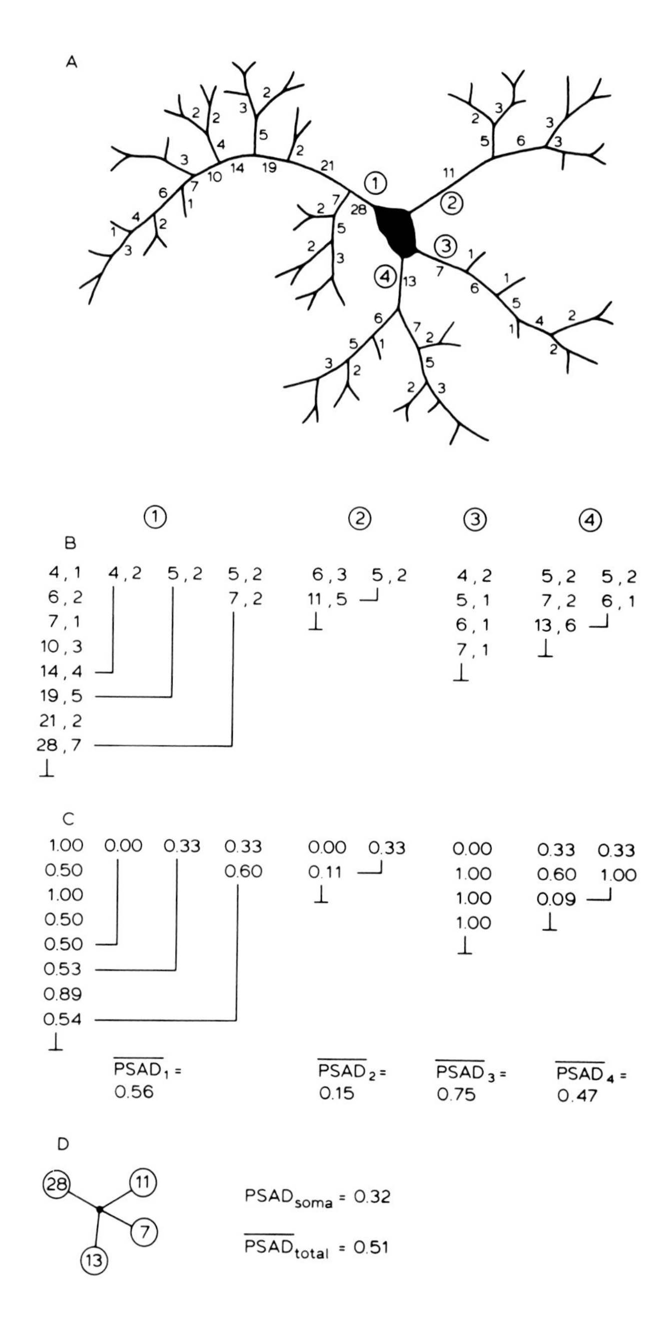

FIGURE 10-7. Illustration of the determination of the proportional sum of absolute deviations (PSAD) of the degrees of (sub)trees, of bifurcations, of dendrites, and of a soma. A: A schematic drawing of an aspiny neuron from the human striatum (Graveland et al., 1985). B: All-order partitions indicated as n,r instead of $n(r,n-r)$. C: PSAD values at all partitions and the mean PSAD values ($\overline{\text{PSAD}}$) of the four dendrites. D: PSAD value of the soma and the mean PSAD of the whole neuron, in which an equal weight is given to the PSAD value of the soma and those of each partition in a tree. Reproduced from Verwer and Van Pelt (1986) with permission.

TABLE 10-3. Standardized Topological Measures for Individual Tree Types (up to $n = 8$)[a]

n	Rel. no.	$\overline{C}_S = p_S = (p_e)_S$	D_S	V_S	$\overline{\text{PSAD}}$
4	1	1.00	1.00	1.00	1.00
	2	0.00	0.00	0.00	0.00
5	1	1.00	1.00	1.00	1.00
	2	0.50	0.00	0.00	0.50
	3	0.00	0.00	0.00	0.33
6	1	1.00	1.00	1.00	1.00
	2	0.75	0.50	0.25	0.66
	3	0.50	0.50	0.25	0.66
	4	0.25	0.50	0.25	0.75
	5	0.00	0.00	0.00	0.25
	6	0.00	0.00	0.25	0.00
7	1	1.00	1.00	1.00	1.00
	2	0.86	0.66	0.25	0.75
	3	0.71	0.66	0.25	0.78
	4	0.57	0.66	0.25	0.83
	5	0.43	0.33	0.00	0.50
	6	0.43	0.33	0.25	0.50
	7	0.43	0.66	0.25	0.87
	8	0.29	0.33	0.00	0.53
	9	0.14	0.33	0.00	0.47
	10	0.14	0.33	0.25	0.60
	11	0.00	0.00	0.00	0.10
8	1	1.00	1.00	1.00	1.00
	2	0.91	0.75	0.33	0.80
	3	0.82	0.75	0.33	0.83
	4	0.73	0.75	0.33	0.87
	5	0.64	0.50	0.11	0.63
	6	0.64	0.50	0.33	0.67
	7	0.64	0.75	0.33	0.90
	8	0.55	0.50	0.11	0.65
	9	0.45	0.50	0.11	0.64
	10	0.45	0.50	0.33	0.73
	11	0.36	0.25	0.11	0.40
	12	0.55	0.75	0.33	0.92
	13	0.45	0.50	0.11	0.67
	14	0.36	0.50	0.11	0.67
	15	0.27	0.50	0.33	0.72
	16	0.18	0.25	0.00	0.39
	17	0.18	0.25	0.11	0.33
	18	0.27	0.50	0.33	0.78
	19	0.18	0.25	0.11	0.44
	20	0.09	0.25	0.11	0.33
	21	0.18	0.25	0.33	0.67
	22	0.09	0.25	0.11	0.33
	23	0.00	0.00	0.00	0.00

[a]Abbreviations used: n, degree; rel. no., relative number within degree; $\overline{C}_S$, standardized mean centrifugal order; p, total path length; p_e, total terminal path length; D_S, standardized diameter; $\overline{\text{PSAD}}$, mean PSAD; V_S, standardized inverted vertex ratio (see Section 10.4). The inverted vertex ratio is equal to the ratio of the number of 1st and 2nd Strahler order branches minus two (Horsfield and Woldenberg, 1986).

indicating topological differences. For this reason we compare, in Section 10.4, the $\overline{\text{PSAD}}$ measure with topological variables that are considered to be measures for tree asymmetry or compactness and have a much lower number of classes per degree than the $\overline{\text{PSAD}}$ variable (Sections 10.3.1c and 10.3.2).

10.3.1c. Vertex Analysis. Recently Berry and Flinn (1984) proposed to distinguish three types of branching points (vertices) for the topological analysis of trees (i.e., vertex analysis). Given the interrelationships of the three types of branching points, vertex analysis is based mainly on the ratio of two types of branching points, bifurcation points from which two terminal segments arise and bifurcation points from which one terminal segment and one intermediate segment arise. This ratio is called the vertex ratio, V_r. The number of values (classes) the vertex ratio can obtain is as small as that of the first-order partitions (Table 10-2). For two growth models (i.e., bifurcations are randomly formed at all segments and only at terminal segments), Berry and Flinn (1984) have determined the value of the vertex ratio using computer simulations of large trees. Verwer and Van Pelt (1985) have derived the exact probability expressions of the vertex ratio for these two models.

By using the Monte Carlo test and the formulas given by Verwer et al. (1985), we have extended the application of the vertex ratio to growth hypotheses other than terminal and segmental growth. You should note that the probabilities of the vertex ratios are derived from the partition probabilities. The statistical test to evaluate the significance of the differences found for the observed and the expected vertex ratio was first described by Verwer et al. (1985). In comparison with the vertex analysis, the all-order partition analysis uses more topological information. In addition, the statistical testing of the vertex ratios requires many computer simulations (Monte Carlo tests), whereas the result of the partition analysis can be tested with the chi-square test, which does not require complicated computational facilities (Verwer and Van Pelt, 1983, 1986). Comparison of two groups of neurons examined with the partition analysis is done with the chi-square test for differences between discrete frequency distributions (Verwer and Van Pelt, 1986). Statistical comparison of two groups of neurons using the vertex ratio has not been described so far. The vertex ratio also represents aspects of symmetry of a tree (Berry and Flinn, 1984). Horsfield and Woldenberg (1986) suggest that the inverse vertex ratio is better and use it to express asymmetry of a tree. In completely symmetrical trees there are no bifurcation points from which both one terminal and one intermediate segment arise, and the inverted vertex ratio is zero. As asymmetry increases, so does the inverted vertex ratio.

10.3.2. Classification Based on Topological Distance from the Root

The use of a single measure for a whole tree that is dependent on the topology and rather easy to calculate is attractive. In this section we discuss single measures for whole trees that are based on the topological distance from the root and also reflect the extent of asymmetry of a tree. The topological distance between the root and, say, a branching point is defined as the number of segments on the path connecting these two points.

10.3.2a. Topological Center of Mass. Topological single measures for whole trees, such as the total topological path length of all bifurcation points and terminal tips, the total path length from all terminal tips (e.g., Werner and Smart, 1973), the mean centrifugal order of all segments (Triller and Korn, 1986; Van Pelt and Verwer, 1987), and the mean degree of all segments, are based on the topological distance from the root. These topological variables, however, are actually all identical. Since this is not immediately clear, we will show you these variables in more detail.

The centrifugal order of a segment in a tree is defined as the number of segments between this segment and the root of a tree. Therefore, order zero is assigned to the root segment. Figure 10-3F shows the order assignments of segments of a degree-5 tree. The number of segments at the successive segment orders 0, 1, 2, and 3 of this tree is 1, 2, 4, and 2 (Fig. 10-3F; Table 10-1). The mean centrifugal order, $\overline{C}$, of this whole tree is the ratio of the sum of the centrifugal order of all segments, C, which is tabulated in Table 10-1, and the total number of segments of a tree with degree n; i.e., $C/(2n - 1) = 16/9 = 1.78$. The mean centrifugal order of a tree is the topological equivalent of the center of mass of a tree. For a given size of tree, it indicates how far the segments are from the root. The farther they are, the higher the "mean centrifugal order." This will correlate with a high asymmetry of the tree. The $\overline{C}$ of any tree can be easily calculated in a sample and averaged for all trees of the same degree. A set of trees of unequal degree can be characterized by the way the averaged $\overline{C}$ depends on the degree. Van Pelt and Verwer (1987) have given different relationships between $\overline{C}$ and the degree of a tree for a whole range of different growth models.

The path length of a terminal or a branching point is defined as the number of segments between that point and the root of the tree. This definition is equivalent to the one for the centrifugal order of branching points and terminal tips (see Fig. 10-3F). A degree-n tree, a tree with n terminal tips, has n corresponding terminal paths, and the sum of the path lengths of all terminal tips is called "total terminal path length," p_e. The sum of all path lengths of all bifurcation points and terminal tips is called "the total path length" of a tree, p. The total number of paths of length 1, 2, 3, and 4 of the tree in Fig. 10-3F is 1, 2, 4, and 2, respectively. This distribution is equal to the distribution of the successive segment orders (Table 10-1). The path length of a point is always 1 plus the centrifugal order number of the corresponding segment that terminates in that point (Fig. 10-3F). Consequently, the sum of the centrifugal orders of all segments, C, is less than but related to the sum of all path lengths of bifurcation points and terminal tips, p. The relationship for this difference is

$$C = p - (2n - 1) \tag{10-3}$$

and so $\overline{C}$ is related to the mean path length of all branching and terminal points, $\overline{p}$, as

$$\overline{C} = \overline{p} - 1 \tag{10-4}$$

Table 10-1 gives you the sum values of the centrifugal orders of all segments (C), the total path length (p), and the total terminal path length (p_e) for trees up to degree 8. The number of different values (classes) per degree of the mean centrifugal order or the "sum of centrifugal orders of all segments" of a tree, C, is equal to the number of different values (classes) of the total path length (Table 10-2). This number also equals the number of classes of the total terminal path lengths, p_e. This equality is evident from the following relationship:

$$p = 2p_e - (2n - 1) \tag{10-5}$$

This in turn follows from the relationship

$$p = p_e + p_i \tag{10-6}$$

in which p_i is the total path length of all bifurcations, and

$$p_e = p_i + 2n - 1 \tag{10-7}$$

(Werner and Smart, 1973).

The degree of a segment is defined as the number of terminal tips of the subtree that arises from the pertinent segment; terminal segments are degree-1 segments (Fig. 10-8). The degree of a segment and path length of terminal tips are closely related variables, since the total path length of terminal tips is equal to the sum of the degrees of all segments (Werner and Smart, 1973). This is evident when we realize that the degree of a segment indicates how many paths from different terminal tips to the root contain that segment. So the mean degree ($\bar{n}$) of all segments of a tree is $p_e/(2n - 1)$, and consequently, the mean centrifugal order is related to the mean degree ($\bar{n}$) as

$$\bar{C} = C/(2n - 1) = 2(\bar{n} - 1) \tag{10-8}$$

From equations 3 to 8 it can be concluded that the following single topological measures for whole trees—the mean centrifugal order ($\bar{C}$), the sum of the centrifugal order numbers of all segments (C), the mean total path length ($\bar{p}$), the mean total terminal path length ($\bar{p}_e$), the total terminal path length (p_e), and the "mean degree of all segments of a tree" ($\bar{n}$)—are actually identical. They all have an equal topological interpretation and an equal number of classes per degree; after standardizing, their value per degree ranges between 0 and 1 (Section 10.4), and the standardized values of these single measures are identical.

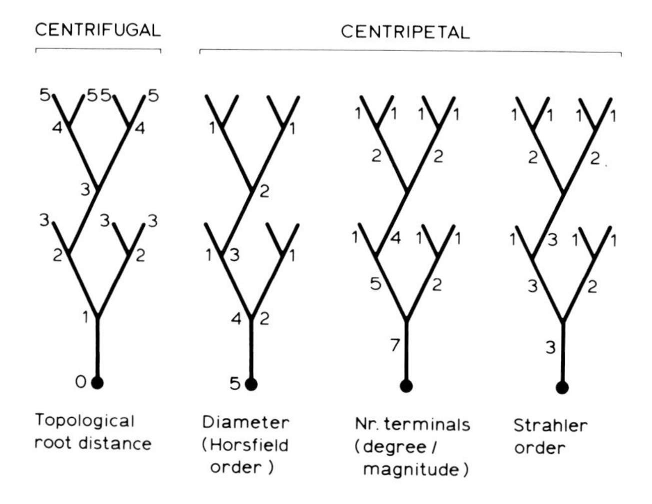

FIGURE 10-8. Illustration of different ordering systems. The centrifugal order or topological path length indicates the topological distance from the root. The topological diameter is the largest topological distance from the root. The centripetal Horsfield order reflects the diameter of all subtrees of a tree. Reproduced from Uylings et al. (1986a) with permission.

10.3.2b. Topological Diameter of a Tree. Each tree type has at least two terminal points with the greatest path lengths in the tree. This maximal path length of a tree, i.e., the maximal topological extension, is called the topological diameter of a tree, D. For a set of trees with a given degree, there is one tree type whose diameter is greater than those of other types, D_{max}, and at least one tree type whose diameter is smaller than those of the other types, D_{min}. The diameter measure, D, is also indicative of the extent of asymmetry, although in a different way from the mean centrifugal order. As can be seen in Fig. 10-4 and Table 10-1,

$$D_{max} = n \quad (10\text{-}9)$$

for the tree that is topologically the most asymmetric or elongated in the set of degree-n trees, i.e., trees with n terminal segments. The value of D_{min} is also a function of n, although a less easy one (see Appendix, Section 10.9) for trees that are topologically most symmetrical. The diameter of trees of a given degree takes all integer values between its maximal and minimal values, and therefore the number of classes of the topological diameter (N_D) is given by

$$N_D = D_{max} - D_{min} + 1 \quad (10\text{-}10)$$

This number of classes is considerably smaller than that of the individual tree types or the mean centrifugal order and is slightly larger than the number of classes of the first-order partitions and the vertex ratios (Table 10-2). The topological diameter, D, is used in river network analysis (see for review Jarvis and Woldenberg, 1984) and has the advantage of being one of the simplest topological measures. The D_{max} and D_{min} values for any given degree n are also parameters in the relations for the extremes of "total path lengths" and sum of centrifugal orders, which are ultimately all functions of n. Werner and Smart (1973) described these functions for the maximal total path length value (p_{max}) as

$$p_{max} = n^2 + n - 1 \quad (10\text{-}11)$$

and the minimal total path length (p_{min}) as

$$p_{min} = 2nD_{min} - 2^{D_{min}} + 1 \quad (10\text{-}12)$$

The extreme sums of the centrifugal order numbers of all segments of a tree for any given n are (Van Pelt and Verwer, 1987):

$$C_{max} = n(n - 1) \quad (10\text{-}13)$$

and

$$C_{min} = 2n(D_{min} - 1) - 2^{D_{min}} + 2 \quad (10\text{-}14)$$

These extreme value relations are given here to show the interrelationships to the extreme values of the topological diameter (Table 10-1).

10.4. COMPARISON OF MEASURES FOR ASYMMETRY OF TREES

10.4.1. Trees with Bifurcations Only (Strictly Binary Trees)

In Section 10.3 we have seen that the single topological measures for whole trees in a way reflect, among other things, the extent of topological asymmetry of trees. In this section we describe how these different measures are correlated with the $\overline{\text{PSAD}}$, the direct and best measure for asymmetry (Section 10.3.1b). As was indicated above, the $\overline{\text{PSAD}}$ is a sensitive measure for the topological structure of trees, which has a very large number of classes per degree (Table 10-2), and has a value range between 0 and 1 (Section 10.3.1b). For this comparison we have standardized the other measures for whole trees to obtain for any given degree and for each measure a range of discrete values between 0 and 1. The standardization for any given degree n used is the ratio of the observed value minus the lowest possible value to the maximal possible difference for the given n. You can see from the definition that the number of classes for the nonstandardized and standardized parameters are equal. They are given in Table 10-2.

From the relations given in Section 10.3.2a and this standardized definition, it follows that the values of the standardized mean centrifugal order, $\overline{C}_s$, for the different tree types are identical to the values of the standardized sum of the centrifugal orders of all segments of a tree, the standardized total path length, the standardized total terminal path length, and the standardized "mean degree." So we compare the $\overline{\text{PSAD}}$ with $\overline{C}_s$, the standardized inverse vertex ratio, V_s, and the standardized diameter, D_s. The standardized diameter values can be easily calculated manually (see Appendix, Section 10.9, for formulas). In addition, we determined the correlation of the $\overline{\text{PSAD}}$ with the first-order partition, using the PSAD of the first-order partition, $(\text{PSAD})_1$. Note that, given the definition of PSAD, the $\overline{\text{PSAD}}$ and $(\text{PSAD})_1$ can only reach a value of 0 if the degree of the tree is even.

Table 10-3 shows the values of the standardized measures for the different tree types displayed in Fig. 10-4. Here you can observe rather clear differences between particular $\overline{C}_s$, V_s, and D_s values. In a first comparison of the different single measures for tree structures, the relative correlation coefficients are determined. First, for trees with the size of degree 9 and 8, respectively, Table 10-4 shows that for particular degrees the D_s measure has the highest correlation coefficient with $\overline{\text{PSAD}}$. For the degree-8 tree types, the correlation coefficient between $\overline{\text{PSAD}}$ and D_s is 0.929; between $\overline{\text{PSAD}}$ and $\overline{C}_s$ it is 0.797, and between $\overline{C}_s$ and D_s it is 0.909. From this comparison it appears that the D_s measure is an interesting topological measure, since it corresponds better with the $\overline{\text{PSAD}}$, although the number of D_s classes is more reduced than that of $\overline{C}_s$. On the other hand, the inverted vertex ratio does not appear to be the best index for the asymmetry of a tree if we consider the lower correlation coefficient (Table 10-4) for the particular degrees.

At present there are no good arguments from graph theory (e.g., Dr. E. Kaczmarek, Poznar, Poland, personal communication) to allow the comparison of the topology of trees of different degrees or to allow the discrete values of topological measures for trees of different degrees to be grouped together. The average of $\overline{\text{PSAD}}$ for a set of trees depends on the frequency of occurrence of different tree types and therefore depends on the mode of growth. It appeared from our analysis that for a whole range of different growth models the averaged $\overline{\text{PSAD}}$ has different values for different models, which, moreover, remain more or less constant for different degrees (tested up to degree 31). The

TABLE 10-4. Correlation Matrix of Standardized Topological Measures[a]

r	$\overline{C}_S$	D_S	V_S	$(\text{PSAD})_1$	$\overline{\text{PSAD}}$
Degree 8 (n_8 = 23)					
$\overline{C}_S$	1				
D_S	0.909	1			
V_S	0.682	0.745	1		
$(\text{PSAD})_1$	0.749	0.671	0.312	1	
$\overline{\text{PSAD}}$	0.797	0.929	0.700	0.578	1
Degree 9 (n_9 = 46)					
$\overline{C}_S$	1				
D_S	0.899	1			
V_S	0.649	0.716	1		
$(\text{PSAD})_1$	0.748	0.568	0.270	1	
$\overline{\text{PSAD}}$	0.763	0.910	0.690	0.449	1

[a] r = correlation coefficient. See Section 10.4.1 for explanation of the abbreviations of the topological measures.

'constant' averaged $\overline{\text{PSAD}}$ value for the different degrees for one growth model indicates that the discrete values of the $\overline{\text{PSAD}}$ measure may be grouped together, irrespective of the degree of the trees.

It has not yet been determined whether the values of the standardized topological measures D_s, $\overline{C}_s$, and V_s can be grouped together irrespective of the degree. However, we did compare the correlation of the $\overline{\text{PSAD}}$ values of all tree types of degree 6 to 9 with the D_s, $\overline{C}_s$, and V_s values, respectively (Table 10-5). In this comparison, too, the $\overline{\text{PSAD}}$ measure correlates better with the D_s measure than with $\overline{C}_s$ and much better than with V_s. In general, the D_s measure correlates well with the $\overline{C}_s$ measure (Tables 10-4 and 10-5). This also holds for tree types that are grown according to segmental growth (i.e., new segments arose at random with equal probabilities from all segments). For such a frequency distribution of tree types there is a high correlation (close to 0.9) between $\overline{C}_s$ and D_s for all degrees (Werner and Smart, 1973).

The power to distinguish between different groups of neuronal trees, using the diameter D or the standardized diameter D_s has not yet been examined. The diameter measure, however, appeared useful in river network analysis (e.g., Jarvis and Werrity, 1975; Werner and Smart, 1973; both articles also appeared in Jarvis and Woldenberg, 1984). The statistical testing of the D or D_s measures has not yet been done extensively. The topological diameter measure itself is one of the simplest measures. For statistical comparison with growth models, however, some simple measures, such as the vertex ratio (Verwer et al., 1985) and the mean centrifugal order, may require extensive computational facilities.

As shown in Tables 10-4 and 10-5, the correlations of the PSAD of the first-order partition with $\overline{\text{PSAD}}$ and D_s are much lower. For the degree-8 tree types the correlations are 0.58 and 0.67, respectively. Apparently, the first-order partition measure is a topological descriptor that is much less sensitive in indicating the asymmetry of a whole tree than the $\overline{\text{PSAD}}$, D_s, $\overline{C}_s$, and V_s measures.

TABLE 10-5. Correlation Matrix of Standardized Measures for Degrees 6 to 9 ($n_{6-9} = 86$)[a]

r	$\overline{C}_S$	D_S	V_S	$(PSAD)_1$	$\overline{PSAD}$
$\overline{C}_S$	1				
D_S	0.891	1			
V_S	0.676	0.742	1		
$(PSAD)_1$	0.772	0.593	0.289	1	
$\overline{PSAD}$	0.783	0.888	0.646	0.546	1

[a] r = correlation coefficient. See Section 10.4.1 for explanation of the abbreviations of the topological measures.

10.4.2. Multifurcating Trees

All neuronal trees are essentially binary trees; that is, nearly all branching points are bifurcations, but occasionally a multifurcation (e.g., trifurcation) occurs. The frequencies of multifurcations differ for the different neuronal cell types (e.g., Percheron, 1979). Nearly all branches of the neocortical pyramidal cell dendrites are bifurcations (Hollingworth and Berry, 1975; unpublished data). The frequency of trifurcations is somewhat higher in hippocampal pyramidal cell dendrites (unpublished data), and in Purkinje cell dendrites 5% to 10% of the branch points are trifurcations (Hollingworth and Berry, 1975).

What appears to be a trifurcation might be observed when the length of an intermediate segment is too small to enable resolution of separate bifurcations (e.g., Verwer and Van Pelt, 1987). In general, both the instrumental error of a semiautomatic measuring microscope system and the thickness of the neuronal tree segments determine the criterion of the computer program to detect tri- and multifurcations. For the neocortical dendrites measured with our system (Overdijk et al., 1978; Uylings et al., 1986b), we use the criterion that two bifurcations within a distance of 2 μm are considered to form a trifurcation. Multifurcations can influence the topological analysis. A general formula for PSAD has been developed to account for the presence of multifurcations in a tree (Verwer and Van Pelt, 1986). In a multifurcation (Fig. 10-3E), the number of terminal tips (n) is partitioned over m subtrees (in Fig. 10-3E, $m = 4$). In a multiple partition the PSAD measures the difference between the degree of each subtree and the mean degree of the m subtrees. The sum of these values is divided by the maximal possible difference, and the equation is

$$\text{PSAD} = [m/2(m - 1)(n - m)] \sum_{i=1}^{m} |r_i - (n/m)| \qquad (10\text{-}15)$$

in which r_i is the degree of the ith subtree.

You can establish topological asymmetry of a tree by calculating the mean PSAD value (Fig. 10-7). As was indicated above, a multifurcation can be observed if the length of an intermediate segment is too small to distinguish adjacent bifurcations. In that case, the multifurcation is an aggregate of bifurcations. This might affect the PSAD value, especially if the number of subtrees in the multifurcation (m) is close to the sum of the degrees of all subtrees, that is, when $n - m$ is small. For instance, a 4(2,2) bifurcation may be distinguishable from a 4(1,1,2) trifurcation, but the PSAD of a 4(2,2) partition is 0, and the PSAD of the 4(1,1,2) partition is 1.

For whole neurons we can regard the soma as one multifurcation and calculate the $\text{PSAD}_{\text{soma}}$ given the degrees of the different dendrites. In Fig. 10-7D, for instance, m is the number of dendrites, 4, n is the sum of all dendritic terminal tips of the neuron, 59, and the r_i, the degree of the ith dendrite, is indicated in the circles. Thus, the $\text{PSAD}_{\text{soma}}$ of the Fig. 10-6D neuron is $[4/(6 \times 55)] \times 26.5 = 0.32$. The $\text{PSAD}_{\text{soma}}$ can distinguish between different groups of neurons, for example, after or without a fiber lesion (Verwer and Van Pelt, 1986). The occurrence of trifurcations also influences the topological diameter measure and its standardized expression D_S. We have expanded the relations for D_{max} and D_{min} for trees containing a number of trifurcations. The standardized topological diameter of essentially binary trees with some trifurcations can be determined fairly easily.

10.5. ORDERING SYSTEMS FOR SEGMENTS

In previous sections we have shown you two topological systems for grouping segments: (1) the centrifugal ordering system starting at the root segment and (2) the ordering according to the degree of segments, that is, the number of terminal tips to which a segment can lead, starting at the terminal segments. The objective of ordering of segments is to group together those segments with similar structural characteristics (Uylings et al., 1975). This appeared to be essential in the analysis of segment lengths to elucidate in which part of the tree length changes take place.

Different systems have been used to order segments. Two main schemes can be distinguished: the centrifugal ordering system and the centripetal ordering system. The centrifugal ordering system starts at the root (i.e., the origin) of a tree and increases the order by one beyond each bifurcation (Fig. 10-8). Given the definition of order, i.e., the number of segments between the tree element considered and the root (Fig. 10-3), order zero is assigned to the proximal or root segment. The centrifugal order of points (branching points and terminal tips) is equivalent to their topological distance or path length to the root of a tree. Van der Loos (1959) and Rall (1959) introduced the centrifugal ordering system into the field of neuronal dendritic analysis, and it has been frequently applied since (Bok, 1959; Jones and Thomas, 1962; for review see Uylings et al., 1975, 1986a). The centrifugal order system is particularly applicable to studying the development of tree structures. When new segments are formed at the terminal segments, the centrifugal ordering of the already existing segments does not change, whereas it does change in the centripetal ordering (Uylings et al., 1975). In the neuroanatomical literature the centrifugal ordering of segments generally starts with 1. When the centrifugal ordering starts at 1, the relation of order of segments is the relation of topological distance of branch points and terminal tips, that is, the path length relation (Section 10.3.2a and Table 10-1).

Different centripetal ordering methods have been used as well (Fig. 10-8; Uylings et al., 1975). Centripetal ordering according to the number of terminal tips, i.e., degree, and the topological diameter are considered in Section 10.3. Until recently, the ordering according to Strahler (1952), a river geomorphologist, was the only centripetal ordering method applied in the analysis of neuronal dendrites after its introduction by Hollingworth and Berry (1975). In the Strahler ordering all terminal segments are called the first-Strahler-order segments. The Strahler order increases by one only when two branches of equal Strahler order are joined. Contiguous segments may have an identical Strahler order (Fig. 10-8). All contiguous segments with the same Strahler order form a so-called Strahler branch.

In all other centrifugal and centripetal ordering systems there are no contiguous segments with a similar order. There a branch is similar to a segment. In the analysis of river networks, the Strahler ordering has been abandoned. Other topological measures, such as topological diameter, path length, and the degree of a network, are applied at present (e.g., Jarvis and Woldenberg, 1984). The main reason for this is that the Strahler ordering in the analysis of river networks did not sufficiently distinguish between different river networks. For analysis of the lung airways (Horsfield, 1980) and some botanical trees (Park, 1985), however, the Strahler ordering and analysis can be meaningful. On the other hand, the Strahler ordering was abandoned for studies of modes of growth by Sadler and Berry (1983), but not completely, since the vertex ratio is related to the ratio of 1st and 2nd Strahler order branches (Horsfield and Woldenberg, 1986). From the analysis of segment length values (Uylings et al., 1986a) it appeared that centrifugal ordering is preferable to distinguish groups of terminal segments of pyramidal and nonpyramidal neurons. With respect to the length values of intermediate segments, however, the centripetal ordering according to the degree of subtrees is preferable (Uylings et al., 1986a; see Chapter 11 of this book).

In comparing control and experimental groups, "number of segments per centrifugal order per neuron" is frequently used (e.g., Greenough and Volkmar, 1973; Uylings et al., 1978a; Juraska, 1984). Segments arising from the same branching point have the same centrifugal order and thus constitute dependent observations (Verwer and Van Pelt, 1986). To avoid this dependence it is probably preferable to compare groups of neurons using the topological measures of topological diameter or mean centrifugal order (Table 10-1 and Section 10.3.2).

10.6. ANALYSIS OF INCOMPLETE TREES

In practice, the dendritic and axonal trees examined may not be completely visible because of incomplete staining, a too complex intermingling with other stained tree structures, or cutting. Therefore, they cannot be analyzed completely. The cutting effect can almost be nullified by reconstruction of the tree structures from a series of contiguous histological sections. Complete reconstruction of neurons, including merging of multisection trees, is often very time consuming (Capowski and Sedivec, 1981). For that reason, and because of the large variability of neurons within the same organism, it is often unfeasible to reconstruct a sufficiently large number of neurons for comparison of control and experimental groups. When multisection merging is not done, the cutting frequency can be reduced both by using very thick histological sections, preferably thicker than 300 μm, and by selecting the appropriate plane of cutting when the neuronal tree structures have a spatially preferential orientation (e.g., Uylings et al., 1986a). In Golgi-stained sections of about 160 μm, about 15% of the stained segments of pyramidal basal dendrites are cut or not measurable for other reasons (unpublished data). If only complete trees are allowed to be analyzed, a drastic reduction in the data will generally occur, which will most probably be biased towards small neurons (Uylings et al., 1983, 1986a). This can lead to wrong conclusions about differences in topological size between groups of neurons. For statistical evaluation of growth models, a number of complete trees may be sufficient for testing if the first-order partitions of trees are lumped per degree in two-class distributions (see Fig. 10-6). These two-class distributions are then tested according to the procedure described by Uylings et al. (1983).

Two situations can be distinguished for incomplete trees: (1) peripheral parts of a tree are invisible, and (2) only the peripheral parts of the tree can be measured and the proximal part, including the soma, is invisible (cf. Verwer and Van Pelt, 1986). When peripheral parts cannot be measured, the tree structures can be analyzed (though in a less complete way) provided that wrong sampling by a spatially preferential orientation of the tree structures is avoided and the incomplete trees are sufficiently large. Then the degrees (topological size) of groups of neurons can be compared to indicate the existence—but not the extent—of topological size differences. In addition, it has been shown by Van Pelt and Verwer (1984b) that if trees have grown according to the random terminal or segmental growth and the cutting of segments was at random, the topological properties in terms of the frequency distributions of the partitions remain unchanged. On the other hand, the effects of cutting on the frequency distributions of most topological measures is as yet unknown (Sections 10.3.2 and 10.4) and needs to be further examined.

When only complete peripheral parts of dendrites or axonal trees can be analyzed, the partition analysis can be applied as though they were normal trees. When only one or a few peripheral parts of large tree structures can be analyzed, the partition analysis can be used in the comparison of groups of neurons by examining the composite partition distribution $(1,n_i - 1)$, $(2,n_i - 2)$, $(3,n_i - 2)$, etc., in which n_i is the degree of a (sub)tree, which will be different for the different (sub)trees, to obtain a distribution with well-filled classes and to eliminate any possible dependence between the subtrees of a peripheral part (see the example of statistical tests in Verwer and Van Pelt, 1986; Verwer et al., 1987). Note, however, that the topological characteristics of peripheral tree parts need not be similar to those of the whole tree.

10.7. IN CONCLUSION

In this review different topological measures for describing tree structures with or without multifurcations were discussed. The topological tree type (which contains all the topological information) can rarely be used in the statistical comparison of control and experimental groups, since a sufficient number of all topological tree types is seldom observed in all groups. Therefore, topological parameters have to be used that provide a grouping of the observed trees into another discrete frequency distribution with a manageable number of filled classes, which necessarily ignore some structural information. The following topological measures appear to be of importance for statistical analysis: (1) the degree (i.e., the number of terminal segments) of a tree or a whole neuron—the degree indicates the size of a tree and can easily be used to test the difference between two or more groups (see, e.g., Verwer and Van Pelt, 1986); (2) the distribution of the partitions of subtrees in neuronal trees at all bifurcations or the first-order bifurcation. Different topological tree types can be grouped or lumped into a smaller number of classes on the basis of the partitions in the trees that retain sufficient topological differences to distinguish between different groups of neurons or to be applied in the analysis of growth models (e.g., Van Pelt and Verwer, 1984a). (3) For indicating the extent of asymmetry of trees, with or without multifurcations, the mean PSAD and the standardized topological diameter are the preferred topological parameters (Section 10.4). The standardized topological diameter can be determined easily and corresponds well with the more precise mean PSAD parameter, which is based on the partitions within a tree but is less easy to calculate. (4) The total path length of a tree (p), the total path length from terminal tips to the root of the tree (p_e), the mean degree of all segments of a tree, the total centrifugal order (i.e., the sum of the centrifugal orders of all segments, C), and the mean centrifugal order (i.e., the total centrifugal order divided by the total number of segments, $\bar{C}$) are all equivalent topological parameters. The relationships among these parameters are given in Section 10.3.2a and Table 10-3. They can be used to distinguish between different growth models and groups of neurons. They indicate indirectly whether or not trees are topologically asymmetric or whether the segments are farther away from or closer to the root.

In this chapter we have emphasized the importance of lumping topological tree types, especially according to the discrete topological variables discussed. One of the main reasons for this was to avoid the occurrence of classes that contain no observations in the discrete frequency distributions analyzed. This is important in statistical testing of discrete variables with, e.g., the discrete chi-square test. However, if the classes of values of topological variables are ordered in a natural way, other statistical tests, the Kruskall–Wallis test or the Mann–Whitney test, can be used (Verwer and Van Pelt, 1986; Conover, 1980). In contrast to the chi-square test, these two tests take the order between the classes into account, and as a further advantage classes without observations do not cause any problem. For detailed examples see Tables 10-3, 10-4, and 10-5 and Verwer and Van Pelt (1986, p. 203).

Topological variables discussed in this chapter that can be ordered in a natural way (i.e., in lower and higher values) are, (1) for a given degree, the mean PSAD, all standardized variables discussed and their nonstandardized variables, and the first-order partition variable, and, (2) irrespective of the degree of a tree, the mean PSAD variable, the "composite partition" distribution $(1,n_i - 1)$, $(2,n_i - 2)$, etc. (Section 10.6; Verwer and Van Pelt, 1986), the degree of a tree or total degree of a neuron, and the number of dendrites per neuron. It has not been established whether or not the standardized variables discussed in this chapter can be grouped together irrespective of the degree of a tree. Whether or not the above-mentioned topological descriptors can reflect functional properties of neurons still needs to be determined. Certainly, relationships exist between neuronal form and function (Horwitz, 1981; Miller and Jacobs, 1984).

10.8. SUMMARY

A review is given of different topological measures that can efficiently characterize different groups of neuronal trees. Several topological classifications are compared to evaluate the merits of simple topological measures as degree, that is, the number of terminal segments of a tree, the partition of terminal segments at a bifurcation, the topological diameter (i.e., the maximal extension of a tree), the total path length of a tree, and the mean centrifugal order (i.e., the topological center of mass of a tree).

In this review, relations of topological measures are given for trees with bifurcations only, and relations are described for multifurcating trees. The mean PSAD measure (Section 10.3.1b) and the standardized diameter measure appear to be the preferred topological measures for the extent of asymmetry of whole trees. In addition, ordering methods for further analysis of segments are discussed, and the preferable ordering is indicated for both terminal and intermediate segments.

10.9. APPENDIX

The topological diameter of a tree is the longest path length from a terminal tip to the root. The standardized topological diameter, D_s, is defined as

$$D_s = (D - D_{min})/(D_{max} - D_{min})$$

D_{max} is the maximum topological diameter possible for a tree of a given degree n, i.e., the tree that is topologically most elongated or asymmetric, and thus

$$D_{max} = n$$

The value of D_{min} (the minimal topological diameter possible for a tree with degree n) is also a function of n (Werner and Smart, 1973):

$$D_{min} = \lceil \log_2(n) \rceil + 1$$

in which $\lceil x \rceil$ is the smallest integer greater than or equal to x. The formula for D_s is therefore

$$D_S = \frac{D - \lceil \log_2(n) \rceil - 1}{n - \lceil \log_2(n) \rceil - 1}$$

With this formula the D_s can be easily determined, even without a computer.

ACKNOWLEDGMENTS. The authors are indebted to Mr. A. Janssen for correcting the English and typing this manuscript.

11

Statistical Analysis of Neuronal Populations

HARRY B. M. UYLINGS, JAAP VAN PELT, RONALD W. H. VERWER, and PATRICIA McCONNELL

11.1. INTRODUCTION

The neuronal tree alters during development and aging and in several cases during disease and experimental treatment (e.g., Mrzljak et al., 1988; Coleman and Flood, 1987; De Ruiter and Uylings, 1987). In quantitative assessment, metric and topological analysis offer us important tools for estimating the type and size of these alterations. Topological analysis, dealing with the number of branchings and connectivity pattern of segments but not the physical size of individual neurons, has been dealt with in Chapter 10. Metric analysis, referring to neuronal size of individual neurons, has been dealt with in Chapter 9 of this book. In this chapter we discuss metric analysis further, with special attention to populations of neurons, and consider the variation between and within animals to estimate the number of animals and the number of neurons required for a statistical comparison.

The metric analysis of three-dimensional tree structures requires measurements of the three-dimensional coordinates of points that characterize the tree structure (Fig. 11-1). Analysis of the two-dimensional, projected images gives an underestimation of the length ranging from 0 to 100%. For spherically oriented trees, this underestimation is on the order of 20% (Uylings et al., 1986a; Sedivec et al., 1986). This underestimation is practically absent in two-dimensional structures such as retina ganglion cells analyzed in whole mounts or Purkinje cells. Because of the very large number of measurements, semiautomatic neuron-tracing systems are usually necessary for the acquisition of these three-dimensional coordinates. For a review of these semiautomatic systems, we refer to Chapter 4 in this book and to Uylings et al. (1986a).

HARRY B. M. UYLINGS, JAAP VAN PELT, RONALD W. H. VERWER, and PATRICIA McCONNELL • Netherlands Institute for Brain Research, 1105-AZ Amsterdam, The Netherlands.

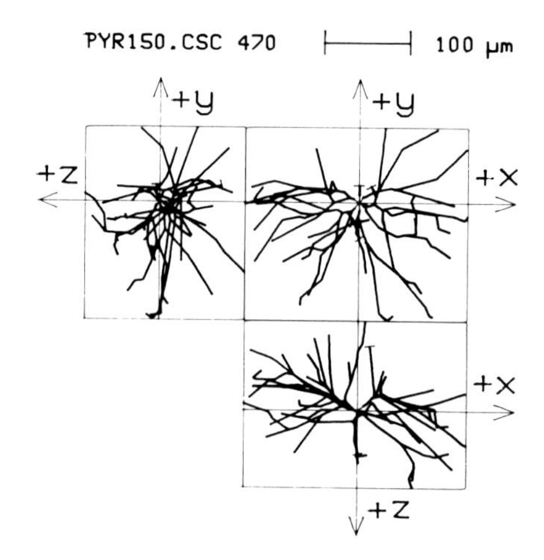

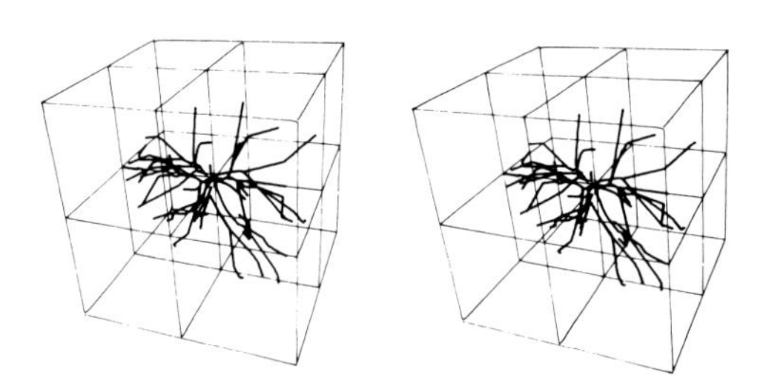

FIGURE 11-1. Computer graphics displays of three-dimensional basal dendritic field of a layer-III pyramidal neuron in the visual cortex of an adult rat, reconstructed with the neuron-tracing system described in Overdijk et al. (1978). The top display is the projection onto three orthogonal planes in which the *y* axis is perpendicular to the pia. The bottom figure is a stereo pair of the same dendritic field. The apical dendrite is left out. Reproduced from Uylings et al. (1986a) with permission.

11.2. METRICS OF SOMATIC SIZE AND DENDRITIC SEGMENTS

In this section, techniques and limitations for some elementary measures of neuron populations are discussed. These techniques are to calculate soma areas and volumes and dendritic lengths and to analyze the lengths of intermediate segments as a function of their distal targets.

11.2.1. Somatic Size

One of the first neuronal variables measured in many morphometrical studies is somatic size. The somatic surface area is one of the variables of interest in establishing electrophysiological properties of a neuron (e.g., Rall, 1977). The somatic size is reported to be moderately but significantly correlated with the summed cross-sectional area of the proximal or root segments of dendrites of a neuron (e.g., Hillman, 1979; Rose et al., 1985; Sedivec et al., 1986), with the total dendritic length (Hillman, 1979; Hull et al., 1981; McMullen et al., 1984), with the dendritic volume (Arkin and Miller, 1988), and with the diameter of a two-dimensional dendritic field (Rodieck et al., 1985) of motoneurons, cortical pyramidal and nonpyramidal neurons, Purkinje cells, neostriatal spiny neurons, and retinal ganglion cells.

Because of the optical shadow effect of Golgi-impregnated or HRP-filled somata, it is not possible to obtain accurate measurements of somatic depth values with normal research light microscopes. Therefore, the somatic size is generally inferred from its projected area (e.g., De Ruiter and Uylings, 1987; Dann and Buhl, 1987; Katz, 1987), or estimated as the surface area on the basis of the measured length of the projected major and minor axes (e.g., Kernell, 1966; Rose et al., 1985) or volume of a prolate (i.e., elongated) spheroid (e.g., Dinopoulos et al., 1988). The projected surface area (S_p) and the volume (V) of a prolate spheroid are calculated with the following equations:

$$S_p = (\pi/4)AB \quad (11\text{-}1)$$

$$V = (\pi/6)AB^2 \quad (11\text{-}2)$$

in which A and B are the lengths or diameters of the major and minor axes, respectively. The minor axis of the soma is the longest axis perpendicular to the major axis.

The determination of the somatic size is more difficult for those types of neurons in which the border between soma and dendrite (i.e., the dendrite origin) is not really clear (e.g., in reticular formation and globus pallidus, some cortical nonpyramidal neurons, and motoneurons). This difficulty may also cause problems in determining the number of dendrites per neuron. Leontovich (1973) gave rather objective rules for the localization of a morphological dendritic origin: the dendritic origin is the point from which a dendritic segment with a more or less constant diameter originates. The dendritic base is relatively thick in comparison to the segments, and as it tapers relatively rapidly, it is considered to be part of the soma. Usually the outline shape of the soma is extrapolated through the dendritic base. To be certain, Ulfhake and Kellerth (1981) and Rose et al. (1985) compared the number of dendrites measured at the somatic outline with the number of dendrites at a distance of about 30 μm from the cell body and found in their studies somewhat lower numbers of dendrites at the somatic outline. For this reason, plus the fact that dendrites can branch within 30 μm from the soma, additional measurements at a short distance of the somatic outline are not really helpful in the estimation of the number of dendrites per neuron.

11.2.2. The Size of a Dendritic Tree

In a three-dimensional metric analysis the size of a dendritic tree is generally estimated by the total length of all segments. Another measure for the size of a dendritic tree is the total membrane area of a dendrite or of all dendrites of a neuron (e.g., Cameron et al., 1985; Rose et al., 1985; Cullheim et al., 1987a). Together with the diameter and segment tapering, this is of importance in the relationship between tree structure and electrophysiological characteristics of the neuron (e.g., Rall, 1977; Horwitz, 1981). The estimation of the membrane area of the dendritic field may only be accurate in neurons that have sufficiently thick dendritic segments such as spinal motoneurons. Several types of neurons, however, have many dendritic segments whose diameters are close to the resolving power of the light microscope. Examples are dentate granule cells (De Ruiter and Uylings, 1987; Flood et al., 1987), cortical nonpyramidal neurons (Parnavelas and Uylings, 1980; McMullen et al., 1988), and basal dendrites of pyramidal neurons.

Studies of motoneurons (Cullheim et al., 1987a; Ulfhake and Kellerth, 1981; Cameron et al., 1985; Ruigrok et al., 1985; Rose et al., 1985; Sedivec et al., 1986) showed that membrane area, total length, and number of terminal tips of individual dendrites are correlated with the diameter of its root segment. For techniques estimating these relationships when measurements of both variables are subject to error or variation, we refer to the papers by Hofman et al. (1986) and Uylings et al. (1986b). Occasionally the size of the dendritic field is estimated by the area of the envelope of the whole dendritic field projected onto the plane of sectioning. This measure is appropriate only if the projection plane is an optimum one, e.g., the sagittal plane for cerebellar Purkinje neurons, if the dendritic field is rather flat, and if the area is densely packed with dendritic segments, e.g., the Purkinje cell dendritic field (Hollingworth and Berry, 1975). This measure is not really informative in other cases.

The "total length of the dendritic field" is a variable that indicates the existence of general size differences between groups of neurons (De Ruiter and Uylings, 1987) and the rate of dendritic growth in developmental studies (Uylings et al., 1986a). Our studies on the neuronal development in rat visual cortex, for example (Parnavelas and Uylings, 1980), showed that in the period of rapid dendritic growth, i.e., between postnatal days 6 and 18, the basal dendrites of the large layer-V pyramidal cells grew about three times faster than the dendrites of the layer-IV nonpyramidal cells. The total-length variable, however, is susceptible to the effect of cutting if the dendritic field is not completely reconstructed from serial sections or not completely contained in single histological sections. If the dendritic field has a preferential orientation in the three-dimensional space, the effect of cutting may be dramatic when the wrong plane is chosen for sectioning, e.g., perpendicular to the preferential orientation. When incomplete dendritic trees are examined in single histological sections with an appropriate plane of sectioning, the age at which the total dendritic length reaches its maximum value can be checked by comparing it with the age at which the maximum average value of "radial distance" or "path length" from uncut terminal endings to soma center is reached (Uylings et al., 1986a).

The metric path length of a terminal tip is the summation of the curvilinear length along the dendritic segments from the terminal tip to the dendritic root or origin (Van der Loos, 1959). A very high percentage of the total dendritic length in uncut trees is formed by the total length of the individual uncut terminal segments. For rat visual cortex layer-IV multipolar nonpyramidal cells, this is about 80%; for layer-II/III basal dendrites, about 85%; and for the human dentate granule cell dendrites, about 75% (unpublished data). The comparison of different experimental groups may therefore be based on the analysis of the length of individual terminal segments instead of the total dendritic length when neurons in single sections are studied and a significant part of their dendritic field has been cut off. The assumption here is that the sample of uncut terminal segments contained in the single section is representative of all terminal segments.

11.2.3. Metric Analysis of Segments

As indicated above, metric analysis of segments is of importance for revealing length changes and especially for the question of which types of segments are altered in length. In Chapter 10 of this book we have discussed the different systems for ordering or classifying segments. From our studies it appears that the centrifugal ordering system is preferable for the analysis of length of terminal segments. The centrifugal ordering indicates the topological distance from the root and starts with order numbering from the root or dendritic origin and increases the order by one beyond each bifurcation (Fig. 11-2).

For tree structures with a main shaft from which dendritic branches arise asymmetrically, the centrifugal ordering has to be adapted to ensure that "equivalent" segments receive the same order (Fig. 11-3). The lengths of terminal segments of cortical pyramidal dendrites and nonpyramidal dendrites differ significantly between different centrifugal orders (e.g., Uylings et al., 1978a,b, 1980). The centrifugal ordering has the advantage that it can also be used if some peripheral parts of the tree are lost by cutting. In addition, when new segments arise at the terminal segments, which is the most common situation in normal development (e.g., Van Pelt et al., 1986), the centrifugal order number of already existing segments will not change (Uylings et al., 1975).

On the other hand, it appears more appropriate to examine the length of intermediate segments in relation to their degree and type (Fig. 11-2; Table 11-1). The degree of a segment is equal to the number of terminal tips to which it leads. Table 11-1 shows the length values of segments of different degrees and types for cortical pyramidal basal dendrites and dentate granule cell dendrites. These values have been described for cortical nonpyramidal neurons by Uylings et al. (1986a). For these types of neurons, it appears that the group of segments called "next to terminal" (Buell and Coleman, 1981) consists of three significantly different subgroups, i.e., the degree-2 segments, the degree-3 (iit-type) segments, and the iit type of segments with a degree equal to or larger than 4 (Table 11-1). The iit-type segment is an intermediate segment branching into a terminal and an intermediate segment.

Terminal segments in particular (Uylings et al., 1978a,b; Juraska et al., 1980; Parnavelas and Uylings, 1980; Flood et al., 1987; De Ruiter and Uylings, 1987), and also the degree-2 intermediate segments (De Ruiter and Uylings, 1987), display length changes resulting from experimental treatment or continued growth in adulthood. These length changes can be obscured if the different types of segments are not distinguished (Uylings et al., 1978a,b; Juraska et al., 1980).

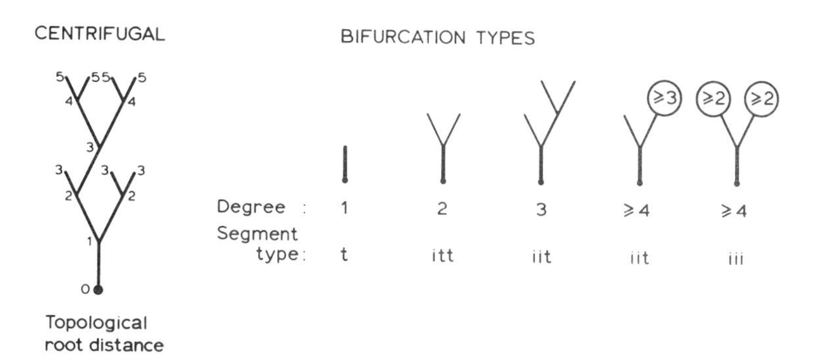

FIGURE 11-2. The centrifugal ordering system is illustrated in the left drawing; the root segment is the order-0 segment. The right drawing indicates the different types of segments distinguished. The degree of a segment is the number of terminal tips to which it leads. t, terminal segment; itt, intermediate segment that bifurcates into two terminal segments; iit, intermediate segment bifurcating into an intermediate and a terminal segment; iii, intermediate segment bifurcating into two intermediate segments. Reproduced from Uylings et al. (1986a) with permission.

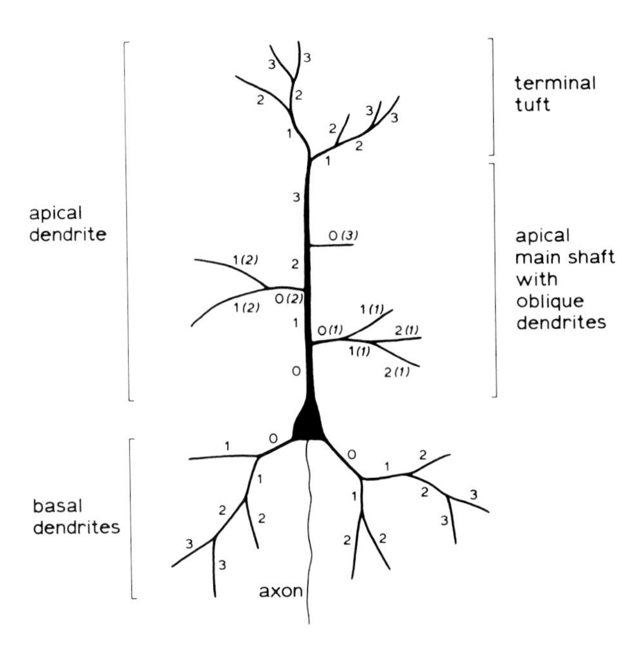

FIGURE 11-3. The adapted centrifugal ordering for an apical dendrite with a main shaft from which oblique dendrites arise asymmetrically and in their turn branch symmetrically. The numbers indicate the centrifugal order of the segments, and the numbers in parentheses in the oblique dendrites indicate the topological distance of the origin of the oblique dendrite to the soma. Reproduced from Uylings et al. (1986a) with permission.

11.3. ANGULAR METRICS OF BIFURCATIONS

Relatively few authors have published data on the bifurcation angles in the dendritic tree. Bifurcation angles have been measured in cortical pyramidal and nonpyramidal neurons, dentate granule cells, and Purkinje cells (Mungai, 1967; Smit and Uylings, 1975; Williams and Matthijsse, 1983; Calvet and Calvet, 1984). From these studies we

TABLE 11-1. Length Segments of Rat Pyramidal Basal Dendrites and Human Dentate Granule Cells[a]

Type of segment	Mean (μm)	CV (%)	Number of animals
Rat pyramidal basal dendrites			
All segments	46.4	5.1	6
Terminal segments	70.5	6.0	6
Degree 2 (itt)	24.9	17.	6
Degree 3 (itt)	15.8	37.	6
Degree ≥4 (itt)	11.7	27.	6
Degree ≥4 (iii)	10.4	7.7	6
Human dentate granule cells			
Terminal segments	130	33	5
Degree 2 (itt)	99	21	5
Degree 3 (iit)	67	36	5
Degree ≥4 (iit)	41	34	5
Degree ≥4 (iii)	29	32	5

[a]CV, coefficient of variation (i.e., ratio of standard deviation and mean) between subjects; itt, intermediate segment branching into two terminal segments; iit, intermediate segment branching into an intermediate and a terminal segment; iii, intermediate segment branching into two intermediate segments.

see that angle values show a very large variability and display no significant changes under the experimental circumstances of an enriched condition (Uylings et al., 1978a) or aging and senile dementia (Williams and Matthijsse, 1986). At present, therefore, it only appears meaningful to analyze bifurcation angles if "real" dendritic trees have to be generated by the computer. It is also noteworthy that nearly all bifurcations of three-dimensional tree structures examined so far have their parent and daughter segments lying within a plane or a thin planar sheet (Uylings and Smit, 1975).

11.4. SPATIAL ORIENTATION OF TREES

In the previous sections, the size of neurons and their parts was analyzed. In this section, the spatial extension of dendrites is discussed. The tissue surrounding the neurons is divided into various shaped regions, and the extent of the dendrites and axons that grow into each region is calculated, thereby showing where a neuron or population extends its influence.

11.4.1. Spherical and Circular Orientation Methods

To examine the spatial extent of neuronal tree structures, Sholl's (1953) concentric-spheres method and its projection variant, Eayrs' (1955) concentric-circles method, are frequently used (e.g., Katz, 1987; Bellinger and Anderson, 1987; Juraska, 1984). In Sholl's concentric-spheres method illustrated in Fig. 10-2, the number of dendrites per neuron is counted, and the number is determined of dendritic intersections with consecutive spheres centered at the neuronal somata and spaced equidistantly at 20 μm. In the concentric-circles method the same dendritic features are determined for a set of concentric circles or cylinders spaced equidistantly at about 20 μm, together with the number of bifurcations and terminal tips within each concentric shell. The concentric-circles method is relatively easy to apply and may allow the detection of great differences between groups of neurons. The concentric-spheres or -circles method quantifies only some aspects of the radial distribution of the neuronal trees. However, these methods are not really very sensitive, since quite different neurons may have the same set of dendritic intersections (Fig. 10-2). These methods do not enable orientation differences to be detected; also, the number of intersections is determined by both metric and topological characteristics of neurons. A classic example is the study of Valverde (1970), in which the concentric-circle method was used but no significant differences were detected. However, an orientation-density analysis (Section 11.4.4) of the same neurons (Ruiz-Marcos and Valverde, 1970) showed significant differences.

The concentric-circles method can be refined so that orientation differences can be studied by dividing the circles into different sectors of about 30° (e.g., Greenough et al., 1977; Coleman et al., 1981; Eysel et al., 1985). Orientation analysis with circular techniques is especially appropriate for neurons that have their dendritic field in a two-dimensional plane, such as the retinal ganglion cells in whole-mount preparations. Because of the physiological orientation sensitivity of retinal ganglion cells, the study of the orientation of the entire dendritic field within the different retinal regions is of great interest (Leventhal and Schall, 1983; Eysel et al., 1985; Rodieck et al., 1985; Schall and Leventhal, 1987). For the statistical comparison of the angular distribution of dendritic intersections between concentric circles or between two or more groups of neurons, the chi-square test described by Batschelet (1981) can be used.

Other but more complicated techniques were also used for this comparison by Glaser et al. (1979b) and Coleman et al. (1981). The circular orientation of (projected) dendritic fields is also examined by superimposing a large circle on the entire dendritic field subdivided into angular sectors (e.g., into 30° intervals or bins). Different analysis procedures are followed in the literature. A polar diagram (Fig. 9-28) determines the total length of dendritic elements present within each angular sector per neuron and per group of neurons (e.g., Borges and Berry, 1978). Another procedure determines the orientation of the mean angle of the terminal tips per neuron (Verwer and Van Pelt, 1986). The whole sample of neurons is tested for randomness of orientation with the Rayleigh test or with the nonparametric test of Moore (for a detailed description, see Batschelet, 1981; Leventhal and Schall, 1983). The orientation of two or more groups of neurons is statistically compared using the chi-square test (Batschelet, 1981).

The orientation preference of three-dimensional neuronal tree structures is frequently examined by means of two-dimensional orientation analysis of these neuronal trees projected onto three orthogonal planes. True three-dimensional techniques (extended from two-dimensional techniques) are also presently available for studying the possible presence of a preferential orientation of complete trees in the three-dimensional space. Such a three-dimensional extension is the extension of the "concentric circles with angular sectors" method given by McMullen et al. (1984). They determined the total three-dimensional length of segments within "spherical sectors" bounded by cones at 30° angular intervals with respect to the vector that points from the soma to the pia (Fig. 11-4) and by concentric spheres at 50-μm radial distance intervals.

Recently, McMullen et al. (1988) applied this spherical sector method without using concentric spheres in their analysis of the effect of neonatal deafening on dendritic orientation in the auditory cortex, having spherical sectors with equidistant angular intervals of 30° and 15°, respectively. This three-dimensional procedure does not have the projection shortcomings present in the circular orientation methods. The three-dimensional space contained within each equidistant angular sector, however, is not the same (Uylings et al., 1986a). This follows from the relationship between the solid angle (Ω) and the apical angle of its cone (2ϕ), the apex of which is at the center of the neuronal soma (thus, Φ is the angle with the vector that points from the soma perpendicular to the pial surface). This relationship is:

$$\Omega = 2\pi(1 - \cos\phi) \tag{11-3}$$

The solid angle is a dimensionless measure for three-dimensional space contained in a conical sector of a sphere. From equation 11-3 it follows that the three-dimensional space for the six equidistant 30° angular sectors can differ by up to a 1 : 3 ratio (see lower figure in Fig. 11-4; Uylings et al., 1986a). In Fig. 11-4 (top) we show the six spherical sectors that contain equal volumes with ϕ values of 48.2°, 70.5°, 90°, 109.5°, and 131.8°. When the sphere is divided into seven spherical sectors, each having the same volume, the ϕ values are 44.41°, 64.6°, 81.79°, 98.21°, 115.37°, and 135.58°. François et al. (1987) used seven spherical sectors in their orientation analysis, which they subdivided in such a way that the sphere is divided in 38 equispatial portions; i.e., each portion has the same

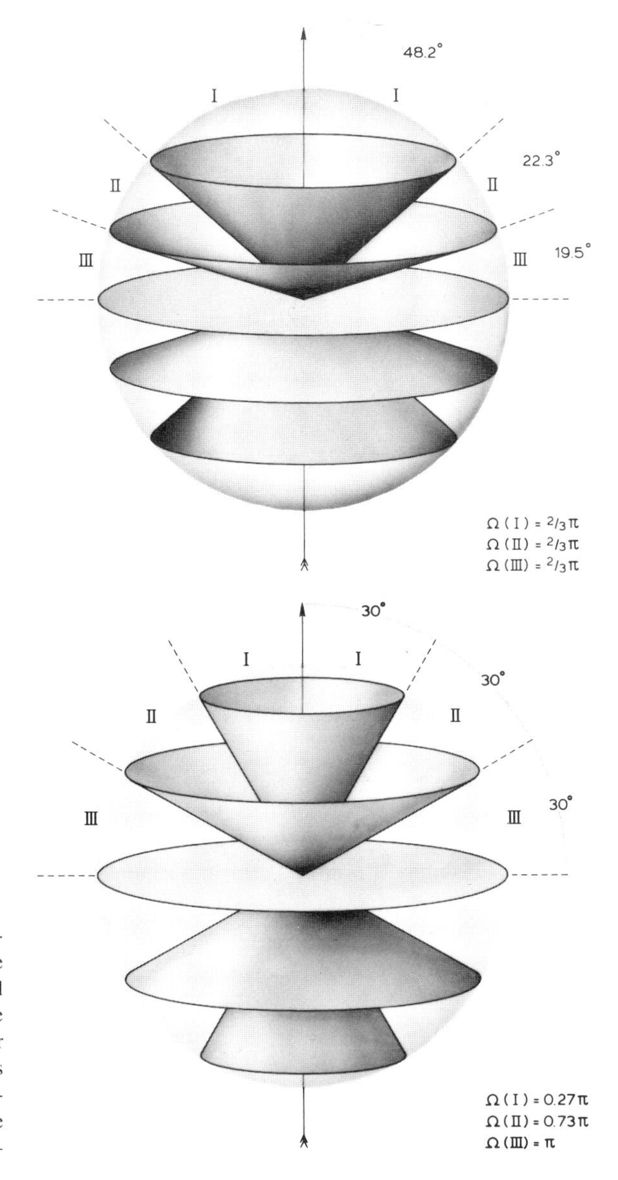

FIGURE 11-4. The spherical sector division of three-dimensional space. In the upper figure the volume of each spherical sector is the same, and the angles of the various sectors are different. In the lower figure the angles of the various sectors are the same, but the volumes of the sectors are not. $\Omega(I)$ is the solid-angle value of spherical sector I, indicative of its relative volume.

volume. They calculated the following ϕ values for the seven spherical sectors using the cone angle equation: 18.68°, 42.54°, 71.59°, 108.41°, 137.46°, and 161.32°. Because of the further subdivision into portions, their seven spherical sectors do not all have the same volume. Note that their angle is equal to $90° - \phi$.

For their spherical sector orientation analysis, Glaser and McMullen (1984) developed a "fan-in projection," i.e., a computer graphic method in which all dendrites are rotated around the soma–pia vector into a half plane, thus minimizing projection artifacts with respect to their original length. This is appropriate for the spherical sector orientation analysis in which the presence of asymmetry perpendicular to the axial vector (e.g., from soma toward pia) is not examined. When the detection of any kind of asymmetry is desired, the fan-in projection around one axis cannot reveal asymmetries in all directions. Fan-in projection around additional main axial vectors (axes) may be helpful, but subdivision of spherical sectors in equispatial portions (e.g., Seldon and Von Keyserlingk, 1978; François et al., 1987; Cullheim et al., 1987b) might then be preferred.

Visualization of orientational differences of the spherical sectors subdivided into 38 equispatial portions is not always very easy (see François et al., 1987). On the other hand, it still has to be established whether or not, in general, 38 equispatial portions are too many for detection of orientational differences. Empty classes (i.e., portions) will make statistical testing more difficult. We expect that in general a much lower number of equispatial portions should be used. Noteworthy in this respect is that in the study of François et al. (1987) the visualization of the orientation of the same dendrites in or projected onto three orthogonal planes by polar histograms with 30° angular intervals, appears to be equally informative, especially because of the very interesting additional information on the direction of the main bundle of axonal afferents (e.g., Fig. 10 in François et al., 1987).

Apart from the orientation of the whole dendritic field, it may also be of interest to know the orientation of each dendrite (e.g., Steffen and Van der Loos, 1980). Glaser et al. (1979b) and Cullheim et al. (1987b) determined the three-dimensional orientation of an individual dendrite by the three-dimensional centroid vector, i.e., the line from the dendritic origin toward the center of mass of the pertinent dendrite. For the calculation of the center of mass of a dendrite, Cullheim et al. (1987b) accounted for the membrane area of the different dendritic segments. The stereoscopic visualization of the three-dimensional orientation of the dendritic centroids and the centroid of all the dendrites of a neuron, weighed for their membrane areas, by Cullheim et al. (1987b) is very informative (see Fig. 11-5). Glaser et al. (1979) also defined a very interesting dispersion measure of the dendrite around the centroid vector, which assigns less weight to the peripheral, outlying dendrites.

Statistical tests for three-dimensional orientations are not always easy and not well known in neuromorphometry. For testing of three-dimensional randomness, i.e., absence of a preferred direction or of clustering around preferred directions, Batschelet (1981) refers to Stephens (1964, 1966), Beran (1968), and Prentice (1978). For comparison of three-dimensional orientations of two or more groups, he refers to Watson and Williams (1956), Stephens (1969), and Mardia (1975).

11.4.2. Cubic Orientation Methods

In the previous section the analysis of dendritic orientation or the spatial arrangement of dendrites of a neuron was done by subdividing the three-dimensional space into spherical or circular sectors. Another procedure for subdividing the three-dimensional space into equal volumes is used by Cullheim et al. (1987b): the space around the soma can be divided into so-called "half-spaces," e.g., dorsal versus ventral half-space, medial versus lateral half-space, and rostral versus caudal half-space. This is obtained from three orthogonal planes passing through the neuronal soma and aligning with the dorsal–

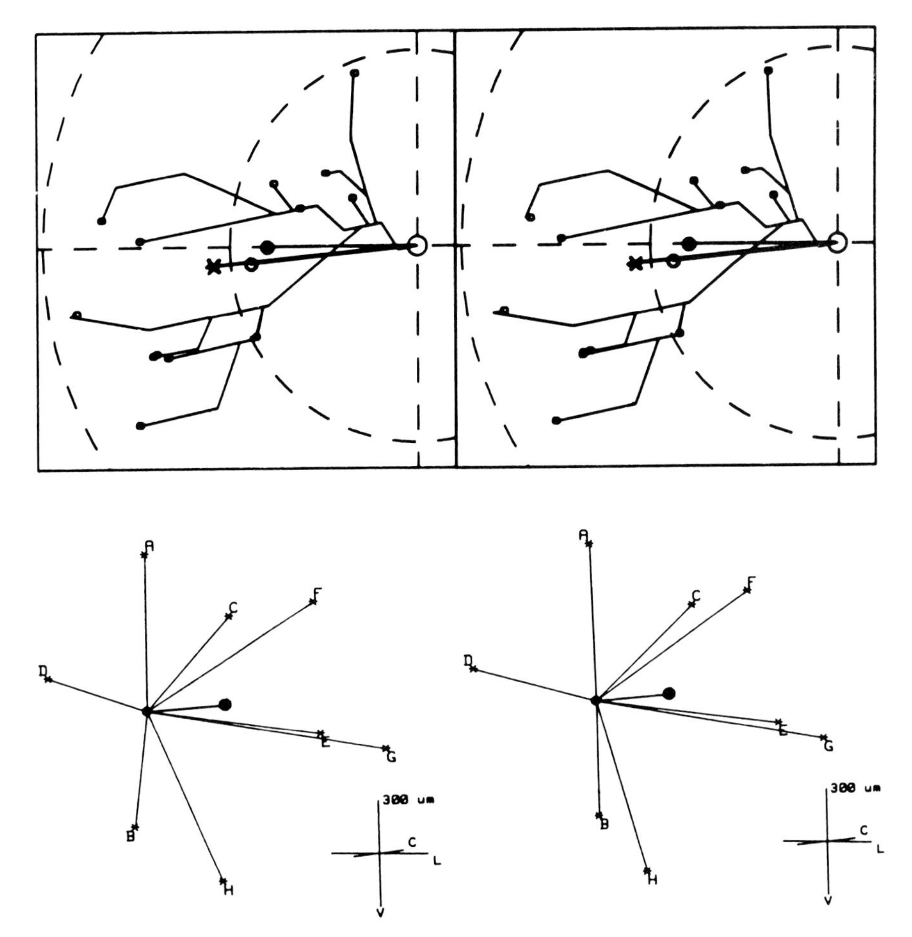

FIGURE 11-5. The upper figure is a stereo pair of motoneuron dendrite with the centroid (i.e. center of mass) vector from the soma in three-dimensional space. The solid circle is the center of mass of the dendritic membrane area. Asterisks and open circles indicate the center of mass of all terminal tips and bifurcations, respectively. The dashed inner circle has a radius of 500 μm, and the outer circle one of 1000 μm. The lower figure is a stereo pair of the centroid vectors of all dendrites of a motoneuron (asterisks are the dendrite centroids) and of the vector to the center of mass of the whole dendritic field (●). The centroids account for differences in membrane area of the different dendritic segments. C, caudal; L, lateral; V, ventral. Reproduced from Cullheim et al. (1987b) with permission.

ventral, medial–lateral, and rostral–caudal axis, respectively (Fig. 11-6B). Cullheim et al. (1987b) also combined these rectangular axes with the radial direction from the soma by dividing the cubic space into six equispatial sectors, i.e., pyramids with the apex at the soma (so-called "hexants") (see Fig. 11-6A, in which the ventral hexant is shaded).

The three orthogonal planes through the soma used for the determination of half-spaces also divide the cubic space into eight equispatial octants (see, e.g., the shaded

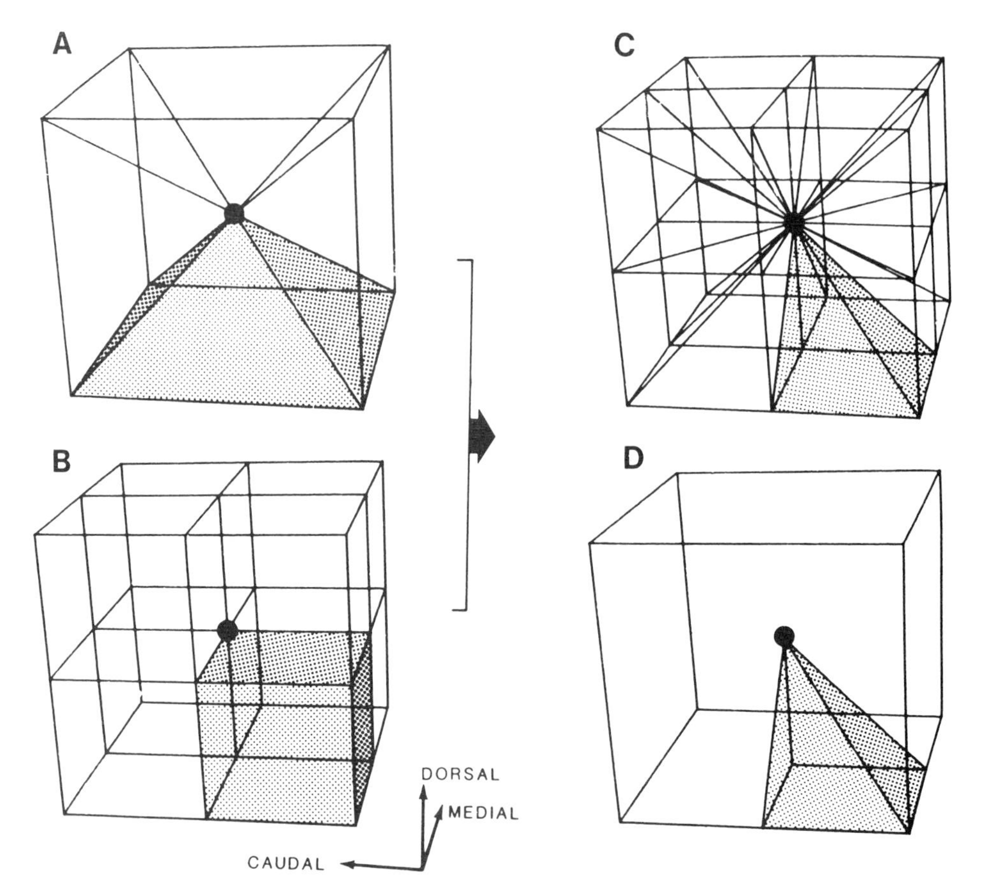

FIGURE 11-6. A: Cubic space divided into six hexants; the ventral hexant is shaded. B: Three orthogonal planes divide the cubic space both into half-spaces, e.g., dorsal and ventral half-space, and into eight equispatial octants. The rostral–ventral–lateral octant is shaded. C,D: Intersection of hexants with octants: the shaded figure is the intersection of the ventral hexant with the rostral–ventral–lateral octant. Reproduced from Cullheim et al. (1987b) with permission.

rostral–ventral–lateral octant in Fig. 11-6B). A further spatial subdivision is the intersection of the cubic hexants and octants, which results in quarter-hexants (see, e.g., the shaded quarter-hexant intersection of the ventral hexant and the rostral–ventral–lateral octant in Figs. 11-6C and 11-6D). Whether or not the further subdivision into quarter-hexants is really necessary still needs to be established. The hexant spatial subdivision appears to be illustrative for orientational differences when the dendritic intersections at certain radial distances from the soma are considered (e.g., Fig. 11-6C; Cullheim et al., 1987b). Thus, we have a combination of the hexant spatial subdivision with some concentric spheres. For statistical comparison of the hexant distribution of two or more groups of cells, the chi-square test described in Batschelet (1981) can be applied, and for testing the randomness of orientation of the hexant distribution, i.e., absence of deviation from a uniform distribution, the chi-square test described can also be used.

11.4.3. Principal-Axes Method

Colonnier (1964) was the first researcher to report a quantitative study of the possible presence of a preferential orientation of the dendritic field of a neuron. He determined the length and width of the projected dendritic field from camera lucida drawings in which the length is defined as the maximal distance between two terminal tips along an axis through the soma. The width was defined as the maximal distance between terminal tips perpendicular to the length axis (e.g., Uylings et al., 1986b). This method considers only the extreme locations of terminal tips and is therefore sensitive to outliers.

In 1974, Stevens developed a procedure for a two-dimensional dendritic field that considers every dendritic segment (Stevens and Van den Pol, 1982). He assigned to each segment a "mass" proportional to its length and centered this mass at its midpoint to determine the center of mass (i.e., the centroid) for the whole dendritic field and the major axis of the whole dendritic field (i.e., the first principal component axis) through this center of mass, i.e., the rotational axis with the smallest moment of inertia. The minor axis through the centroid is orthogonal to the major axis. As a measure of the size of the major and minor axes, the roots of the mean (mass-weighed) squared distances of the segment midpoints to both axes, respectively, were taken. This two-dimensional principal component analysis was applied by Borges and Berry (1978) and later by Van den Pol and Cassidy (1982). The geometric center of mass of an entire planar dendritic field was studied in retinal ganglion cells by Leventhal and Schall (1983) and Schall and Leventhal (1987).

Brown (1977) extended Stevens's two-dimensional procedure to a three-dimensional method. In this three-dimensional principal component analysis of a neuronal dendritic field, the orientation and size of three orthogonal principal axes are determined. The three-dimensional principal component analysis provides the three-dimensional dendritic field of a neuron with the best-fitting lines and planes (Pearson, 1901, cited in Brown, 1977). In this type of analysis the trees are decomposed into disconnected points located in the three-dimensional space. Brown (1977) and Yelnik et al. (1983) take equidistant (15-μm) points along the dendrites. The first principal axis is defined as the axis that minimizes the sum of squared distances from each point to this axis and runs through the center of mass. The principal plane (which is determined by the first and second principal axis through the centroid) is defined as the plane that minimizes the sum of squared distances from each point to this plane. Thus, the dendritic field shows the greatest (projected) extension in this principal plane (Calvet and Calvet, 1984). The third principal axis runs through the centroid perpendicular to the principal plane. Brown (1977) advised taking the extreme length between two projected dendritic points on the three axes, respectively, as the length of these three axes, as was also done by Yelnik et al. (1983). This length measure is, however, susceptible to dendritic outliers. The principal component analysis results in a very concise description of the three-dimensional dendritic field (Fig. 11-7).

The ratios of the length of two principal axes may indicate the spatial shape of the three-dimensional dendritic field, but see the measures used for axialization and flatness in Fig. 11-7. The principal axes are rectilinear axes, but the dendritic field of a neuron may in some areas be centered around curvilinear axes (see, e.g., Van den Pol and Cassidy, 1982).

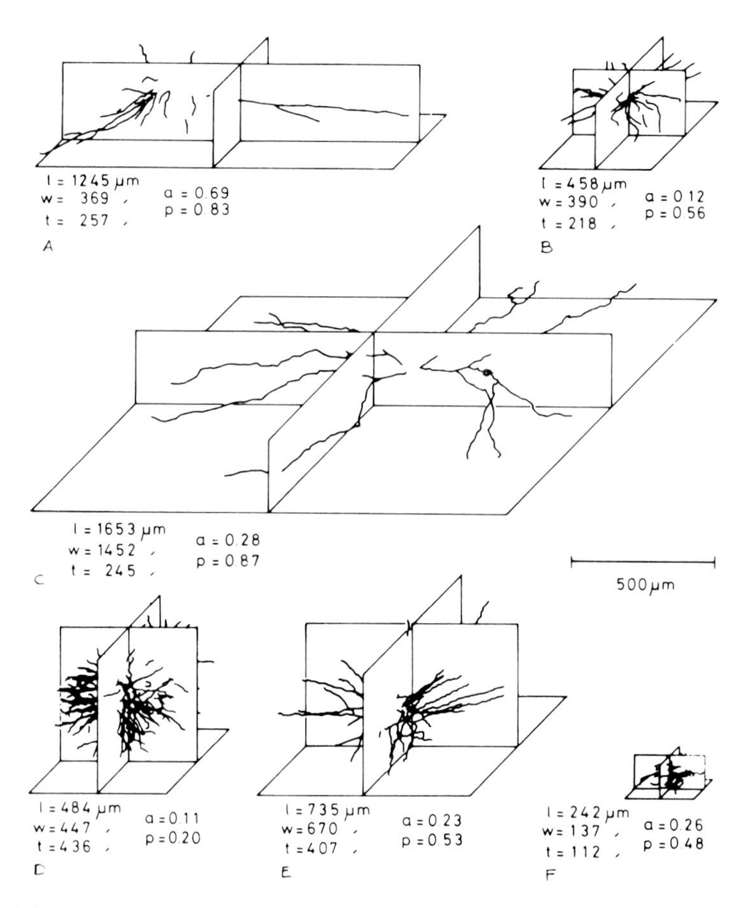

FIGURE 11-7. The first, second, and third principal axes through the centroid of the whole dendritic field of six neurons from different brain regions. *l*, length of first principal axis; *w*, length of second principal axis; *t*, length of third principal axis; *a*, index of axialization, i.e., $1-(S_2/S_1)$, in which S_2 and S_1 are the standard deviations of the second and first principal axes, respectively; *p*, index of flatness, i.e., $1-(S_3/S_1)$, in which S_3 is the standard deviation of the third principal axis. Reproduced from Yelnik et al. (1983) with permission.

In the case of a group of neurons with elongated dendritic fields, it is important that you know whether or not the orientations have a random distribution. An interesting feature of the principal component analysis is that the significance of elongation of the dendritic field along one of the principal axes can be statistically tested (Coleman et al., 1981; Batschelet, 1981, Chapter 7; see Batschelet, 1981, Chapter 4 for testing of two-dimensional randomness and for three-dimensional randomness). In addition, it is of interest to know whether or not the centroid of the dendritic field of a neuron significantly deviates from the center of the soma (Borges and Berry, 1978; Tieman and Hirsch, 1982; Cullheim et al., 1987b).

11.4.4. Cartesian Grid Analysis for Dendritic Orientation and Density

Another type of analysis for the orientation and density of the dendritic field of neurons is the method of Ruiz-Marcos and Valverde (1970), which has been improved by

Ruiz-Marcos (1983). At this moment the method is still applicable only to two-dimensional dendritic fields or projected dendritic fields, but it indicates the orientation and density in a visually attractive way and can be extended to a rectangular three-dimensional method. In this two-dimensional method the whole dendritic field is projected onto a Cartesian grid of square elements of equal size (e.g., 15 × 15 μm). In each square element the (projected) dendritic length is determined for each neuron. Before projection onto a Cartesian grid the dendritic field of a neuron is rotated and tilted in such a way that the soma projects onto the center of the grid and the vertical grid lines point in a standard direction for each neuron, e.g., the direction perpendicular to the pial surface. In principle, the fan-in projection images, which do not foreshorten their dendrites (see Section 11.4.1; Glaser and McMullen, 1984), can also be examined with the Cartesian grid analysis. However, in that case lateral asymmetries cannot be determined. The mean dendritic length and its variance in each square or matrix element are calculated for a group of neurons.

Groups of neurons are compared in a different way in the literature. Ruiz-Marcos and Ipiña (1986) compared the mean dendritic densities per matrix element with the "least significant difference" (LSD) as a measure for detecting significant differences (Andrews et al., 1980; Sokal and Rohlf, 1981). In planned multiple comparisons the use of LSD values is preferred to the multiple *t*-test for comparing homologous pairs of matrix elements. A condition for using the LSD test, however, is that the mean dendritic density values of neighboring matrix elements of a group of cells be independent (Uylings et al., 1986a). This condition is more or less related to the size of the square matrix element. A short-range correlation of dendritic densities can be assumed to be present, but the larger the size of the square matrix element, the lower the correlation between the dendritic densities of neighboring matrix elements. Another comparison method is the descriptive use of the two-tailed Student's *t*-test for each pair of homologous matrix elements (Uylings et al., 1986a). The comparison of two groups of neurons at a two-tailed $P = 0.05$ criterion will only indicate significant differences if sufficiently large clusters of matrix elements differ significantly in the same direction. This estimate of local density differences will be strengthened if the estimate at $P = 0.01$ still shows significant clusters (Fig. 11-8) and if, after an increase in size of the matrix elements, there are still significant differences at the same location and in the same direction. On the basis of patterns of negative and/or positive differences according to *t*-tests, the minimum size of a matrix element is taken to be about 15 × 15 μm for the dendritic field of rat cortical pyramidal and nonpyramidal neurons.

11.4.5. Comparison of Orientation Methods

For three-dimensional tree structures, all three-dimensional methods discussed in Sections 11.4.1–11.4.3 require the use of very thick (300-μm) histological sections in which almost the entire dendritic field is included or the reconstruction of a neuron from consecutive sections. It is still extremely time-consuming to reconstruct a complete dendritic field of a neuron from consecutive sections, even with semiautomatic measurement systems. For the three-dimensional reconstruction of one spinal cord neuron within a tissue slab whose size is 1800 × 1500 × 400 μm, approximately 40 hr was needed in a study carried out by Capowski and Sedivec (1981); for the present state of the art, see

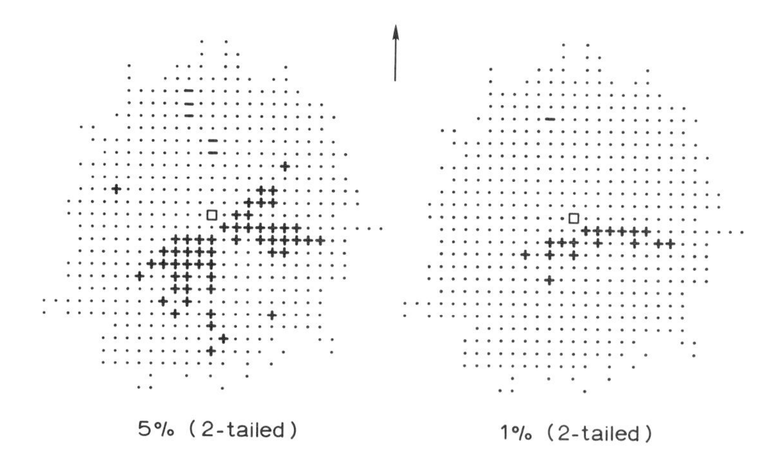

FIGURE 11-8. Comparison of the "Cartesian grid" density and orientation of the basal dendritic field of layer-III pyramidal neurons in 30-day-old control and underfed rats. The plane of projection is the coronal plane, and the arrow is perpendicular to the pial surface. The size of each square element is 15 × 15 μm; +, significantly higher density; −, significantly lower density of the control group. Reproduced from Uylings et al. (1986a) with permission.

Chapter 4 in this book. A statistical comparison of dendritic orientation of cortical neurons in two different conditions requires the reconstruction and measurement of about 15 neurons per animal per neuronal type per layer (Section 11.5). Therefore, it still seems preferable to use, if possible, very thick mercury Golgi–Cox sections (200, 300, or 500 μm: Glaser and Van der Loos, 1981) cut in such a plane that nearly the entire dendritic field can be contained within a section. In our experience it appears that if uncut neurons are selected that are located in 120- to 160-μm-thick sections, biased measurements (e.g., an underestimated total length) will result. For rapid Golgi-stained tissue, sections of 200 μm and thicker are generally not useful, as they are too dark from the heavy impregnation, whereas Golgi–Kopsch staining can be sectioned in 300-μm-thick sections. The Golgi–Kopsch technique, however, stains many fewer neurons than the Golgi–Cox technique.

The abovementioned considerations indicate that a fair comparison of the different three-dimensional orientation techniques is hampered by the practical difficulty of complete three-dimensional measurements. Because of a lack of data we can give only a preliminary comparison based on a few studies (Borges and Berry, 1978; Coleman et al., 1981; François et al., 1987). Borges and Berry (1978) examined the projected dendritic fields of cortical nonpyramidal neurons in control and dark-reared rats with two-dimensional principal axes analysis, Cartesian grid analysis, two-dimensional centroid vector analysis, and circular sector analysis. They found the most obvious differences with the Cartesian grid analysis. Statistically significant differences were also detected with the circular sector analysis and with the location of the centroid.

Coleman et al. (1981) examined the three-dimensional dendritic field of layer-IV nonpyramidal neurons in the visual cortex of kittens reared in control and vertically and horizontally striped cylinders. To analyze dendritic orientation, they applied a three-dimensional principal component analysis, a spherical orientation analysis, and a two-dimensional concentric circles method with angular distribution of projected dendritic intersections both in the coronal plane and the tangential plane parallel to the pial surface. The Golgi-stained sections were 200 μm thick and cut parallel to the pial surface of the medial bank of the lateral gyrus containing area 17. They did not report in this paper on the frequency of cut segments within the dendritic field of a neuron. Coleman et al. (1981) obtained the clearest results with the concentric circles method combined with the angular distribution of dendritic intersections. In control kittens they found that the neurons analyzed had a preferential orientation to the direction of the pial surface and the white matter. In the experimental kitten groups, this preferential orientation was disturbed. The three-dimensional principal component analysis did not reveal clear differences, since no statistically significant elongation of the dendritic field was detected. It is not clear whether this was obscured by the cutting plane of their histological sections.

In their three-dimensional analysis of the complicated neuronal organization of the primate substantia nigra, François et al. (1987) noted that the orientation of the two principal axes (i.e., the principal plane) was not the most suitable method for analyzing the dendritic orientation. For the nigral dendrites, they selected the three-dimensional spherical sector method, subdivided into 38 equispatial portions. In addition, their two-dimensional visualization of dendritic orientation in three orthogonal planes appears to be very informative.

In view of the abovementioned limitations regarding the availability of three-dimensional data, and in view of the abovementioned studies, it is not possible at present to give solid arguments in favor of one of the orientation methods discussed in Sections 11.4.1–11.4.4. At this moment, we still use the two-dimensional Cartesian grid analysis for dendritic orientation and density because of its visual attractiveness in comparing different groups and because the cells we analyzed were not complete because of cutting or, occasionally, excessive Golgi precipitation. The preferential plane of cutting with regard to the orientation of the dendritic field must of course be taken into account then.

11.5. STATISTICAL EVALUATION OF GROUPS OF NEURONS

In the statistical comparison of neurons from two or more experimental conditions, the mean and variance of an experimental or control group are frequently based on the total number of neurons or the total number of segments studied per group. This can lead to erroneously small group variances because of the large number of observations within a group (e.g., 1500 terminal segments or 120 neurons per group) and, consequently, erroneous conclusions on group differences. The group observations based on the total number of terminal segments or the total number of neurons per group are not independent. Segments are found within a neuron located within an animal, which in its turn constitutes only one of the examined conditions. Therefore, experiments to test significant differences between groups defined by experimental conditions have a hierarchy of sampling levels. The first level is the group level, in which two or more groups are different in a fixed, circumscribed condition. The second level is that of the "random" sample of individuals or animals that belong to a group. The third level can be the brain regions studied. The fourth level is, for example, the "random" sample of neurons within a brain region. And the fifth level is the measurements of each neuron. The number of levels is smaller when only one measurement per neuron is analyzed and/or the neurons examined are all part of one particular brain region. The abovementioned experimental design is called a "nested sampling design."

The statistical analysis of such a design is called the "nested analysis of variance" or the "hierarchic analysis of variance" (e.g., Sokal and Rohlf, 1981). This parametric analysis of variance requires normally, continuously distributed variables and cannot be applied to topological, circular, or spherical variables. The nested analysis of variance takes into account the variance components of each hierarchical level and gives procedures to examine the presence of significant differences and to establish the optimum sample sizes of the different levels.

Since sampling efficiency and the test of statistically significant differences are of great importance in any study on the neuronal branching structure, we give a short introductory description. If we have a nested design with levels (1) animals, (2) neurons within animals, and (3) measurements within neurons, the variances are related in the following way. The total observed variance among animals ($O_{s^2_a}$) is a summation of the "true" or actual variance among animals (s^2_a) and the observed variance of the mean between neurons within animals ($1/n_n \times O_{s^2_n}$). The number of neurons examined within each animal is n_n. In its turn, the observed variance among neurons is a summation of the actual variance between neurons within animals (s^2_n) and the observed variance of the mean of the measurements within neurons ($1/n_m \times O_{s^2_m} = 1/n_m \times s^2_m$). This leads to the equation

$$O_{s^2_a} = s^2_a + s^2_n/n_n + s^2_m/(n_n \times n_m) \tag{11-4}$$

If only one measurement per neuron is used, the total observed variance between animals for the nested design becomes

$$O_{s^2_a} = s^2_a + s^2_n/n_n \tag{11-5}$$

For computation of the variance components s^2_a and s^2_n, see the clear descriptions by Sokal and Rohlf (1981, Chapter 10, including unequal sample sizes) and Shay (1975). In biomedical morphometric research, the relative contribution of the true variance among animals to the observed overall variance is generally much higher than that of the nested lower levels, e.g., measurements within animals (see for examples Gundersen and Østerby, 1981; Gupta et al., 1982). In such a situation the statistical test of group differences can be based only on the mean value of each animal and does not require the complete nested analysis of variance, which also takes into account the variance component(s) within animals. The approximate sample size of number of animals per group (n_a) can be estimated from the ratio between s^2_a and d^2, in which d is the smallest difference between the means desired to be detected, by (Cruz-Orive and Hunziker, 1986; Shay, 1975):

$$n_a \geq 2k^2\, s^2_a/d^2 \tag{11-6}$$

in which k is a constant depending on the power and the level of significance of the test. The power is the probability that a difference as small as d will be found to be significant (see, e.g., Sokal and Rohlf, 1981, p. 263). Usually the power is taken at 0.80 (see Keppel, 1982, p. 70), which means that for a significance level of 0.05, $k^2 = 6.2$. Note that in equation 11-6 both s_a and d can be divided by the mean of the control group. The ratio of s_a and d becomes the ratio of the coefficient of variation (CV_a) and the relative difference of means. For CV_a it is assumed that s^2_a is the same in both control and experimental groups.

In Table 11-2, we give examples for the number of animals or number of subjects per group on the basis of data of De Ruiter and Uylings (1987) and unpublished data of P. McConnell and H. B. M. Uylings on 30-day-old rats. This table shows that with respect to the dendritic variables considered, the animals show less variation than the human subjects examined. The neurons examined were partially cut (about 10% of the segments), but it is not expected that this cutting interferes with the data concerning the variables "somatic surface," "number of dendrites per cell," and "length of individual terminal segments" and, as a consequence, with the estimated number of subjects per group for these variables. From Table 11-2 we can expect that if the sample size is taken at approximately six rats per group, differences of about 15% between a control and an experimental group should yield a significant result.

TABLE 11-2. Estimated Number of Subjects per Group to Detect Relative Differences between Groups (with $k^2 = 6.2$)

	Relative difference to be detected				
	10%	15%	20%	25%	30%
Layer-III pyramidal cells (rat)					
Somatic surface	21	10	6	4	3
No. basal dendrites/cell	7	4	2	2	1
No. basal segments/cell	22	10	6	4	3
Total length dendrites/cell	28	13	7	5	4
Length ind. terminal segments	11	5	3	2	2
Radial distance terminal tips	11	5	3	2	2
Metric path length terminal tips	10	5	3	2	2
Layer-IV multipolar nonpyramidal cells (rat)					
Somatic surface	24	11	6	4	3
No. dendrites/cell	7	4	2	2	1
No. segments/cell	12	6	3	2	2
Total length segments/cell	14	6	4	3	2
Length ind. terminal segments	17	8	5	3	2
Radial distance terminal tips	12	6	3	2	2
Metric path length terminal tips	11	5	3	2	2
Human dentate granule cells					
Somatic surface	49	22	13	8	6
No. dendrites/cell	14	6	4	3	2
No. segments/cell	42	19	11	7	5
Total length dendrites/cell	129	58	33	21	15
Length ind. terminal segments	105	47	27	17	12
Radial distance terminal tips	104	46	26	16	12
Metric path length terminal tips	82	36	21	14	10

In the statistical textbook of Sokal and Rohlf (1981), the relative magnitude of the true variance among subjects (animals), s^2_a, is frequently expressed as the percentage of the sum of the "random" variance components (e.g., $s^2_a + s^2_n$) and is called the percentage variation between subjects. On the other hand, recent morphometric articles (e.g., Gundersen and Østerby, 1981; Gupta et al., 1982; Cruz-Orive and Hunziker, 1986) express the relative magnitude of the variance component among animals, s^2_a, as the relative contribution to the observed overall variance among animals, i.e.,

$$P = s^2_a/(s^2_a + s^2_n/n_n) \tag{11-7}$$

in which n_n is the number of neurons examined per animal and in which one measurement per neuron is analyzed. The last-mentioned percentage, P, will increase to 1 if n_n increases. In Table 11-3 we tabulate both percentages.

Table 11-3 shows that for the dendritic variables of the human dentate granule cells evaluated, the relative contribution of s^2_a to the observed variance between animals (P) is higher than 70%, so that group differences can be tested only on the basis of the mean value of each animal. The rat cortical pyramidal and multipolar nonpyramidal data, however, require a complete nested analysis of variance (Table 11-3). There the P is generally lower than 50%, while the percentage variation between animals is generally lower than 10%. This has also been shown for the length of individual terminal segments of nonpyramidal and pyramidal neurons in adult rats (Uylings et al., 1981).

For an estimate of the optimum number of measurements or number of neurons within animals, the following equation is generally used (Shay, 1975; Cochran, 1977; Sokal and Rohlf, 1981):

$$\text{optimum } n_n = [(s^2_n/s^2_a)(c_a/c_n)]^{\frac{1}{2}} \tag{11-8}$$

in which (c_a/c_n) is the relative cost of an animal versus the measurement of a neuron, to be calculated in working hours with or without financial costs. In fact, equation 11-8 indicates that the more the neurons vary, or the more expensive the animals or subjects are, the more neurons have to be measured. Table 11-3 shows the estimated optimal number of neurons to be sampled for each variable. In view of the difference in working hours, c_a/c_n in this table is 12 for rats and roughly estimated at 48 for human subjects. For procedures for determining the optimal number of sampling units for several variables simultaneously, we refer to Waters and Chester (1987). The optimal numbers of cortical neurons per rat and of dentate granule cells per human subject would be about 15–24 and 10–15, respectively.

11.5.1. Parametric and Nonparametric Testing

In biomedical research, many variables are clearly not normally distributed. An example of this is the distribution of length values of intermediate segments (Uylings et al., 1978a). In such situations one of the following statistical approaches is applied: (1) the data are transformed (e.g., log transformation, square root transformation) in such a way that their distribution more closely resembles a normal one so that parametric statis-

TABLE 11-3. Percentage Variation between Subjects and Relative Contribution to Observed Variance between Subjects

	Percentage variation among subjects	P^a	Optimal number of neurons[b]
Pyramidal neurons (rat)			
Somatic surface	17	78	8
No. basal dendrites per cell	2	28	24
No. basal segments per cell	18	79	7
Total length dendrites per cell	17	77	8
Length individual segments	1	13	34
Radial distance terminal tips	6	50	14
Metric path length terminal tips	4	43	17
Multipolar nonpyramidal neurons (rat)			
Somatic surface	21	81	7
No. dendrites per cell	10	63	10
No. segments per cell	2	20	24
Total length dendrites per cell	2	22	24
Length individual segments	1	14	34
Radial distance terminal tips	1	15	34
Metric path length terminal tips	1	15	34
Dentate granule cells (human)			
Somatic surface	39	90	9
No. dendrites per cell	3	29	39
No. segments per cell	22	80	13
Total length dendrites per cell	45	92	8
Length individual segments	14	71	17
Radial distance terminal tips	53	94	7
Metric path length terminal tips	33	88	10

[a]P is percentage contribution of "true" variance among subjects to observed variance among subjects; n_n (pyramidal cells) = 16.67; n_n (nonpyramidal cells) = 16.5; n_n (dentate granule cells) = 14.9.
[b]Estimated per subject with equation 11-8.

tical methods (e.g., analysis of variance, ANOVA) can be applied, or (2) a distribution-free (i.e., nonparametric) procedure is applied (e.g., Kruskal–Wallis test). The nonparametric methods, however, are not sufficiently developed to be applicable to many experimental designs.

Conover and Iman (1981) proposed combining the parametric and nonparametric methods by applying the parametric techniques (e.g., a type of ANOVA) to rank-transformed data. For (nested) ANOVA the entire set of data is ranked from lowest to highest value, the lowest value having rank 1, the next-to-lowest rank 2, etc. Average ranks are assigned in case of ties. The advantage of this combination of parametric and nonparametric statistics is that an extensive set of statistical computer programs becomes available for nonparametric analysis (Conover and Iman, 1981). Although this procedure is only conditionally distribution-free, it is robust, which means that the true level of significance is usually fairly close to the calculated level of significance, irrespective of the type of distribution of the rank-transformed data (Conover, 1980).

11.5.2. Geometric Dendritic Size and Body Size

In studies on the effect of experimental treatment on a particular brain region, the treatment sometimes alters the size of the entire body. To examine whether or not a brain region is disproportionately affected by experimental treatment, you must take into consideration the relationship between brain region and body weight of control animals (e.g., Uylings et al., 1986b, 1987). This might also be necessary for neuronal size differences. For interspecific comparison, a relationship between dendritic complexity, indicated by the number of dendrites per cell, and body weight was reported for superior cervical ganglion neurons by Purves and Lichtman (1985). Voyvodic (1987) found a correlation during development of the dendritic tree between body size of rats and the total dendritic length of superior cervical ganglion cells. If this relationship is taken into account, there is no significant difference in dendritic growth of these neurons between male and female rats, which differ substantially in weight.

For the dendritic variables examined in rat cortical neurons, however, we have not found significant correlations so far between the size of cortical neurons and body weight in control and experimental groups of rats (A. Kalsbeek and H. B. M. Uylings, unpublished data). The variables examined were "somatic size," "number of dendrites," "number of dendritic segments," and "total dendritic length per neuron." Therefore, in case of intraspecific comparison, the cortical neuronal trees of different groups of rats are tested without taking into consideration the differences in body size.

11.5.3. Multivariate Comparison of Sets of Dendritic Variables

As indicated in previous sections, several variables describe different features of the neuronal tree. So far, groups of neurons have been compared by testing single variables. However, the application of multivariate analysis techniques may be of importance. Pearson et al. (1985) and Harpring et al. (1985) studied a group of neurons with six variables, somatic size, dendritic field shape, dendritic field extent, density of somatic spines, density of dendritic spines, and location of each neuron within the brain region examined. They could not separate the group of neurons into different classes if each of these variables was tested separately, whereas three distinct morphological classes could be designated by a principal component analysis that considered all variables simultaneously. The multivariate principal component analysis is a descriptive method that transforms a set of observed variables to a new set of variables, which are uncorrelated and arranged in decreased order of variance (Chatfield and Collins, 1980). The first principal component accounts for as much as possible of the variation in the original data.

When the original variables are all uncorrelated, principal component analysis is not meaningful since it will not reduce the number of variables. For neuronal trees, however, at least some of the variables appear to be correlated (e.g., Sections 11.2.1. and 11.2.2.). The principal component analysis was also applied by Yelnik et al. (1987), considering 13 variables per neuron. A subdivision of nigral neurons into neuronal types, however, was less clear (see their Fig. 9). On the other hand, it showed that the types recognized were also grouped together according to the principal component analysis.

In principal component analysis, no assumption is made about the type of distribution of the original variables. An analytical multivariate technique for examining the existence of differences among means of a number of groups (especially when there are more than two groups) is the canonical variate analysis (CVA). Unlike the principal component analysis, CVA is invariant when origin and scale of measurements are changed. On the other hand, CVA assumes that the multivariate samples have normal distributions and homogeneity of the variance–covariance matrices (Chatfield and Collins, 1980). So far CVA has only been applied to dendritic analysis by Ipiña and Ruiz-Marcos (1986) and Ipiña et al. (1987) to study the effect of neonatal thyroidectomy, aging, and thyroxine replacement therapy on cortical pyramidal neurons. These analyses showed that hypothyroidism affects both basal and apical dendrites, and no recovery was found after thyroxine replacement therapy. Ipiña (1987), however, used principal component analysis instead of CVA in the multivariate comparison of groups of neurons from three different regions of the cerebral cortex of a mammal, a bird, and a reptile, since he could not assume all neurons were of a similar type, and the descriptive principal component analysis does not require a specific underlying statistical model.

11.6. IN CONCLUSION

In this review we consider metric variables that are used to describe the neuronal branching structures, and we indicate what statistical tests can be used for analysis of the metric variables discussed. The problems that arise in analyzing tree structures that are incomplete because of cutting, excessive precipitation from staining, and/or the inextricability of segments of several tree structures, all of which make the usefulness of this analysis questionable, have been described in Chapter 10 of this book.

For metric analysis, tissue shrinkage is a factor to be considered. Studies including data on shrinkage of neurons by the histological treatment of the tissue are relatively rare. Some methods for estimating the extent of shrinkage are described by Uylings et al. (1986b). On the basis of osmolality of different Golgi fixations (Uylings and Feirabend, 1983), we expect relatively low shrinkage differences for Golgi rapid immersion and Golgi–Cox immersion (290 and 245 mOsm/liter, respectively). The method of embedding also influences the shrinkage. We can expect celloidin-embedded tissue to differ from plastic-embedded tissue (Blackstad et al., 1984). The linear shrinkage of the cerebral cortex in Golgi–Cox-stained celloidin sections is similar to that in rapid Golgi sections and has been reported to be generally between 5% and 10% (Lund, 1973; Jones, 1975; Haberly, 1983; McMullen et al., 1984). The Golgi–Kopsch solution has a much higher osmolality (1315 mOsm/liter, Uylings and Feirabend, 1983). Linear shrinkage from Golgi–Kopsch is estimated to be approximately 20% (Lund, 1973), but it can be reduced to 5–10% (Jones, 1975) if the dichromate is dissolved in 1% glutaraldehyde solution.

11.7. SUMMARY

A series of metric variables describing size, orientation, and density of neuronal tree structures is discussed. The metric analysis of segments is considered, indicating into which groups segments have to be classified and examined. Different types of orientation analysis of planar and three-dimensional tree structures are discussed. In addition, statistical tests that can be applied in metric analysis are indicated. The nested sampling design inherent in studies comparing the metrics of neuronal trees under experimental conditions is explained. Sample sizes of number of rats per group and number of neurons per brain area per animal are given for cortical pyramidal and nonpyramidal neurons in rats and dentate granule cells in humans.

ACKNOWLEDGMENTS. The authors are indebted to Mr. A. Janssen for correcting the English and typing this manuscript.

12

Controlling the Computer System
The User Interface

ELLEN M. JOHNSON

12.1. INTRODUCTION

In the days before personal computers, very little commercial software was available for neuroanatomical applications. Researchers who used computers either wrote the necessary software themselves or hired professional programmers or students to write it for them. As a result, a great deal of time and effort was spent writing highly personalized application programs. Because these programs were designed to be used by the author of the program or someone in close proximity to the author, the user benefited directly from the author's mental model of the program's operation as well as the author's knowledge of computers and the programming process. Program effort was spent largely on providing the problem-solving functions. The extra work required to allow ease of learning and use was usually not done. If program structure allowed and time was available, enhancements might be added later in the process to make life easier for the user.

Now, a vast array of commercial software products are available. It is much more likely that today's neuroanatomist can find an "off-the-shelf" program to acquire, analyze, and display data. For some applications, customized software may still be needed; but as more and more commercial software is developed and enhanced, that situation becomes the exception rather than the rule. In addition, a detailed knowledge of computer hardware and software is no longer required to make use of the powerful research tools provided by computer technology. Because the user need not have great computer expertise and is usually remote from the software developer, however, more effort must be made to ensure that programs are easy to learn and use.

This chapter discusses current techniques for implementing the "user–computer dialogue." The area through which this dialogue takes place has become known as the "user interface." The goal of this chapter is to provide you, the neuroanatomist, with some criteria for evaluating the user interfaces of user-written as well as commercially available computer software programs designed for neuroscience applications.

ELLEN M. JOHNSON • Department of Physiology, The University of North Carolina, Chapel Hill, North Carolina 27599.

In describing and evaluating the user interface of any computer program, we must be concerned with the manner in which the functions of the program are presented to you and with the options you have for selecting these functions. We are also concerned with the ways in which data pass through the interface as output and input. We must consider the interactive devices used to enter and manipulate data, the way in which output data are stored and presented to you, and the responses that a computer program makes to user errors and hardware malfunctions.

If a computer program, be it a word processor or a data analyzer, performs the desired functions for you, how much does the form of that program's user interface really matter? Experience has shown that it matters a great deal (Foley et al., 1984). The quality of the interface determines the amount and kind of learning that you will have to do to be able to use the program. It will strongly influence how much you are able to accomplish with the program once you have learned how to use it. A poorly designed interface can produce the kind of frustration that makes you want to destroy (or at least punish) the computer (Fig. 12-1).

A well-designed interface is expensive and requires a great deal of programming effort that usually produces no additional user functions. Probably, we must double the time to write a program if we also make it easy to use. Whether or not this doubling of time is worth the investment often depends on the size of the user population and the lifetime of the program's usability. If a program is intended for use by one person over a short period of time, that person might be willing to tolerate an inelegant interface as long as the program is usable and produces the desired results. If many people are going to use a program over a long period of time, on the other hand, the extra time spent by a programmer diminishes in relative cost. In either case, time spent enhancing usability will be time that does not have to be spent helping you deal with the limitations of a crude user interface.

12.2. PRINCIPLES OF USER INTERFACE DESIGN

Those of us who used computers in the days of punched cards and teletype machines often think the video display terminal is such an improvement in the user interface that we need not be much concerned with the issues discussed in this chapter. In the past, we submitted a stack of punched cards at a dispatcher's window and often waited for hours before seeing our output. We could see the computer in the background, but we were rarely allowed to touch it. A little later, we might have been able to use a teletype terminal; computing becomes a little more interactive, but we shared the use of one computer with others. Response time became a big issue, but programmers spent very little time designing the user interface. Computers were large mysterious machines! We did not expect them to be friendly. Who ever heard of "user friendliness"?

Today, user-interface management is a growing subspecialty of computer science. There are interface software designers and human-factors engineers. The principles of interface design are becoming well known. The most important of these principles, from Hansen (1971), Norman (1981), Foley and Wallace (1974), and Shneiderman (1987), are summarized in Fig. 12-2.

Norman elaborates on Hansen's first principle by admonishing the software designer

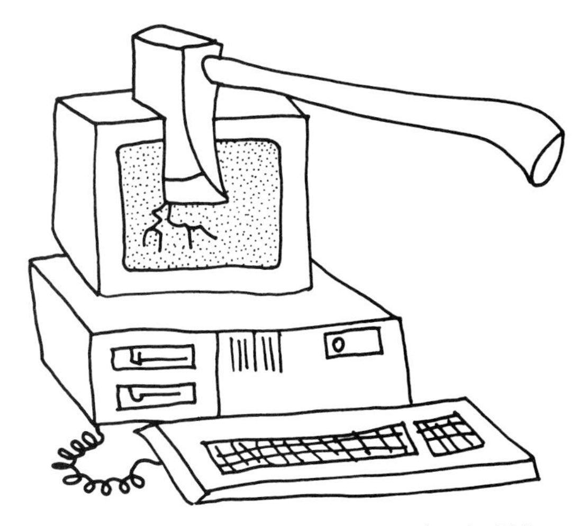

FIGURE 12-1. Have you ever wanted to do this?

to "design the system for the person, not for the computer, not even for yourself." Unfortunately, the commercial software developer cannot know everything about you. For example, the developer cannot know your level of computer expertise. As was pointed out in the introduction to this chapter, knowledge of the user is one major advantage of user-written or customized software.

Norman also points out that the "most powerful memory aid is understanding." That is why it is important to provide the user with a general understanding, a mental model, of how a program works. You form some picture of a program's inner workings. The programmer wants that picture to be an accurate one. Immediate access to this model is an advantage of software that is written by the user or by someone in close proximity, but it is not possible for the remotely located programmer to document a completely adequate description of how a program works.

First principle: Know the user
Minimize memorization
Provide simple, logically consistent commands
Provide quick, polite, and meaningful response to every user action
  Provide immediate, informative feedback
  Let the user know when a task is completed (closure)
Handle errors smoothly
  Be forgiving—allow user to easily try again
  Be robust—avoid program crashes
  Detect common errors
  Allow reversible actions
Protect the user from destructive acts
Provide short cuts for knowledgeable users
  Adapt to the user's ability
Keep the user in control of the program's actions
Provide an explicit model of how the program works

FIGURE 12-2. Principles of user interface design.

The impossibility of completely knowing the user makes it essential that software developers be able to apply the principles of good user-interface design. If you are buying an off-the-shelf application program or specifying the functions for customized software, you should expect, if you are willing to pay for it, a product that not only does the job intended but is easy to learn and a pleasure to use. Such a program is more likely to be an effective research tool.

In the following descriptions of various types of user interfaces, we refer to the principles presented above and discuss how each of the types adheres to them.

12.3. SELECTING COMMANDS

Using a program is interactive. The human gives a command to the program, either by hitting a key or pushing a button, and the program obediently executes the command, perhaps providing a result, and then waits for the human to give another command. This is the computer–human dialogue.

12.3.1. Introduction

Before describing some of the many ways commands are presented to the user and the options for selecting these commands, we would like to stress the following principle: once a command is selected, you must get an immediate response from the computer. Few things are more frustrating than staring at an apparently dead computer. If you do not get some sort of feedback from the computer, you may try to enter the command again, which would probably interfere with the operation of the program. (Of course, if enough attention is paid to the user interface, it should be very difficult to confuse a program.) If the execution of a command is not instantaneous, you should receive some message such as "Writing to disk . . . please stand by" to assure you that things are going well. You should be able to abort commands that take a long time to execute so you will not have to wait if you have made a mistake or change your mind about a command selection.

It should also be stressed that the principle of "closure" is very important for user satisfaction. You need to know that a command has been successfully executed. When successful execution involves a graphics task such as joining two points with a straight line, task completion is often obvious. For other actions, a message from the computer may be needed. If a command has not been successfully executed, you should be told why as well as what is required for successful execution. You should never be stranded without an option for a next step.

These qualities of the user interface—immediate response, feedback messages, the ability to abort commands, the existence of an obvious next step, and closure—all help make you feel that you, not the computer, are in control of the operation of the program.

12.3.2. Menus

The most common method of presenting the functions of an application program to its user is to display a menu of commands on the computer screen. In the computer world a menu is a list of command options, just as in the restaurant setting a menu is a list of culinary choices.

You can select a command from a screen menu by typing a single keyboard character that represents the command (Fig. 12-3) or by moving a marker on the screen to the position of the command on the list (Fig. 12-4). This marker is called a "screen cursor" and can take many forms. In Fig. 12-4, it is a box surrounding the selected command.

In Figs. 12-5 and 12-6, it is an arrow pointing to the command. The keyboard keys with the up, down, left, and right arrows are called "cursor control keys" and can be used to control the movement of the screen cursor. After these keys are used to move the cursor to the position of the desired command, you press the "enter" key to select the command.

Because it eliminates the need to type and is faster than using the cursor keys, a device called a "mouse" is very popular for pointing to and selecting commands from a menu. The movement of the mouse guides the movement of the screen cursor. You position the cursor at the desired command and press a button on the mouse to signal command selection. Most mouse buttons make a clicking sound when they are pressed; thus, mouse-button pressing is often referred to as "clicking." The mouse does have some disadvantages. It takes up desk space, the wire that connects it to the computer can be distracting, and it has to be picked up and repositioned for long movements. However, it can be used to position the screen cursor rapidly and accurately once the necessary eye-hand coordination has been developed.

The light pen is a direct-pointing device that is sometimes used for command selection. You simply point the pen at the desired command in the menu displayed on the computer screen. While this seems a very natural way to select commands, many users find that their arm gets tired after using the light pen for long periods because direct pointing at the vertical screen requires holding the pen up in the air. Other disadvantages of a light pen are that you must move your hand from the keyboard and pick up the light pen to use it, and the hand holding the light pen partially obscures the screen.

Just as there are various ways to select menu items, there are several ways to display menus on the screen. Until recently, the most common was to fill the screen with the display of the currently active menu. Many commercially developed programs, especially those that make heavy use of graphics, now have menus that appear on the screen only temporarily. There are two main types of these "dynamic" menus: pull-down and pop-up. A pull-down menu appears on the screen when you select it from a "menu bar" across the top of the screen. It appears to be pulled down from the menu bar as a window

Optical Density Measurement Program

Current frame identifier is RH005

I—IMAGE operations
W—WRITE frame results into the disk file
R—READ frame results from the disk file
M—MOVE to the next frame
T—TRACE and measure regions in this frame

X—EXIT the program

Ready for command . . . .

FIGURE 12-3. Menu of commands to be selected by a single keystroke. For example, the user presses the "R" key to execute the command "READ frame results from the disk file."

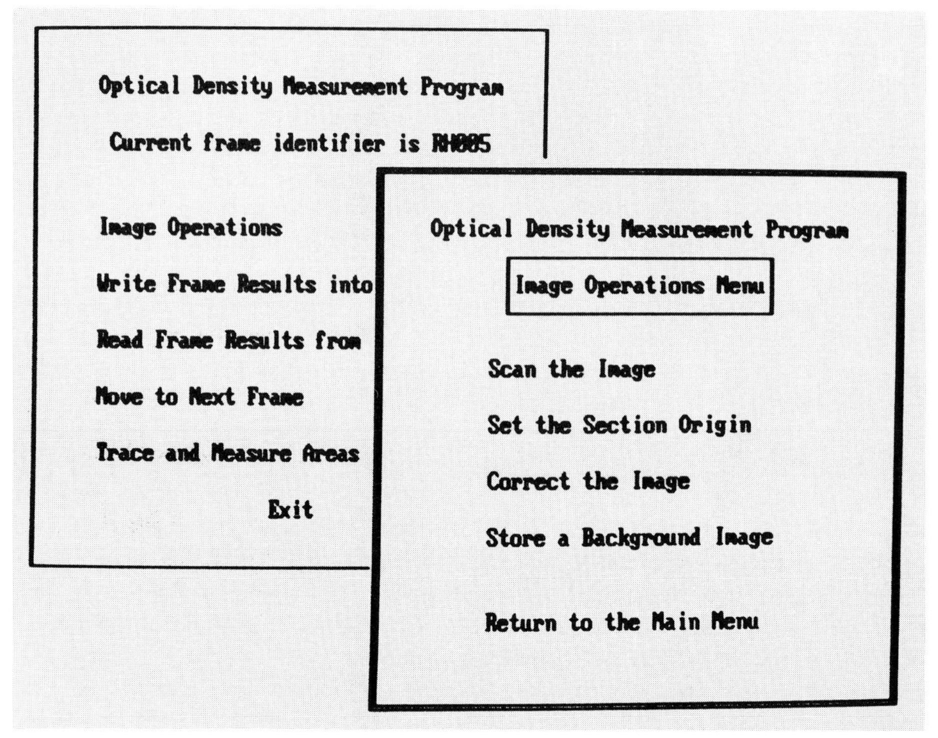

FIGURE 12-4. Menu and submenu of commands to be selected by moving a screen cursor to the position of the command in the list. In this example, the mouse is used to guide the cursor from command to command. When the user pushes a mouse button, the command in the box is executed. The command ''Image Operations'' causes the submenu to appear. Here, the submenu is displayed in a window that will disappear on return to the main menu.

shade is pulled down from the top of the window (Fig. 12-5). A pop-up menu appears at the screen cursor and helps even more to preserve visual continuity; you do not have to look away from the areas of immediate interest on the screen to see the menu options (Fig. 12-6).

In graphics and image-processing applications in which an image is manipulated on a display screen that is separate from the screen of the computer monitor, it is especially desirable to have menus that are overlaid temporarily on the image screen when the user demands. Then, you do not have to look away from the image screen to the computer monitor to select commands and operands.

Menus are often hierarchical; the selection of a command at one level causes a menu of commands for a sublevel to be displayed. Figure 12-4 shows the main menu of a program for measuring optical density as well as the submenu that appears when ''image operations'' is selected. As in this example, submenus are often displayed in a bordered area of the screen that appears to open as a window in the main menu area. Because it allows portions of previous screens to remain visible, this ''windowing'' can remind you of where the current menu is in the menu hierarchy.

Menu systems with many levels can be slow and awkward if you must go through the selections of many submenus before reaching a desired command. This problem can be somewhat alleviated if the program has a "macro" or "scripting" capability that allows you to store a group of commands at one time and recall them for execution on demand and with few keystrokes at any time thereafter. In addition, programs that require you to type a keyboard character for command selection should queue the characters of a frequently used sequence in a buffer that will allow you to enter the sequence rapidly, without waiting for each command to execute. There is a trade-off here, however; the ability to enter commands before the next one is entered so quickly that each command's feedback is not considered also presents the opportunity to make compounding and difficult-to-correct mistakes.

12.3.3. Command-Line Interfaces

Experienced users, especially those who use a program almost daily, often prefer programs that require the typing in of commands rather than selection from menus with

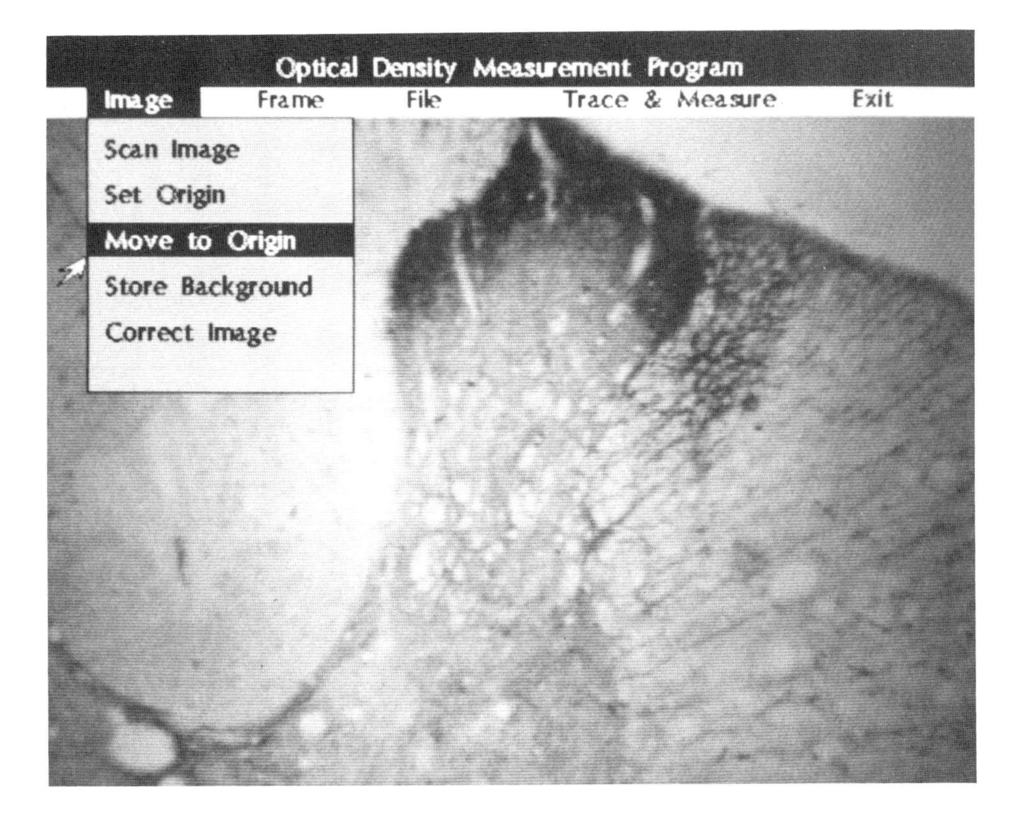

FIGURE 12-5. Pull-down menu. The main menu appears in the bar across the top of the screen. Here, the submenu for "Image" has been pulled down over an image. The cursor has been moved with the pointing device to the command in the black bar. A button or key press will cause the command to be executed. If the user wishes to return to the main menu, he moves the cursor back to the menu bar to select another item. To exit the program, the exit symbol in the upper right-hand corner is selected.

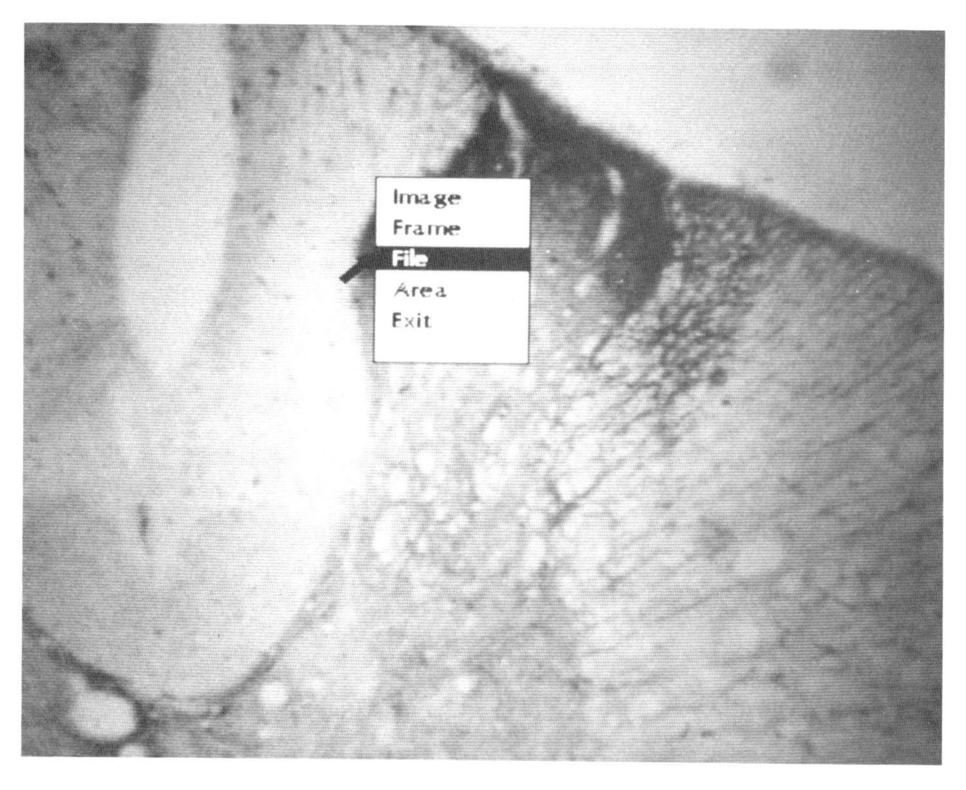

FIGURE 12-6. Pop-up menu. The main menu appears at the cursor position on the image when the user clicks a button or hits a key twice. Clicking once causes the "File" submenu to appear. Clicking when the cursor is outside the menu box causes the menu to disappear.

pointing devices or a single keystroke. Some command interfaces allow you to enter the command and its operand at the same time. To plot data from a file named DATATABLE, for example, you can type:

PLOT DATATABLE

and press "enter." Other command interfaces prompt you for the operand after accepting the command. For example, you can first type:

PLOT

and press "enter." The computer responds with:

Enter name of file to be plotted:

Many operating systems, such as IBM's PC-DOS, have command-line interfaces. In the days before video display terminals, command-line interfaces were widely used for interactive systems.

Because all commands are available at any time, command-line programs offer more flexibility and control of the sequence of function execution than do programs restricted to multilevel menu systems. The command-line interface provides you with a box of tools to use in any order that proves productive in your particular application. However, it takes time to learn the syntax of a command language and use it most efficiently. Memorization is often required because the command list is not usually on the screen for reference; however, the use of help screens can reduce the amount of memorization required. From a programmer's point of view, a command-line interface is the easiest type of interface to implement. It allows a simpler program structure that is easier to debug and maintain.

There will always be a trade-off between the flexibility provided by the loose structure of a command-line interface and the information provided by the rigid structure of a menu interface. The experienced user wants the flexibility, and the less experienced or intermittent user needs the information. It is possible to design programs with both a menu structure and a command-line interface. One way to do this is to give each command in the menu structure a unique and unambiguous name. Either type of user can then select a command from any submenu while in the main menu or any other submenu. Of course, this will add to the programming task. It will create the need for checking whether or not a command from another level of the menu structure is valid in the context of the level from which it is selected.

Some command-line interfaces are programmable. That is, they provide you with all the elements of a program language: arrays and variables, assignment, conditional input/output and control statements, as well as arithmetic and logic expressions that can be combined to form procedures to be executed in the context of the program. Here, for example, is an application program for plotting data using a fairly sophisticated programmable command language. The user/programmer defines a procedure for combining and plotting two data files:

```
COMBPLOT
READ table1
READ table2
ASSIGN table3 := table1 + table2
IF (SIZE(table3) NEQ 0)) PLOT table3
                    ELSE DISPLAY "no data in combined tables"
```

This procedure, COMBPLOT, is added to the program's list of commands and is invoked whenever the user types COMBPLOT. The procedure reads two tables of data from the disk, combines them into a third table, and, if indeed there are data, plots this newly generated third table.

12.3.4 Direct-Manipulation Systems

The Apple Macintosh computer popularized a new method of controlling a computer program, the "direct-manipulation system." With this system, both commands and data are represented visually. Systems utilizing this concept are very popular because they adhere so well to the principles of good interface design. You directly access the data to be operated on as well as the commands to be selected; you are thus able to see immediately the results of command and operand selection. An "undo" command is usually available so that actions can be reversed. All this gives most users the feeling that they, rather than the software, are directly controlling the actions of the computer. To the new user, direct manipulation should seem a more natural way to control the computer than learning the syntax of a command-line sequence. The visual representation by itself often provides the user with an adequate understanding of the program's internal operation—the necessary mental model discussed earlier.

Direct-manipulation interfaces in which the commands and operands are represented graphically are often called "graphical interfaces." The graphic representations, for example, a picture of an eraser, are called "icons." Although the Apple Macintosh system is a well-known example of a graphical interface, most spreadsheet programs are examples of direct-manipulation systems that do not make heavy use of graphics. An example of a graphical interface is shown in Fig. 12-7, the Media Cybernetics HALOvision Image Editor.

Direct-manipulation systems minimize the need for memorization if the graphic representations are meaningful to the user. The choice of the best icons for commands and

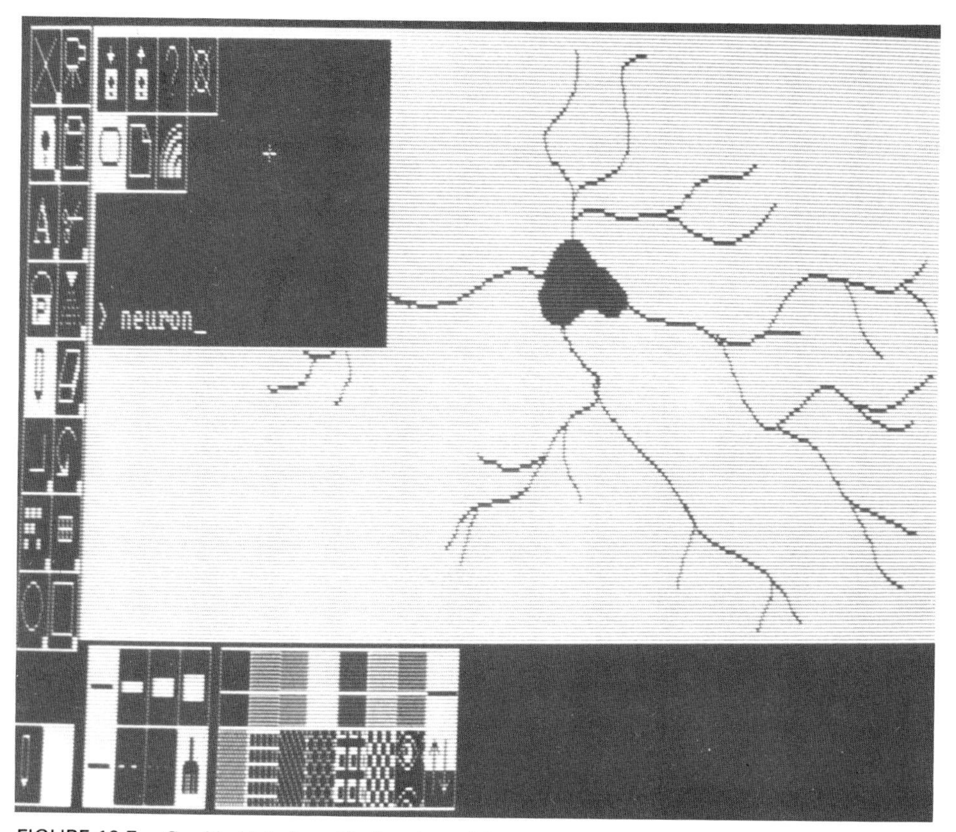

FIGURE 12-7. Graphical interface. The functions of the system are represented by the icons in the boxes on the left-hand side of the screen. Selecting the paint-can icon, for example, allows the user to fill an area with a color or hatch pattern selected from the representations at the bottom of the screen.

operands is a difficult task for the system designer. Icons that have an obvious meaning to the designer may make no sense at all to the user.

There are other disadvantages to direct-manipulation systems. Because they are usually hierarchical menu systems with a graphical interface, experienced users may be impatient if they are forced to go through many levels of the menu structure to access a command. A cluttered and poorly arranged layout of the commands and data on the screen will be confusing to the user. Icons may take up a good deal of screen space, leaving the user inadequate space for manipulating an image. Pop-up and pull-down icon menus alleviate this last problem.

12.3.5. Command Operands, Default Values, and Parameters

User interfaces that minimize the use of the keyboard are attractive to users who are nontypists, especially in applications where the user's hand is required to be on another interactive device. However, typing cannot be eliminated. The execution of many commands requires the entry of data on which the command operates. Some applications use a pointing device to select values from a list, table, graph, or scale displayed on the monitor, but numeric data such as parameters and textual data such as graphs and labels usually must be entered by typing. The user is prompted for such data at the appropriate points in program execution. The following prompt might appear after a user selects a command such as "Adjust Gray-Level Range." The current value of the parameter is displayed after the prompt.

ENTER MAXIMUM GRAY LEVEL VALUE (range 128 → 255) : 200
Press ESC to return to command selection if you do not wish to change the current value.

Figure 12-8 is an example of a dialogue-box format used with pop-up and pull-down menus. The user has selected an option for opening a data file. To enter the file specification, the user may use a pointing device to select a specification from the list of existing files displayed in the box. Thus, the user does not have to enter data that the computer already has. Typing would be necessary only if the user specifies a file not on this list. The

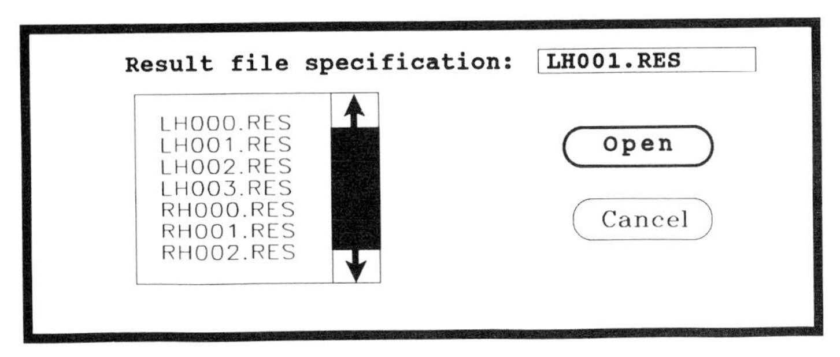

FIGURE 12-8. Dialogue box. The user can use a pointing device to select the "OK" option if he wants to open a file with the specification displayed after the prompt. A new specification could be typed before selecting "open." The user could scan the directory of ".RES" files to select the specification of an existing file.

current file specification is displayed so the user can elect to use it with a button click to select the "open" box.

The application program should assign reasonable initial values to all program parameters and file specifications; then it should give you commands for changing these values to fit the application. These initial values are called "default values" and are to be used by the computer if you do not select or enter other values. Once you set them, these values are not changed until program termination. If a program has a facility for storing the current parameters and file specifications in a file at program termination and then retrieving them when the program is restarted, you will not have to reset them. If a program is used by more than one person, the ability to create personalized parameter files is helpful.

12.4. USING INTERACTIVE DEVICES TO CONTROL TRACING

12.4.1. Introduction

Menus and command-line interface provide the user with the means for selecting and entering commands. Data also must pass through the user interface. The neuroanatomist often enters data into a computer by tracing structures—for example, capturing points along branches of neurons and the boundaries of tissue. To capture or enter a point means to store, in computer memory, its X,Y,Z coordinates relative to some predefined origin. Tracing can be controlled by several interactive devices. The usual interactive devices used are described in detail in Chapter 2 of this book. Here, we describe three types of tracing systems and discuss how interactive devices are used to control them.

Because devices that can be used to control tracing may also be used as pointing devices to select commands, programs that use the same device for both tasks must provide a mechanism for switching back and forth between tracing mode and selection mode. The design of the user interface should make the current mode obvious to the user. The advantage of using a pointing device to control tracing as well as to select commands is that tactile continuity is preserved. You do not have to grope for the keyboard to enter a command while tracing.

The design principle of preserving visual and tactile continuity is especially applicable to the tracing task. Ideally, your hand should remain on the tracing device. Your eye should not be distracted from the structure that is being traced. A third kind of continuity should also be preserved if possible—"concentrational." In addition to your hand and eye, your attention may be displaced from the tracing process. This may happen in poorly designed neuron-tracing systems when a branch point is encountered and it is necessary to specify that this is a branch point to the computer. The program should allow you to identify this point without losing your concentration.

12.4.2. Tracing with a Data Tablet

A typical tracing technique is to place a photograph or drawing of a structure on a data tablet (Figs. 5-4, 5-5, and 5-7). The image of the structure may also be projected onto the tablet, either from above or from below. You move the hand cursor, sometimes called a "puck," of the data tablet along the outline of the structure to capture a series of points.

Data-tablet cursors with buttons allow you to perform the basic tracing operations—capturing and deleting points and signaling the completion of an outline—without removing your hand from the cursor; thus, tactile continuity is preserved. If the outlines are displayed on a graphics screen as they are traced, you get immediate feedback. To increase visual continuity, place the graphics screen as close as possible to the tablet.

When tracing with a data tablet, point entry can be handled in three ways by the application program. It can allow continuous distance-based entry, continuous time-based entry, or point-by-point entry. For continuous distance-based point entry, a spatial filter is applied to the data. That is, the computer captures points that are more than a specified distance apart along the outline. It adds to the flexibility of a system if you are able to vary the size of this filter to accommodate the shape of the outline to be traced. A jagged or sharply curved outline requires a shorter interval than a smooth one.

In the case of continuous time-based entry, the points are captured at specified time intervals as you move the cursor along the outline. Usually, you move the cursor more slowly along a complex outline, thus capturing more points than you would when you move the cursor quickly along a smooth outline.

For point-by-point entry, you click a button when the tablet cursor is aligned with a point to be captured. This method is slower than continuous entry, but you have more control over which points are entered. For many applications, the ability to use either continuous or discrete point entry is desirable.

Tracing with the data tablet requires calibration. When the calibration sequence is invoked, you are prompted to enter two individual points in the tissue and the distance between them in your units, usually micrometers or millimeters. The computer then computes the scale factor for converting from data-tablet units to your units.

12.4.3. Tracing by Moving the Microscope Stage

Many tracing systems are designed so you view the tissue containing the structure to be traced through a microscope. The microscope is equipped with a motorized stage that is controlled by the computer. You direct the movement of the stage to align the points that you want to capture under a cross hairs in the eyepiece of the microscope or under a cursor overlaid on the microscope image. The overlay is done by using a camera lucida to mix a cursor drawn by the computer on a graphics display with the microscope image. The outline of the traced structure is also displayed on the screen so that it too is mixed with the microscope image. This gives you feedback so you will know that the tracing is proceeding correctly.

Stage movement in the *XY* plane may be controlled with a mouse, a data-tablet cursor, a joystick, a trackball, or the keyboard. The movement of the device is mapped to the movement of the microscope stage. When the keyboard is used, use the cursor keys for up, down, right, or left movement. Holding down a key causes the stage to move continuously in the selected direction. With the use of any of these devices, hand–eye coordination is preserved; when the movement of the device or the pressing of the keys is stopped, stage movement stops. Focusing—stage movement in the *Z* direction—can be controlled with buttons (on the mouse, data-tablet cursor, or trackball), with the keyboard, or with a rotatable center shaft on a joystick.

Of the mouse, data tablet, joystick, and trackball, the mouse is the least expensive and the most easily obtained. The tablet, mouse, and trackball are excellent for controlling two-dimensional movement, but they are awkward when movement in a third dimension is required. Only a joystick with a rotatable center shaft allows simultaneous and smooth control of microscope focus and movement. Often outlines are traced as input to a three-dimensional reconstruction system. In that case, the joystick can be used to control both the microscope movement and the rotation of the graphic display of the reconstruction.

The data tablet may be the most expensive of the three devices, but it can also be used for tracing without the use of a microscope if you desire a choice between the two methods. It is an absolute-positioning device that may be used in other applications to simulate a relative-positioning device such as the mouse. However, a mouse or any other relative-positioning device cannot simulate an absolute-positioning device.

12.4.4. Tracing on a Digitized Video Image

If a computer with a video-image digitizer is used for the analysis of data, boundaries may be traced on the digitized image of the tissue to be analyzed as it is displayed on a television screen. This is often done to outline regions in which optical density is to be measured or in which particles are to be counted.

In Fig. 6-6, the operator is using a mouse to direct the movement of a cursor on the screen. When one of the mouse buttons is held down, the computer program captures points on a tissue boundary. The operator gets immediate feedback as the computer connects points as they are entered. The cursor and connecting lines are a graphics display on the digitized image. The gray-level values of pixels in the image frame buffer are not changed. With some video hardware, lines may only be drawn on an image by changing the gray value of the pixel; if a tracing error is made, a new image must be digitized. This may be undesirable in some applications.

Any of the devices mentioned in the previous section may be used to guide the movement of the cursor. Positioning with the cursor keys on the keyboard is the slowest and most difficult method, but it is adequate and does not require the purchase of an additional device. The use of the data-tablet hand cursor is very similar to the use of the mouse; however, the mouse is less expensive and takes up less space.

12.5. DATA STORAGE

12.5.1. Introduction

Most, if not all, data-entry programs provide a mechanism for transferring the data from computer memory, which is volatile, to a nonvolatile mass storage device. The most commonly used devices are fixed or removable (floppy) magnetic disks. The data are written into files on these devices. Magnetic tape is sometimes used to provide additional backup.

12.5.2. Disk Files

In most operating systems, a disk file is identified by its specification, which consists of a device name, a file name, and a file extension name. An example would be

"A:DATAFILE.DAT." The device name is the identifier of the mass-storage device on which the file is located. In the above example, "A:" is the name of a floppy disk drive, "DATAFILE" is the name of the file, and ".DAT" is the file extension name. If a device is not specified, most operating systems assume that the file is on the default device, which in many cases is a fixed disk. The file extension may be used to distinguish types of files. For instance, the file specified as "A:DATAFILE.DAT" may specify an input file to a data-analysis program, whereas "A:DATAFILE.RES" may specify an output or "result" file from the program.

File specifications provide a powerful tool for identifying and keeping track of data. An application program should allow you to enter customized file specifications for files read or created by the program. Default specifications should be provided if you do not care to customize. You should be able to read files from and write files to any valid device in the system. You should only be required to enter the minimum amount of a filename. Once that minimum amount of filename is entered, the program should automatically supply the rest of the name.

There are a number of ways for saving data that may be implemented in a program. The program may provide a command that you may invoke at any time. How often you invoke it then depends on how much work you are willing to lose in the event of a power failure or computer malfunction. This strategy gives you flexibility, but you are also liable for data loss. A program that employs this method should warn you when a program function could destroy data in memory that you may wish to save; it should also remind you to store data before terminating the program.

Some programs automatically store data after some prespecified amount of data has been entered or some prespecified amount of time has passed. Other programs store when some event occurs. An example would be a program that stores coordinates of the vertices of a polygon when you close the polygon. For most applications, it is probably good to have a combination of automatic and user-invoked storage of data.

The format of the data storage files created by an application program is a consideration if they are to be the input to other programs such as spreadsheets or word processors. If you are using LOTUS 1-2-3 for a statistical analysis, it is very handy to have a data acquisition program that produces files in LOTUS format.

12.5.3. Backup

Because fixed disks have been known to fail or to be inadvertently erased, it is absolutely essential to back up the data files that are created on them. This is done by copying the files to floppy disks or other appropriate media. Some programs do this automatically; others leave backup entirely to you, without so much as a reminder of its importance. If backup is not automatic, it is often convenient to be able to invoke the operating-system command for copying files from within the application program.

Floppy disks can be damaged, lost, or destroyed. If a file that has been backed up to a floppy disk is to be erased from the fixed disk, it is a good idea to make an additional copy of it on another floppy disk. Some cautious users even make three copies of important files that would be difficult or impossible to replace and store them in three different locations. Today's computer technology makes it possible to store a great deal of data in a very small amount of space. But it is easy to lose a small thing like a floppy disk or to wipe out a small area like the surface of a fixed disk. Backing up the backup may seem like the action of a paranoid person; however, losing a week's worth of data in a microsecond may destroy other aspects of your mental health.

12.6. ERROR HANDLING

In evaluating a computer system or program, one of the most important considerations is how the software responds to and copes with errors. The errors may result from invalid input from the user or from occasional hardware problems.

The entry of data through the keyboard is an error-prone process. The user interface should be designed to detect many of these errors. First, you need to know what the valid input is. That information should, of course, be provided in the user manual—an important part of the user interface. Many programs provide an online user manual in the form of help screens that can be viewed on demand. (These are described in detail in Section 12.9.) Prompts that include the required syntax for a file specification or the accepted range for a parameter help to guide you. When entering responses to prompts, you should be able to use the backspace or delete keys to make corrections as you type. You should also be given the option of continuing without changing the current value.

If you enter an incorrect file specification or a value that is out of the accepted range, the software should respond with an understandable message and an opportunity to try again. If invalid data are not detected on input, they should be easy to delete when detected later. For example, it should be possible for you, when tracing with a data tablet, to delete the most recently entered point or to restart the tracing of an outline.

Some programs respond to user errors by beeping. Although this does attract your attention, it is not by itself an understandable message and should not be the computer's only response. It should also be possible to disable beeping. Many people find it annoying when a beep calls attention to an error or indicates the successful completion of a command.

The most common kinds of hardware errors involve the reading or writing of disk files. Such errors should be handled smoothly within the program. If the program does not check for these errors, they either go undetected or are handled by the operating system software. When the operating system takes over, a message such as "file read/write error" is displayed after the application program is terminated. It is much better if the program displays a message such as:

Error reading file—b:samples.dat—press ENTER to continue or ESC to return to the main menu.

If the program's detection of input errors is adequate, arithmetic errors, such as attempts to divide by zero, will usually not occur. Such errors may cause a program to "crash," that is, either return control abruptly to the operating system or completely stop the operation of the computer.

The interface should protect you from destructive acts such as writing over an existing file. If in response to a prompt from the computer for the specification of a file to be created, you enter a specification for an existing file, the computer should respond with a warning such as:

File datapts.dat exists. Write over it? (Y/N) . . .

A negative answer returns the user to the point where the file-creation command was selected. A new attempt may be made with a new file specification, or you may exit the program to copy the file to a different disk or to rename it. A sophisticated program allows you to invoke the operating system functions for copying or renaming without exiting the program.

In summary, an application system that handles errors well should do the following:

- Check the validity of input data.
- Display understandable error messages.
- Allow the user to delete invalid data.
- Detect and report file-reading and writing errors.
- Allow the user to retry after an error.
- Protect against the destruction of data.
- Not crash.

12.7. PRESENTATION OF OUTPUT

One of the functions of a fully developed application program is to allow you to manipulate the presentation of the program's output. The quality of the presentation and the ease of manipulation are features of the user interface that should be considered. This is especially true in neuroanatomical applications, where much of the output is best presented in graphic rather than tabular form. For instance, neuron and serial section reconstructions can be used most effectively if they are displayed dynamically using the techniques of interactive three-dimensional computer graphics. Dynamic display allows you to experiment with various orientations of the reconstructed object and to select the most useful ones for hard copy.

If histograms and other graphs are produced by an application program, you should have some control over their format. Many programs allow you to do such things as vary the number and size of the class intervals in a histogram, change the scaling of graphs, and enter text for labels. You can also experiment with various axis limits. Some programs automatically scale the data for graphs by making the lengths of the axes correspond to the maximum and minimum data values. This produces good-looking graphs, but you cannot compare one automatically scaled graph to another.

If you are able to manipulate the format of graphs, you should be able to display them on the computer screen before producing hard copies of them. Some programs allow you interactively to edit a dynamically displayed graph.

Hard copies can be made on an X–Y plotter or a graphics printer. Some of these devices produce low-quality lots quickly; others produce high-quality plots slowly. An application program should be able to produce hard copies on several different kinds of devices. Commercial programs do this by supplying a software item called a ''device driver'' for each of the most common devices. The availability of a large variety of device drivers is probably a good indication of a program's maturity.

12.8. SUBROUTINE LIBRARIES

Application programs can often be purchased with additional software, sometimes called ''programming toolboxes'' or ''subroutine libraries,'' that can be incorporated into programs you write. If you use an image-processing program, for example, you might purchase a library of subroutines for accessing the frame-grabber board that is used with the program. You then write programs that call the subroutines provided. These programs may be used separately, or, in some cases, they may also be installed as user modules that are invoked from the application program.

Subroutine libraries, like programmable command-line interfaces, appeal to users who have or wish to acquire programming skills. They are a good choice if you cannot find a commercial application program that does the job at hand and you are willing to devote time to programming or pay a professional programmer. If source code for the subroutines is available, it may be worth the extra money. If necessary, the subroutines can then be modified to fit the application. If the programmer is able to refer to the source code for the subroutines that the program calls, program design and debugging are often a good deal easier.

12.9. USER MANUALS AND OTHER DOCUMENTATION

Writing down the instructions for using a program was not so important in the days when users wrote their own programs. Now, it is a necessary part of the user interface. The manual or some other documentation that comes with a program should provide more than the instructions. It should give you a general description of both how the program works and general strategies for how to use the program in common applications. A detailed description is not necessary as long as an accurate, understandable model is provided.

It has become very common in commercial programs to make information from the user manual available on help screens that are displayed on demand during program operation. This online help can be very convenient to access. It does not take up desk space, and it can be updated cheaply and electronically. If it is possible to search through it in the same manner in which a word-processing program searches for text, the desired topic can be located quickly. On the other hand, computer screens are more difficult to read than printed pages, and one screen holds far less information than a sheet of paper. The time necessary to go from screen to screen may be longer than the time it takes to page through a manual. You can also keep the relevant manual page in sight. Although in some programs a small portion of the screen is dedicated to online help, the help screen usually disappears when the user switches from help mode back to the program. Also, you can read a manual in book form away from the computer to familiarize yourself with the program. It seems best to have both help screens and a printed manual available.

User manuals are often only reference manuals. If they are complete, they provide all the information you need for operating a program. However, this information may not be organized in a manner that helps you learn to use a program. Tutorial information and examples of program use are necessary to get started. If they are not provided in the user manual, they should be available in some other document. Keep in mind that the quality of the user manual and the other program documentation is another good indicator of the maturity of a program. A user manual that is well written and serves as a strategy guide on how to apply the program to a problem as well as a reference for specific details is a difficult-to-write and thus expensive document.

ACKNOWLEDGMENTS. Any insightful and useful information presented in this chapter is partially the result of years of sometimes painful experience designing and writing computer programs for use in the Department of Physiology of the School of Medicine of the University of North Carolina at Chapel Hill. Much encouragement and help has come from Joseph Capowski, Edward Perl, Elizabeth Bullitt, Paul Farel, and Nicholas Moss of the Department. The author is grateful to these people. The cartoon of the "axed" computer (Fig. 12-1) was drawn by David Johnson. Financial support over the years has been provided largely by grant NS14899-10 from the National Institutes of Health.

13

Video Enhancement Techniques

DAVID G. TIEMAN

13.1. INTRODUCTION

Image enhancement transforms an existing picture into a new picture that is suitable for a specific application. Enhancement is often used to make objects stand out from their background and become more detectable, as Fig. 13-1 illustrates. The object of enhancement is to increase the signal and decrease the noise. In many cases, the enhanced image is the desired end: the experimenter is seeking a "clean" picture to present to an intended audience. In other cases, the enhanced image is used for counting, measuring, or tracing the objects represented: the experimenter creates a simplified representation that eliminates the need for subjective judgments during analysis. Such simplified representations can facilitate analysis by inexperienced human observers or by computer.

Different applications require different types of image enhancement. A transformation that successfully emphasizes fine detail in a high-magnification picture of a neuron filled with nuclear yellow will unfortunately emphasize artifacts and irrelevant details in a low-magnification picture of ocular dominance columns demonstrated with the 2-deoxyglucose technique. Likewise, a transformation that simplifies boundaries and increases coherence for the ocular dominance columns will decrease visibility for the fine processes of the individual neuron. In fact, a single transformation can have both good and bad consequences for the same picture. For a neuron filled with horseradish peroxidase, a transformation may increase contrast between the background and fine axonal arbors; this will be of little advantage if the transformation also creates a "salt and pepper" background by increasing the contrast for small cells that have phagocytized HRP-containing residues. Thus, the useful transformation or series of transformations must be determined for each new application. In general, image-enhancing transformations are most useful when the spatial or gray-scale characteristics of the "signal" are distinctly different from those of the "noise," allowing enhancement of one without enhancement of the other.

Image enhancement requires specialized equipment (Section 13.2). Specific techniques are discussed later with examples. These examples are meant to be illustrative, not exhaustive. Additional information concerning both equipment and techniques can be found in source books on image enhancement (Section 13.6).

DAVID G. TIEMAN • Department of Biological Sciences, State University of New York, Albany, New York 12222.

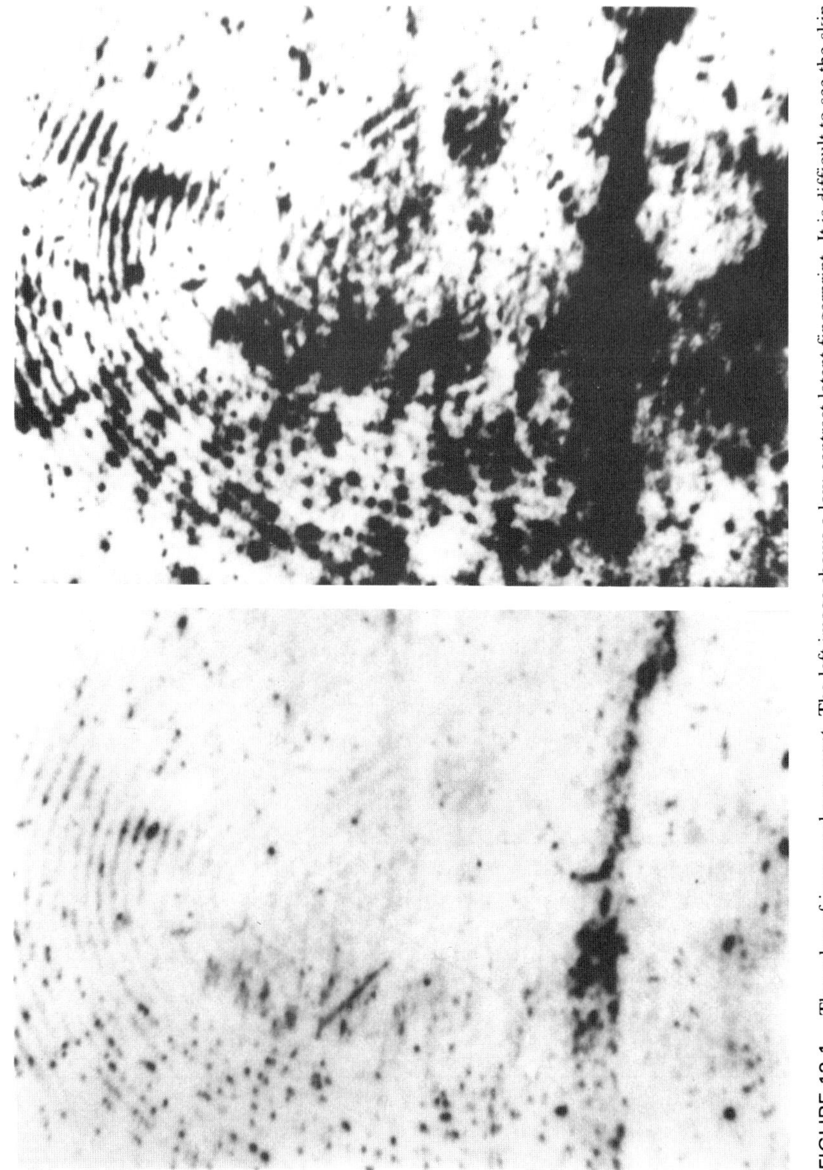

FIGURE 13-1. The value of image enhancement. The left image shows a low-contrast latent fingerprint. It is difficult to see the skin ridges. When the left image is enhanced with a contrast-stretching histogram equalization computer program, the right image results. The ridges are now much more visible. Courtesy of Barry A. J. Fisher, the Los Angeles County Sheriff's Department, Scientic Services Bureau.

13.2. HARDWARE

The hardware of the basic image-processing system is diagrammed in Fig. 13-2. This hardware can cost as little as $4000, but the capabilities of the system can increase drastically as additional money is spent. Any attempts to upgrade a system later can meet with frustration because the new hardware is likely to be incompatible with existing hardware and software. Thus, it is important to anticipate the future applications of a system before purchase.

A system should be tested on its intended applications before purchase. An inexpensive system may appear to do everything when it is demonstrated, but that is often an illusion created for the demonstration on data selected by the manufacturer specifically to show off the capabilities of the system. You can be sure that each system, no matter how inexpensive, does some things well. You cannot, however, be guaranteed that the system will do anything useful when presented with the preparations from a real experiment.

13.2.1. Image-Processing Boards

Image-processing boards for standard personal computers vary in price from $1000 to $5000. A typical inexpensive board has one video input, one input look-up table (LUT;

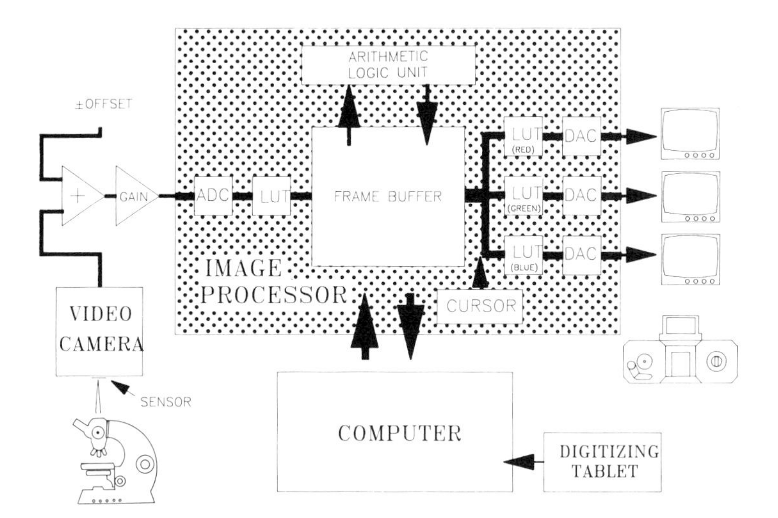

FIGURE 13-2. A schematic diagram of an imaging system. Images are entered into the system by the video camera indicated on the left. The final negatives are obtained with the 35-mm camera indicated on the right. Most of the processing is performed by the image processor and computer. The image processor contains 8 bits of information about each location in a picture that has been digitized to 512 × 480 resolution. On input, the gray values of the video signal are linearly transformed to 8-bit numbers by an analog-to-digital converter (ADC), and these numbers can be mapped to any other set of numbers by the input look-up tables (LUT); on output, the 8-bit numbers are transformed by the LUTs and digital-to-analog converters (DAC) to determine the gray values of the video output. The three output LUTs allow each of the three video outputs to contain different gray-scale representations of the same picture or to provide outputs for the three primary colors of a color picture. Video images can be rapidly manipulated by image processors that include an arithmetic and logical unit.

Section 13.2.1b), one video frame buffer with eight bits per pixel (Section 13.2.1a), three output look-up tables, three video outputs for RGB color coding, and an interface to the computer. More expensive boards add multiple video inputs that can be selected from software, multiple LUTs, a hardware cursor, zoom, real-time arithmetic and logical operations, special interfaces for high-speed array coprocessors, and multiple frame buffers that can be independently selected for input, manipulation, and display.

13.2.1a. Frame Buffers. In the frame buffer, a two-dimensional scene is represented by a matrix of numbers. Some characteristic of the scene, usually gray scale, is sampled at discrete points across the scene and digitized. Each of the sampled points is called a picture element or pixel. The frame buffer contains one number per pixel. Typically, frame buffers contain an array of 256 × 256, 512 × 512, or 1024 × 1024 numbers. As the number of pixels increases, the resolution of the digitized image also increases. For a frame buffer attached to a microscope using a 20× objective, a 512 × 512 frame buffer has pixels with diameters of approximately 1 μm. The numbers that encode values at each pixel are written serially to and read serially from the frame buffer. For a typical 512 × 512 frame buffer, these operations must occur at a rate of about 8 MHz (512 × 512 × 30/sec). The delay between input and output for the system is 1/30 sec or less.

Most moderately priced imaging systems do not use ideal frame buffers. The ideal frame buffer has a large number of pixels, and each is square. Because of requirements of the fast Fourier transform, which is used in frequency-domain analyses (Section 13.4.5), it is desirable for the dimensions to be powers of 2. In most practical situations, however, the use of the frame buffer is dictated either by television standards or by computer video standards. American television standards (RS-170) specify that 525 horizontal scans occur during each 1/30 sec. Since time is required to reset the scanning beam from the lower right to the upper left of the field in a video camera or TV screen, only 480 horizontal scans, and hence 480 vertical sampling points, are reliably presented. Thus, a standard TV video input fills only part of a 512 × 512 frame buffer. Since the dimensions of the standard television screen do not match the dimensions of the frame buffers, the dimensions of each pixel are in a ratio, called an "aspect ratio," of about 4 : 3. Therefore, $n \times n$ spatial-domain filters (Section 13.4.3b) have slightly different characteristics in the horizontal and vertical directions.

Various video standards allow higher resolution. For example, European TV uses 625 horizontal scans, from which it is easy to obtain a 512 × 512 image without using special equipment. Thus, standards for European TV (CCIR and PAL) are better for image processing but are based on the 50-Hz alternating current used in Europe. Additional standards that allow 512 scan lines to be visible (e.g., 559 scan lines) are available on some imaging boards. The difficulty with the proliferation of standards is that the same standard must be supported by each piece of video equipment—monitor, frame buffer, camera, and VCR. As a result, vendors of lower-priced systems have emphasized U.S. television standards (RS-170, RS-330, and NTSC). In the near future, an international standard for high-density television (HDTV) will allow 1100 lines of vertical scan for broadcast TV. At this writing, a version of HDTV exists on some scientific imaging systems, a standard has been proposed by the Japanese electronics industry, and a European consortium is developing an alternative standard. Once a single standard is accepted and the associated equipment is manufactured, the 1024 × 1024 frame buffer and display will become more common.

The display screens on personal computers can sometimes be used with frame buffers, but only rarely do the pixel arrays that are defined for the computer graphics screen correspond to the pixel arrays in the frame buffer. Unfortunately, computer graphics screens do not improve on frame buffers. Frequently used standards for these screens, such as the IBM color graphics adaptor (CGA) standard (640 × 200), the Hercules monochrome standard (720 × 350), the IBM enhanced graphic adaptor (EGA) standard (640 × 350), and the IBM video graphics array (VGA) standard (640 × 480) all employ pixel arrays that are not powers of 2. Only the VGA standard has pixels with a 1 : 1 aspect. Because of the differences in video and computer graphics standards, two types of frame buffers are evolving, one to mimic the computer graphics screen and use 640 × 480 pixel arrays so that computer graphics can be merged with the video, and the other to work independently of the computer graphics.

13.2.1b. Look-up Tables. Look-up tables (LUTs) determine the mapping between numbers in the frame buffer and intensity at each pixel in the scene or display. The input is received by the board as a voltage and converted to a digital representation by a fast analog-to-digital converter. Black is usually converted to 0 and white to 255. This number is then passed to the input LUT, which defines a new transformation. Each of the 256 input numbers can be mapped to any number in the range 0 to 255 with any one-to-one or many-to-one mapping. The normal LUT would map each possible input value to itself. A negative LUT would subtract the input value from 255 to determine the new value. Thus, LUTs can be used for transforming scales such as transmittance to density or to linearize a measured quantity such as autoradiographic label. Figure 13-3B shows the effect of a logarithmic transformation on the image of an autoradiogram of cat visual cortex with ocular dominance columns labeled with [^{14}C]2-deoxyglucose.

One of the most common uses is to stretch a limited range of grays so that it covers the entire range, thus increasing the contrast in the picture (Fig. 13-3C). The value from the input LUT is entered in the frame buffer and then passed to the output LUTs. Here the value is transformed independently by three LUTs working in parallel. The values from the output LUTs are then converted to video voltages by three independent digital-to-analog converters. These video outputs can be used independently either to control three black and white monitors, each of which shows the same scene with a unique gray scale rendition, or to control the red, green, and blue beams of a color display monitor. In this latter case, the three output LUTs map the numbers in the frame buffer onto 256 colors in much the same way as a ''paint-by-numbers'' picture from the hobby shop. This type of presentation in which arbitrary colors are presented for gray values is called a pseudocolor display. It is discussed in Section 13.4.2d and illustrated by Fig. 13-8. The 256 colors can be chosen from a palette of 16 million possible colors (256 × 256 × 256), all the colors the image-processing system could present.

Although either the input LUT or an output LUT can be used to manipulate the video display, each differs in its effect on the data available to the computer. Any transformation implemented with the input LUT affects the data available to the computer; any transformation implemented with the output LUT does not. Thus, the input LUT is normally used

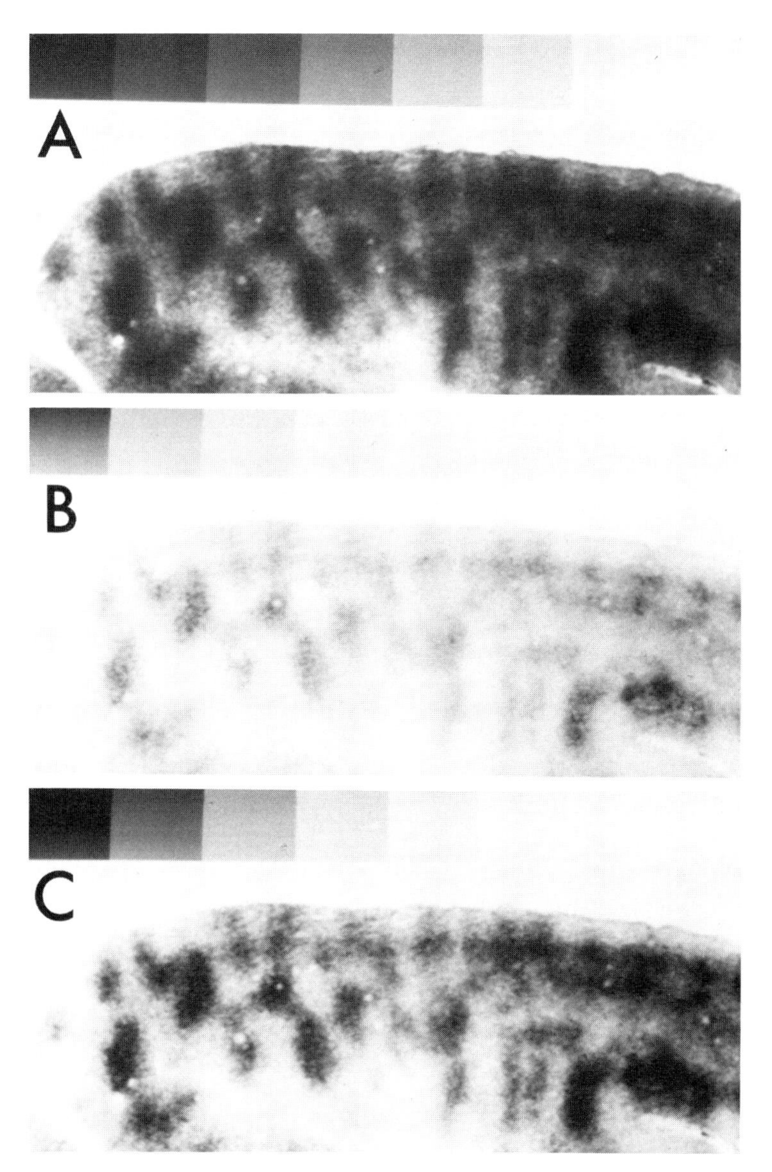

FIGURE 13-3. An image showing the effect of gray-scale transformations. Panel A shows a direct video image of an autoradiogram of cat visual cortex. The dark areas represent [^{14}C]2-deoxyglucose labeling of ocular dominance columns in a cat that was reared with alternating monocular deprivation (Tumosa et al., 1989). The offset and gain of the video signal were adjusted so that an opaque region of the film gave a frame buffer reading of 0 and so that the unexposed film gave a reading of 255. The scale at the top is continuous from the upper left and continues from top to bottom in each of the eight subregions of the scale. Since the video camera encodes light transmission through the film, a more accurate representation of the label is given with a logarithmic transformation into a density scale. This is shown in panel B. Note the difference in the gray scale at the top of A and the top of B. The gray scale in B was created by logarithmic transformation of the gray scale in A. A more pleasing picture of panel B can be obtained by stretching the gray scale so that the lower half of the logarithmic scale is better displayed. The grays in the range 128 to 248 have been expanded to fill the full scale from 0 to 255. Values from 0 to 127 have been set to 0, and values from 249 to 255 have been set to 255. The gray scale that is shown at the top of C represents a linear transformation of the logarithmic values in B. Although this stretching of the gray scale provides a more pleasing picture, the values that result from it no longer provide a simple measure of a physical property of the autoradiogram.

for transformations that are necessary for quantitative operations, and the output LUTs are normally used for transformations intended to affect the appearance of the image.

13.2.1c. On-Board Arithmetic or Logical Processing. Image-processing boards for personal computers can be divided into three categories: (1) those that have only frame-grabbing and display functions, (2) those that provide additional arithmetic processing, and (3) those that are combined with high-speed coprocessor boards. Although these boards are widely advertised to function with IBM PC/XT/AT and compatible computers, you should be wary. These boards often use high-speed data transfer techniques that are not a routine part of "compatibility" tests. You should verify compatibility before purchasing a computer and image-processing board.

The least expensive boards ($1000–$1500) are the frame grabbers available for IBM PC-compatible computers. These boards, which are often used in the simpler image analysis systems provide no real-time manipulations except the gray scale and pseudocolor manipulations made possible by the LUTs. All additional enhancements must be performed by the computer.

The medium-priced boards ($3000–$5000) are available for IBM AT-compatible computers and contain an arithmetic and logic unit (ALU). The ALU allows the boards to perform operations on the picture in real time, for example, to subtract a previously digitized background image from a live image, pixel by pixel. These boards can typically integrate the input image to create and display a running geometrically weighted average.

More expensive ($8000–$25,000) sets of boards are most frequently used in more powerful computers, those containing a Motorola 68000 series CPU and the VME bus. This high-speed bus or even dedicated cables are used to transmit images between the boards. One board has the capabilities of the medium-priced boards, and another board can perform high-speed spatial-domain filtering and can perform arithmetic operations with greater precision. Enhancement operations involving 3×3 filters (Section 13.4.3b) can take a minute when performed by the computer but can take less than 1 sec when the picture is transmitted to the co-processor board, the operation is performed by the on-board hardware, and the enhanced picture is transmitted back to the frame buffer. The least expensive of the coprocessor boards are limited to spatial-domain filtering; the most expensive provide arithmetic array processing that can be used to implement frequency-domain filtering.

13.2.1d. Multiple Frame Buffers. Multiple frame buffers allow a user to store more than one picture on the image-processing board. The most elegant implementations of multiple frame buffers provide overlays, scrolling, keying, and independent read, write, and display operations.

An overlay is a line drawing that appears to be on top of the video image but does not destroy the pixel values under it. Typically, one of the eight bits assigned to each pixel is allocated to the overlay and protected from being overwritten. When an overlay bit has a zero value, its pixel is unaffected. When the bit is set to 1, the pixel is displayed in whatever color has been assigned to the overlay. If additional bits can be assigned to the overlay, the overlay can be multicolored or have several levels of gray. When sufficient bits are available, the overlay can display the full range of gray values, and a special case exists—the frame buffer has two independent pictures that are stacked. Enabling or disabling the overlay at each pixel determines which of two independent values is displayed, and any combination of pixels from one picture can be displayed simultaneously with the complementary pixels from the other picture. The overlay bit is said to be "keying" between the two independent pictures. This keying can be used to display part of one picture overlaid on another or, when half of the pixels are keyed at random, to create an apparent double exposure that allows the two pictures to be compared and aligned.

If multiple frame buffers are adjacent rather than stacked, then they can be treated as either independent buffers or adjacent locations. For instance, a board that contains enough memory for four 512 × 512 frame buffers can store four separate pictures or it can create one 1024 × 1024 frame buffer. With the 1024 × 1024 buffer, the video display, which is limited to 512 × 512, can be used as a window to scroll around the large picture. With the four 512 × 512 frame buffers, different functions can occur simultaneously in separate buffers. Thus, the input might be storing an image in frame 1 while the output is displaying frame 2 on the screen and the computer is enhancing frame 3.

The various configurations of frame buffers are not mutually exclusive. One processor board may be capable of treating the available video memory as one large frame buffer, as several adjacent frame buffers, as overlaid frame buffers, and as one frame buffer with additional bits per pixel for arithmetic operations. These possible configurations provide flexibility to the programmer.

13.2.2. The Video Input

Although the most common video input to a laboratory image-processing system is a video camera that uses television standards, there are many alternative input devices with their own advantages and disadvantages.

13.2.2a. Scanning Stages. An input device that uses a scanning stage moves the object rather than the sensor. Thus, one fixed sensor is used, and the pixels are created when the object is moved in discrete steps. The most common implementation of this technique in neuroanatomy uses a microscope with a scanning stage. A single-diode densitometer is attached to the trinocular head of a microscope, and a preparation is moved so that all the pixel positions are scanned and recorded. This system has several advantages. First, the sensor is inexpensive and can be customized for a particular application. Many single-diode detectors are available with superb sensitivity. Second, because all digitizing occurs at the same location in the microscope, identical optics and lighting conditions occur at every pixel. Variations across the image must be a function of the preparation. Third, depth of focus can be controlled. A large field scanned with a 100× objective will have a much shallower depth of field than the same field scanned with a 10× objective, and even less depth of field can be obtained by using a point light source focused on a small diode or a small aperture in front of the diode. Unfortunately, the scanning-stage system also has several disadvantages. First, high-resolution scanning stages for microscopes are expensive. Second, the time required to scan a single image is great. For a 512 × 512 image that is scanned at 1 msec per pixel, over 4 min is required to create a full picture. Scanning-stage systems are used most often when accurate gray-level measurements are required from photographic negatives or from autoradiographs.

13.2.2b. Scanning Linear Arrays. A compromise between sensors that "see" a complete two-dimensional field and sensors that "see" only a single pixel is provided by linear diode arrays. A row of diodes is scanned across an image either by rotating a mirror or by moving the object in one dimension. High-resolution linear arrays of up to 4096 elements are available, and cameras based on this technology are often employed in systems designed for quantification. Although linear array cameras can scan a field with very high resolution much faster than a single diode, they are not adequate for real-time work. In addition, resolution beyond 512 × 512 is often lost because of the practical constraints of the memories in frame buffers. The time required to scan a single field is measured in seconds.

13.2.2c. Two-Dimensional Diode Arrays. Two-dimensional diode array cameras (CCD or charge-coupled devices) have recently increased in resolution and speed. Diode array sensors are now available in standard 640 × 480 arrays that can be incorporated into very small cameras compatible with TV standards. These cameras are rapidly overtaking the more standard models for most applications. Unfortunately, few models of CCD cameras are available for special applications.

13.2.2d. Video Cameras. Standard video cameras have two main components, a sensing tube with a light-sensitive face and electronics to control the capture of an image that is focused on the face and to generate a standard electrical video signal that represents the image. Different sensing tubes can be purchased for the same camera body. Different camera bodies have different noise tolerances and features. The most commonly available sensing tubes are the vidicon, the plumbicon, the chalnicon, and the newvicon. The vidicon is the least expensive and is generally used in inexpensive cameras. The vidicon tube does not have a linear response but does have very good spatial resolution. Newvicon and plumbicon tubes are more sensitive than vidicon tubes and have a linear response. Because plumbicon tubes retain an image for a short time, they are frequently used to observe moving images.

The electronics in many video cameras implement automatic offset and gain adjustments and automatic γ correction. (The γ correction provides a nonlinear transformation to counteract the nonlinear response characteristics of the phosphors on the video screen.) With these automatic features, the cameras provide adequate pictures over a wide range of light levels and contrasts. Unfortunately, these automatic features change the transformation between the gray values present in the object and the digitized values presented to the computer. And these changes can occur at inopportune times. Cameras that provide manual override switches with potentiometer controls of offset and gain are desirable. In fact, it is useful to replace the standard potentiometers with clock-face, 10-turn potentiometers and to increase the available gain and offset to double that normally provided.

Parameters of interest in the specifications for video cameras include minimum light sensitivity values, lines of resolution, intensity range, linearity of response, and signal-to-noise ratio. The minimum sensitivity is important for low-light viewing, the number of lines of resolution is important for fine detail (the number of lines of resolution specifies how many distinct vertical lines in a frame the camera can distinguish under standardized conditions), the intensity range is important for making the camera usable in more situations, the linearity of response is important for quantification, and the signal-to-noise (S/N) ratio is important for a noise-free image. Additional parameters that can be important for a specific application are the time constant for decay of the image from the sensor and the extent to which the sensor can be damaged by bright light or a nonchanging pattern (''burn-in'').

It is difficult to use specification sheets and product descriptions to make a rational choice of the best camera. First, gains for one parameter frequently result in losses in another. Second, comparable test values are difficult to obtain. Third, cameras that are good at everything can cost more than the entire imaging budget. The best technique for purchasing a camera is to borrow a cheap one while you are learning about your image-processing system. When you understand the strengths and weaknesses of your imaging system applied to your data and what you want to do with it, then you will be better able to obtain useful information from the available sources: the representative who sold you your imaging system, exhibitors at conventions, technical support staff of the camera manufacturers, other scientists or their publications, and any technical staff available at your site. Vidicon cameras for nondemanding applications can be obtained from \$200 to \$1000. Newvicon cameras, which are more sensitive and have a linear response, cost from \$1500 to \$4000. Plumbicon and SIT cameras often cost \$5000 to \$20,000.

13.2.2e. SIT Cameras. Silicon intensifier target (SIT) tubes are designed for applications that involve the detection of extremely low light levels. They are essentially two-dimensional photon counters and are often 100 times more sensitive than other sensors. They are sometimes necessary, for example, when viewing a very-low-light-level image of a neuron stained with a fluorescent dye, but are expensive and provide a noisy image. Integration or averaging of successive images is needed to obtain a good image (Fig. 13-10). In all but the most demanding applications, the newvicon tube, pushed to its limits, or a diode array provides a practical alternative. Cooled diode-array cameras with the sensitivity of SIT cameras will be available in the near future.

13.2.3. Video Displays

An image-processing system should have both a monochrome and a color monitor. The video monitor should have a ''reduced scan'' or ''underscan'' option, which will allow the entire image to appear on the screen. This option is necessary because the normal television presents only the central 85% of the image. Little advantage is gained by obtaining a monitor with spatial resolutions greater than that of the frame buffer. Computer monitors that have only composite video inputs (RS-170) can be used with image-processing boards, but they provide poor renditions of the gray scale. RGB computer monitors can be used with most boards if the monitors can accept color (RGB) input as three independent analog signals; most RGB computer monitors, however, accept only RGB/TTL inputs, which are not compatible with analog video. Some of the newer ''multisync'' monitors, intended for use with a wide range of computer graphics outputs, support both RGB/TTL and RGB analog inputs.

For 512×512 resolution, acceptable monochrome video monitors cost about \$300 to $600, and acceptable color monitors cost about $500 to $1000. As the number of pixels increases above 512 × 512, prices increase dramatically because the manufacturing quantities of consumer TV are no longer applicable.

13.2.4. True Color Systems

Most imaging systems available for laboratory computers do not implement true color images. That is, when a color picture is displayed, it does not represent the colors that appeared in the scene. Systems that represent true color can be constructed and are available, but they are expensive. Such a system requires a separate video input and a separate frame buffer for each of the three primary colors (red, green, and blue). The primary colors must be stored independently and displayed simultaneously. An imaging system that provides true color can be configured simply by installing three single-color image-processing boards into the same computer and synchronizing their displays. Many manufacturers of boards provide this capability, but you should be wary because the expansion capability of most laboratory computers is limited by the size of the power supply as well as by the number of expansion slots.

True color systems often do not use a color video camera. Although real-time applications require a color camera, non-real-time color images can usually be obtained by triple exposing the image through filters onto a black-and-white camera. Several sets of Kodak Wratten filters are available for color separation work.

13.2.5. Storage of Images

Once an image has been digitized into a computer memory, we may want to store it on a disk for later recall and mathematical processing. Unfortunately, storing an image requires a large amount of disk space, so image-processing facilities must spend money and effort on this practical problem. Some of the various methods and media for storing images are now reviewed.

13.2.5a. Computer Mass Storage. Images from frame buffers can be stored on the standard devices for small computers. The primary considerations are amount of storage and speed—there is no loss of resolution. Since one 512 × 512 picture requires 256 kilobytes of storage, storage devices with large capacity are preferred. A standard 360-kilobyte IBM PC floppy disk can store only one picture. A 40-megabyte hard disk can store 160 pictures. Even with such a large hard disk, it is critical to have a removable storage medium, especially when several investigators use the same equipment. Hard disks that seem infinitely large can fill rapidly with images, especially color images. The readily available devices that could be used by an investigator to store working images include removable hard disks, Bernoulli boxes, and magnetic tape. Magnetic tape drives, though slowest, are the cheapest and have removable media that are inexpensive and easy to store. Some computer software techniques are available to compress an image for storage on disk. Two of them, run-length encoding and pixel chaining, are discussed in Chapter 6. The amount of space saved is dependent on the data and may vary from significant to none.

13.2.5b. Video Cassette Recorders. Video cassette recorders (VCRs) are often used to record analog video information for later analysis with image-processing systems or for lecture presentations. Any imaging system that uses standard RS-170 video signals can use a VCR either for input or for storing output. Most imaging systems can synchronize on regular or stopped-frame modes. They cannot, however, synchronize on input that was created with a time-lapse mode. In this case, the timing signals are too inexact for the computer circuitry.

The disadvantage of VCRs is their lack of adequate frequency response to store fine detail in a picture. A VCR can resolve about 300 lines, and this means that much of the resolution of a 512 × 512 imaging system is lost.

A recent development is the use of VCRs to store digital data. For this application, the VCR is used with an inexpensive interfacing board and replaces other kinds of magnetic tape drives. The VCR stores data with no loss of resolution but cannot provide real-time display of video images.

13.2.5c. Laser Video Disks. Because of their recent popularity for audio systems, optical compact disks (CDs) are being widely developed as a medium for mass storage with computers. The standard CD players that are used with audio systems have been adapted for computer use and can store 400 megabytes (1600 video pictures). Although CD drives are readily available and inexpensive ($1000), they have not become popular because you cannot write new information to a CD. They are "read only," and therefore, you can only retrieve information provided by the manufacturer of the disk. This problem can be overcome with the more expensive WORM (write once/read many) drives ($15,000), which allow storage of approximately 800 video images per optical disk. The disks, which cost about $125 each, cannot be erased but can rapidly store and retrieve high-resolution pictures. Drives are available to store and display in real time.

13.3. SOFTWARE

The normal principles of writing versus buying lab software as described in Chapter 3 apply to image-processing software. However, the real cost of writing your own software deserves emphasis. The temptation to buy hardware and write your own software is great, but you must avoid it because the real cost of the software is too high. Furthermore, you will always underestimate the time involved in creating the software. For these reasons you should seriously reconsider any decision to write your own software; it is rarely cost-effective.

If you should decide to write your own software, then at least buy a "toolbox." This is a package of already-written subroutines that perform the standard image-processing functions. By using these, you or your programmer may concentrate on higher-level functions unique to your application rather than on the details of image processing.

13.4. ENHANCING THE IMAGE

The techniques for image enhancement can be divided into three broad categories: analog techniques, spatial-domain techniques, and frequency-domain techniques. Analog techniques manipulate the video signal before it is digitized or after it is returned to analog form. Spatial-domain techniques manipulate individual pixels and regions of the digitized image. Frequency-domain techniques treat the image as a two-dimensional waveform and manipulate it by modifying the Fourier transform of the wave. The first two techniques are relatively easy to implement in an inexpensive interactive video system for the laboratory and thus are emphasized in this chapter. The third technique is described briefly. It is much more computationally intensive and, therefore, has been relegated to mainframe applications and expensive dedicated systems. This relegation may be temporary, however, as the speed of ''personal'' computers increases dramatically. The 20-MHz 80386 computer is 20 times as fast as its forerunner, the 4.77-MHz 8088 computer (the classic IBM PC), and new coprocessor boards that are specifically designed for two-dimensional Fourier processing now exist.

13.4.1. Analog Adjustments of the Gray Scale

The standard composite RS-170 video signal has two components, the continuous waveform that encodes the light intensities present in the scene and a series of pulses that provide horizontal synchronization of the scan lines. To represent these two components, the voltage range from 0.0 to 1.0 V is subdivided by a ''pedestal'' voltage of 0.35 V. The synchronizing pulses, which occur at approximately 15.7 kHz, are downward pulses from 0.35 V to 0.0 V. The analog signal is represented from 0.35 V (black) to 1.0 V (white). Manipulations of the analog video signal are applied before combination with the pedestal voltage. Thus, we can treat the analog signal as a separate waveform with a maximum range of 0.0 to 0.65 V.

The offset of the analog video signal can be defined as the extent to which its minimum voltage deviates from 0.0 V. The gain of a video signal can be defined as the extent to which the range of its voltages covers the maximum allowable range from 0.0 to 0.65 V. Offset correlates with the overall blackness of a scene, and gain correlates with the contrast present. The offset and gain controls that are incorporated into video cameras change the sensitivity of the sensing tubes. An optimal use of the video signal is one with an offset of 0.0 and a gain of 1.0, so that the full scale of available gray values is used. These are the settings that automatic offset and gain controls attempt to approximate.

Manual manipulations of offset and gain allow resolving power, the ability to separate two items within an image, to be increased. This phenomenon was first discovered by Allen (Allen and Allen, 1983) and Inoué (1986). The reason that offset and gain manipulations can increase resolving power can be understood by considering the resolution of two points. Figure 13-4 shows the video signal as the camera's electron beam makes a horizontal pass across two closely spaced spots of light that have been presented to the camera's face. When the video signal is fed into a TV monitor and the points are displayed on the screen, the viewer's ability to resolve the two points will be a function of the ratio of b, the dip between the two points, divided by a, the brightness of the points. As the two points move closer together, the dip in brightness between them becomes smaller (i.e., the ratio b/a becomes smaller), and the viewer has problems discriminating between the two points. When the two points are close enough together, the viewer sees only one point. While the viewer is seeing just one point, however, the analog video signal may still show a dip (Fig. 13-4B). It is this discrepancy between what the viewer perceives and what the

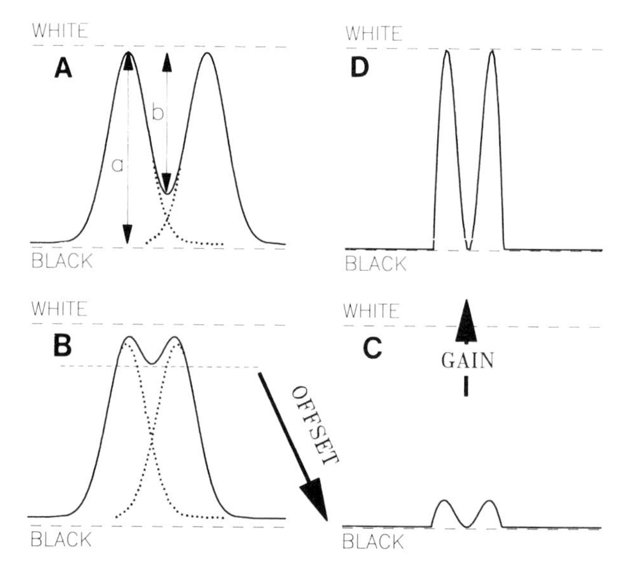

FIGURE 13-4. Diagram of light intensity values for scans passing through the centers of two point sources of light. The blur circle is represented schematically by a Gaussian distribution. In A the two point sources of light are far enough apart to be distinct. The critical value for determining that two points, rather than one point, are present is the ratio of *a* to *b*. In B the two point sources of light are close together and perceived as a single point. In C the offset of the video signal has been adjusted so that the dip between the two peaks has a value of zero. This scan would be perceived as two very dim points of light. In D the gain of the video signal has been increased so the gray values cover the full available range. The two points would be perceived as two distinct bright spots. Note that the intensity scans are now very different from the Gaussian distributions that were used for the starting conditions.

analog signal shows that allows for analog enhancement. First, the offset can be manipulated so that the waveform is moved lower in the available voltage range (Fig. 13-4C). As this is done, voltages below 0 V are set to 0, and consequently discriminations between the dark gray values are lost, and a very faint image results. Second, the image can be made more visible by using the gain control to increase the brightness. In the final image (Fig. 13-4D), the ratio b/a is greater than the ratio in Fig. 13-4A, and the two points become perceptually distinct.

Analog enhancements are limited. First, the technique allows only the better detection of objects. Because of distortions, inferences about the sizes and shapes of the objects relative to other objects with gray values in ranges less affected by the offset and gain adjustments are difficult. This can be seen by comparing the contributions of individual points in Figs. 13-4A and 13-4D. Also, because offset and gain are controlling the video sensor, the technique is limited for light objects on a dark background. When the background is light and the points are black, then the blackest values can be offset to 0.0 V as diagrammed in Fig. 13-5B. Subsequent attempts to increase the resolution by increasing the gain will be defeated, however, because a point on the sensor that is made overly bright (>0.65 V) will build up excess electrical charge that will spread or ''bloom'' into adjacent regions (Fig. 13-5C). This blooming obscures all nearby structures and sets serious limits on the extent to which offset and gain can be employed. The maximum brightness can be increased only to the point at which blooming does not obscure adjacent structures of interest.

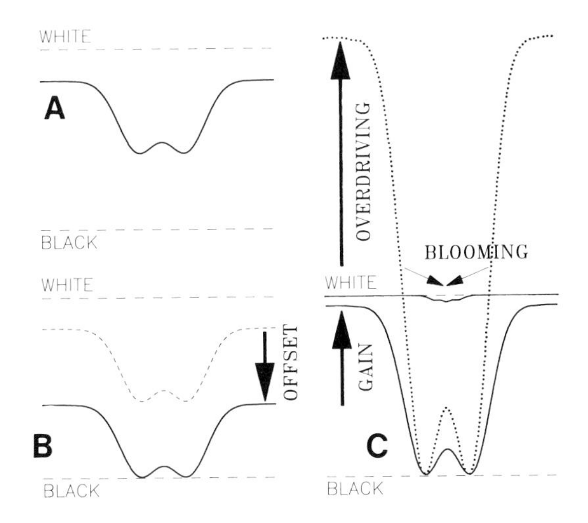

FIGURE 13-5. Diagram of light intensity values for scans passing through the centers of two dark spots on a light background. The scan in A is an inverted version of that in Fig. 13-4B; the two dark points are perceived as a single dark spot. In B the offset of the video signal has been adjusted so the peaks are offset to zero volts. In C the gain has been used to increase the signal so the grays cover the full available range, but the effect on differentiation is minimal. Further attempts to overdrive the video signal are ineffective because blooming, which results from charge spread on the face of the sensor, invades the dark region and causes the image to turn white.

13.4.2. Digital Manipulation of the Gray Scale

13.4.2a. Bits of Resolution. Because imaging systems have limited memory, and that memory can be assigned to purposes other than storing the gray scale of an individual picture, it is important to know the minimum adequate number of bits of resolution. Most imaging systems use an eight-bit analog-to-digital converter to transform the analog signal into digital form. With such a converter, black corresponds to 0, white corresponds to 255, and a total of 256 shades can occur. A video image with 256 discrete shades of gray appears to have continuous variation in the gray scale. As a general rule, an image with 64 shades of gray appears to have a continuous variation of gray; an image with 16 shades of gray looks like the object but has definite planes of uniform gray. Figure 13-6 shows video images with two, four, eight, and 16 shades of gray, corresponding to one, two, three, and four bits of memory.

When few discrete values are available to encode the gray scale, improved images can result if the values are not spread evenly over the entire range. In many images, important information is represented within one region of the gray scale, and other regions contain little information. For instance, a low-power image of an autoradiogram of a section of rat brain might contain a very bright background (255), relatively bright and unlabeled myelinated pathways (180–200), and a relatively dark cortex that contains the gray values of primary interest (0–63). If only three bits were available for storage, it would be inefficient to allocate the eight discrete values equally across the gray values 0–256: only two discrete values would occur within the region of interest. Rather, an unequal assignment by the input LUT (e.g., the gray range 0–9 to the value 0, 10–19 to 1, 20–29 to 2, 30–39 to 3, 40–49 to 4, 50–99 to 5, 100–224 to 6, and 225–255 to 7) could better allow the available discrete values to encode the information of interest. The output LUT can then provide gray values for the picture by assigning the eight discrete values to gray values (e.g., 0 to a gray of 5, 1 to 15, 2 to 25, 3 to 35, 4 to 45, 5 to 55, 6 to 200, and 7 to 255).

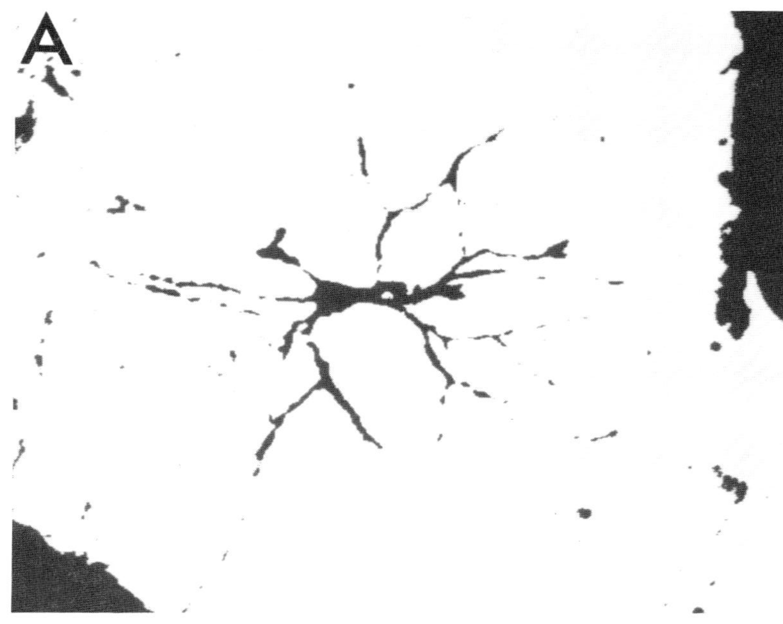
A

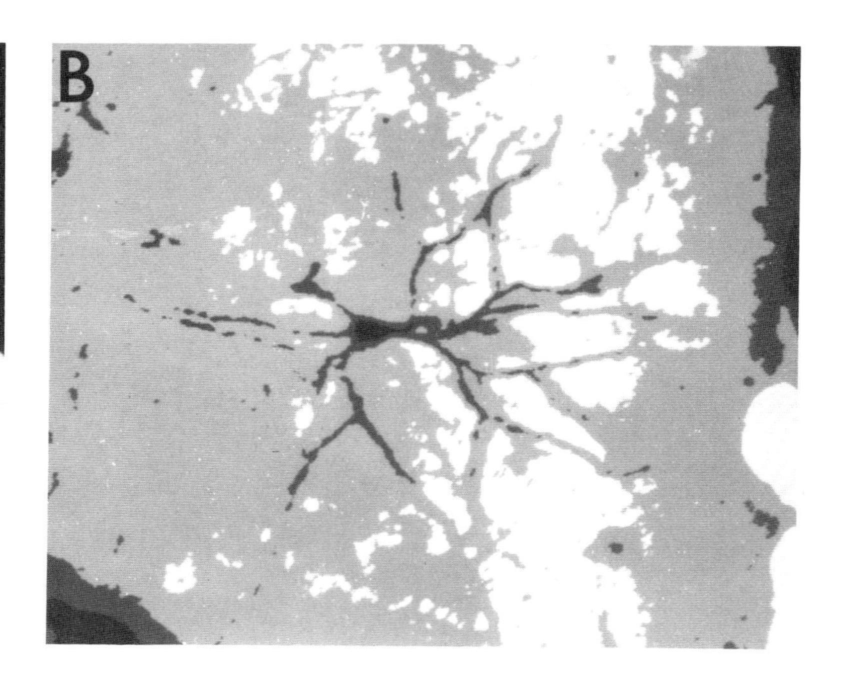
B

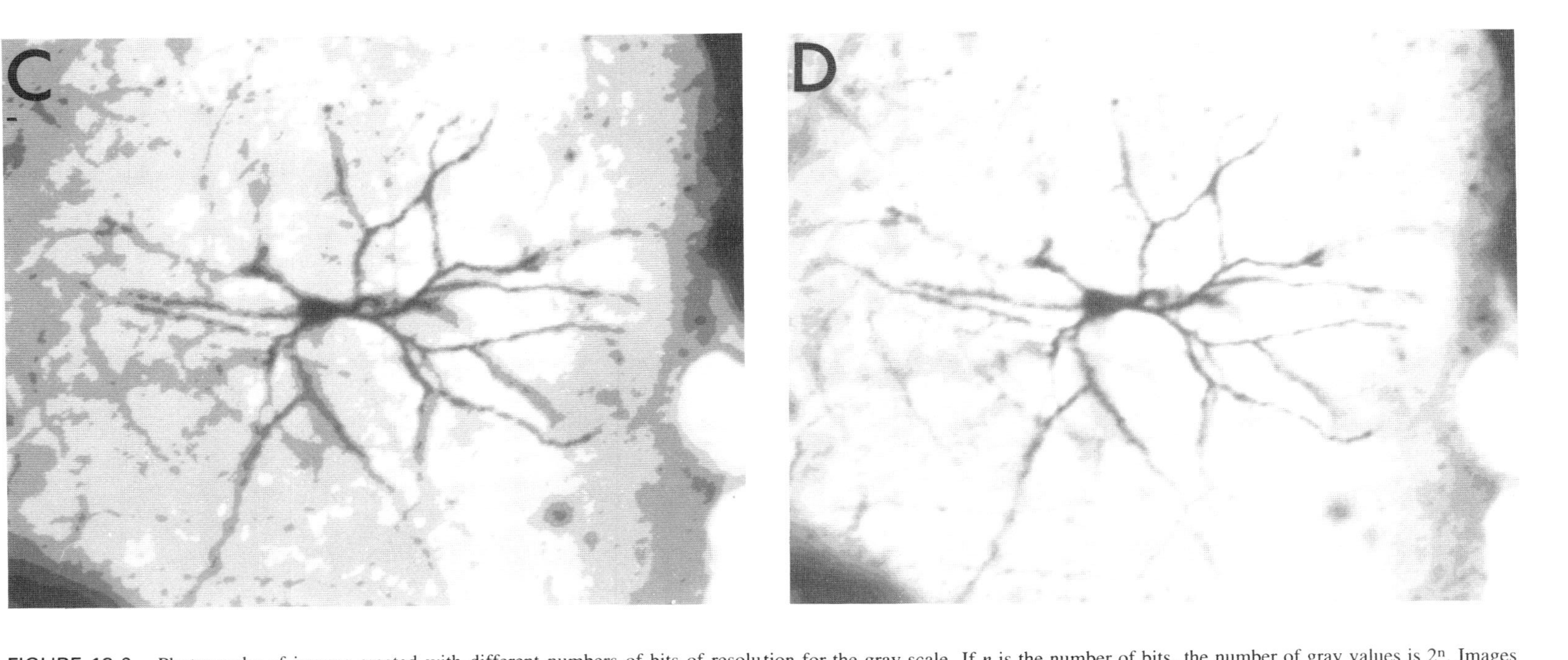

FIGURE 13-6. Photographs of images created with different numbers of bits of resolution for the gray scale. If n is the number of bits, the number of gray values is 2^n. Images with 2 (A), 4 (B), 8 (C), and 16 (D) shades of gray are illustrated. The image shows a layer-II pyramidal cell in the visual cortex of the cat. The neuron has been stained by the Golgi–Kopsch method. (Preparation courtesy of S. B. Tieman.)

13.4.2b. Scale Transformations. The LUTs allow any input gray value to be effortlessly mapped onto a different value. This mapping can be designed to transform inputs by a log or γ function and can be used to transform, for example, transmittance into density. To linearize the gray values with respect to some underlying variable, such as a radioactive label that may not be linearly related to density, the gray values from radioactive standards are often digitized and an appropriate transformation implemented in the LUT (Altar et al., 1986).

13.4.2c. Gray-Scale Stretching. Expanding one part of the scale linearly is a common method for enhancing image contrast. One offsets the scale values by a constant and expands a limited range of values so that it covers the full range (Fig. 13-3). This has effects similar to the analog manipulations of gain except that blooming is not a problem; hence, the technique is applicable to both black-on-white and white-on-black images. The primary limitation of digital gray-scale stretching is that the number of gray values remains fixed. Thus, if 256 grays are digitized and 25% of the gray range is then stretched to full scale, as in Fig. 13-3, then the resulting gray scale can have only 64 shades of gray. Therefore, gray-scale stretching is limited by the number of grays required for an adequate picture (Section 13.4.2a).

Gray-scale stretching and scale transformations both modify the histogram of gray values. Another common histogram modification technique is histogram equalization. The transformation used is one that will convert the observed distribution of gray values into a uniform distribution, that is, one for which every value of gray is equally probable. This is accomplished by first determining the number of pixels that have each gray value. This value is then divided by the total number of pixels in the image to determine the probability of a particular value. The necessary transformation, T, for an eight-bit LUT is then

$$v = T(w) = 255 \cdot \sum_{i=0}^{w} \text{Prob}(i) \text{ for each } 0 < w < 255. \tag{13-1}$$

Transformations to convert the observed histogram to other histograms, whether defined a priori or interactively, can be derived. For a discussion of these transformations, see Gonzalez and Wintz (1977).

13.4.2d. Color Transformations of the Gray Scale. The human eye is much more sensitive to differences in intensity and hue available with a color monitor than to differences in intensity alone that are available with a monochrome monitor (Gonzalez and Wintz, 1977). For this reason, pseudocolor transformations, which map the gray scale onto an arbitrarily chosen color continuum, are implemented in almost all image-processing systems. In addition, it is easy to implement and is demanded by the marketplace. In my opinion, mapping the continuum of gray values onto a color continuum has much more effect on the differentiation of the grays than does any attempt to represent the gray scale more accurately with expensive sensors or expensive monitors.

The colors available on video monitors are obtained by additively mixing three primary colors in varying intensities. Additive mixing of colors is different from the subtractive mixing that occurs with paints. For paints, a red pigment absorbs all light except red, and a green pigment absorbs all light except green. The mixture of the two pigments absorbs all wavelengths, and any light reflected from the paint appears gray. In contrast, a red light projected onto a white screen is reflected and a green light projected onto a white screen is reflected. When the red and green lights are projected onto the same white screen, both wavelengths are reflected, and the additive mix appears yellow. Additive color mixing occurs on video screens, where three separate phosphors are present for each pixel. When excited, one phosphor glows red, a second green, and a third blue. These three phosphors are excited independently by three separate electron beams that are controlled from the three separate outputs of image-processing boards. Since dots or lines of the three color phosphors are closely spaced, the viewer does not normally resolve the individual colors but perceives the screen as if the colors were overlapping.

The color space can be represented as a cube with three orthogonal edges representing the three primary colors (Fig. 13-7). The intensity of the color is represented by the distance from the "black" corner of the cube. If the cube is considered in three-dimensional polar coordinates with black at the origin, then each point in the cube has an intensity defined by its distance from black and a hue or chromaticity value defined by the angle. The standard two-dimensional chromaticity diagram is defined by the colors present on a plane passing through the "red," "green," and "blue" corners. Projection lines that are perpendicular to this plane define sets of points that differ from each other in "saturation," which can be roughly defined in terms of the amount of white light diluting the color. Thus, a pink light and a dim red light differ in their saturation. For the major, black-to-white diagonal of the cube, saturation is meaningless since no color exists to be saturated.

Continua for use in pseudocolor mapping can be constructed along paths through the color cube. Figure 13-8 shows the result of several choices of continua that are diagrammed in Fig. 13-9. For "ideal" pseudocolor transformations, equal quantitative differences in the image in the frame buffer translate to equally perceived differences on the color scale. Unfortunately, there is no simple mapping between the psychological attributes of perceived color and brightness and the physical quantities that the experimenter can control. Thus, constructing an ideal color continuum is almost impossible—first, because it depends on the experience of the viewer (even ignoring various degrees and kinds of color blindness and states of dark/light adaptation), and second, because it depends on the exact phosphors used on the viewing screen, on the exact intensity and contrast settings of the monitor, on the type and exposure of film used for reproduction,

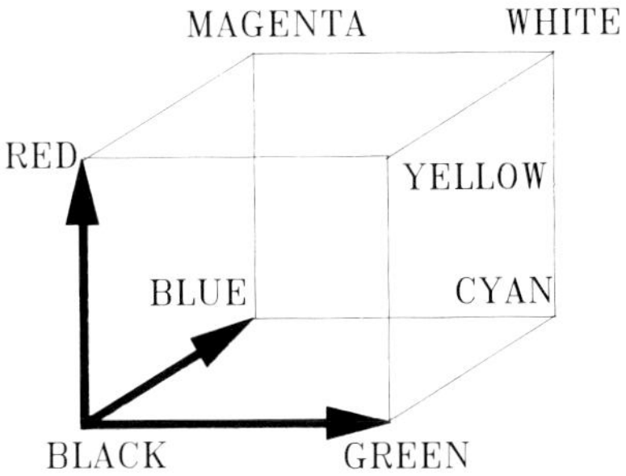

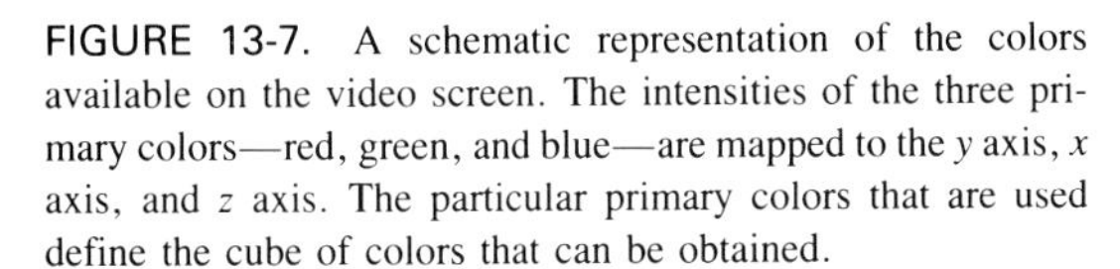

FIGURE 13-7. A schematic representation of the colors available on the video screen. The intensities of the three primary colors—red, green, and blue—are mapped to the y axis, x axis, and z axis. The particular primary colors that are used define the cube of colors that can be obtained.

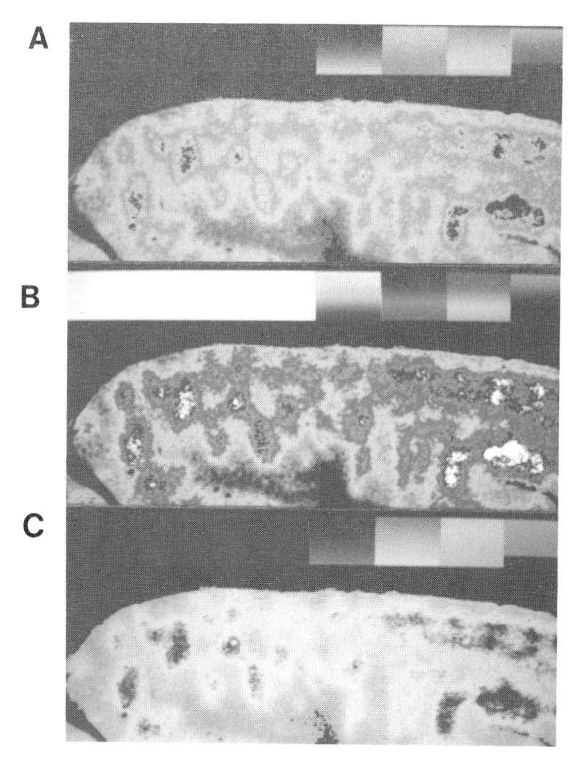

FIGURE 13-8. Three pseudocolor representations of the image in Fig. 13-3C. In each panel, the gray scale has been represented in the upper part. The color continuum has been applied within the gray region from 128 to 248. Hence, the color images are analogous to the images in Fig. 13-3. Values below 128 were set to the same color as 128, and the values above 248 were set to black. A shows the color continuum in Fig. 13-9A. B shows the color continuum in Fig. 13-9C. C shows the color continuum in Fig. 13-9B. A color reproduction of this figure appears following p. xx.

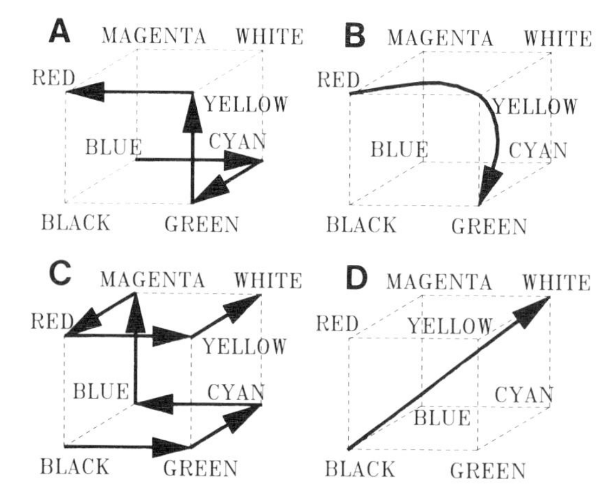

FIGURE 13-9. A schematic representation of four different color schemes that map the numbers in the frame buffer to pseudocolors. These color schemes are used in Fig. 13-8 and are created only for illustrative purposes. The pseudocolor scheme in panel A of Fig. 13-8 correlates with the physical wavelength. The color scheme in panel B was chosen to have black at one extreme and white at the other and otherwise to have the colors as saturated as possible and to have the colors change as rapidly as possible. The bipolar color scheme in panel C is specified by the equation: $c(t) = \frac{1}{2}(-t + t^{1.5})*(255,0,0) + \frac{1}{2}(-t + t^{1.5})*(0,255,0) + (1 - t^{1.5})*(180,180,180)$, where $t = (\text{gray}-127)/127$ and gray $= 0,1,2,3,\ldots,254$ (Carr et al., 1986).

on the inks and quality of the printed publication, and on room lighting. Because of these problems, the researcher can strive only to obtain a pseudocolor continuum that shows the distinctions of interest and does not mislead the audience. When appropriate, the researcher then provides a legend showing the mapping between the quantitation and the color scale. Examples of color scales can be found in numerous publications on quantitative autoradiography (e.g., Boast et al., 1986).

13.4.3. Decreasing the Noise

The signal in any image can be made more visible by decreasing the noise. Although it is not always easy to determine what is signal and what is noise, in most situations reasonable decisions can be made, and the characteristics of the signal and the noise then dictate which noise-reduction techniques should be used. These techniques include temporal averaging, spatial smoothing, frequency-domain filtering, and subtraction.

13.4.3a. Temporal Averaging. Noise that occurs randomly in time can be eliminated by averaging successive images. Temporal averaging is available on any system, but it is only practical on a system that has hardware that implements arithmetic operations. The averaging is generally accomplished either with integration and rescaling to provide standard arithmetic averages or with exponentially weighted averaging (see below).

Standard arithmetic averages of successive pictures can reduce noise that is independent from one frame to the next. Such averages are calculated for each pixel separately. Unhappily, the signal-to-noise (S/N) ratio is increased in proportion to only the square root of the number of pictures that are averaged. For example, if 16 pictures are averaged, the S/N increases by a factor of 4. Figure 13-10 shows the effect of such averaging.

On most image processor boards, the implementation of arithmetic averaging has a major disadvantage—the averaged picture can be displayed only at fixed intervals. If n pictures are averaged, then the average can be displayed every nth picture because averages are calculated by integrating the input picture and dividing by the number of pictures integrated. If n pictures are integrated, then the system can calculate a new average every $n/30$ sec. The number of pictures that can be integrated depends on the number of bits used for the input picture and the number of bits available for the sum. If an eight-bit image (256 values) is to be integrated into a 16-bit buffer (65,536 or 64K values), then as many as 256 frames can be integrated; if an eight-bit image is to be integrated into a 12-bit buffer (4096 values), then only 16 frames can be integrated.

A system that could display a real-time image showing the average of the last n pictures would be too expensive to be practical for values of n that are useful. A system that could display a moving average of the prior 16 pictures would need a minimum of 16 frame buffers to store the needed data. A system that could display a moving average of 256 pictures would need 256 frame buffers. The cost of a system that could access this amount of memory would be prohibitive.

An averaging method that allows real-time display and some control of the averaging interval is exponentially weighted averaging, which does not treat each frame equally. Rather, the average is calculated by the formula:

$$PIXEL(t + 1) = (1 - 0.5^n) \cdot PIXEL(t) + 0.5^n \cdot PIXEL_{\text{input}} \qquad (13\text{-}2)$$

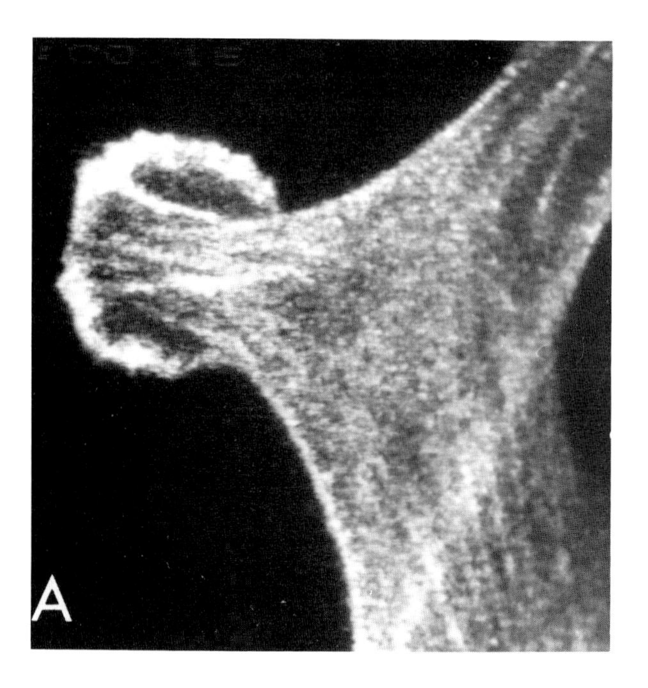

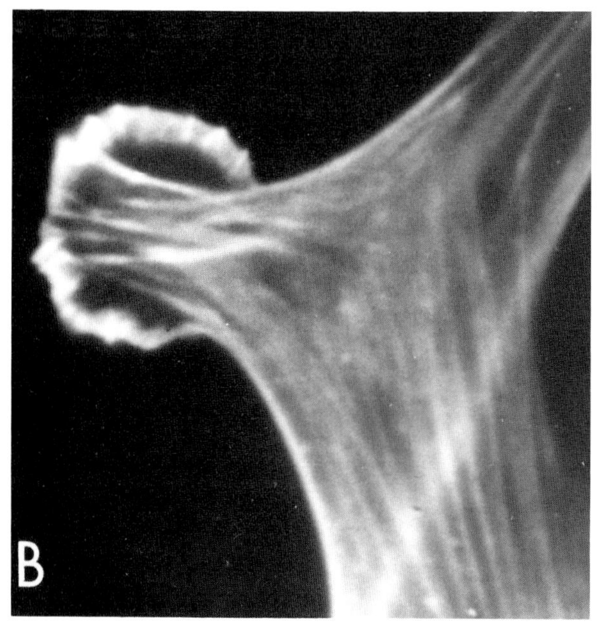

FIGURE 13-10. A fluorescent specimen imaged with an ISIT video camera. The top image (A) shows a single frame and illustrates snow, which can obscure single ISIT images. The bottom image (B) shows the same image with temporal averaging of 256 frames to increase the S/N. The preparation is a chick embryo fibroblast labeled with NBD-phallacidin. The fluorescence demonstrates the distribution of F-actin in the lamellipodium at the cell margin and in stress fibers within the cell. The preparation was chosen from tapes made available by C. Izzard and demonstrates a moderate level of noise. Images of F-actin-containing structures from similar preparations can be found in DePasquale and Izzard (1987).

where $PIXEL(t + 1)$ is the pixel value in the new average, $PIXEL(t)$ is the pixel value in the previous average, $PIXEL_{input}$ is the pixel value from the image that is currently being input (the "live" image), and n determines the time constant for the average. This type of averaging can be implemented easily in hardware for real-time display. The values of n that are practical are determined by the speed with which important changes occur in the image. When changes occur rapidly, n must be small; when changes occur slowly, n can be large.

The noise that is present after exponential averaging does not necessarily decrease with n, as it does for a standard average. For exponential averaging, n and the number of frames included in an average are not the same and can be set independently. The amount of noise reduction is related to both. The relationship is complicated because geometric averaging starts with a noisy picture in the frame buffer. When 0.5^n is small, the contribution of the noise from this initial picture decays slowly; when 0.5^n is large, the contribution of the noise from the initial picture decays rapidly. Since the averaged noise will be smallest when all pictures contribute equally, there is a trade-off between the size of the contribution from the first picture and the size of the contribution from the most recent picture. When 0.5^n is large, the noise from the most recent picture weighs heavily; when 0.5^n is small, the noise from the initial picture weighs heavily. Figure 13-11 shows, for various values of n, the relationship between the expected noise and the number of frames (and time) during averaging. The time over which averaging occurs is defined as the time between the enabling of averaging and the use of the averaged image.

Real-time averaging is one of the primary reasons that systems with on-board arithmetic processing are desirable. Many situations arise in which electrical noise cannot be eliminated. Averaging provides one of the simplest ways to minimize the effects of the noise. An example of the usefulness of averaging is shown in Fig. 13-10. Additional examples become apparent when a 3 × 3 feature-enhancing filter (Section 13.4.3b) is used on almost any video input. Attempts to use these filters to enhance fine detail are often impractical unless the electrical noise, which is enhanced by the filters, is minimized first.

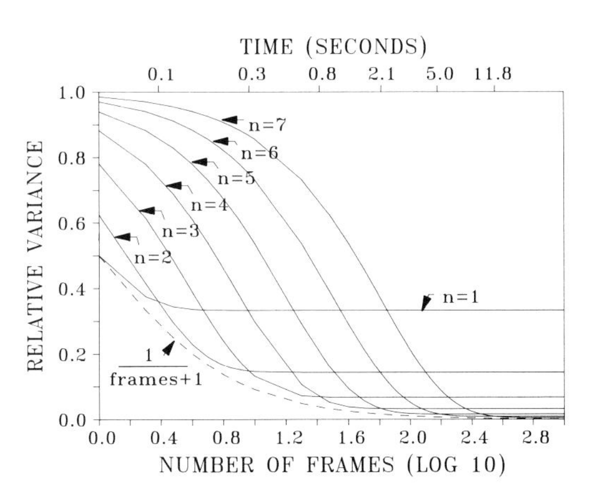

FIGURE 13-11. A plot of the relative variance of noise as a function of the number of frames over which averaging occurs. The noise variance for the average is expressed as a proportion of the noise variance in a single picture. For each of the solid lines, the averaging is by exponential temporal averaging with the incoming picture weighted by 0.5^n. The value of n is given for each curve. For the dashed line, the averaging assumes an equally weighted average. The model of noise assumes all noise is random and independent from one frame to the next and, further, that averaging begins with the previous image in the frame buffer. The approximate times at which successively higher values of n become "better" are given across the top of the graph. Unweighted averaging is always best but can not be easily implemented in hardware.

13.4.3b. Spatial Smoothing. Noise can often be reduced by averaging several adjacent pixels in the frame buffer. Each pixel in the new, smoothed picture is assigned the average value for the pixels in a rectangular array centered on that pixel. This is accomplished by the equation:

$$PIXEL'_{i,j} = \sum_{r=-m}^{+m} \sum_{s=-n}^{+n} \frac{PIXEL_{i+r,j+s}}{(2m + 1)(2n + 1)} \tag{13-3}$$

where $PIXEL_{i,j}$ is the value of a pixel in the *i*th row and *j*th column, $PIXEL'_{i,j}$ is the corresponding averaged value, and m and n define the size of the region over which averaging occurs. If $N = 2n + 1$ and $M = 2m + 1$, then averaging is occurring over an $N \times M$ region. For a 3×3 average, $n = m = 1$. At the edge of the picture, a border of width m or n is undefined.

A more general formula for spatial filtering is described by the equation:

$$PIXEL'_{i,j} = \sum_{r=-m}^{+m} \sum_{s=-n}^{+n} a_{r,s} \frac{PIXEL_{i+r,j+s}}{|\Sigma\Sigma a|} \tag{13-4}$$

where the values are the same as above, except that an additional array of $N \times M$ constants, $a_{r,s}$ is defined and $\Sigma\Sigma a$ is the sum of these constants. The $N \times M$ array of $a_{r,s}$ values is called an $N \times M$ "kernel" or "mask," and the operation of applying this kernel across the image is called a "convolution" of the kernel and the image. The unweighted averaging function presented first is simply equation 11-4 with all the $a_{r,s}$ equal to 1. Table 13-1 shows the 3×3 kernel for unweighted averaging and a 3×3 kernel for averaging that gives greater weight to the center pixel than to the surrounding pixels.

The convolution process has become a standard technique in image processing and, with different masks, is used to perform the many functions described below. With most imaging systems, 3×3 filters can be completed by the computer in a few seconds and give useful results in most applications. The time to complete the convolution on the computer, however, increases greatly as the size of the filter increases. When speed is essential for 3×3 filters, or when larger filters are employed, arithmetic coprocessor boards are used. For some coprocessor boards, 3×3 convolutions can be completed in a fraction of a second, and 7×7 convolutions in less than 2 sec.

Spatial averaging reduces the resolution for fine detail and sharp edges. Figure 13-17A (Section 13.4.4d) shows the slight blurring that is caused by using a 3×3 spatial smoothing. This reduction is an advantage when fine detail is noise and sharp edges are not important; the reduction is a disadvantage when small objects and sharp edges are important. Figure 13-12A demonstrates the effect of an averaging filter for a one-dimensional scan. The effect is analogous in two dimensions. The arrows indicate two points of light that are sharp and close together (dashed line). After averaging, the points become much less distinct (solid line). A similar effect occurs at an abrupt transition (asterisk) but does not occur when the transition is gradual. Figure 13-17A shows the results of applying this spatial averaging to Fig. 13-17B. Averaging that gives greater weight to the center of the kernel has the same advantages and disadvantages, but to a lesser extent.

Spatial smoothing based on medians rather than means is advantageous when sharp edges are important but fine detail is not. In median filtering, the median or middle value

TABLE 13-1. Useful 3 × 3 Convolution Filters

Average			Weighted average			Inverting		
1	1	1	1	2	1	0	0	0
1	1	1	2	3	2	0	−1	0
1	1	1	1	2	1	0	0	0
sum = 9			sum = 15			sum = −1		
Feature extraction			**Feature enhancement**			**Horizontal line**		
−1	−1	−1	−1	−1	−1	−1	−1	−1
−1	8	−1	−1	9	−1	2	2	2
−1	−1	−1	−1	−1	−1	−1	−1	−1
sum = 0			sum = 1			sum = 0		
Vertical line			**Diagonal line**			**Diagonal gradient**		
−1	2	−1	−1	−1	2	−1	−2	0
−1	2	−1	−1	2	−1	−2	0	2
−1	2	−1	2	−1	−1	0	2	1
sum = 0			sum = 0			sum = 0		
Vertical gradient			**Horizontal gradient**			**Identity**		
1	2	1	1	0	−1	0	0	0
0	0	0	2	0	−2	0	1	0
−1	−2	−1	1	0	−1	0	0	0
sum = 0			sum = 0			sum = 1		

of the surrounding pixel array, rather than the average, is chosen as the new value. The effects of this kind of filtering can be seen in Fig. 13-12B. The isolated points are lost, and the sharp transition is unaffected.

Median smoothing is especially useful when noise spikes cause very large fluctuations but affect few pixels. Figure 13-13 shows hypothetical scans with this problem. These scans, which are on the left and represent adjacent horizontal scans on the monitor,

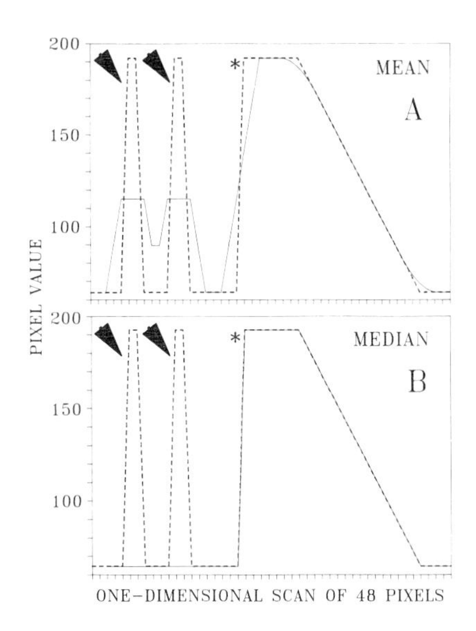

FIGURE 13-12. A schematic one-dimensional scan showing the effects of a spatial averaging filter (A) and a spatial median filter (B). The median filter can be seen to follow better the transitions between large objects, but to eliminate completely isolated points. The averaging or mean filtering can be seen to blur both transitions and isolated points.

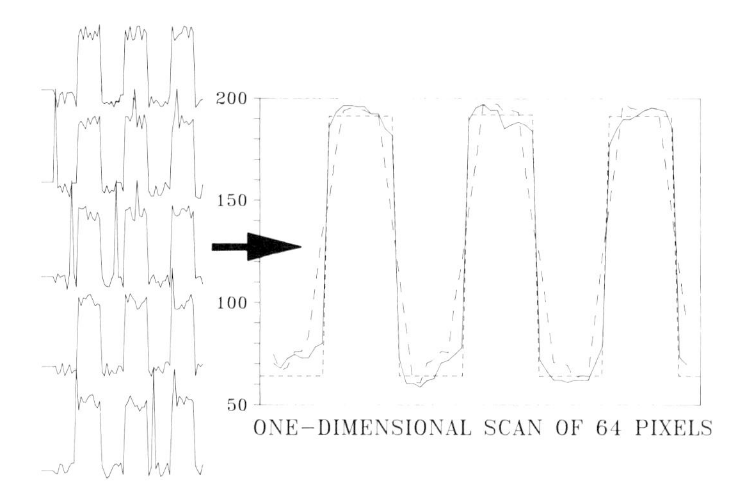

FIGURE 13-13. A comparison of spatial smoothing techniques. The underlying signal is a square wave on which low-amplitude random noise and high-amplitude noise spikes have been superimposed. On the left are five scans, which represent the data from five consecutive horizontal scans. The center scan is smoothed by considering its neighbors within a 5 × 5 matrix. Hence, the two scans above and the two scans below modify the central scan during the filtering. Even in a noisy image, the median of the 25 neighbors can be seen to follow better the shoulders of the square wave than does the mean of the 25 neighbors. Both the median and mean filters minimize the noise spikes. If the number of neighbors is decreased (e.g., 3 × 3), the superiority of the median filtering at eliminating noise spikes becomes more apparent, and the difference between the two methods for representing sharp transitions becomes less apparent.

were created by superimposing two kinds of noise onto a square wave. One kind of noise causes both positive and negative deviations and is relatively small; the other kind of noise causes occasional very large deviations in the positive directions. These latter deviations would create white spots or "snow" on the image. The traces on the right are filtered. The line composed of short dashes shows the underlying square wave. The line composed of long dashes shows the middle trace on the left as it would appear after transformation with a 5 × 5 mean filter. The solid line shows the middle trace transformed with a 5 × 5 median filter. Although both types of spatial filters partially reconstruct the square wave, the data from the median filter do a much better job of fitting the transitions and eliminating noise spikes. For noise spikes that affect single pixels, even a 1 × 3 median filter can smooth an image; a 1 × 3 averaging filter would be much more limited.

13.4.3c. Logical Operations with Spatial Filters. When a threshold is applied to discriminate objects by isolating regions that are above or below a particular quantitative value (see Section 13.4.4a), noise in the original image is likely to cause random fluctuations above and below threshold, creating discontinuities in the objects, isolated points in the background, and a very complicated border between the two. One solution to this problem is to smooth the image before thresholding. Another is to use filters that employ logical operations to simplify the borders and eliminate discontinuities.

Logical operators employ an image that has been thresholded so that each pixel can be considered to have either value 1 (true) or 0 (false), and an $N \times M$ kernel that is an array of ones and zeros. Then the summations and multiplications of equation 13-4 can be replaced with the logical operators AND, OR, and XOR (exclusive or). For example, the equation below will set $PIXEL_{i,j}$ to zero if any of the $N \times M$ pixels used for the computation are zero or if any of the values in the kernel is zero.

$$PIXEL'_{i,j} = \underset{r=-m}{\overset{+m}{\text{AND}}}\ \underset{s=-n}{\overset{+n}{\text{AND}}}\ a_{r,s} \text{ AND } PIXEL_{i+r,j+s} \tag{13-5}$$

The most common logical operations are called ''erosion'' and ''dilation.'' Erosion removes the outside border of pixels from any object composed of pixels equal to 1. This has three effects: (1) large fields of ones are made smaller, (2) large fields of zeroes are made larger, and (3) small fields of ones (that is, those that are no more than one or two pixels across) are eliminated. Erosion is defined by equation 13-5 above and an $N \times M$ kernel with the $a_{r,s}$ all equal to 1. Larger values of N and M yield more severe erosions.

Dilation adds a row of pixels to any object composed of pixels equal to 1. This has three effects: (1) large fields of ones are made larger, (2) large fields of zeroes are made smaller, and (3) small fields of zeroes are eliminated. Dilation is defined by equation 11-6 below and an $N \times M$ filter with the $a_{r,s}$ all equal to 1.

$$PIXEL'_{i,j} = \underset{r=-m}{\overset{+m}{\text{OR}}}\ \underset{s=-n}{\overset{+n}{\text{OR}}}\ a_{r,s} \text{ AND } PIXEL_{i+r,j+s} \tag{11-6}$$

Erosion and dilation operations are most useful when used in combination with each other and with other operations. ''Opening'' and ''closing'' operations (Fig. 13-14) eliminate small objects and simplify borders, although they have little effect on the size and shape of large objects. ''Opening'' is defined as an erosion followed by a dilation, and ''closing'' is defined as a dilation followed by an erosion. The opening and closing operations differ only in that one eliminates isolated black points and the other eliminates isolated white points. The terms ''opening'' and ''closing'' are somewhat arbitrary and

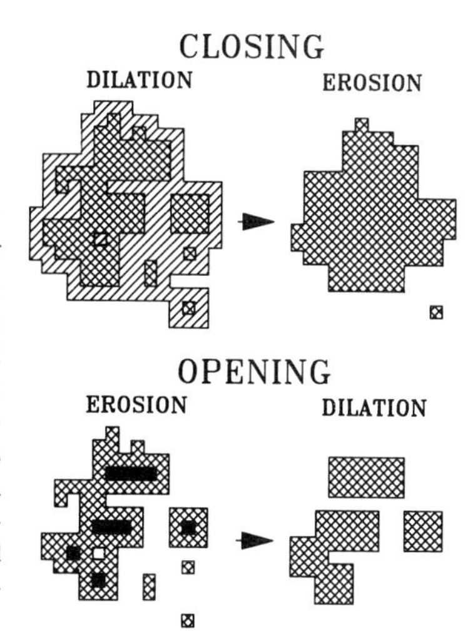

FIGURE 13-14. A diagram illustrating the effect of ''dilation,'' ''erosion,'' ''closing,'' and ''opening'' operations. The top pair of diagrams illustrates a closing operation, which consists of a dilation followed by an erosion. On the left, the set of pixels that were above threshold in the original picture is indicated by cross-hatching, and the result of the initial dilation operation is indicated by diagonal shading. On the right, the result of the subsequent erosion is indicated by the cross-hatching. The bottom pair of diagrams illustrates an opening operation, which consists of an erosion followed by a dilation. On the left, the original pixels are again indicated by cross-hatching, and the result of the initial erosion is indicated by solid shading. On the right, the result of the subsequent dilation is indicated by the cross-hatching.

depend on the user's perception of a black point as an object or a hole. An operation to specify the edges of objects occurs when a picture is compared with an eroded or dilated version of itself. This comparison can use an XOR operation on corresponding pixels from the two pictures. The picture that results from the comparison will have pixels set to 1 on edges and set to 0 elsewhere.

Although erosions and dilations are applied to binary images, variations on the techniques can be applied to images with several shades of gray. For instance, an image with eight shades of gray could be simplified by seven separate applications of the procedures. First the threshold can be set so that the binary image discriminates the darkest gray from all the other grays, and the erosion/dilation operations applied. Next, the threshold can be set so that the binary image discriminates the two darkest grays from the others and the operations again applied. This can be continued until each of the seven borders between gray values has been used as a threshold and its border simplified. Other algorithms besides dilation and erosion have been developed for filling in discontinuities and eliminating isolated points. One that is particularly suited for filling in linear arrays has been described by Ford-Holevinski et al. (1986).

13.4.3d. Correcting for Background Fluctuations. Under many viewing conditions, especially in high-power microscopy, the field of view does not appear evenly illuminated. The unevenness can result from the light source, the optics, and the characteristics of the light sensor and can be extremely troublesome for quantitative densitometry and video-enhanced microscopy. In most cases, the various sources of unevenness can be treated as if they are related to unevenness of the light source. Uneven illumination can be corrected either subtractively or multiplicatively. The multiplicative correction is more correct, but the subtractive correction is more easily implemented on medium-priced imaging boards.

For a subtractive correction, the object to be imaged is first placed in the field, and the desired *manual* gain and offset conditions are found. Next the object is removed from the field and replaced with a neutral-density filter, and the resulting blank field is stored in a frame buffer. Finally, the real-time subtraction mode can be activated so that the live visual output is subtractively corrected. The advantage to this subtractive correction is that it can be effected in real time.

Subtractive corrections need not use a blank field to create the background image. For instance, in surveillance work it is often necessary to detect change or movement. To do this a scene is stored and then subtracted from the live input. If the scene does not change, the screen remains blank; if the scene changes, only the changed objects appear on the screen. Since the effectiveness of subtracting an image from itself requires the incorrect assumption that illumination conditions do not change, systems that are designed for this type of surveillance work often include the capability of using a temporal average (see Section 13.4.3a) for the background image. The time constant of the average can then be chosen to be faster than the time constant for illumination changes but slower than the time constant for the changes to be detected. Applications of this technique in neuroanatomy could allow for the subtraction of autofluorescence, nonspecific labeling, or extraneous features.

Multiplicative operations better correct for uneven illumination. First, a background image with no object is stored in the same way as for subtractive corrections. Then, for each pixel in the background image, a gain factor is determined in such a way that the application of the gain would raise or lower the pixel value to the mean (or median) of all the pixels in the background image. Finally, the object is placed in the field, and each pixel is multiplied by its gain factor.

Multiplicative corrections calculated in this way can be very time consuming because of the number of operations that must be performed. The calculations can be somewhat simplified by creating tables of values rather than computing them for each of the 256,000 pixels. After the mean or median has been calculated, the transformation has two inputs, the value in the background picture and the value in the picture to be transformed. Hence, if 256 values of gray are used, a 256 × 256 array can be calculated once and then used as a LUT to determine the transformed image. The advantage of the LUT increases as the number of gray values decreases and the number of images with the same background correction increases.

Although real-time multiplicative corrections are rarely implemented, most image-processing boards that can subtract one image from another can also perform an approximation to multiplicative corrections by using the input and output LUTs for logarithmic and antilogarithmic transformations. To accomplish real-time multiplication, the pixel values of the background picture are transformed onto a logarithmic scale with an appropriate input LUT. These logarithmic values then become the background image. If the live input is transformed onto the logarithmic scale and the background subtracted, then a subsequent antilogarithmic transformation by the output LUT completes a multiplicative correction.

In general use, subtractive background corrections are used for speed, and multiplicative background corrections are used for accuracy. In theory, however, both types of corrections might be necessary simultaneously. For instance, a diode array camera could have poorly matched diodes with different offsets and gains. The problem is to decide what combination is needed. Trial and error usually works best. Transformations can be tested with different neutral-density filters in the light path. Theoretically, you should be able to gather data with various filters and from these data determine a correction function that employs both multiplicative (slope) and subtractive (intercept) components. In practice, the necessary calibrations would be sufficiently laborious that they would be warranted for only the most precise equipment. For most systems and situations, either subtractive or multiplicative corrections are adequate. It is usually easier simply to defocus the specimen and capture a background image than to remove the specimen and insert a neutral-density filter. This may work if there are no ghosts of the preparation remaining in the defocused image.

13.4.3e. Correcting for Artifacts with Known Characteristics. Occasionally applications of image processing create pictures that have distortions with known characteristics. If the transformation that created the artifact is known, then the reverse transformation that eliminates it can be calculated. These reverse transformations are beyond the scope of this chapter but are frequently discussed in books on image processing (Andrews and Hunt, 1977). One particular application of such reverse transformations has been developed (Agard, 1984) for eliminating the blur of out-of-focus objects in microscope images. When a plane of focus is imaged from a thick preparation, the gray values represent the sum of the in-focus images from the current plane and the out-of-focus blur from nearby planes of focus. Since we know the characteristics of blur, the out-of-focus image can be calculated for any nearby object. The problem arises because the objects are not known until the blur is eliminated, and the blur is not known until the objects are known. This problem is solved by storing optical sections through the tissue. As a first approximation, it is assumed that all the gray values in one plane contribute blur to the adjacent planes. From this assumption, the hypothesized blur can be predicted and subtracted from each plane. The result is a first approximation to the objects present. This first approximation can then be used to recalculate the blur and subtract this new prediction of blur from the original images. This generates a second approximation, and this procedure can be continued through more iterations. In fact, the first approximation seems to create a significant improvement in image quality (Weinstein and Castleman, 1971).

Corrections for known noise are rarely discussed with regard to laboratory imaging systems. The corrections are heavily dependent on the characterization of the noise source, and the calculation of the reverse transformations is computationally intense. Little work has been devoted to procedures that could be implemented on inexpensive systems.

13.4.3f. Editing the Image. Imaging systems allow the user to access individual pixels, or groups of pixels, for read and write operations. This allows the user to change the gray value for any pixel and thus to "edit" the image. Such editing can be useful when obvious artifacts exist in an image. For example, in darkfield autoradiograms, reflections from objects other than silver grains can be edited to the background black, or in 2-deoxyglucose autoradiograms, the black rings caused by bubbles under the tissue can be edited to a mean gray value so that averaging of several adjacent sections is little affected by the artifact.

Editing is best accomplished with a cursor that is controlled by a digitizing tablet or mouse. The user edits either points or areas. When editing points, a point size and gray level are specified, and the cursor is used to position these points on the screen. By pressing the cursor button, the user can replace the pixel values in the frame buffer. When editing an area, the user first delimits the area by drawing its outline in the overlay plane and then filling the outline with pixels of a specified gray. For either method of editing, it is important that the user be able to determine nearby gray values, so that the edited regions will blend in.

Although either a digitizing tablet or a mouse is appropriate for imaging, the best choice is a digitizing tablet that is capable of emulating a standard mouse. The digitizing tablet is preferable for most neuroanatomical applications, whereas the mouse is preferable for many commercially available software packages. Even the cheapest digitizing tablets are adequate for image processing because precision is limited by the number of pixels (usually 512 × 512). Good digitizing tablets with four-button cross-hair pointers can be purchased for under $600 and should be chosen to emulate a standard, such as that used in Summagraphic BitPads, that is supported by general graphics software.

13.4.4. Delimiting the Object

In many situations it is advantageous to divide the image into subregions. Individual objects need to be discriminated from each other and from the background as objectively as possible. The localization of spots on two-dimensional gels, the measuring of labeled structures in brains, the counting of red blood cells or silver grains, the measuring of parts of intracellularly filled neurons, and the characterization of the shapes of chromosomes all require that the objects of interest be first discriminated from other objects, from artifacts, and from the general background. This task seems deceptively easy to the novice for two reasons. First, vendors of image-processing equipment often illustrate trivial cases. A standard picture in brochures shows brightly colored pill capsules on a white background. For such a simple image, application of gradient-detecting or feature-extraction filters creates distinct lines at the edge. Second, the human visual system is extremely efficient at delimiting objects. Differences between objects on the video screen are often so obvious that the viewer assumes that a simple algorithm will delimit them. Real biological images are, unfortunately, not so simple, and because of this, the problem of "segmentation" of images and "feature detection" are continuing areas of research in digital image processing.

13.4.4a. Segmentation by Threshold. To segment regions of a picture is to find those areas (regions) with similar gray levels. The simplest method for segmenting regions is to set a global threshold (i.e., one with a constant value over the entire image) for gray level. Techniques for noise reduction are first used, and then all pixels with values above the threshold are assigned to one region and all pixels with values below the threshold are assigned to a different region. Regions can then be simplified with the opening and closing operators defined in Section 13.4.3c.

Thresholds are effective when transitions between objects are abrupt relative to the noise level and the gray scale of the objects is uniform across the field. In other situations, either the necessary threshold changes with position in the field or the edges of objects are sufficiently blurred that slight changes in the threshold make significant changes in the extent of an object. Both of these conditions arise when thick preparations are viewed in the microscope. The blur from out-of-focus structures causes the background grays to vary from one point to the next. When global thresholds are used, parts of isolated structures are often lighter than the background regions between more closely packed structures (Fig. 13-15). Numerous techniques have been developed to determine local thresholds to deal with these problems in images with relatively large objects (Rosenfeld and Kak, 1982; Gonzalez and Wintz, 1977). Applying local thresholds to images that contain fine detail or rapidly varying background illumination remains difficult.

13.4.4b. Outlining a Region Defined by Threshold. The border of a region that has been defined with a threshold can be determined by either erosion or dilation as described in Section 13.4.3c or by edge-following routines. Edge-following routines first search for a place in the image where one pixel is below threshold and an adjacent pixel is above threshold. The position of the latter pixel is stored and serves as a starting point for an irregular line that circumscribes the object. The line goes to each successive point either by horizontal and vertical movements (Fig. 13-16A) or by movements to any of the eight adjacent positions (Fig. 13-16B). The process terminates when the line returns to its starting point. To simplify the data storage, each successive point is specified relative to the preceding point rather than by its own coordinates. The 3×3 matrix at the top left of

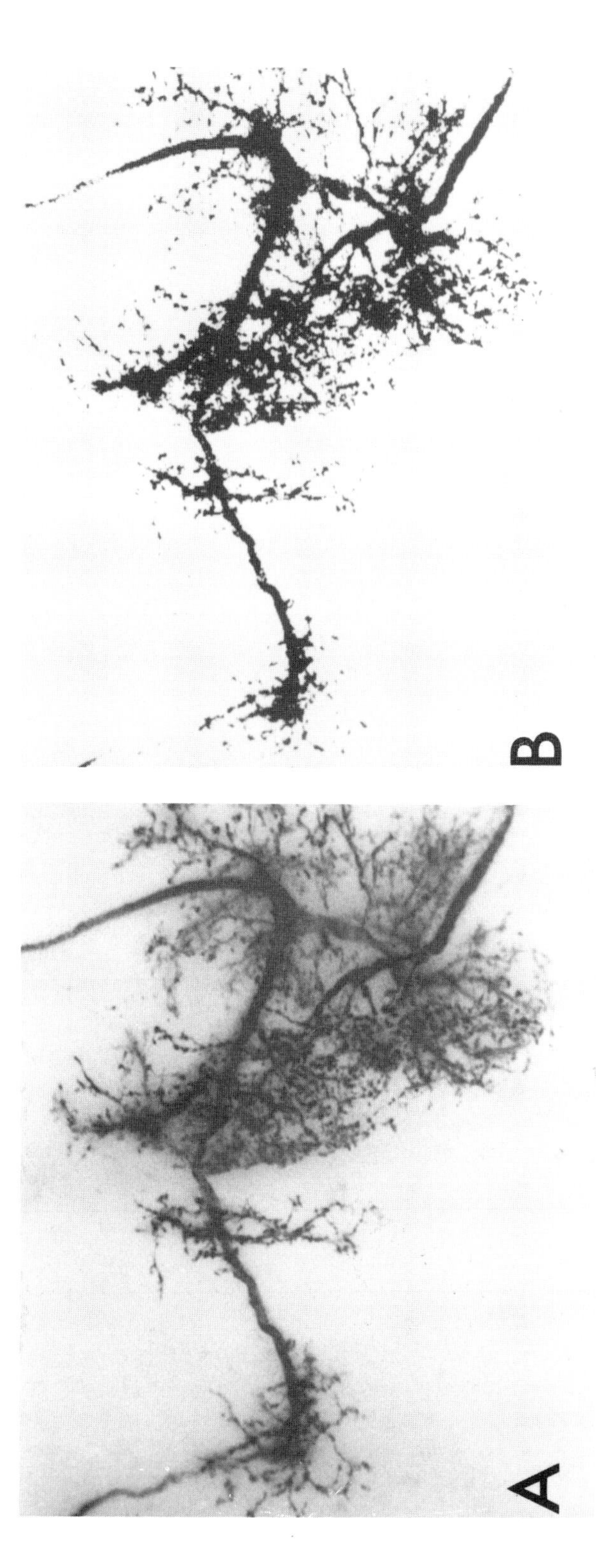

A
B

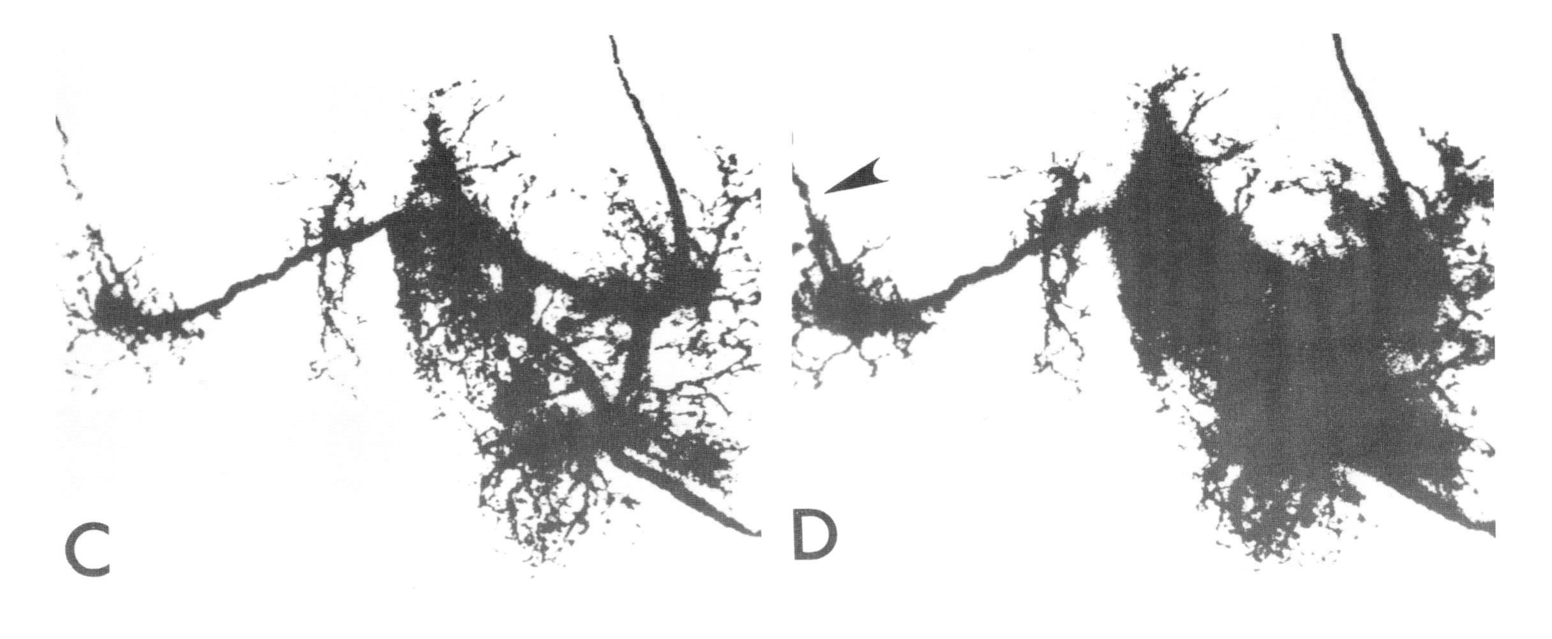

FIGURE 13-15. Several images of the same preparation with different thresholds set on the gray scale. There is no threshold that will cleanly separate the neuron from the background. A shows the neuron with a continuous gray scale. B shows a threshold that maintains the lacy appearance of the cell but creates obvious gaps in the dendrites. C shows a threshold that begins to contain the fine processes but has obvious filled-in regions. D shows a threshold that closes the break in the neurite (arrow). By the time this break is filled, the main dendritic structures are almost totally obscured.

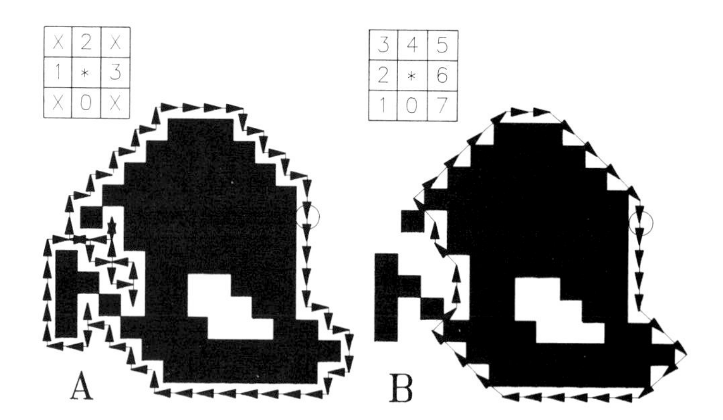

FIGURE 13-16. A diagram showing the effect of two different following algorithms for drawing a line at a threshold value. In A the line is allowed to move in the four primary compass directions. In B the line is allowed to travel in diagonal paths and therefore can jump across staggered pixels and create a simpler outline. The matrix in the upper left of each drawing shows the coding numbers that are used to store the relative movements along the line. This coding allows the line to be stored in a minimum of computer memory.

Fig. 13-16A and B is used to code the directions of movement. With the asterisk centered on the preceding point, the code for the next movement is determined from this matrix. The codes of Fig. 13-16A can be stored in two bits (eight pixels per 16-bit memory word), and the codes of Fig. 13-16B can be stored in three bits (five pixels per 16-bit memory word). This saves memory and does not lose any information about the border.

13.4.4c. Contour Map Generation. If each gray level is treated as a separate threshold, then gray-scale contouring can be accomplished by the above procedure. The threshold is first set for the lowest gray, and a contour is generated. Then a higher threshold is set, and a new contour is added. This can be continued until a contour map is created.

13.4.4d. Edge and Feature-Enhancing Filters. An alternative to defining the characteristics of the pixels that compose an object is to define the characteristics of the edges of objects. The most common techniques for determining edges use $N \times N$ filters with the same mathematical operations as were used for spatial averaging, but with different weighting values. Table 13-1 shows some commonly used kernels. Intuitions about the filtering characteristics of these can be gained by observing effects of different kernels on the same image. Figures 13-17 and 13-18 show the effect of several kernels on an image of a Golgi-impregnated neuron. In Fig. 13-17, one-dimensional scans are shown for a picture (B) that has been transformed by smoothing (A), by feature enhancement (D), and by feature extraction (C). An extracting filter is designed to select an object from its background; i.e., it will eliminate the background and pull out the object of interest. An enhancing filter does not eliminate the background; it only makes the object of interest more distinct on its background.

Feature-extracting filters and feature-enhancing filters differ in the extent to which they retain the overall gray values of objects in the image. For the feature-extraction filter, only transitions of gray that occur within a local $N \times N$ region contribute to the trans-

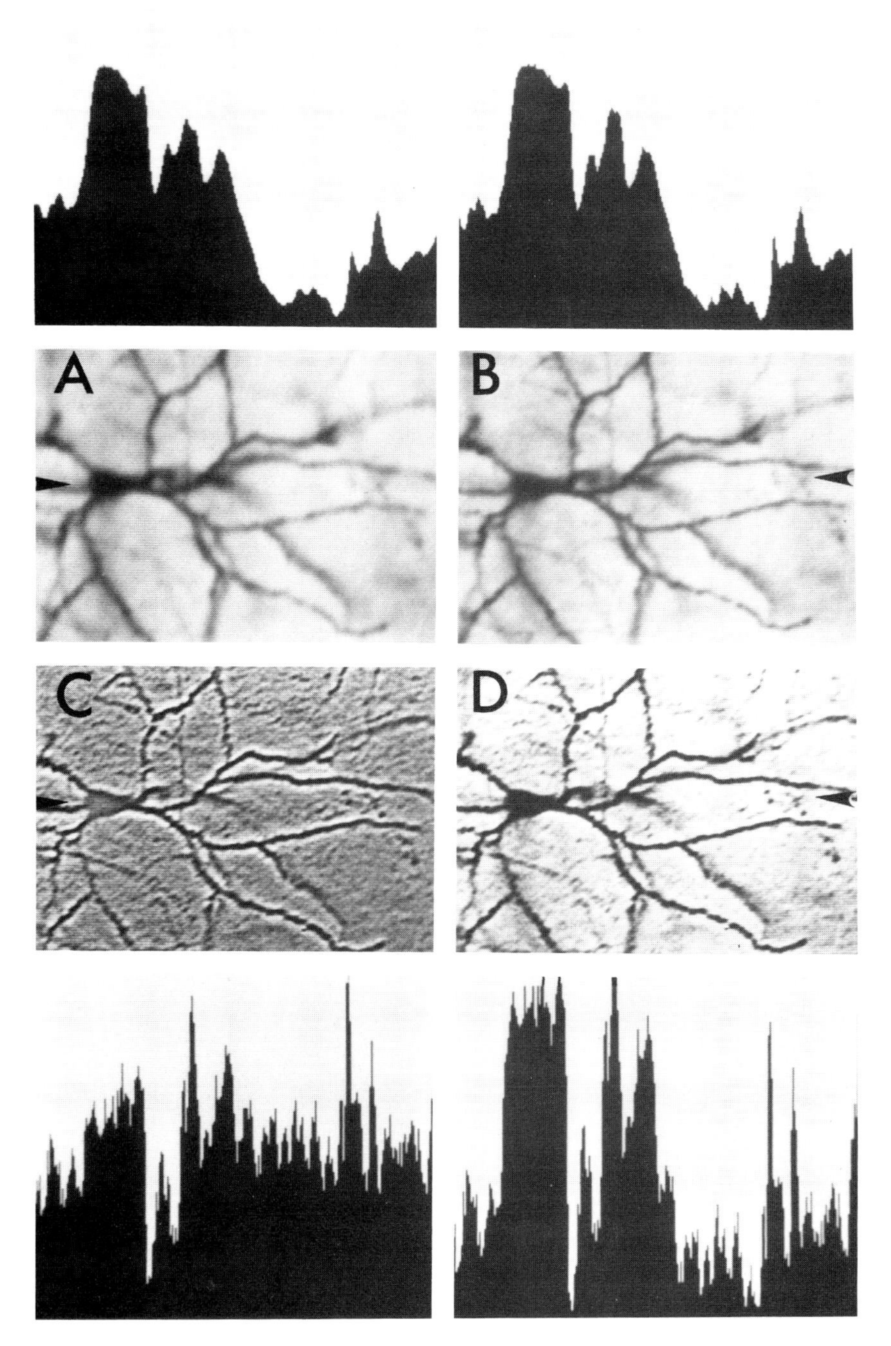

FIGURE 13-17. Transformations and one-dimensional scans of a central quadrant of the video image in Fig. 13-19A. The one-dimensional scans are obtained at the arrows. B shows the nontransformed video image. A shows the image with 3 × 3 smoothing. D shows the image with 3 × 3 feature enhancement. C shows the image with 3 × 3 feature extraction. The arrows at the right and left edges of the figure show the horizontal lines across which the intensity scan was taken. In this figure and in Fig. 13-18, the area presented is a 240 × 256-pixel region of the 480 × 512 image in Figs. 13-6, 13-19B, and 13-19D.

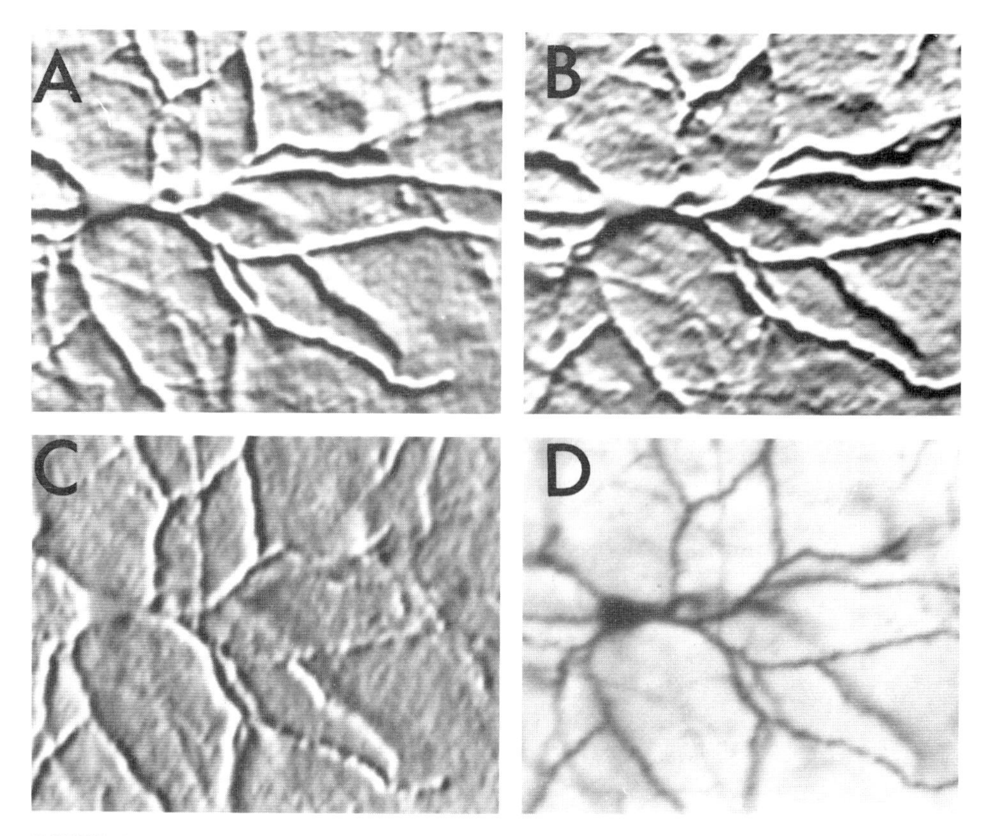

FIGURE 13-18. The effect of 3 × 3 gradient extraction transformations on the neuron in Fig. 13-17. D shows the nontransformed video image. C shows the effect of vertical gradient extraction. B shows the effect of horizontal gradient extraction. A shows the effect of diagonal gradient extraction.

formed picture; the average gray of the local $N \times N$ region is ignored. As a result, objects that are a uniform gray are distinguished only by their edges and their texture and not by their brightness relative to other objects; all objects in the extracted image are the same brightness (i.e., have the same gray value) (Fig. 13-7C). For the feature-enhancing filter, both the transitions and the average gray contribute to the transformation. Thus, objects that were a uniform gray in the original image are distinguished by their brightness as well as by their edges and texture in the transformed image (Fig. 13-7D). Kernels used for feature extraction have values that sum to zero; those used for feature enhancement have values that sum to one. Adding the identity kernel to a feature-extraction kernel typically yields an appropriate enhancement filter.

The independence or orthogonality of different kernels can be tested by applying the kernels to each other. When the result is zero, the kernels are orthogonal. For instance, the various line-detecting filters (horizontal, vertical, and diagonals) are orthogonal to each other and to the spatial averaging filter but are not orthogonal to the feature-extracting or -enhancing filter (Table 13-1).

Line filters are reminiscent of simple cells in visual cortex and selectively enhance edges having particular orientations (Table 13-1). To obtain a general edge or line filter, the absolute values for each of the individual filters must be averaged, and this time consuming. As an alternative, the feature-enhancement and feature-extraction filters (Laplacian filters) enhance edges of any orientation in a single convolution. Such filters are reminiscent of the center-surround organizations found in many ganglion cells of retina and in neurons of the lateral geniculate nucleus. These filters are especially powerful for emphasizing in-focus objects, since their edges are sharp but those of out-of-focus images are gradual. The extraction filter is more commonly used when features are to be isolated; the enhancement filters are more commonly used when a sharper picture is desired.

The advantage of these filters can be seen in Figs. 13-19 and 13-20. Figure 13-19 shows video pictures of a Golgi-impregnated neuron viewed with a 16× objective (Fig. 13-19A) and a 40× objective (Fig. 13-19B). Figures 13-19C and 13-19D show the effect of enhancement. The benefits of enhancement in overcoming some of the limitations of video imaging can be seen in Fig. 13-20. This figure shows video images for one plane of focus for the cobalt-filled neuron in Fig. 13-23. The top two panels show enlargements of part of the video screen when the neuron is viewed with a 16× objective. A full reconstructed view of the neuron with this objective is shown in Fig. 13-23. Figure 13-20A shows the normal video image and illustrates the loss of resolution by the video systems. Figure 13-20B illustrates the recapture of sharpness with enhancement. For comparison, Fig. 13-20C shows a direct photomicrograph of the neuron.

Anything that causes a loss of high frequencies in an image will cause it to look blurred. Therefore, all video images, because they were spatially sampled from the original, will be blurred to some degree. An out-of-focus image is quite blurred. To recover the best original image or the best focused image, filters are used to deblur the video image. Figure 13-21 diagrams the effect of a simple enhancement (here applied one-dimensionally) when used to recover a blurred image. In this case a point of light that would have had a height of 1 (to the dotted line) has been blurred with Gaussian noise. The filter has the greatest effect when the blur is least and can even enhance the gray value of the object beyond its true value.

The problem with the feature-extraction and feature-enhancement filters is that they are extremely sensitive to isolated points such as single pixels that differ in value from the pixels that surround them. Since such isolated pixels are characteristic of noisy images, it is important that temporal averaging or some form of spatial median filtering be applied before feature enhancement. The use of spatial averaging filters in combination with feature-enhancing and -extracting filters is usually counterproductive because smoothing creates shallow gradients for which the feature-extracting and -enhancing filters are ineffective.

13.4.5. Fourier Analysis of Images

The use of Fourier's theorem provides a powerful method for processing images. Although a detailed description of Fourier analysis is beyond the scope of this chapter, it is briefly described. Complete descriptions are found in the references.

When an image is digitized into discrete samples, the pixels are said to be in the "spatial" domain; that is, they are gray values specified for each location in the image.

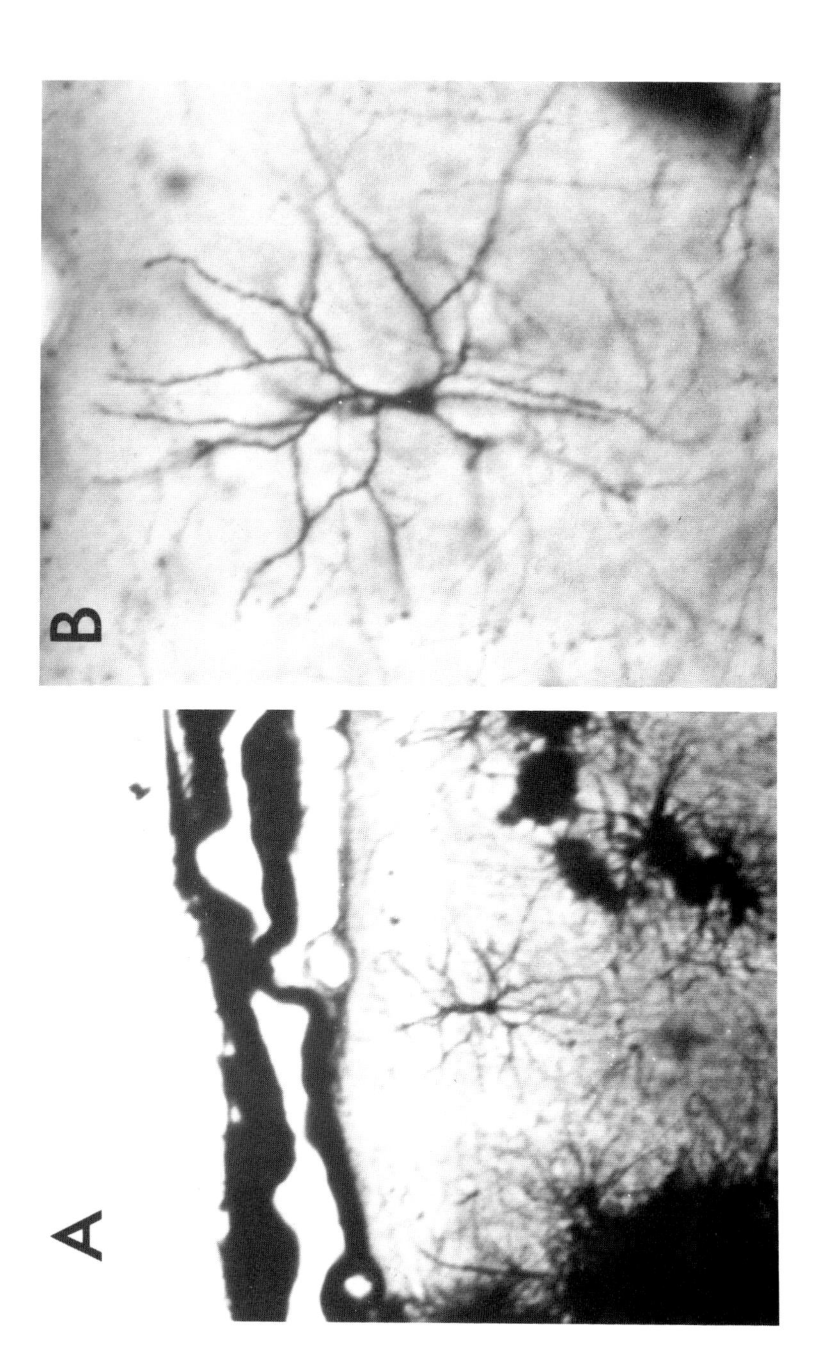
A
B

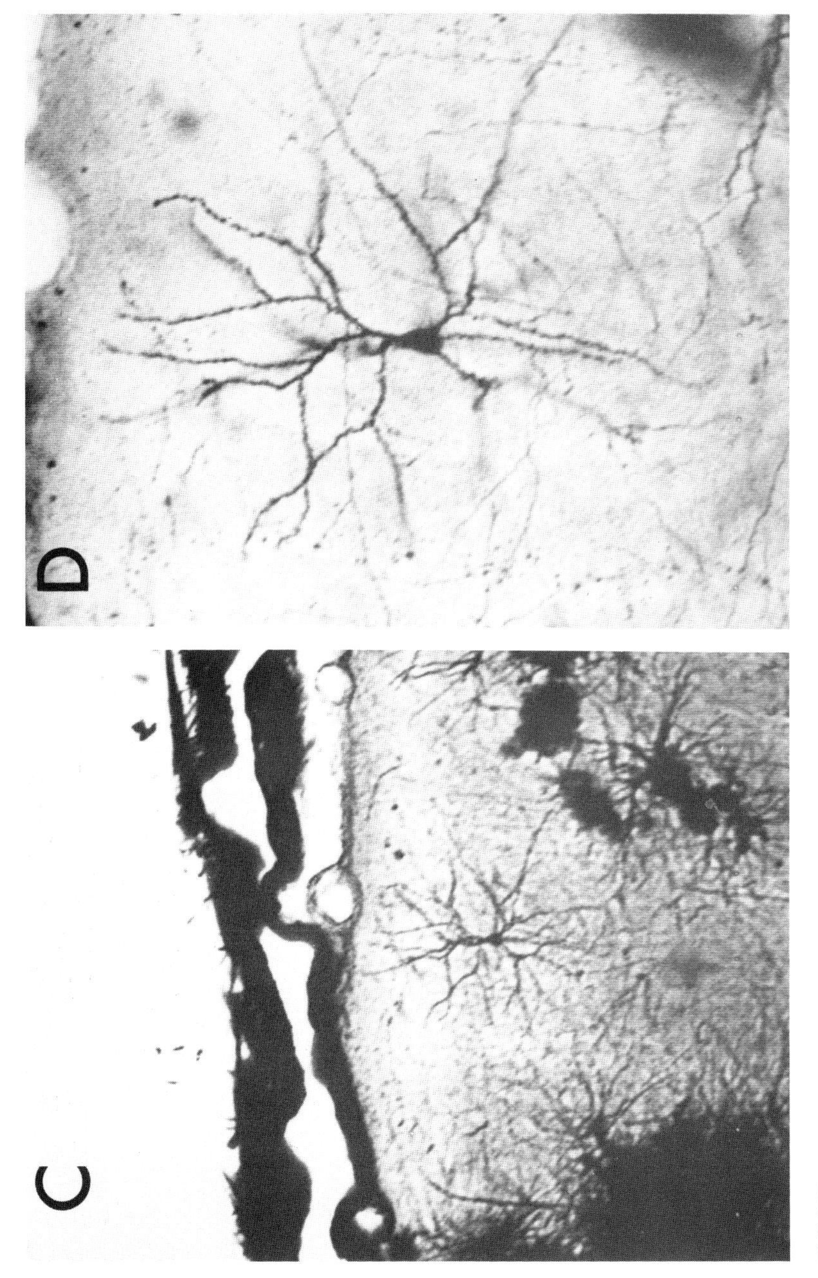

FIGURE 13-19. Video images of a layer-II pyramidal cell from cat visual cortex. A and B are nontransformed video images. C and D show the effect of feature enhancement. Out-of-focus blur is little affected by feature enhancement, whereas in-focus edges are made darker.

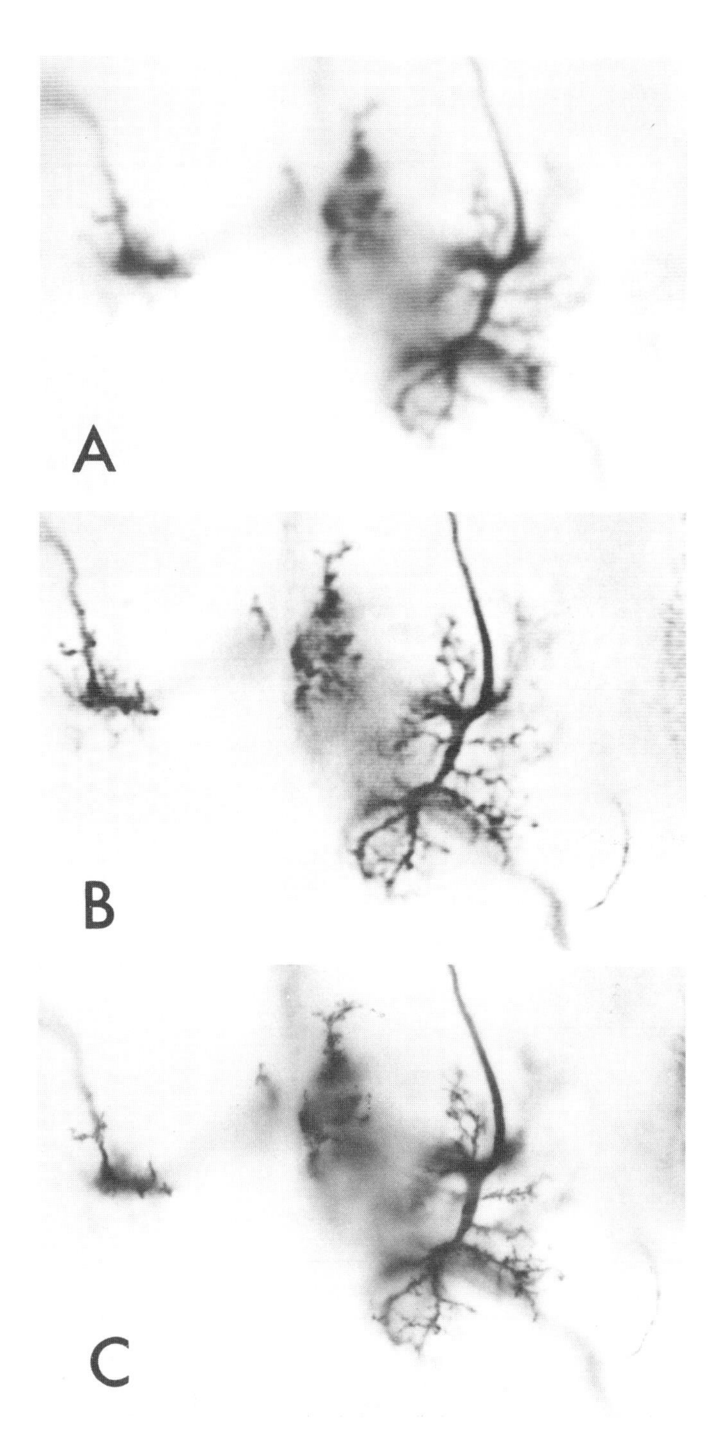

FIGURE 13-20. Three illustrations of the same region of the cobalt-chloride-filled neuron that is illustrated in Fig. 13-15 and Fig. 13-23. A is a nontransformed video image obtained with a 16× microscope objective. In the picture shown, the video scan lines are large because the picture is an enlargement of a portion of the video screen. The entire video image is the same magnification as that in Fig. 13-23A. B shows the image in A with feature enhancement. Clearly, the enhanced image is sharper. C shows a photomicrograph of the same region. This photomicrograph shows that video imaging is limited in its resolution and sharpness.

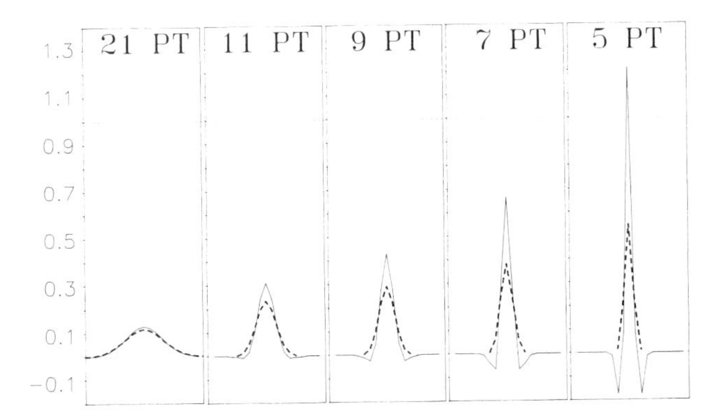

FIGURE 13-21. A diagram illustrating the effect of enhancement on a blurred image created by a point of light. If no blur occurred, the point of light would have a height of 1.0. The dashed curve is Gaussian, and the area under it is 1.0, but the number of points over which it spreads becomes smaller from left to right. The spread is indicated at the top of each graph. The solid curve indicates the effect of enhancement on the dashed line. Clearly, when the blur is limited to a few points, enhancement can restore the intensity of the point of light. When the blur is great enough, as it is for a spread of 21 points, the enhancement has little effect.

According to the Fourier theorem, the pixels can be converted into the "frequency" domain using a forward transformation. Since some mathematical operations are easier to apply in the frequency domain than in the spatial domain, images are manipulated in the frequency domain and then converted back to the spatial domain by a reverse transformation so that the results can be displayed. The Fourier theorem can be expressed by a forward transform and a reverse transform. Letting $j = \sqrt{-1}$, the forward transform into Fourier space is

$$F(u,v) = \frac{1}{MN} \sum_{x=0}^{M-1} \sum_{y=0}^{N-1} f(x,y) \exp\left[-j2\pi \left(\frac{ux}{M} + \frac{yv}{N} \right)\right] \tag{13-7}$$

for $u = 0, 1, 2, \ldots, M - 1$; $v = 0, 1, 2, \ldots, N - 1$. The reverse transform is

$$f(x,y) = \frac{1}{MN} \sum_{u=0}^{M-1} \sum_{v=0}^{N-1} F(u,v) \exp\left[j2\pi \left(\frac{ux}{M} + \frac{yv}{N} \right)\right] \tag{13-8}$$

for $x = 0, 1, 2, \ldots, M-1$; $y = 0, 1, 2, \ldots, N-1$. Since it is known from Euler's equations that

$$\exp[-j2\pi ux] = \cos 2\pi ux - j \sin 2\pi ux \tag{13-9}$$

the variables u and v are referred to as "frequency variables," and the less sophisticated version of Fourier theorem can be stated: Any function can be expressed as a series of weighted sine and cosine functions. The extent to which each frequency contributes to $f(x,y)$ is expressed the "frequency spectrum." Determination of the transforms is computationally intense. The fast Fourier transform (FFT) is easy to calculate but requires too much time to be useful for interactive analysis on small computers without special hardware coprocessors.

Fourier frequency filtering normally changes the frequency spectrum before the reverse transform is applied. Images with high-frequency noise can be smoothed by decreasing the high-frequency components (low-pass filtering). Images with uneven lighting can be evened by decreasing the low-frequency components (high-pass filtering). Images that are not sharp can be enhanced by increasing the high-frequency components (frequency compensation). Images with artifacts occurring at particular frequencies can have those frequencies decreased (notch filtering). Images with critical information occurring in limited frequency ranges can have all other frequencies decreased (band-pass filtering). Unfortunately, the calculations for any one of these operations would take minutes on an IBM PC.

The Fourier transforms can also be used for operations such as rescaling or rotating an image. These operations are extremely difficult when performed for individual pixels but are simple as Fourier transforms. The advantage to the Fourier formulation derives from its treatment of pixels as discrete samples of a continuous function. Because the continuous function is computed, there is no problem with sampling the transformed function at points that correspond to pixels. When the data are treated as a discrete array of pixels rather than as a continuous function, there are problems defining pixel values. Figure 13-22 illustrates the problem for rescaling an object by a factor of 1.5 and rotating an object by 45°. When the object is shrunk or the axes are rotated, the distance between the sampled points also shrinks for both the x and y axes. After the operation, the available values no longer fit one to one into the memory locations available. Attempts to interpolate intermediate values for the pixels are as computationally intense as the Fourier analysis.

In summary, the Fourier transforms provide an elegant formulation of image-enhancement techniques. The computational complexity, however, makes Fourier analysis impossible in real time and cumbersome for interactive systems. It is rarely implemented on laboratory imaging systems that are commercially available for personal computers.

13.4.6. Combining Images

Images that have been stored individually can be combined by the computer. Combinations are useful when a group of pictures are acquired in such a way that they

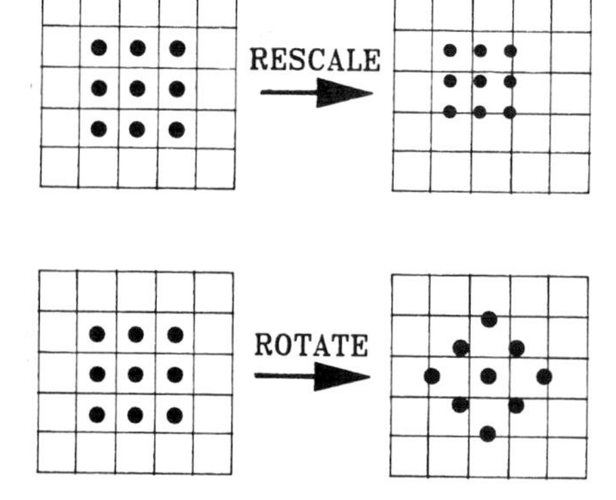

FIGURE 13-22. A diagrammatic demonstration of the difficulties encountered when pixel-defined images are rotated and rescaled. The squares represent the locations corresponding to each pixel in the frame buffer. The black dots represent the original sampling points, which correspond to the pixels. On the right are rescaled and rotated versions of the sampling points, demonstrating that these operations destroy the correspondence between the pixels and sampling points. When a rotated or rescaled image is placed in the frame buffer, there is no longer a simple 1 : 1 relationship between pixels and data values.

constitute a "stack" of images with pixels that correspond directly "above" each other. Thus, the x and y positions are encoded by the pixel location within the plane of the picture, and the z position is encoded by the picture's location in the stack. Stacks of pictures for combination can be obtained either by optical sectioning or by alignment of adjacent histological sections as discussed in Chapter 5. Alignment can be accomplished either by Fourier transforms (Section 13.4.5) or by rotating and shifting a mechanical stage and displaying both images simultaneously by keying (Section 13.2.1d) or real-time differencing.

Optical sectioning involves focusing through a thick specimen and storing pictures at specific focusing points. Typically, the optical sections are obtained at equally spaced intervals, with the distance between sections depending on the depth of field provided by the viewing conditions. Figure 13-23 shows a neuron and a sensory afferent that were intracellularly filled with cobalt chloride. The two views were prepared from separate stacks of optical sections, one gathered at 16× and the other at 40×. The procedure is first to section optically, then to enhance the individual optical sections using the feature-enhancing filter (Table 13-1) to increase the difference between in-focus and out-of-focus objects, and finally to combine the stacks of pictures into a single image by one of the techniques described below (Tieman et al., 1986).

The obvious combination techniques are to determine averages (or medians) or maxima (or minima). The combination algorithm treats the data as a three-dimensional array with elements (i,j,k), where i and j correspond to values on the x and y axes for individual pictures and k corresponds to section number and, hence, depth in the tissue. Thus, $512 \times 480 \times n$ data points are typically used, where n is the number of sections. The combination algorithm for each pixel (i,j) in the final image can be expressed as in equation 13-10 for the maximum algorithm. Since this algorithm must be performed 245,760 times and all of the necessary data cannot be in the computer's fast memory simultaneously, the creation of the final picture takes a significant amount of time. On a relatively slow processor and a single frame buffer, the algorithm takes 2 or 3 min per optical plane. Some of the newer boards, however, can perform pairwise maximum, minimum, or averaging functions in a fraction of a second by using multiple frame buffers and on-board processing units.

$$(i,j)_{\text{normal}} = \text{MAX}[(i,j,1), (i,j,2), (i,j,3), \ldots, (i,j,n)] \tag{13-10}$$

Maximum and minimum are useful when the aim is to view an object that is in focus in some images and out of focus in others. When focusing at various planes through a preparation, you can notice that maximum and minimum gray values occur when objects are in focus. It does not follow that peaks in a z-axis scan of gray values indicate an object. The peaks for objects, however, are sharper (especially with enhancement) than those created by the blur of the same objects. A dark object, such as that in Fig. 13-23, will appear in the combination as long as the background granularity does not accumulate to obscure it. For deeper planes of focus, light scatter causes a loss of contrast. This can have beneficial effects for creating a cue for depth ("haze effect") but can also cause deeper objects to be dim enough that they become lost in the background of the combined image. Because of this, it is sometimes necessary to adjust the analog gain and offset for deeper planes of focus in thick specimens.

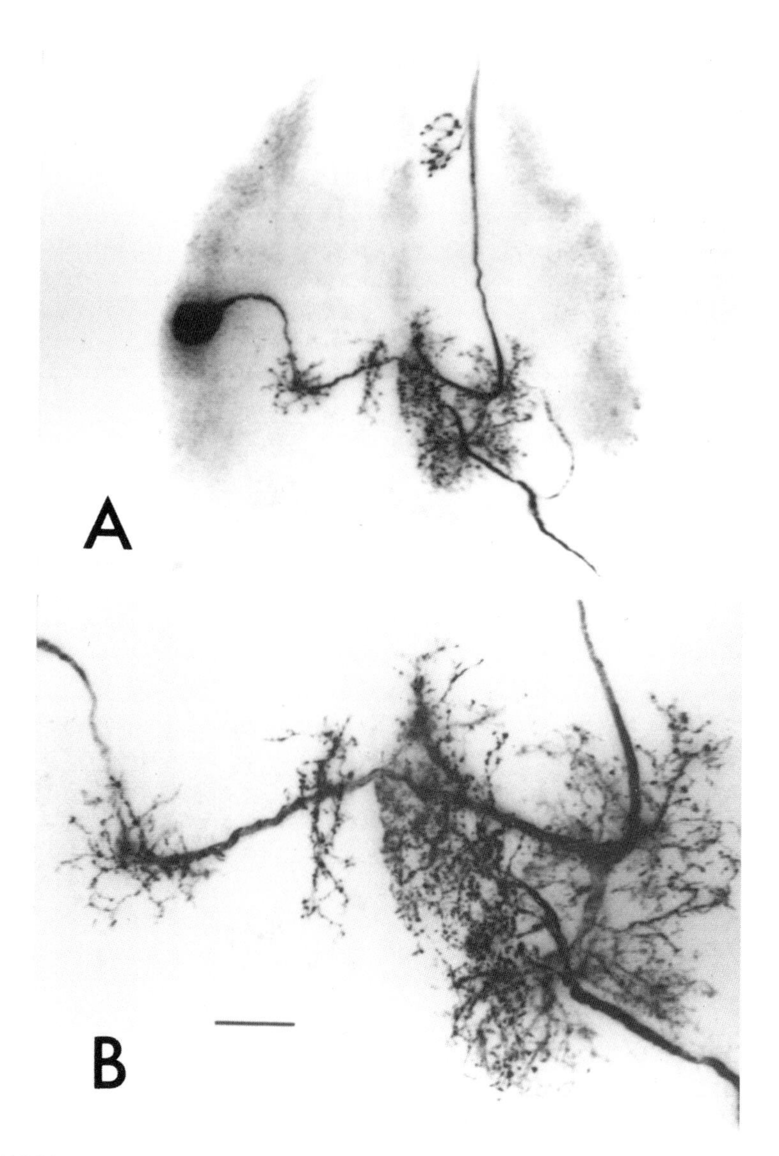

FIGURE 13-23. Pictures of an interneuron and a sensory afferent in the abdominal ganglion of the cricket. A was viewed with a 16× objective. The afferent enters the ganglion through the cercal nerve (lower right) and terminates in the ganglion. The axon of the interneuron exits the ganglion at the top of the picture. This neuron was intracellularly filled with cobalt chloride that was injected into the cell body. The optical sections were sampled every 16 μm, and 11 optical sections were needed to traverse the depth of the neuron. In B the same neuron was viewed with a 40× objective. Optical sections were sampled every 6 μm, and 26 were needed to traverse the neuron. The scale bar is 25 μm for the 40× image. (Preparation courtesy of R. K. Murphey.)

Stereoscopic pairs for three-dimensional viewing can be created by modifying equation 13-10 to equations 13-11 and 13-12. The rationale for this is shown in Fig. 13-24. If the optical sections were a stack of transparencies with space between them, then the view through the stack would be slightly different for the left and right eyes. These different views provide the depth cues for stereograms and are obtained from the optical sections by creating two pictures, one with equation 13-11 and the other with equation 13-12.

$$(i,j)_{\text{right}} = \text{MAX}[(i,j,1), (i-1,j,2), (i-2,j,3), \ldots, (i-n+1,j,n)] \quad (13\text{-}11)$$

$$(i,j)_{\text{left}} = \text{MAX}[(i,j,1), (i+1,j,2), (i+2,j,3), \ldots, (i+n-1,j,n)] \quad (13\text{-}12)$$

Combination by averaging or computing the median can be important in situations such as quantitative autoradiography. The investigator can average the images from several adjacent samples of tissue and thus obtain a more representative and cleaner image for analysis. In this situation, averaging is preferable to the maximum or minimum functions, since the statistical characterization and predictions from maximum or minimum functions are of limited use.

Subtracting one image from another is a useful way of combining information from two images and is used in quantitative autoradiography to emphasize the differences between two pictures. For instance, Altar et al. (1986) have employed subtractive image

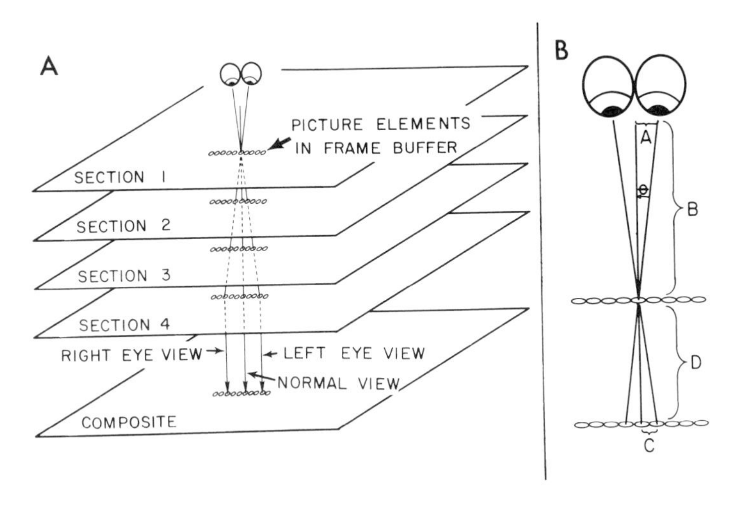

FIGURE 13-24. A diagram showing the relationships critical to the construction of stereo pairs. A: The stereo pairs are created as if a stack of transparencies were being viewed binocularly. The diagram shows that the visual angle is shifted by one pixel in each successive section. B: The values necessary for obtaining proper depth relationships in the stereo pairs are shown. Length A is the distance between the adjacent picture elements and can be determined with a stage micrometer; length D is the distance between adjacent sections. In order to keep the proper relationship such that $A/B = C/D$, distances between optical sections must be adjusted according to the magnification used. For example, with a 20× objective, the horizontal field of view is approximately 570 μm for video, and we would sample optical sections every 16 μm, which yields an angle θ equal to 4° and is appropriate for a 45-cm viewing distance.

processing to identify dopaminergic and serotonergic sites in the brain. First, they incubated brain sections in only [^{3}H]spiroperidol, which labels dopamine (D_2) sites, serotonin (S_2) sites, and some nonspecific binding sites. Next, they incubated adjacent brain sections in [^{3}H]spiroperidol and unlabeled competitors for the D_2 and S_2 sites. In this case some of the label should be displaced from the sites where specific binding for the competitor occurs. Therefore, the last step is to scale and linearize the images of the autoradiograms from the adjacent sections and to align and subtract their digital images to show the sites that were displayed by the competitor.

Although none of the techniques involved in such digital subtractive autoradiography is complicated, all are difficult to implement on the most inexpensive imaging systems for two reasons. First, such systems have no adequate mechanism to allow the two adjacent sections to be aligned so that corresponding points in space occur at corresponding pixels in the frame buffer. Second, the inexpensive systems have only minimal techniques for averaging noise. When pictures are subtracted, the noise variance in the resulting picture is the sum of the noise variances of the individual pictures, but the signal to be detected is often smaller. Because of this, it is important to minimize the noise in the individual pictures. Temporal averaging, one powerful technique for this, is not implemented on the inexpensive imaging boards.

13.5. IS IMAGE PROCESSING LEGITIMATE?

Skepticism about video-enhanced images is not misplaced. The number of manipulations possible and the difficulty of determining from a picture when these manipulations have been done yields tremendous potential for abuse. Furthermore, the video digitization of the original images causes an initial loss of information, and it may not be possible by processing the digitized images to recover fully from this loss. The skeptic must realize, however, that similar potential exists in almost every method of presenting data. The photographs presented are selected, the photographic montages are judiciously cut, and the photographs are manipulated in color filtering, contrast, and evenness. Ultimately, the reader must assume that the author is not misrepresenting the data. With image processing, however, the reader must be sensitive to an additional issue—image-processing techniques can blur the distinction between photographs and schematic diagrams. The author must make clear what techniques have been employed, and the reader must realize the implications.

The researcher who is presenting data must be careful not to employ image processing to make a result seem more important than it is. The danger is analogous to that in statistics where a result can be statistically significant because an abnormally large sample was used, yet have no significance in the real world. Statistics can distort our definition of "significance," and image processing can distort what is "obvious" in a picture. An extremely obvious difference in an image that has been processed and presented with pseudocolor may have been relatively trivial in the raw data. Image processing cannot create a difference where none exists, but it can create an exaggerated visual impression of a small difference.

13.6. FOR FURTHER READING

The topics in this chapter can be divided into those on video-assisted microscopy, digital image processing, and applications in neurobiology. Additional information on video-assisted microscopy is contained in Inoue (1986), which includes detailed information on video systems and their various components. Additional reading on digital image processing includes Andrews and Hunt (1977), Gonzalez and Wintz (1977), Pavlidis (1982), Rosenfeld and Kak (1982), and Wegman and DePriest (1986). Some applications in biology have been presented by Boast et al. (1986), Brown (1976), Geisow and Barrett (1983), and Mize (1985a,b,c). Although not specifically mentioned above, the work of Blaisdel (1986) demonstrates many techniques to acquire and enhance images of the light from voltage-sensitive dyes.

14

The Analysis of Immunohistochemical Data

R. RANNEY MIZE

14.1. INTRODUCTION

Neuroscientists have been interested for many years in the anatomic distribution and concentration of neurochemicals in the brain. Particular attention has focused on the localization of neurotransmitters. Biochemical techniques are now used routinely to assay for the concentration, uptake, or evoked release of neurotransmitters in brain. Microdissection techniques allow one to make these measurements in relatively small brain regions. However, the spatial resolution of biochemical assays is extremely coarse relative to the anatomic complexity of the brain, and it has rarely been possible to measure concentrations at the level of single cells and fibers.

Much better resolution of neurotransmitters can be obtained with anatomic staining methods. Among the earliest of these was the formaldehyde-induced fluorescence technique, which was first used to visualize catecholamines in the CNS (Falck et al., 1962). More recently, the antibody immunocytochemistry technique revolutionized the ability to localize neurochemicals at the single-cell and membrane level. Immunocytochemistry employs antibodies that have been raised against an antigen that one wants to identify in the brain. The antibody binds to the antigen and is then visualized by being coupled to a secondary antibody and one or more other molecules, one of which is tagged with a substance that can be seen with the light microscope. The final reaction product can be clearly localized within single cells and fibers, often within specific subcellular organelles. The technique has been thoroughly reviewed by Sternberger (1979) and Vandesande (1979), among others. The immunocytochemistry technique has vastly expanded our knowledge about the location of neurotransmitters in brain. However, the technique is used rather infrequently to quantify the amount of reaction product in the tissue or to estimate antigen concentration. In addition, relatively few immunocytochemical studies make quantitative measures of the geometries or laminar positions of immunocytochemically

R. RANNEY MIZE • Department of Anatomy and Neurobiology, University of Tennessee Health Science Center, Memphis, Tennessee 38163.

labeled cells and fibers. This is unfortunate because quantitative data of this type can dramatically increase the information obtained from the technique. In addition, computer and imaging technology is available to perform quantitative measures at relatively modest cost.

Computer-based image analyzers are best suited to the task of quantifying immunocytochemistry because they are capable of measuring simultaneously the labeling intensity, the location, the size, and the shape of labeled profiles. Image analyzers accomplish this by imaging the specimen, digitizing the image, and then processing the image to extract and measure the profiles that have been immunocytochemically labeled. To the author's knowledge, no commercial image-analysis systems have software specifically designed to analyze immunocytochemically labeled tissue. However, several university laboratories have developed software for quantitative immunocytochemistry (Benno et al., 1982a; Cassell et al., 1982; Briski et al., 1983; Agnati et al., 1984a; Gross and Rothfeld, 1985; Holdefer et al., 1988).

This chapter describes the principles involved in image analysis and the imaging techniques used to quantify immunocytochemistry. Both the hardware required and the image-processing procedures used to analyze the tissue are discussed. The chapter also describes methods for estimating antigen concentration by comparing the concentration of an antigen coupled to a nonbiological standard with the optical density produced by antibody labeling of that standard. Finally, specific examples of the analysis of immunocytochemically labeled brain tissue are shown. This information should be of interest to both neuroscientists and cell biologists who use immunohistochemistry in their research.

14.2. HARDWARE FOR IMAGE ANALYSIS

Image analyzers used for quantitative immunocytochemistry should include the following components: an image-acquisition device (scanner); an image-digitizing device (A/D converter); an image memory; an image processor; display devices (monitors); and a general-purpose computer. A typical image analyzer is illustrated in Fig. 14-1. Its components are diagrammed in Fig. 14-2. Components of image analyzers have been described in detail previously (Gallistel and Tretiak, 1985; Ramm and Kulick, 1985; Feingold et al., 1986) and in Chapters 2 and 13 of this book. The author's comments on hardware are therefore confined to recommendations specific to immunocytochemistry applications.

14.2.1. Image-Acquisition Devices

High-resolution scanners are commonly used for this type of image analysis. They use a scanning "head" that contains a linear array of approximately 2048 photodetectors, each of which senses the light level incident on it. The linear array is used to scan electronically one axis of the specimen, and the specimen is somehow physically moved in the other axis to cover the entire image and digitize a matrix of pixels from it. Two types of scanners use this concept, drum and flat bed. In a drum scanner, the specimen is wrapped around a rotating drum that passes under the scanner head. In a flat-bed scanner, the specimen is placed on a table, and the scanning head is translated above it. Flat-bed and rotating-drum scanners can have very high spatial and gray-scale resolution. How-

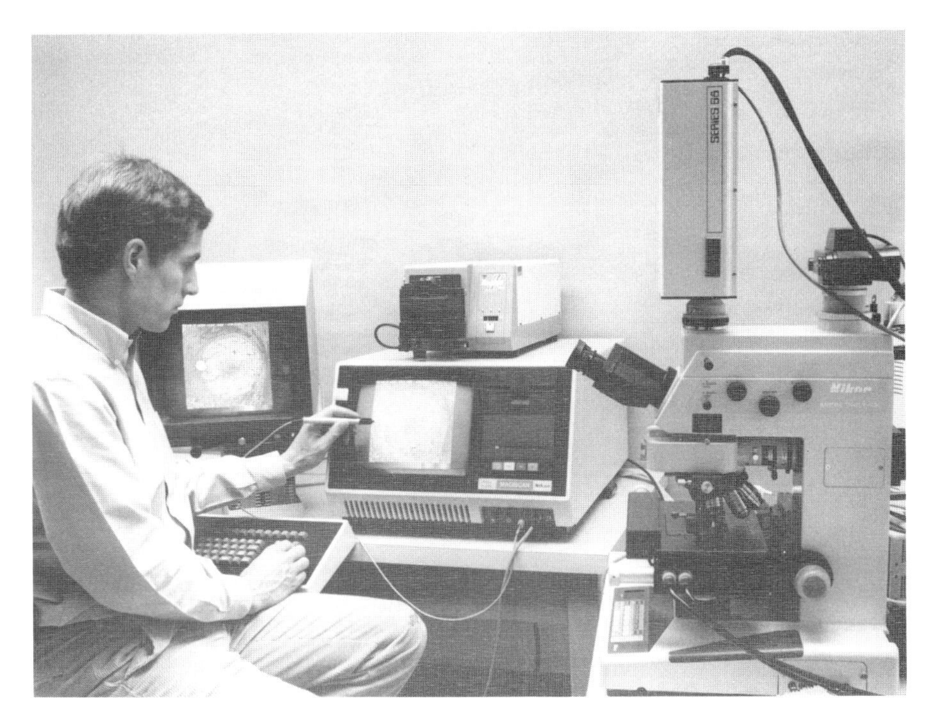

FIGURE 14-1. A typical image-analysis system for quantitative immunocytochemistry. (A) Light microscope; (B) motorized stage; (C) newvicon TV camera; (D) image-analyzer console; (E) color monitor; (F) graphics film recorder. Note how the light pen is used to track individual pixels on the monitors.

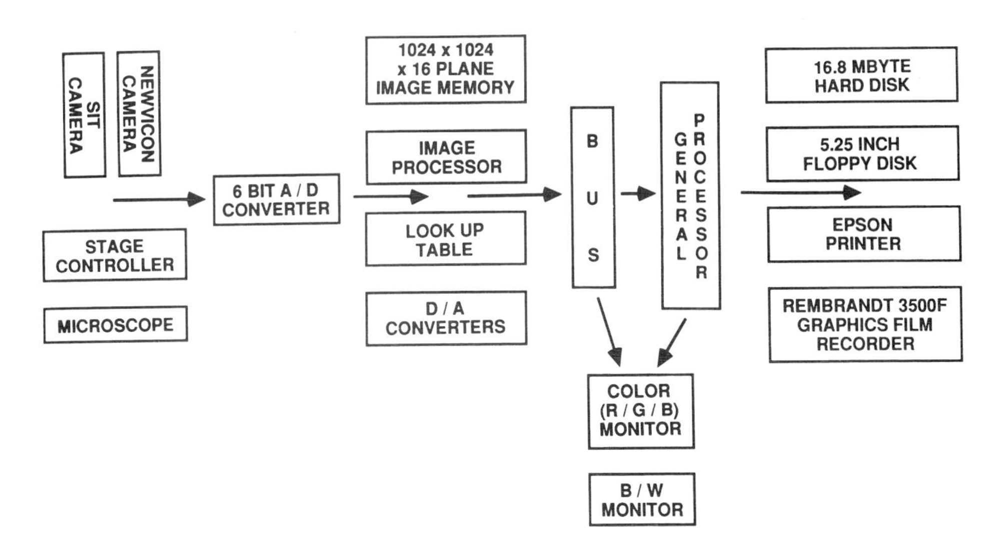

FIGURE 14-2. Block diagram of the components of an image analyzer for quantitative immunocytochemistry. The components particularly important to imunocytochemical labeling are covered in the text. The remaining components are described in Chapters 2 and 13. Reproduced from Mize et al. (1988) with permission.

ever, they are not real-time devices; they require several seconds to several minutes to acquire an image. They are also quite expensive.

Solid-state cameras generally also have an array of photodetecting elements. These cameras, which include charge-coupled devices (CCD) and charge-injection diode (CID) devices, have moderate spatial resolution and sensitivity. The CCD cameras used in current commercial densitometry systems have less spatial resolution and lower sensitivity than high-quality newvicon tube cameras (Alexander and Schwartzman, 1984; Feingold et al., 1986; McEachron et al., 1988). The CCD cameras have less spatial resolution and a lower signal-to-noise (S/N) ratio than the scanners because (1) for technical reasons, the sensor chip cannot be as dense (resolution), and (2) they use U.S. TV video standards, so they count photons at each pixel for a short time (S/N). Newvicon tube cameras are more sensitive and may have better resolution.

Vidicon-type TV cameras are by far the most common input devices used with image analyzers. High-quality cameras of this type are moderately priced ($2000 to 5000), have good temporal stability, have reasonable spatial resolution, and have good sensitivity (Ramm et al., 1984). Both high-quality TV cameras and solid-state cameras are good choices for quantitative immunocytochemistry. For those using fluorescent-tagged antibodies, it is useful to have two camera systems: a newvicon or chalnicon tube for normal light conditions and a silicon-intensifier target (SIT) camera for low-illumination work with fluorescent dyes.

It is obvious that only cameras, not scanners, can be used to digitize data directly from a microscope image. Scanners can be used to digitize a photograph or a microscopic gel.

14.2.2. Image Memory and Processor Boards

Most high-quality image analyzers include special image memory and processing boards. These allow for much faster image storage and processing than is possible using the central processing unit (CPU) of a general-purpose computer. To achieve acceptable performance, only image analyzers with these special components are recommended.

Most image-memory devices, called "frame buffers," consist of a spatial matrix of bits called a "bit plane." These typically range from 128×128 up to 2048×2048. Each bit plane holds an array of binary values. To represent gray-level information, one must have a stack of bit planes. Six planes allow storage of 64 gray levels (2^6); eight planes store 256 (2^8). In practice, 10 to 12 bit planes are desirable so that one can store a picture with 256 gray levels as well as several binary and text overlays (which are contained in the remaining bit planes).

The image or array processor is used to manipulate the contents of image memory. The image processor usually includes logic for boolean (and, or, nor) and arithmetic operators (add, subtract, multiply, divide) so that you can somehow combine two images in various ways. In addition, image processors have procedures for image filtering (mean, Gaussian, Laplacian), thresholding, and binary transformations (erosion, dilation, thinning). These are discussed below.

14.2.3. General Information

Most image analyzers communicate with (or even plug directly into) a general-purpose computer. The computer is the interface to the image analyzer and is used to run the operating system, to execute programs, and to store data. Most image analyzers have a facility for storing complete images on the mass-storage device, usually a hard disk. However, image storage can take time and is rarely necessary for quantitative immunocytochemistry. It is only necessary to store the data file containing the measurements of the image, not the image itself. For this reason, a hard-disk capacity of 10 to 20 Mbytes should be adequate. As optical-disk technology improves, multiple images will be stored rapidly and inexpensively, possibly encouraging the off-line manipulation of stored images after initial image capture.

14.3. CHARACTERISTICS OF IMAGE ANALYZERS

A large number of image-analysis systems are available to analyze immunocytochemically labeled tissues (Ramm and Kulick, 1985; Mize, 1984, 1985a). The specifications of these systems vary widely. Some amateur systems are suitable only for crude image digitization, whereas others are capable of handling sophisticated images such as those collected from satellites. At a minimum, a system for quantitative immunocytochemistry should have adequate spatial and gray-scale resolution, an appropriate dynamic range, and acceptable acquisition and processing speed. These characteristics are described in detail below to help you choose appropriate hardware.

14.3.1. Spatial Resolution

Ideally, the spatial resolution of the image-analysis system should match or exceed the resolution of the labeled tissue. The spatial resolution of a system is determined by the resolution of all of the following: (1) the imaged material (light microscope slide or electron micrograph), (2) the camera and A/D converters, (3) the image memory, and (4) the display monitors. The spatial resolution of commercial image-analysis systems usually ranges from a low of 128 × 128 to a high of 2048 × 2048.

It should be emphasized that the final resolution of the system is only as good as the resolution of any of its components. Thus, an image memory with 1024 × 1024 pixels usually cannot improve resolution if the input imaging device has a resolution of only 800 lines. Similarly, it is not possible to display each pixel of a 1024 × 1024 image on a monitor that has only 512 × 512 resolution.

In general, 512 × 512 pixel resolution should be adequate for quantitative immunocytochemistry. Input devices, image memory, and monitors with that resolution are modestly priced. Systems with 1024 × 1024 resolution are much more expensive and rarely offer significantly greater advantages.

An inexpensive method for testing system resolution is to use a carbon grating replica. These replicas are commonly used in electron microscopy to calibrate magnification. They consist of a series of lines or gratings that have been etched and shadowed with carbon. Fullam, Inc. (Schenectady, NY), for example, sells a grating replica with 2160 lines per millimeter with a 0.463-μm line separation (0.3% error). The replica can be mounted on a glass slide and coverslipped. You should be able to resolve some of the lines of a Fullam replica using a 100× plan apochromat oil objective, an N.A. 1.4 oil condenser, and a 2× auxiliary lens. You can then measure the line-to-line separation with a light pen on the screen monitor.

14.3.2. Gray-Scale Resolution

The gray scale is a series of digital steps that represent different light intensities called "gray levels." The number of gray-level steps specifies the gray-scale resolution of the system. Like spatial resolution, gray-scale resolution depends on several components of the system. It is determined primarily by camera response, A/D converter resolution, and the number of bit planes in image memory. Gray-scale resolution in commercial systems generally ranges from 64 to 256 levels.

14.3.3. Dynamic Range

The dynamic range of a system refers to the range of light intensities between the threshold for detection and saturation. Dynamic range depends almost entirely on camera sensitivity and light intensity. Light intensity is usually expressed as optical density. Optical density is a measure of the amount of light that is attenuated by the specimen. It is the natural logarithm of the reciprocal of transmittance, or

$$\text{density} = \log_{10} (1/\text{transmittance})$$

Transmittance is the ratio of incident light to transmitted light. Optical density is a commonly used measure of light intensity because brightness is perceived by the human visual system along a logarithmic scale. Optical-density units (ODU) range from low values (0.01 OD = 80% transmittance) to middle values (0.03 OD = 50% transmission) to high values (2.0 OD = 1% transmission).

Image analyzers typically express transmittance by calculating integrated optical density (IOD) over the field. The IOD is calculated by the following formula:

$$\sum_{i=1}^{n} \{\text{graylevel} \cdot \log_{10} [63.5/(\text{grayvalue} + 0.5)]\}$$

The IOD can be normalized for area by dividing the IOD by the area over which it was integrated. When more than one sample is taken of a field, the samples can be averaged to yield an average optical density (AOD) value. Vidicon camera image-analysis systems are typically sensitive over a range of 1.5 to 2.0 optical-density units. An image-analysis system should be linear over much of this range. However, it is possible to correct for nonlinearities by transformations that can be stored in a look-up table (Tayrien and Loy, 1984).

14.3.4. Image Input and Processing Speed

These are especially important specifications of image analyzers because they will determine how rapidly you can collect and analyze data. The time required to digitize an image will be determined by the characteristics of the input device, the conversion rate of the A/D converter, and the time required to store the image in memory. Some systems are real-time devices; that is, they will digitize a video image in 1/30 sec. Others require seconds, minutes, or even hours to digitize an image. Acquisition speed depends on scanner speed, A/D conversion rate, and storage and display rates. Mechanical scanners do not provide real-time acquisition.

Processing speed is determined by many factors, including the cycle rate of the image processor and the efficiency of the processing algorithms. Slow image processing can dramatically impede efficient data collection. The best way to test this is to describe a series of processing steps that are likely to be used during analysis of your data. During a demonstration, you should ask the vendor to demonstrate processing steps that will actually be used in analyzing tissue. You should keep track of the exact time required to perform each of these steps on a specimen. One method is to use a test image composed of circles and triangles of different sizes (produced by transfer letters) pasted onto a blank sheet of paper. Processing speed will vary with the type of processing being performed and, in some cases, with the number and size of the objects measured. You should be able to process and measure one image completely in 1 to 5 min. Otherwise, the data collection becomes inefficient, and you will be exasperated.

14.4. HOW TO EVALUATE AND RESOLVE PROBLEMS WITH AN IMAGE-ANALYSIS SYSTEM

Many factors influence the accuracy with which image analyzers measure tissue. These include variations in photometric uniformity, variations in temporal uniformity, nonlinearities in the transfer function of the system, and inaccuracies in geometric and photometric measurements. This section describes methods for evaluating and correcting these nonuniformities.

14.4.1. Photometric Uniformity in Space

When no specimen is present, the image analyzer should detect and store a perfectly uniform field of light. In reality, this is rarely achieved. The light source may have local "hot" spots produced by imperfections in the filament and the light's projection to the specimen. There may be slight irregularities in the microscope optics, especially at the outer edges of the field, which may transmit light less efficiently. The camera tube may be less sensitive in local regions of the field because of variations, or even blemishes, in the light-sensitive phosphor coating on the tube face. Any nonuniformity in the field produced by these imperfections will affect the accuracy of densitometric measurement. Spatial nonuniformities can be corrected easily so long as they remain constant in time.

You can make corrections using a procedure called "shading correction." Shading corrections usually consist of two routines. The first measures the nonuniformities in a background field and calculates the correction coefficients necessary to correct them. The second applies the correction coefficients to the real image to eliminate the nonuniformities produced by the background field.

One of the algorithms for shading correction that has been used is illustrated in the flow diagram shown in Fig. 14-3. The algorithm first captures an image of the background field. This field should be a blank region of a slightly defocused microscope slide. The algorithm then builds a histogram of the gray levels within the background field. The gray level of peak frequency (i.e., highest number of occurrences) is determined from this distribution, and this value is used as the reference gray level of the field.

The gray level of each pixel in the background field is then compared to this reference value, and the difference is stored in a correction image. Both positive and

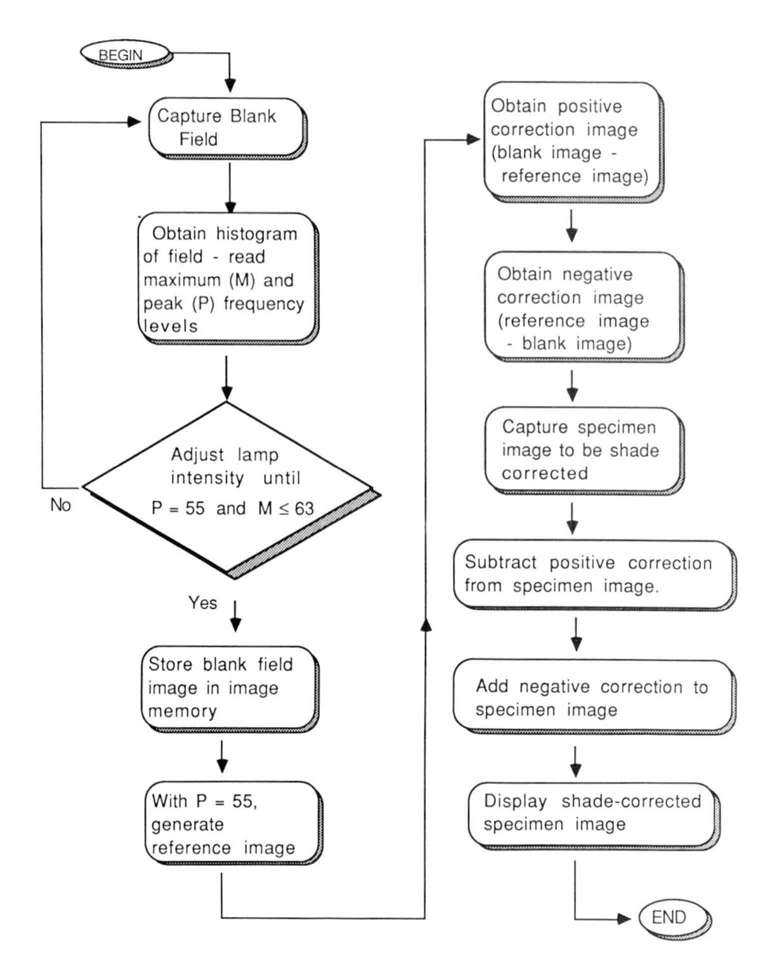

FIGURE 14-3. Flow chart of a shading-correction algorithm used to correct spatial nonuniformities in a background field image.

negative correction images must be generated if the image memory on the instrument does not store negative numbers. The positive correction image is subtracted from the image when it is corrected. The negative correction image is then added to this positively corrected image. The corrected image becomes the new real image, which is stored in image memory and used in all subsequent image manipulations.

Other approaches have been used to correct for photometric nonuniformities. For example, Jones and Lu (1988) rectify shading errors by calculating the mean gray-level value of the background image. Each pixel of the real image is corrected by dividing it by the pixel value of the shading image, multiplied by the mean gray-level value of the background field image. Agnati et al. (1984a) also use the mean gray level of the background field for their shading correction, but the correction coefficients are computed and applied over 4 × 4-pixel blocks of the image rather than pixel by pixel. A commercial program distributed by the Joyce–Loebl Corporation uses the maximum gray-level value in the background field as the reference gray-level value for the corrected field. The correction is applied to 16 × 16-pixel blocks.

Shading correction algorithms that operate on blocks of pixels rather than single pixels do so to conserve computation time and memory space. They are not as accurate as shading corrections that operate on single pixels because they either ignore large variations in single pixels within a block (in the case of using the maximum gray-level value) or distribute the effect of the single pixel variation over the entire block (using the mean gray-level value). Where possible, it is always preferable to correct single pixels rather than blocks of pixels.

14.4.2. Photometric Uniformity in Time

The gray levels of a background image should also remain constant in time. In practice, however, gray levels will fluctuate slightly from time to time. This can be caused by variations in line voltage, gradual fogging of the microscope lamp, variations in camera-tube sensitivity, as well as other factors. The variability can be of two types—short term and long term. Short-term temporal variability (milliseconds to a few seconds) is usually caused by electrical transients. Long-term temporal variability occurs over hours or days and can be caused by a variety of factors.

Short-term temporal variability can be minimized by applying a noise-correction algorithm. These algorithms average a series of images by summing the gray-level values of each pixel in the image and then dividing each summed pixel value by the number of images used to produce the average. The averaged image is stored in image memory and used for all subsequent image manipulations.

Unfortunately, the reduction in noise in an image increases with only the square root of the number of images averaged, but each additional image capture requires a constant amount of time. So there is a disappointing trade-off between the amount of noise reduction obtained and the amount of time that the researcher waits to obtain the noise-corrected image. Figure 14-4 illustrates the effect of increasing trials on reducing the noise of an image on an image analyzer.

The percentage error is used to evaluate the reduction in noise (Jones and Lu, 1989). The time required to collect each image is 2 sec. As can be seen, there is a nearly continuous reduction in noise over the 50 trials illustrated. The most dramatic drop in noise occurs over the first four to eight trials, which require 10 to 20 sec to collect. For work requiring very high gray-level resolution, noise averaging over 30 or more trials is recommended (Jones and Lu, 1988). For routine work not requiring such accuracy, averaging over four or eight trials reduces the effect of noise by over 50%. The time required to collect multiple images creates a strong argument for purchasing a real-time digitizer (i.e., one with which each image capture takes 1/30 sec).

Long-term noise (drift) occurs because of slow changes in line voltage or changes in intensity of the light source or the sensitivity of the camera tube. Long-term noise is not easily corrected, so it is important to evaluate and minimize this error. Line-voltage transients to the light source and camera can be reduced by using constant-voltage transformers. Severe fogging of the light source can be corrected by changing the light bulb periodically. Long-term temporal changes in the system can be measured by taking repeated readings of a background field over the course of several hours. Figure 14-5

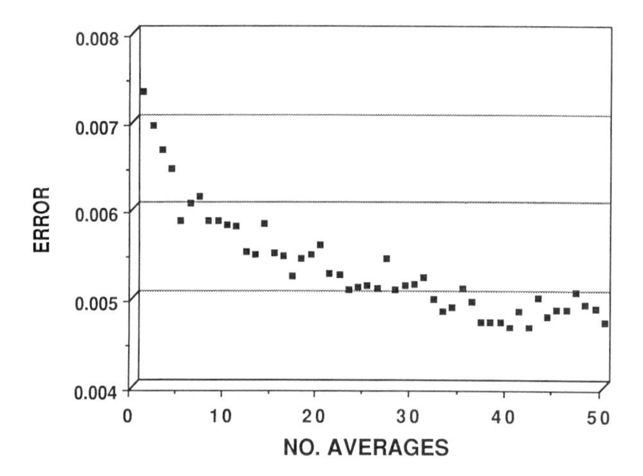

FIGURE 14-4. Effect of averaging on the noise in an image. Note the sharp reduction in error over the first eight trials, followed by a more gradual reduction over the remaining 50 trials.

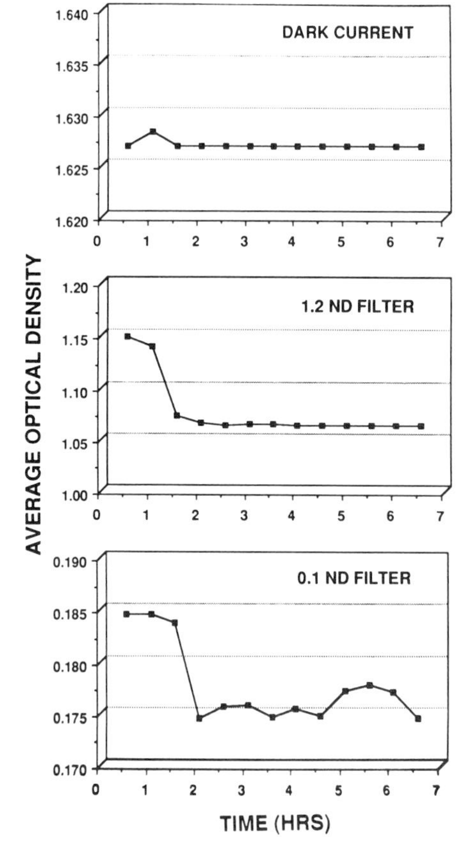

FIGURE 14-5. Measurements of long-term temporal variability in an image-analysis system. The average optical density (AOD) of a blank field was measured every 30 min for a period of 6.5 hr. Light conditions were: no incident light (camera dark current, top); low-light condition (1.2 neutral density filter, middle); high-light condition (1.2 neutral density filter, middle); high-light condition (0.1 neutral density filter, bottom). Reproduced from Mize et al. (1988) with permission.

illustrates the temporal variability found in our system. The integrated optical density (i.e., the measured gray level summed over each pixel of the image) of a square 18 × 18-pixel field was measured very 30 min for a period of 6.5 hr. Three illumination conditions were used: no light (camera dark current), moderate light (1.2 ND filter), and bright light (0.1 ND filter). Figure 14-5 shows that the dark current varied little over the entire 6.5-hr period. In the two light conditions, however, there was a significant decrease in average optical density over the first 1.5 to 2 hr after the system was turned on. This result emphasizes the importance of evaluating long-term drift and correcting for it. To control for transient variability over the first 1.5 to 2 hr, you should turn on all components of a system at least 2 hr before collecting data.

14.4.3. System Sensitivity and Linearity

Image analyzers vary in their sensitivity and linear response to light. Sensitivity is largely determined by the characteristics of the scanner (camera tube). Linearity can be affected by the scanner, the A/D converter, and the computation of integrated optical density. The comparison of input to output of the system is called the "densitometric transfer function" of the system. It can be measured by comparing a calibrated input to the optical-density values output by the system software.

To measure the transfer function neutral-density filters can be used to reduce in defined steps the amount of light reaching the camera. To accomplish this, one can (1) set the tungsten light source of the microscope so that the digitized value has a gray level of 55 (just below saturation of 63), (2) use a 40× plan apochromat objective to image a slightly defocused blank microscope slide, (3) place Kodak Wratten gelatin neutral-density filters over the field diaphragm in the base of the microscope, and (4) measure the integrated optical density (IOD) of the imaged background field.

The data can be normalized for area by dividing the integrated optical density by the area over which the readings are obtained, expressed either as pixels or square micrometers; the normalized value is called "average optical density." Figure 14-6 shows the relationship found between light input (expressed in lux) and the output of a system in average optical-density units. This function illustrates both the sensitivity of the system and its linearity. The dynamic range is approximately 1.5 optical-density units. There is a fall-off in sensitivity at both the highest (0.1 ODU) and lowest (1.5 ODU) light intensities. The system is relatively linear within the midrange (approximately 0.2 to 1.2 ODU). The intensity of the light source should be set so that all immunocytochemical labeling falls within this linear range.

Note that although the video response is always nonlinear (Inoué, 1986, p. 331), these tests are demonstrating the linearity of the entire system.

14.4.4. Evaluation of Measurement Algorithms

The accuracy of the geometric measurements made by image analyzers also varies because the measurements are pixel-oriented and based on vectors calculated between pixels that compose the outline of the object or by the geometry of the pixels within the object being measured. Measurement algorithms can therefore introduce significant errors in geometric computation.

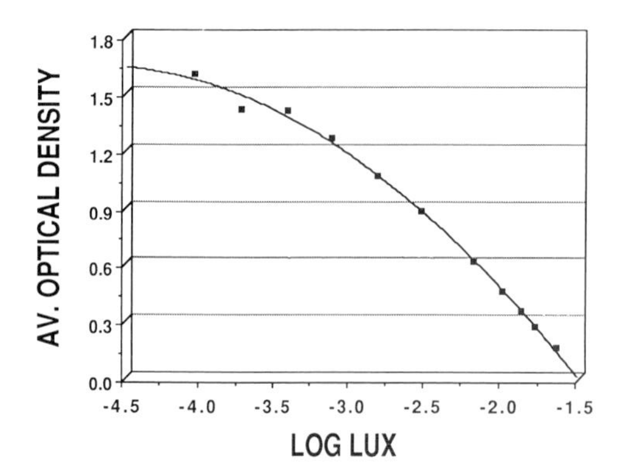

FIGURE 14-6. Densitometric transfer function of an image-analysis system. The output (average optical density, AOD) is plotted against light input (measured in lux). The relationship is linear between approximately 0.3 AOD and 1.5 AOD units.

These errors can be evaluated by making repeated measurements of an object whose size has been established by other techniques. For example, you can measure the cross-sectional area, integrated optical density, perimeter, length, and shape factor of a circle whose exact diameter had been measured with a ruler. An error of 5% or less should normally be adequate for quantitative immunocytochemistry. Length measurements in particular should always be checked, for they may err up to 41% unless proper programming techniques are employed (Cornelisse and Van den Berg, 1984).

14.5. HOW TO MEASURE IMMUNOCYTOCHEMICALLY LABELED TISSUE

The goal of quantitative immunocytochemistry is to obtain simultaneous measurements of the optical density (labeling intensity), geometry, and position of labeled fields, cells, and fibers. The Neuroscience Imaging Center at the University of Tennessee Health Science Center uses a series of image-processing subroutines that extracts the objects to be measured and measures them. Routines are available for calibration, stage movement and position, gray-level analysis, segmentation, editing, binary operations, and measurement. These procedures are described in detail below. Some of these procedures have been described previously by others (Agnati et al., 1984a; Benno et al., 1982a; Briski et al., 1983; Cassell et al., 1982; Gross and Rothfeld, 1985; Hillman et al., 1977; Millar, 1982; Mize, 1985c, 1987; Mize et al., 1988).

14.5.1. Setting up the System

To obtain accurate measurements, careful attention must be given to setting up the image-analysis instrumentation. Before collecting data, it is important to align the optics and adjust the light source of the microscope, to calibrate the magnification, and to enhance the image properly.

14.5.1a. Setting up the Microscope. Precise setup and alignment of the light microscope is essential for accurate quantitative analysis. The Neuroscience Imaging Center uses a 100-W tungsten filament light source to analyze immunocytochemically stained tissue. The power supply for the light source is connected in series to a constant-voltage transformer to stabilize the line voltage. The brightness of the tungsten light is adjusted to match the sensitivity of the camera. The light is increased until the maximum gray-level value of the image is 55 (just below saturation of 63). The intensity of light producing this gray level is read from a photometer built into the microscope. If there is a change in the photometer value, the light source is adjusted until the photometer and peak gray-level values match those set at the beginning of a measurement session. In practice, the values rarely change over the course of a measurement session.

Once the light microscope is set up, the immunocytochemical measurement program is executed. The program first requests the name of a data file and associated file information (user name, date, title of data, reference number, comments). Once this information is entered, the microscope image appears on the screen.

14.5.1b. Spatial Calibration. The image must be calibrated to make measurements in meaningful units. On the Center's system, this is accomplished by placing a stage reticule on the microscope stage and imaging it on the monitor. A light pen is then used to mark the distance between tic marks on the monitor. The actual distance between those tics is entered into the computer, which then computes the magnification factor.

14.5.1c. Enhancing the Displayed Image. Sometimes labeled profiles are difficult to distinguish because the specimen is lightly labeled. At other times, there may be intense background label that obscures specifically labeled profiles. In these cases, it is possible to enhance the gray-level image to better visualize the labeled profiles. There are a number of methods for accomplishing this. One can (1) adjust the sensitivity of the input device, (2) adjust the displayed gray-scale range of the image, (3) use filters to enhance wavelengths that match those of the label, and/or (4) display the image in pseudocolor. The sensitivity of the TV camera is adjusted by changing gain and offset. This allows you to match the gray-level range of the camera to the gray-level range of the specimen. The procedure improves contrast in the specimen but has several disadvantages. The controls for adjusting gain and offset on TV cameras are usually analog potentiometers, which are difficult to readjust to the same settings from experiment to experiment. If each specimen is measured using a different gain and offset, each data set must be normalized to compare it to other data sets. Thus, an additional step is introduced into the computation process, and it is more difficult to compare different sets of data.

Contrast can also be improved by adjusting the gray scale of the digitized image after it has been captured by the camera. The display of the gray-level information is manipulated rather than the digitized image itself. To improve contrast, the gray-scale range is compressed to match that of the specimen. The improved contrast makes it easier to manipulate the image when using binary and gray-level operators, which are described below.

A third approach is to use filters with a wavelength that enhances the contrast of the label. The maximum absorption of diaminobenzidine occurs at a wavelength of roughly 480 nm (Frasch et al., 1978). Green filters with a range of 400 to 500 nm will dramatically improve contrast in immunocytochemically stained tissue. Note, however, that filters will alter the relationship of photometer and gray-level readings because the chromatic characteristics of photometers and the camera differ.

The detection of gray-level values by the viewer can also be markedly enhanced by representing the gray scale with colors. These are referred to as "pseudo-" or "attributed" colors. Colors can be assigned, one to each gray level in the image. The enhancement of an image by pseudocolor display is shown in Fig. 14-7. In this example a color look-up table has been adjusted so that red–yellow represents the lightest labeling in the tissue while blue–black represents the darkest labeling. Although these color assignments enhance the contrast between different gray levels, they also obscure some of the subtle differences seen in the histological specimen. In many cases, differences in labeling may be more obvious in a monochromatic image. The arbitrary color assignments can also be deceptive in that they may not accurately reflect numerical differences in labeling intensity. Despite their frequent use, pseudocolor maps are often of little use except as attractive illustrations.

14.5.2. Collecting Data

Once the microscope, image analyzer, and specimen have been set up, data can be collected. To accomplish this, you need to sample the tissue, capture the image, process the image, and make appropriate measurements.

14.5.2a. Sampling the Tissue. To sample the tissue systematically using the motorized stage, a reference position must be established. To accomplish this, the microscope stage is moved to the left and above the tissue section, and the stage is zeroed (x,y coordinate positions = 0). The stage is then moved until a field, cell, or group of fibers is imaged below the cross hairs of the microscope binoculars. The image is then captured, noise averaged, and shade corrected, and gray-level operators are applied to enhance the image. The image is segmented as well as edited, and binary operators are applied. The extracted profiles are then measured. These steps are described in detail below.

14.5.2b. Capturing the Image. When a field of interest has been moved under the cross hairs of the microscope, the image can be stored in image memory. At this point, the noise and shading corrections are applied to the image automatically, and the corrected image is displayed on the monitors. The image is then processed to extract and measure labeled profiles.

14.5.2c. Gray-Level Operators. Gray-level operators enhance or reduce the contrast of objects in an image by altering the gray-level values of adjacent pixels. The effect is either to enhance or to reduce the contrast found at edges, thus sharpening or smoothing boundaries between areas of more contrast and areas of less contrast. Gray-level operators are used to enhance an object relative to background, thereby improving object detection. Most gray-level operators also alter the size of the object because they add or subtract gray levels from its edges. They should be applied with caution when exact measurements of the size and shape of an object are essential. Because gray-level operators alter the gray levels of pixels near the edges of an object, they also can change the IOD of the object. It is important to assure that optical-density measurements are taken using the gray-level values of the original image, not of the image modified by the gray-level operators.

Gray-level operators include edge enhancement (sharpening), several filters (mean, median, Laplacian, Gaussian), and x and y gradients. A mathematical description of these

FIGURE 14-7. Immunocytochemically labeled tissue section. (A) Section through the cat superior colliculus labeled with an antibody to BSA-conjugated γ-aminobutyric acid. The chromagen is diaminobenzidine, which produces a brown reaction product. (B) Pseudo (attributed) color digital image of the section. Colors are assigned to different gray levels. Black–blue, densest immunoreactivity (lowest gray-level values); red–gray, least dense immunoreactivity (highest gray-level values). A color reproduction of this figure appears following p. xx.

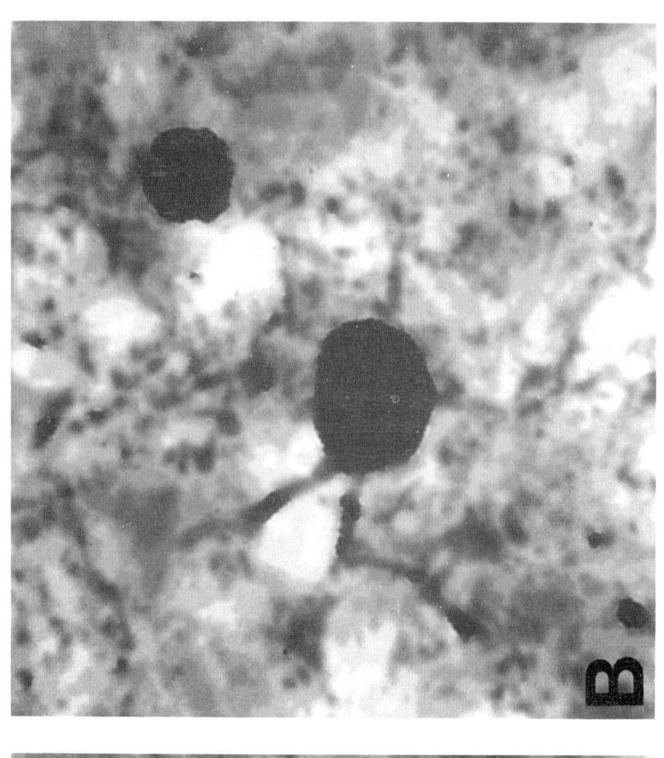
B

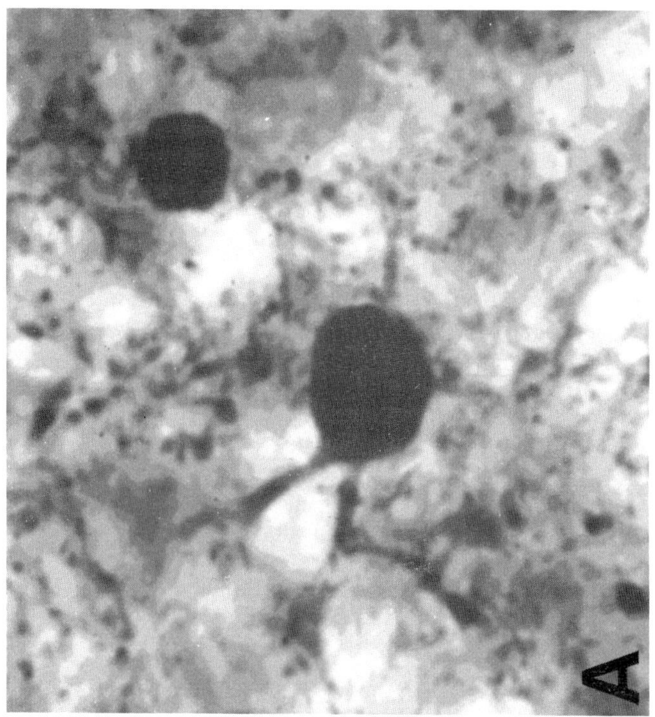
A

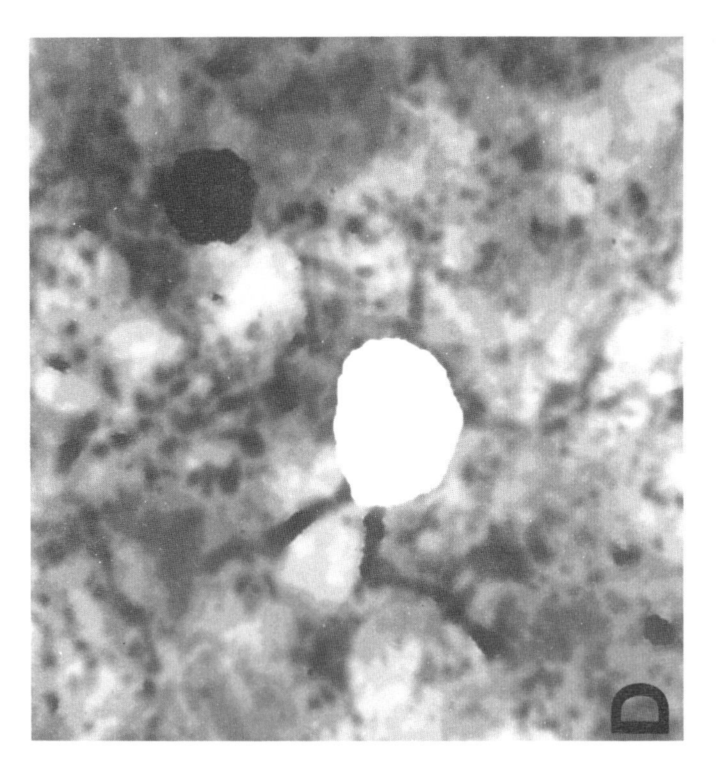

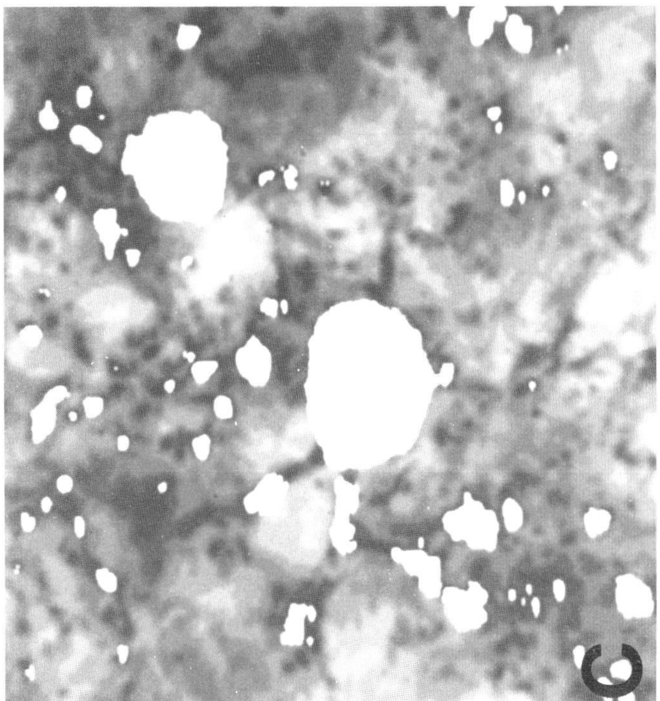

FIGURE 14-8. Image processing of a neuron in the cat lateral geniculate nucleus labeled by an antibody to γ-aminobutyric acid (GABA). (A) On-line digital image of the labeled cell; (B) edge-enhancement operator applied; (C) whole field segmented; (D) interactive binary editing applied to extract the cell and remove all other segmented neuropil. The edge of the cell has also been smoothed using an opening operator.

operators may be found in Chapter 13 of this book. Use of these operators in immunocytochemical analysis is described below.

Edge enhancement is used to sharpen the contrast at edges while smoothing the remaining field. It is used in immunocytochemistry to help define the edges of cells and fibers and to remove artifacts along the borders of these objects (compare Figs. 14-8A and 14-8B). The edge-enhancement operator on the Center's system works by finding the maximum and minimum gray-level values present in a subset of picture elements (pixels). The subset is a matrix of pixels, called a "kernel." On the Center's system, the kernel is 5 × 5 pixels in size. The edge-enhancement algorithm takes the difference between the maximum and minimum values of each kernel in the image and compares that difference to a threshold gray level of 10. If the difference is greater, the center pixel is assigned the maximum or minimum value, whichever is nearer to its original value. If the difference between the maximum and minimum values of the kernel is equal to or less than 10, a mean linear filter is applied that smoothes the kernel.

Other gray-level operators can be used to enhance edges. Gradient operators work on a single plane of the image, either the x or y axis. With an x-gradient operator, boundaries between dark and light that have a vertical orientation are enhanced. With a y-gradient operator, horizontal boundaries are enhanced. These operators work on adjacent pixels along a single axis. Adjacent pixels that have large differences in gray value are further differentiated by adding one or more gray levels to the higher value and subtracting one or more gray levels from the lower value. This increases the difference in gray level at the boundary. The practical effect of gradient operators is to enhance objects with vertical or horizontal orientation. The enhancement looks similar to the raised effect produced using differential interference contrast microscopy (Nomarski optics). Other sharpening operators, such as Laplacian filtering, enhance edges in any direction.

Some gray-level operators are used to smooth an image (reduce contrast). Their use in immunocytochemical analysis is to reduce random background noise. Smoothing operators include mean and Gaussian filters. Mean operators sum the gray-level values of each pixel in a region (kernel), compute the average of these values, and replace each original gray-level value with a value weighted by the mean.

14.5.2d. Segmentation. Segmentation is a gray-level operation that allows the researcher to select a range of gray levels to be discriminated from background. This operation is sometimes called "thresholding." The image is sliced so that objects with a given gray-level range are extracted. These are stored as a binary image in one bit plane of image memory (Figs. 14-8C and 14-9B). The extracted gray levels are displayed on the monitor in white.

This technique is used to extract immunocytochemically labeled cells and fibers. The upper and lower thresholds of the segmentation window can be adjusted on the Center's system by manipulating a sliding scale bar on the screen with a light pen until the best visual fit of the fibers or cells is obtained.

14.5.2e. Setting the Measurement Frame. Often the investigator is interested in measuring only a portion of the specimen imaged on the monitor. Sometimes single cells are measured. At other times, one wants to measure the immunoreactivity of a brain nucleus or lamina. In other cases, one may want to know the integrated optical density of an entire section. To measure subregions of the field, the subregion must be defined for the computer.

On the Center's system, a researcher can specify one of three field types: a frame, a boundary, or a whole field. A frame is a square or rectangular subregion whose size and shape are determined by the user (Fig. 14-10A). The user defines the upper left and lower right corners by pointing to them with the light pen. The frame can then be moved with the light pen until it is properly positioned on the screen.

Irregular objects (cells, laminae, brain nuclei) can be defined by specifying a boundary (Fig. 14-10B). The boundary is traced with a light pen. The whole field is the entire image displayed on the monitor (Fig. 14-10C). Once a frame, boundary, or field is defined, the user has the choice of measuring all objects contained within the frame or measuring the optical density and area of the entire frame. If objects are to be measured, they can be further manipulated using binary operators before they are measured.

14.5.2f. Binary Operators. Binary operators modify the binary image that was created during segmentation. Binary operators are used to connect related objects, disconnect separate objects, and remove artifacts. The Center uses five basic operators to modify extracted cells and fibers. These are erosion, dilation, closing, opening, and thinning.

Erosion is the subtraction of pixels from the boundary of an enclosed object. The erosion operator used on the Center's system sets up a kernel of amendment pixels around the inside edge of the detected object. The width of the amendment kernel is determined by the researcher and can vary from one to ten pixels. Pixels in the amendment kernel that overlap detected pixels of the object are removed.

The effect of erosion is to contract the size of the object equally around its perimeter. The erosion operator can be used to remove edge artifacts or to separate two objects that are in very close proximity. To avoid altering the size of the object, a dilation operator can subsequently be applied to the object field to recover most of the lost pixels.

Dilation is the addition of pixels to the boundary of an object. Like the erosion operator, the dilation operator sets up a kernel of amendment pixels, in this case around the outer boundary of the object. Pixels of the amendment kernel that are not currently part of the object are filled by the algorithm so that they now form the edge of that object and enlarge it.

The effect of dilation is to expand the size of the object a constant amount around its perimeter. This operator is useful for attaching regions of an object that belong together but have become separated, perhaps because of inadequate labeling of one region of the object. It is also useful for recovering the size of objects that have been separated by prior erosion. If this is the first operator to be applied, then you can apply a subsequent erosion operator to decease the object to its original size.

Opening is a function that combines erosion and dilation. It allows you to separate objects and then return them to their original size. It operates just like an erosion followed by a dilation, except that objects smaller than the pixel size of the erosion are erased and are not recovered during the subsequent dilation. The practical effect of this operator is twofold: to separate two closely spaced objects and to remove small artifacts that you do not want to measure. As with erosion and dilation, you can specify between one and 10 pixels to be removed.

Closing is a dilation followed by an erosion. It attaches objects by expanding pixels

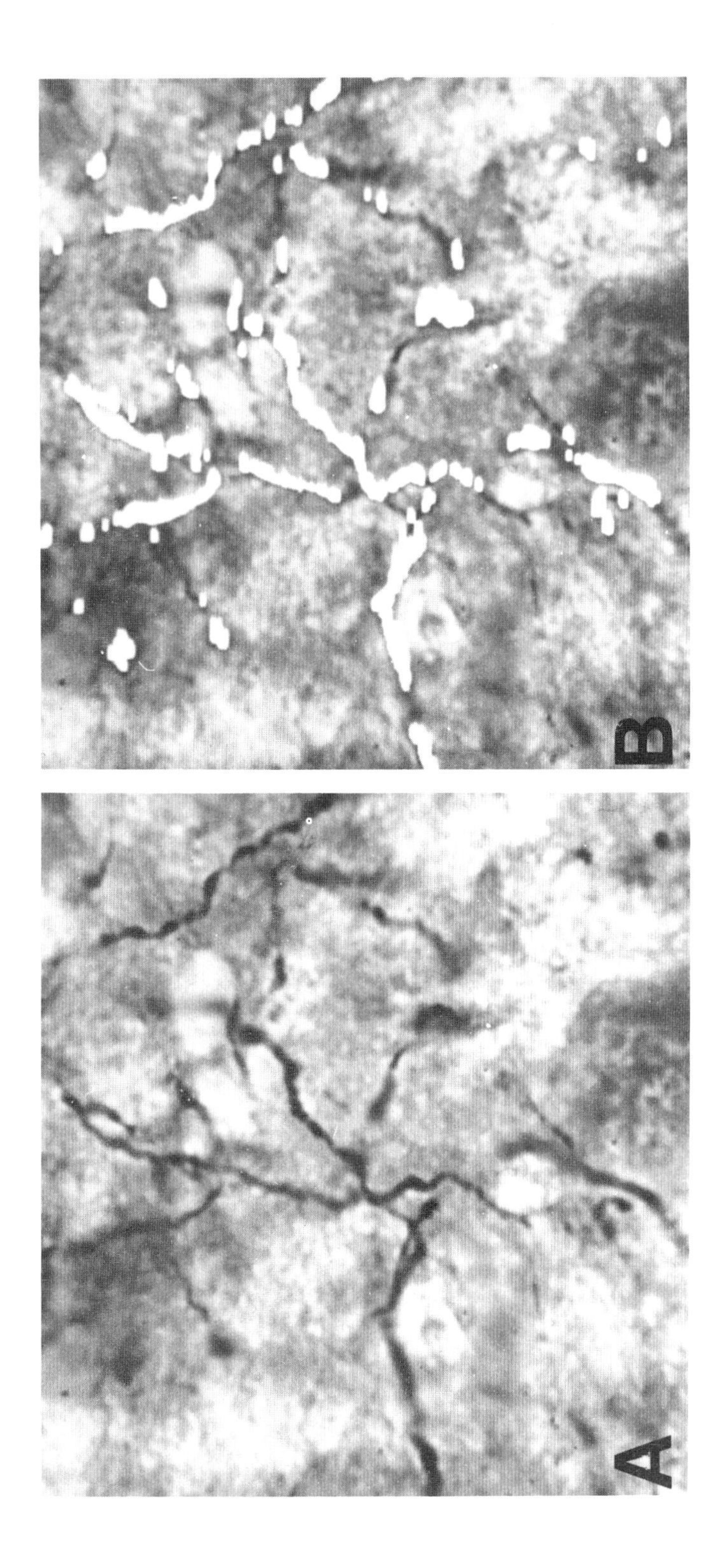
B
A

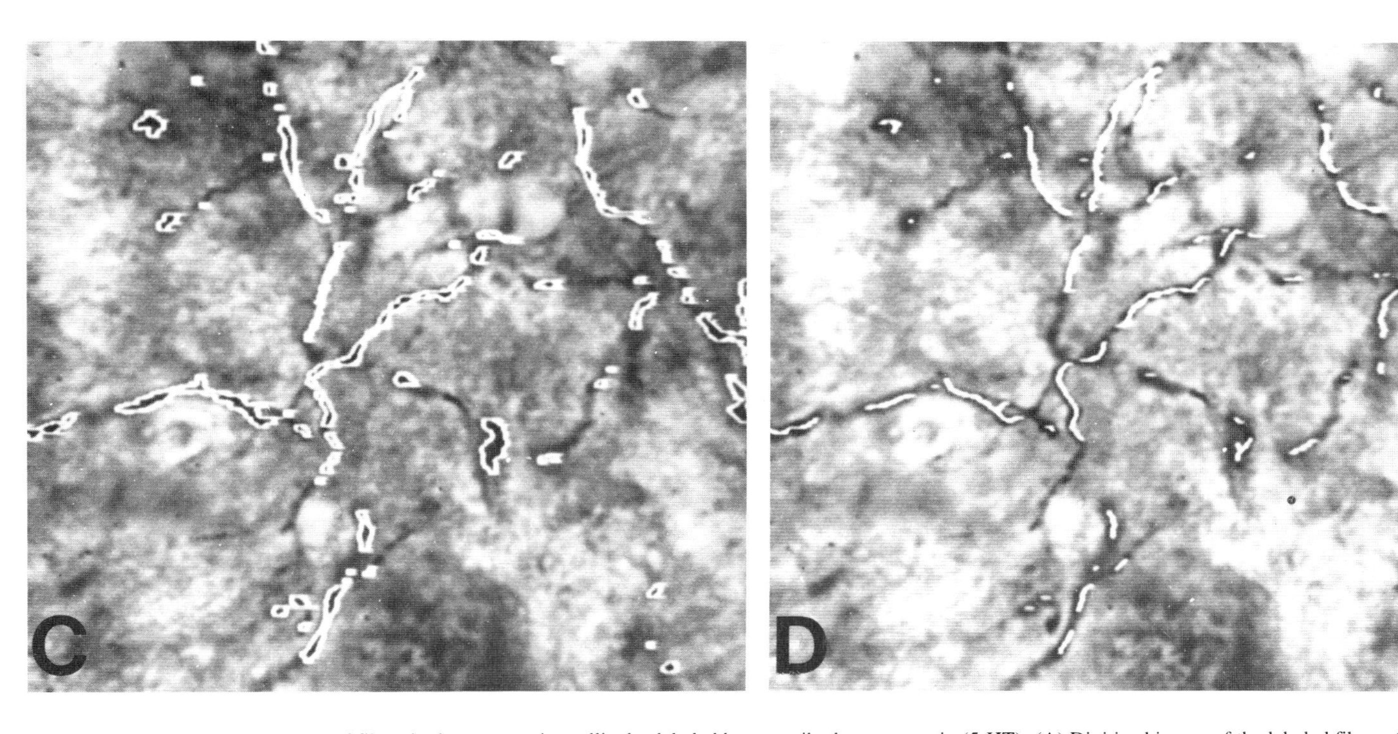

FIGURE 14.9 Image processing of fibers in the cat superior colliculus labeled by an antibody to serotonin (5-HT). (A) Digitized image of the labeled fibers; (B) field segmented to extract densely labeled, focused fibers. Interactive binary editing has also been applied to fill in labeled fibers and to remove artifacts; (C) detected fibers outlined to illustrate regions being measured; (D) thinning operator applied to reduce fibers to pixel-wide lines in order to measure total fiber length.

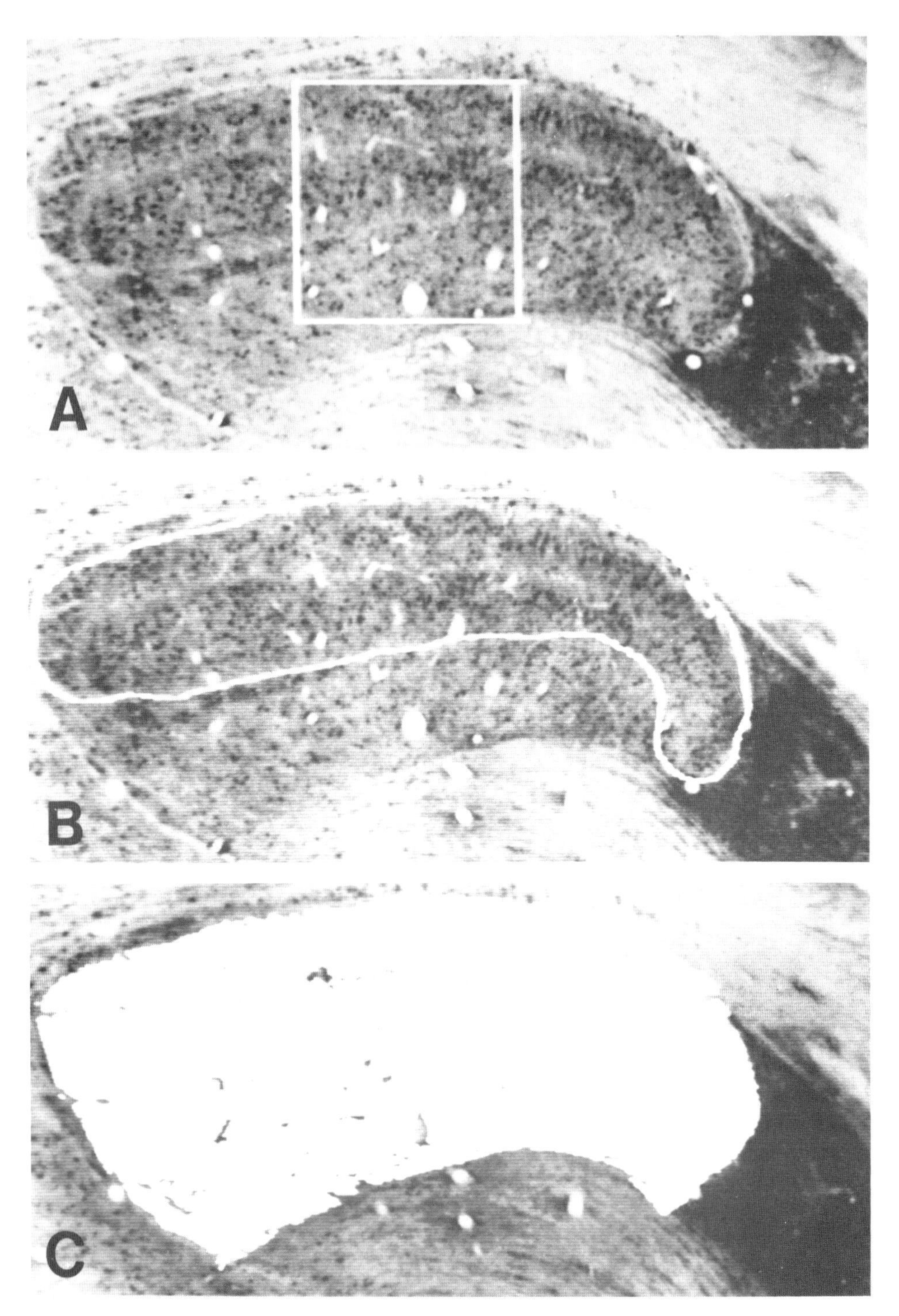

FIGURE 14-10. Field measures of immunocytochemically labeled tissue in the cat dorsal lateral geniculate nucleus (DLGN). (A) A square frame has been set to measure the optical density within the frame. The size, shape, and position of the frame can be adjusted with the light pen on the monitor. (B) A boundary has been traced around several laminae of the DLGN with the light pen. (C) The DLGN has been segmented to measure the optical density of the entire nucleus.

at the edge of the object. Those pixels that do not touch as a consequence of the operation are then removed, returning the object to near its normal size.

Thinning is an extreme case of erosion. This operator erodes objects until their smallest diameter is reduced to a single pixel width. The practical effect of this operation is to produce a series of single-pixel lines. The process is also called "skeletonization." With this procedure, you can measure the length of a field of fibers without regard to fiber diameter (Fig. 14-9D).

Interactive binary editing is also important for extracting labeled profiles. This feature allows the user to add or subtract pixels from the binary image. On the Center's system, the light pen is used to amend the image pixel by pixel or object by object. The user can remove pixels or objects by pointing to them with the light pen, add to the image by drawing in pixels with the light pen, and select or remove regions of the image by drawing a boundary around them. High-resolution pixel editing is facilitated using a zoom feature that magnifies the image 8 to 16 times. Figures 14-8D and 14-9C show some of the effects of binary editing.

Binary editing is very useful for removing artifacts in the binary image that you do not want to measure. You can also separate objects that touch by drawing a single-pixel-wide path between them. Finally, you can fill in regions of reaction product whose gray level was not captured during segmentation.

14.5.2g. Measurements. The purpose of each of the above operations (enhancement, segmentation, binary manipulation) is to produce a faithful binary representation of the immunohistochemically labeled objects that you want to measure. Once an accurate binary overlay of the histochemistry has been produced, image analyzers can measure the objects automatically. The Center's image analyzer has the capability of measuring area, detected area, length, breadth, perimeter, height, width, orientation, and center of gravity. It is also possible to make the following field measures: total area in the field and the total number of objects in the field. These measures can be expressed in pixels or in calibrated units (e.g., the number of objects per square micrometer in the field.)

For measuring single immunocytochemically labeled objects, researchers at the Center compute area, length, perimeter, orientation, and shape factor. Area is calculated by integrating along the boundary of each object, using a calculus formula for area (Fig. 9-4). Area can also be calculated by counting the number of pixels and multiplying this value by the area of each pixel. Length is calculated as the maximum chord of the object and is computed by finding the longest distance between any two pixels in the image. Perimeter is calculated by summing the length of all vectors drawn between adjacent points on the surface of the object. Orientation is determined by finding the angle between the maximum chord length and a line parallel to the y axis of the image. Shape factor, or form factor, is the ratio of the area of the object to its perimeter squared (Fig. 9-5) and is a measure of its "roundness."

14.6. ANALYSIS OF IMMUNOCYTOCHEMISTRY DATA

Three types of measurement can be made on immunocytochemically labeled tissue using the procedures described above. These are field, cell, and fiber measures.

14.6.1. Field Measurements

The most global measure is field area. The field-area measure calculates the integrated optical density of all pixels found in the field. As described earlier, the field can be a frame, an irregular boundary that you draw, or an entire tissue section. Figure 14-10 illustrates the use of a frame, a boundary, and a whole nucleus. Frames can be drawn to take a series of samples from a region of tissue (Fig. 14-10A). This procedure is useful if you want to take random or systematic samples of optical density through different regions of tissue.

Irregular boundaries can also be drawn to define a lamina or brain nucleus. In the example in Fig. 14-10B, a boundary has been drawn around several laminae of the cat dorsal lateral geniculate nucleus (DLGN). The integrated optical density of these laminae is then calculated. Boundaries can be drawn around single cells, single laminae, a brain nucleus, or an entire brain section.

The integrated optical density of complete brain nuclei or entire brain sections can also be measured using segmentation. Figure 14-10C illustrates a coronal cross section through the cat DLGN that has been extracted using segmentation plus binary editing. Because the labeling of the DLGN is darker than the surrounding tissue, it is possible to extract the edge of the nucleus by segmenting it. However, it is usually also necessary to edit the segmented image to remove regions of the field that are not part of the nucleus. In most cases, therefore, it is actually quicker to draw around the boundary with a light pen.

Regional variations in optical density within these boundaries can be illustrated in pseudocolor (Fig. 14-7B) or expressed as a regional map of numerical values. The displayed values can be gray levels or optical density units. It is also possible to sample the optical density along a given strip of tissue. This procedure is particularly useful where there are obvious bands or sharp boundaries of label in different laminae of a brain structure. Figure 14-11 illustrates a section of cat visual cortex that has been reacted with an antibody that recognizes fixative-modified glutamate (Hartman et al., 1988). A single-pixel-wide scan across different cortical laminae has been made using a light pen (Fig. 14-11A). The image analyzer then samples the optical densities within a patch of tissue centered along the scan line and extending some distance above and below the line, as shown in Fig. 14-11B. The optical densities sampled from this patch are displayed in histogram form (Fig. 14-11C) and also treated numerically. This method allows you to examine differences in labeling along any axis of the tissue. The scan line can be straight or curved and can be oriented at any angle.

14.6.2. Cell Measurements

The easiest method for measuring the optical density and geometry of a single cell is to segment it. To accomplish this, the labeled cell must first be enhanced and then edited to remove artifacts (Fig. 14-8). The cell is first imaged on the screen (Fig. 14-8A), and its position is recorded. An edge-enhancement operator is then applied to sharpen the outer membrane of the cell (Fig. 14-8B). The cell is then segmented (Fig. 14-8C). A window of gray-level values is chosen that includes all gray levels within the cell but excludes gray levels in the remaining tissue. The binary overlay of the segmented image is then edited with the light pen to remove artifacts and other labeled processes that fall within the segmentation window (Fig. 14-8D). A closing procedure (erosion followed by dilation)

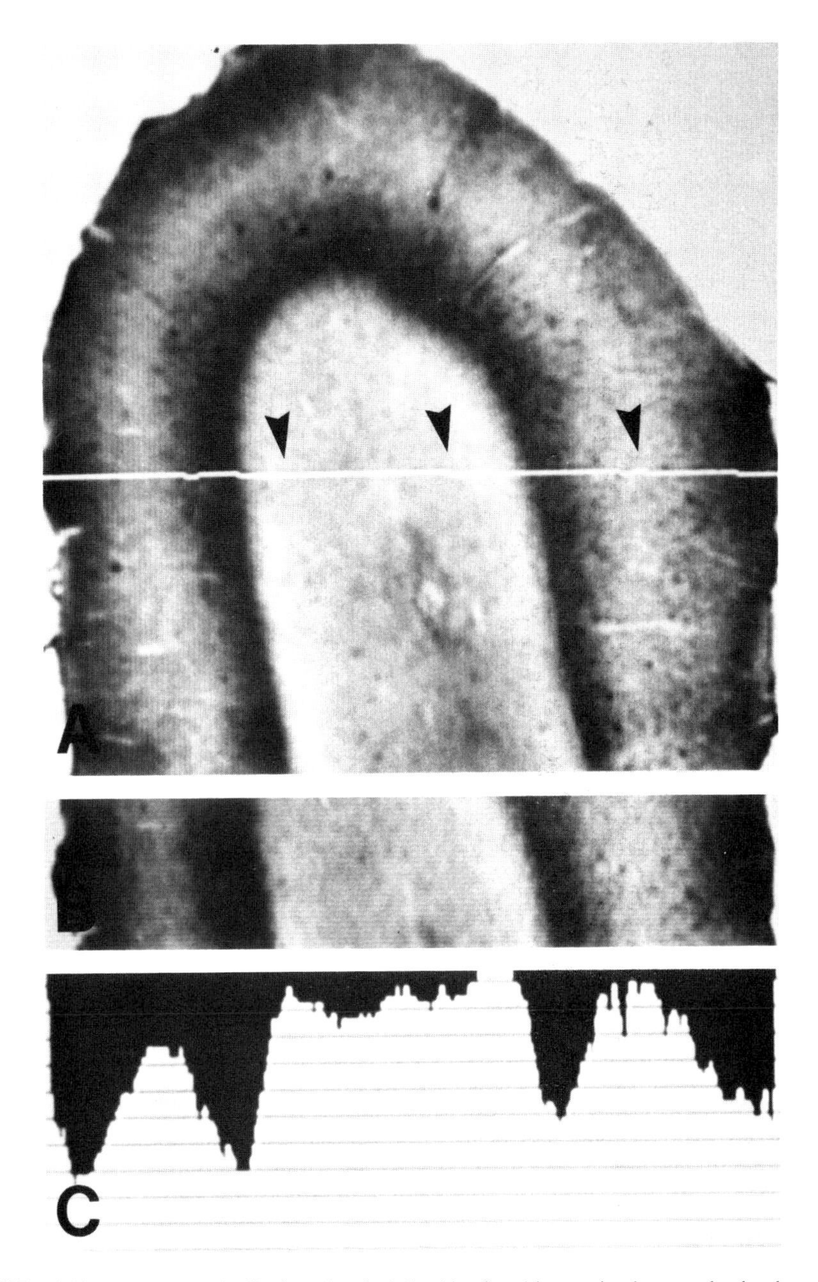

FIGURE 14-11. Frequency distribution of optical densities found in cat visual cortex that has been reacted with an antibody to γ-L-glutamyl-L-glutamic acid. (A) Areas 17–18 (marginal gyrus) of visual cortex. Arrows point to a scan line traced with a light pen on the monitor of the image analyzer. (B) A patch through the marginal gyrus showing the region over which optical densities are calculated. (C) Frequency histogram of the optical densities found in the patch illustrated in B.

can be used to separate cells that are next to each other. An object-selection routine can also be applied to eliminate all objects smaller or larger than a window of sizes defined by the user. The area, perimeter, optical density, and other parameters of the cell are then measured.

Another method for measuring cells is to trace around the outer contour of the cell body with the light pen. The same information can be obtained with this technique, but it is generally slower and less accurate because of the time and care involved in tracing with the light pen. Segmentation is faster, is usually more accurate, and can be applied simultaneously to more than one cell in a single field. By contrast, each cell in a traced field must be drawn separately.

The Center's researchers have used the segmentation procedure to measure cell size and labeling intensity in the lateral geniculate nucleus (LGN) of the cat using an antibody to γ-aminobutyric acid (GABA) (Mize, 1987). The LGN complex includes the perigeniculate nucleus, which contains GABAergic projection neurons, and the main nucleus, which contains GABAnergic interneurons. The Center's researchers compared cells in the two nuclei to determine whether or not there were differences in their morphology and labeling intensity.

The perigeniculate nucleus (PGN) and the lateral geniculate nucleus were systematically sampled by measuring every labeled cell located in a series of scans made through the two nuclei. The area and average optical density of each cell were measured. The position of each cell was recorded and later compared with traces of the nuclei to determine whether the cell was in the PGN or LGN. Differences in cell size were found between the two nuclei. Almost all the LGN cells were small. By contrast, the PGN contained a number of larger neurons (Fig. 14-12). Although the cells in the LGN were smaller, they were more darkly labeled than the larger neurons of the PGN.

14.6.3. Fiber Measurements

It is also possible to segment immunocytochemically labeled fibers and thus to estimate the innervation density of specific transmitter pathways. The procedure is similar to that for cells. A digital image of the fibers is displayed on the monitor (Fig. 14-9A). The fibers are then segmented (Fig. 14-9B), which captures most but not all of the fiber field. Interactive binary editing is used to fill in obvious fibers that were not detected by the segmentation procedure. The fiber field is then measured (Fig. 14-9C).

The fibers can also be thinned to a single-pixel width by using the skeletonization procedure (Fig. 14-9D). This is done to compute total length of the fiber field unaffected by the minor diameter of individual fibers.

The Center's researchers have used this procedure to measure the density of immunoreactive processes in the lateral geniculate nucleus of the tree shrew using an antibody that recognizes GABA (Holdefer et al., 1988). The LGN contains both anti-GABA-labeled neurons and several types of anti-GABA-labeled processes, including presynaptic dendrites and axon terminals. To estimate the number of these processes in different laminae of the LGN, researchers at the Center measured 12 fields of fibers in each of six laminae from five sections in two animals. The total area occupied by immunoreactive processes above a threshold optical density was used as the measure. The Center's

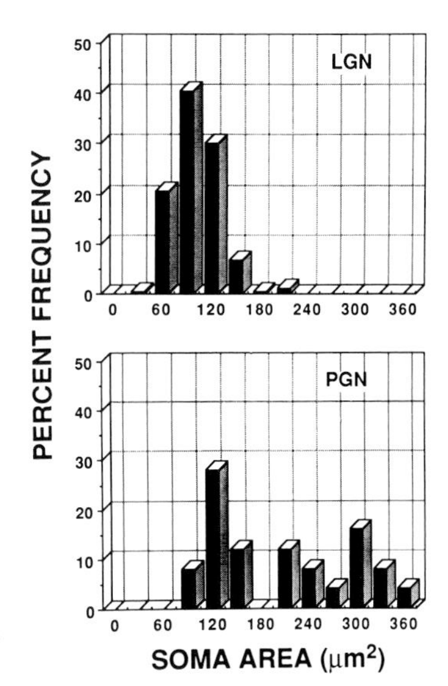

FIGURE 14-12. Histogram illustrating differences in cell size (cross-sectional area) in the cat dorsal lateral geniculate nucleus (LGN) versus the perigeniculate nucleus (PGN). Reproduced from Mize et al. (1988) with permission.

researchers believe this to be the best measure because individual processes overlap and are therefore sometimes difficult to detect.

Figure 14-13 shows that there are measurable differences in total area of immunoreactive processes in different laminae of the LGN. The least labeling is found in the medial laminae (1,2,3), the most in the lateral laminae (4,5,6). Differences between laminae 1,2 and 4,5 are statistically significant. This corresponds to differences in function in these two pairs of laminae, where laminae 1,2 contain on-center cells whereas laminae 4,5 contain off-center cells (Holdefer et al., 1988).

In addition to measuring cells and fibers with the light microscope, you can also apply the technique to electron micrographs. Synaptic profiles, for example, can be

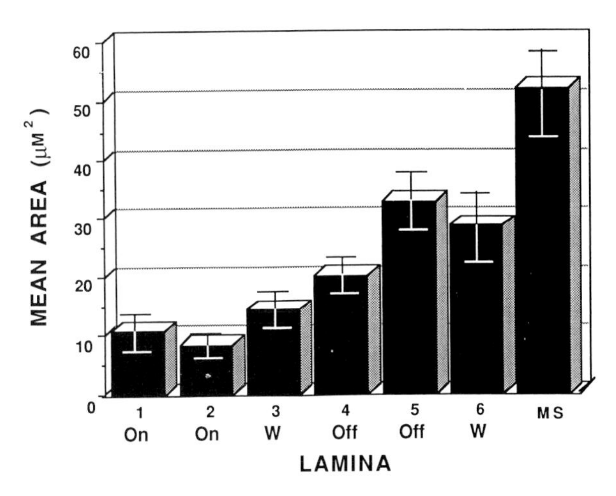

FIGURE 14-13. The total area of GABA-immunoreactive processes in the six cell layers of the tree shrew LGN. The data for each lamina are the average of the total area of labeled neuropil found in 12 sampling fields. The neuropil in the monocular segment (MS) was also labeled and measured. Modified from Holdefer et al. (1988) with permission.

segmented just like immunocytochemically labeled cells. In cases where the profiles are not contrasted enough to segment, they can be measured by tracing the outer contour of the membrane with the light pen (Fig. 14-14). In Fig. 14-14, the EM negatives were imaged with the newvicon camera. A zoom lens was used to adjust the magnification. The negatives were back-lit using a fluorescent light box. The digital image of the negative was converted to a positive image using an inverse-image function available on the Center's system. The digital image is shown in Fig. 14-14A. The light pen was used to draw manually the outline of the synaptic terminal on the TV monitor. The region within the enclosed profile was then filled (Fig. 14-14B), and its size and shape were measured. The time required to image and trace the outline of the synaptic terminal is comparable to that of tracing the outline on a digitizing tablet (Mize, 1983a, 1985b). The image-analysis procedure does not require photographic prints, which saves both time and expense.

Synaptic terminal size can sometimes reveal important relationships between structure and function. For example, researchers at the Center have found statistically significant differences in the size and synaptic connections of retinal synaptic terminals in the medial interlaminar nucleus (MIN), the dorsal and parvicellular C laminae of the lateral geniculate nucleus, and the ventral lateral geniculate nucleus (VLG) (Mize and Horner, 1984; Mize et al., 1986). These structural differences correlate with known physiological differences in the various nuclei. Thus, for example, MIN contains predominantly Y-like cells, whereas VLG contains only W-type cells. The retinal terminals in VLG are smaller and make fewer contacts with F2 presynaptic dendrites. These differences appear to be related to the discharge rate of the afferents as well as to differences in the inhibitory characteristics of the two nuclei (Mize et al., 1986).

14.7. HISTOCHEMISTRY PROCEDURES AND CONTROLS

To quantify immunocytochemistry, it is important to understand the kinetics involved in the labeling reaction. Immunochemistry is a complex procedure involving a number of processing steps. Each of these steps can introduce nonlinearities into the procedure, making quantitation of labeling difficult. In both the indirect peroxidase (PAP) and avidin–biotin complex (ABC) labeling procedures, a layered complex of molecules is introduced between the antigen and the visualized label. With ABC, for example, the labeling complex consists of the antigen, the primary antibody, a biotinylated secondary antibody, a peroxidase-labeled avidin–biotin molecule, and a chromagen (usually diaminobenzidine, DAB). Each of these steps involves kinetic reactions that lead to the desired coupling. It is therefore important to control rigidly each step in the immunocytochemical procedure. Some important controls are listed below. These have been discussed in greater detail by Sternberger (1979) and Vandesande (1979).

14.7.1. Factors Affecting Labeling

A number of variables can affect the labeling intensity produced by antibodies in immunocytochemistry. These variables include the concentration and exposure time to the antibody and other reagents, including the chromagen. Some of these factors are discussed below.

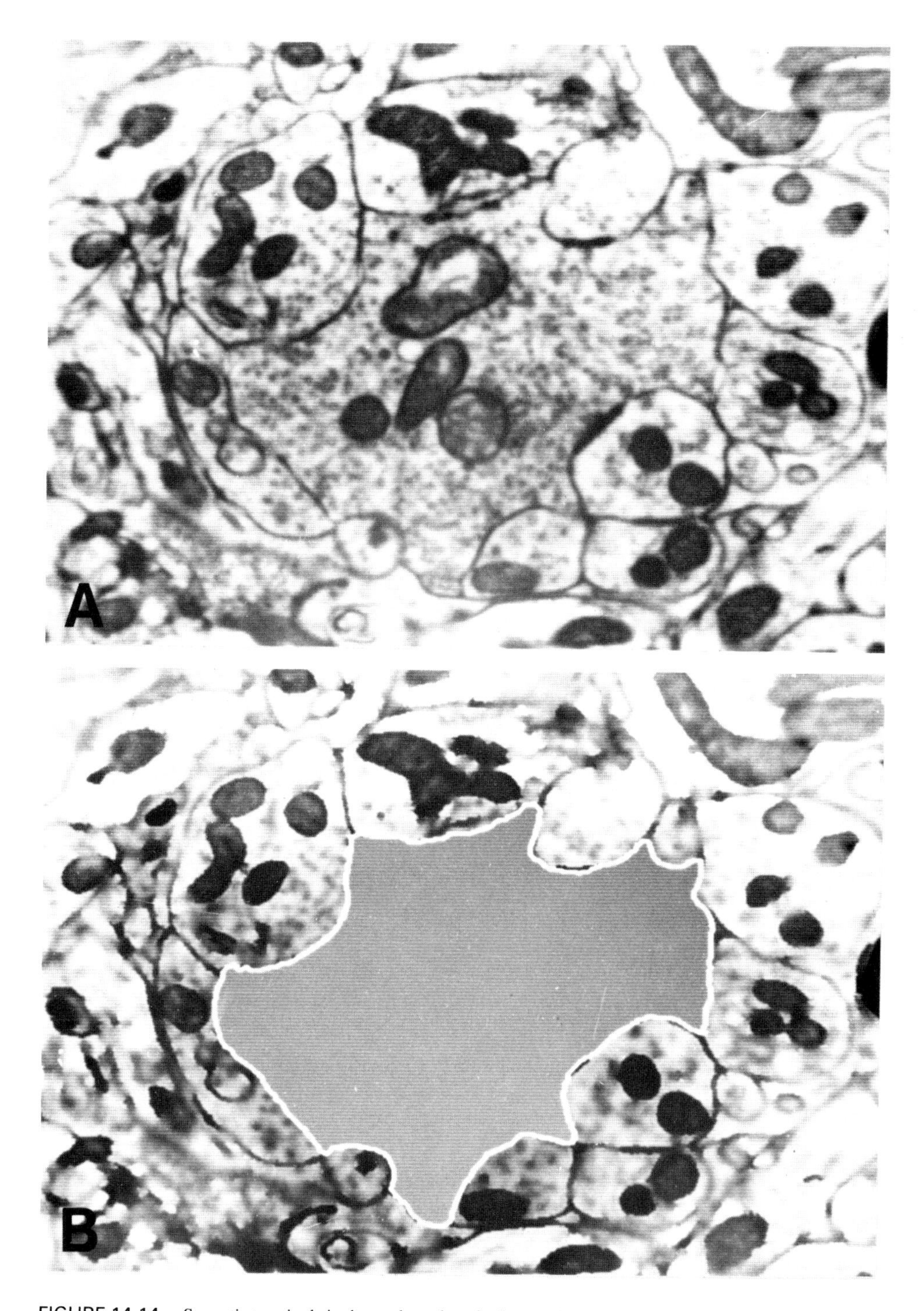

FIGURE 14-14. Synaptic terminals in the cat lateral geniculate nucleus measured with an image analyzer. (A) Digitized image of a retinal terminal in the medial interlaminar nucleus (MIN). (B) Image edge enhanced and outlined using the light pen of the image analyzer and then filled automatically for measurement. Software will compute the cross-sectional area, perimeter, length, etc. of the synaptic terminal.

14.7.1a. Antigen Concentration. In brain tissue, antigen concentration is the variable being measured. It is unknown and cannot be controlled. Antibody concentration and other processing variables may have to be altered depending on the antigen concentration in brain. Large amounts of antigen may require higher antibody dilutions, whereas very small amounts of antigen may not be detected without increasing antibody concentration.

14.7.1b. Antibody Concentration. It is important to establish the optimal dilution of the primary antibody (Bigbee et al., 1977). Antibodies that are too dilute will produce false negatives, whereas those that are too concentrated will produce false positives because of high background labeling. The optimal dilution can vary dramatically for different antibodies as well as for different techniques. In the Center's lab, an anti-GABA antiserum works best at a dilution of about 1 to 5000 using the ABC technique. An antiserum to γ-L-glutamyl-L-glutamic acid, on the other hand, needs to be diluted as much as 1 to 80,000. The ABC technique is more sensitive than PAP and requires five to ten times more dilute antibodies (Hsu et al., 1981; Madri and Barwick, 1983). For these reasons, it is essential to establish the appropriate antibody concentration for each set of experiments. The Center's researchers usually vary the range tenfold in four to five equal steps. A sharp rise in AOD is found up to a certain concentration, followed by a plateau in labeling at slightly higher concentrations. At yet higher concentrations, there is a secondary increase in AOD, much of which represents nonspecific background labeling (Millar and Williams, 1982; Gross and Rothfeld, 1985; Nabors et al., 1988).

The Center's researchers expose the tissue overnight (16 to 20 hr) at room temperature. Variations of 3 to 4 hr seem to have minimal effect on antibody labeling. It is nevertheless important to control exposure time to the primary antibody within any group of experiments in which results will be compared. Antibody concentration, incubation time, temperature, agitation force, amount of incubation medium, and number of sections incubated should all be kept constant. In addition, the exposure times and concentrations of the remaining reaction agents (secondary antibody, ABC solution) should be identical for the entire set of experiments.

14.7.1c. DAB Concentration. The concentration of the chromagen can also affect labeling intensity dramatically. The use of 3,3′-diaminobenzidine tetrahydrochloride to label HRP was originally described by Graham and Karnovsky (1966). The appropriate DAB concentration varies with incubation time and with the amount of H_2O_2 (hydrogen peroxide) used to catalyze the reaction. Gross and Rothfeld (1985) report a linear increase in reaction product with concentrations between 0.01% and 0.05% with an H_2O_2 concentration of 0.016%. Higher concentrations of DAB resulted in visible background labeling when an antibody raised against gonadotropin-releasing hormone was used. In the Center's laboratory, a concentration of 0.05% yields good results for a variety of antibodies. As with other processing steps, an optimal DAB concentration should be determined empirically and then used in all experiments using a given antibody.

14.7.1d. DAB Reaction Time. The optimal reaction time for DAB can also vary with different antibodies. There is a linear increase in AOD for exposure times between 1 and 10 min using a DAB concentration of 0.05% (0.003% H_2O_2 added) and the GABA antibody (Nabors et al., 1988). Gross and Rothfeld (1985) report a linear increase between 6 and 24 min also using a DAB concentration of 0.05%. On the other hand, Reis et al. (1982) show a linear increase in DAB incubation time only over the first 4 to 6 min of the reaction using an antibody to tyrosine hydroxylase. The variability in these results emphasizes the importance of running control experiments to determine the optimal reaction time for each antibody.

14.7.1e. H_2O_2 Concentration. H_2O_2 will also affect staining intensity. Gross and Rothfeld (1985) have shown that the optimum H_2O_2 concentration varies with DAB concentration. There is a linear increase in labeling only between 0.0001% and 0.001% H_2O_2. On the other hand, Streefkerk and van der Ploeg (1973) report no change in reaction absorbance between 0.002% and 0.03% H_2O_2. As with other reagents, an empirical determination of the optimum concentration should be made for each antibody.

14.7.1f. Other Variables. Other variables will affect optical-density readings of an immunocytochemically labeled section. Section thickness should be identical to allow comparison of antibody labeling of different sections. Antibody penetration should also be equivalent in those sections. Reagent incubation times and dilutions should be identical in all experiments where data will be compared. Incubation times can be timed precisely with a stopwatch. Reagents should be measured with precision micropipetters. The pH and temperature of these solutions should also be kept constant.

14.7.2. Controls

Three types of controls can be used to measure background AOD.

14.7.2a. Adsorption Controls. Adsorption controls are produced by preadsorbing the antibody with a known amount of antigen. Adsorption controls reduce labeling because the antibody has already bound to the antigen mixed with it. Residual labeling reflects antibody labeling that is not specific to the antigen of interest. The optical density of the residual labeling can be subtracted from total immunoreactivity to eliminate the effect of this nonspecific labeling. The optical-density readings from the adsorption control sections should be taken from the same brain region in both control and experimental sections.

14.7.2b. Normal Serum Controls. Normal serum controls are also used to measure nonspecific labeling of the antibody to components of tissue other than the antigen. Normal serum controls are produced by incubating tissue sections in nonimmune normal serum from the same species in which the primary antibody was raised. The AOD of these control sections can also be subtracted from the AOD of the experimental sections.

14.7.2c. Negative Controls. Negative controls test for reactivity produced by endogenous peroxidase or other components in brain in the absence of the primary antibody. The primary antibody step is deleted, but the tissue is otherwise processed normally. This control can be used to subtract optical density produced by endogenous peroxidase or other background not related to the primary antibody.

14.7.3. Quantitative Standards

When experimental conditions are rigidly controlled, optical-density differences found in different sections should reflect differences in antigenicity. These optical-density differences are relative. They do not allow you to estimate the actual concentration of the antigen in brain tissue. However, you can estimate concentration if you compare the optical density of a standard of known antigen concentration with the optical density of a brain section of unknown antigen concentration. Standards are substrates that contain a known amount of antigen. The standard is incubated in primary antibody and processed for immunocytochemistry just like brain sections. The optical density produced by different concentrations of antigen in the standards is then plotted to generate a standard curve of optical density versus concentration. If the standard is processed in parallel with brain sections, one can estimate the antigen concentration in the brain section from the known antigen concentration that produced that optical density in the standard.

Both homogenized brain paste and nonbiological substrates have been used as standards. Brain paste is an excellent substrate in some experiments because it has the same protein composition as brain sections (Biegon and Wolff, 1986). However, it is impossible to process brain paste free floating in parallel with brain sections because the paste will not remain intact. It is also an inappropriate medium if it contains unknown amounts of endogenous antigen, as is the case for many widely distributed transmitters.

Nonbiological substrates include polyacrylamide films (van Duijn et al., 1967; Streefkerk and van der Ploeg, 1973), gelatin (Millar and Williams, 1982; Schipper and Tilders, 1983; Schipper et al., 1984), and agarose beads (Capel, 1974; Pool et al., 1984; Streefkerk et al., 1975; Streefkerk and Deelder, 1975; van Dalen et al., 1973). All these substrates can be processed with brain sections. Antigens of known concentration are coupled to the substrate by several methods. In the case of gelatin, antigens can be cross-linked directly to the gelatin with glutaraldehyde (Millar and Williams, 1982; Schipper and Tilders, 1983; Schipper et al., 1984). In the case of agarose, coupling can be achieved by activating the substrate. For example, researchers at the Center use 100-μm sections of 3% agar activated with cyanogen bromide/acetonitrile (Axen et al., 1967; March et al., 1974; Porath et al., 1967). The cyanogen bromide activates hydroxyl groups that then form bonds with the lysine residues of proteins. This technique works well with antigens that are conjugated to proteins such as thyroglobulin or bovine serum albumin (Nabors et al., 1988). Agarose beads also operate on this principle.

Activated agar works well because it can be cut into sections on a vibratome just like brain sections. Agar sections are inert and virtually colorless. They can also be processed in parallel with brain sections. They readily couple to transmitter antigens that have been conjugated with carrier proteins such as BSA. Using agar standards, researchers at the Center have shown that there is a linear relationship between the concentration of BSA-conjugated GABA coupled to agar sections (measured by scintillation counts) and the optical density produced by labeling the sections with an antibody directed against BSA/GABA (Fig. 14-15) (Nabors et al., 1988).

Others have used similar approaches. For example, Millar and Williams (1982) have developed a gelatin step-wedge standard that can be cut and mounted onto the same slide as a tissue section. The step wedge is produced by adding layers of gelatin with different IgG concentrations on top of one another. Both the standard and brain section are pro-

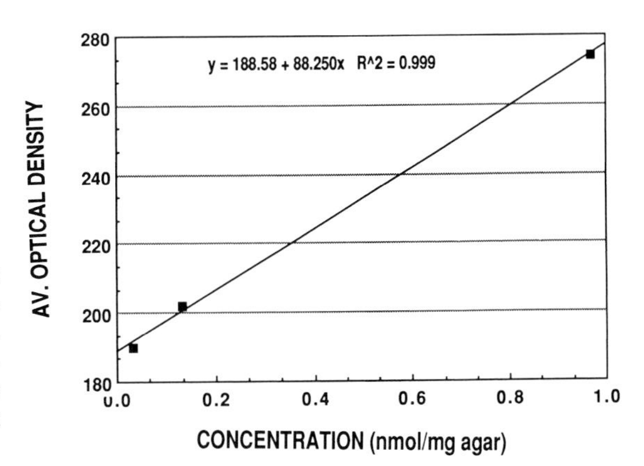

FIGURE 14-15. Relationship between average optical density (AOD) and the concentration of an antigen to γ-aminobutyric acid that was coupled to activated agar sections. The data are fit by linear regression. Modified from Nabors et al. (1988) with permission.

cessed identically. Schipper and Tilders (1983) have also used gelatin sections to compare the antigen concentration of standards and brain sections.

14.8. SUMMARY AND ADDITIONAL INFORMATION

This chapter has shown how image analyzers are used to measure both the geometries and labeling intensities of immunocytochemically labeled tissue. The technique is valuable for comparing the laminar position, morphology, and antibody labeling of cells and fibers in the CNS. Nonbiological standards provide a method for estimating the actual concentration of the antigen in these same cells and fibers. This comprehensive, quantitative approach to immunocytochemistry is a significant advance in the analysis of neurotransmitter distributions in brain.

14.8.1. Hardware Considerations

Systems for quantitative immunocytochemistry should have acceptable gray-scale and spatial resolution, appropriate image-analysis operators, acceptable processing speed, good-resolution monitors, and adequate user interface devices. It is important when purchasing a system to consider the gray-scale and spatial resolution of the entire system. A system with 512 × 512 resolution should be adequate for most immunocytochemistry applications. This size most closely matches the spatial resolution of other reasonably priced components. Thus, high-quality vidicon-type TV cameras usually have resolutions between 600 and 800 lines. Moderately priced solid-state cameras typically have resolutions in the neighborhood of 512 × 512, as do reasonably priced color and black and white monitors. The cost increases considerably for a system with 1024 × 1024 resolution.

Gray-scale resolution on most systems ranges from 64 (2^6) to 256 (2^8) gray levels; 256-gray-level resolution can be obtained at moderate cost and is therefore recommended. However, this resolution exceeds the resolution of some specimens. For example, the gray-level information present in x-ray film autoradiographs is estimated to be 50 to 85

(Lear et al., 1986; Jones and Lu, 1988). A system with 64 gray levels should therefore be adequate for that type of application.

An image analyzer for quantitative immunocytochemistry should also have sufficient image-processing operators and processing speed. This chapter has shown that some gray-level and binary operators are very useful, although others are unnecessary. In general, gray-level operators are most valuable for enhancing images in which labeled objects are difficult to detect. Because there is usually a high signal-to-noise ratio in immunocytochemically labeled tissue, it is often not necessary to apply gray-level operators to extract labeled profiles. Segmentation, of course, is essential for extracting labeled profiles to be measured. Among the other gray-level operators, edge enhancement is the most useful but is by no means indispensable.

The binary operators are far more useful for quantitative immunocytochemistry. This discussion has shown how erosion and dilation operators can be used to separate adjacent profiles or combine binary images belonging to a single object. Thinning is useful for examining the skeletal framework of a field of fibers unencumbered by differences in the thickness of their varicosities.

Image-processing speed varies dramatically in different systems. In general, a field of tissue should be processed completely within several minutes. In the Center's system, noise and shading correction takes about 9 sec, and the edge operator takes 12 sec. Segmentation takes less than 1 sec. An opening procedure (erosion plus dilation) requires around 5 sec. Measuring the area, length, breadth, and optical density of a field of 30 objects requires around 12 sec. Thus, a single field of cells or fibers can be captured, extracted, and measured in about 40 sec (Mize et al., 1988). In practice, it takes longer (1 to 2 min) because you must make multiple passes to obtain the best segmentation window and to interactively edit the binary image. This time frame is acceptable to students in the Center's group who measure fields for 3 to 4 hr a day. However, single gray-level or binary operations that take longer than 15 to 20 sec can be annoying. Before buying a system, you should carefully assess the time required to perform the operations needed in the specific research project. Fortunately, the equipment you can buy today is faster, and often less expensive, than what has been described in this chapter.

14.8.2. Choice of Image Analyzer

There are a large number of commercial image analyzers that can be adapted for quantitative immunocytochemistry (Ramm and Kulick, 1985; Mize, 1984, 1985a). Three different approaches can be taken: you can buy flexible, self-contained systems, specific application systems, or component systems. Self-contained systems are often the most expensive and can cost from $30,000 to $100,000. These systems come complete with scanner, ADC, image memory and processing boards, monitors, a general-purpose computer, a printer, and other peripherals. They are often mounted in an enclosed console. The total resolution of these systems is generally quite good, and the special-purpose image processors are usually high speed. They also usually include extensive software for a variety of gray-level, binary, and measurement operations. The systems can usually be programmed in one or more high-level languages so that it is relatively easy for a programmer to write additional code to accommodate special applications. Many of these systems include specific application packages for densitometry, particle counting, gel electrophoresis, and other uses. To the author's knowledge, however, no system of this type has a commercial package specifically designed for immunocytochemistry analysis, and none provides source code for its software.

An example of this type of system is the Joyce-Loebl Magiscan. This system has a general-purpose program called MENU that serves as a framework program to call a variety of binary and gray-level subroutines contained in the system library. You can link together a text list of subroutines to meet your specific needs. Program-generator software is provided to incorporate other user-programmed application programs. In principle, the system is quite flexible for designing specific applications. In practice, the solution is less simple—largely because no source code is provided. It is thus impossible to make slight modifications in existing library routines. Rather, an entirely new routine must be written from scratch. Much of the difficulty in generating special-purpose applications would be eliminated if source code were made available to the user.

Another type of image-analysis system is designed for specific applications. A large number of these systems have been developed for quantitative receptor autoradiography (Alexander and Schwartzman, 1984; Feingold et al., 1986; Goochee et al., 1979; Palacios et al., 1981; Ramm and Kulick, 1985; Tayrien and Loy, 1984; Unnerstall et al., 1982; Lear et al., 1986). One example is the Loats Associates RAS-R1000, distributed by Amersham. This is a turnkey system with software specific to ligand–receptor analysis. It comes with a light box designed for x-ray images. The software has a facility for entering the optical density of calibration standards specific for radioligand work. Subregions of interest can be extracted to measure localized receptor density or ligand concentration. The gray-level data can be transformed and expressed as disintegrations per minute, quantity of bound ligand, and other units specific to autoradiography. Many of the display features (e.g., pseudocolor display, line profiles of optical density) are especially useful for receptor analysis. The statistical–graphics capabilities are also specialized for receptor autoradiography. The RAS-R1000 now comes with software packages for other applications as well, but each configuration is designed for a specific application.

These special-application systems lack flexibility and are less readily adapted to more general image-analysis applications. For example, receptor autoradiography systems have excellent densitometric capabilities but usually include only rudimentary gray-level and binary operators. They are thus less suitable for extraction of cells and fibers for quantitative immunocytochemistry. However, the turnkey approach makes these instruments very user friendly for their specific purpose and allows for quick startup.

A third approach is to buy component image-processing boards to be used with a microcomputer. Specific software packages for these boards can be purchased from third-party vendors. The scanner, motorized stage, stage drive, monitors, and other peripherals must also be purchased separately. Several companies manufacture modular board systems at a surprisingly modest cost. The author has been favorably impressed with the speed and capabilities of these board systems. For example, Imaging Technology (Woburn, MA) and Matrox (Dorval, Quebec, Canada) sell a family of processing boards that can handle all the processing subroutines reported in this chapter. The top-of-the-line Imaging Technology system, the Series 150/151, includes ADC interfaces, a frame buffer with a $512 \times 512 \times 32$-bit resolution, a pipeline image processor with extensive processing functions, a real-time convolver for real-time gray-level transformation, and a histogram/feature extractor for real-time thresholding. Other manufacturers, listed in Chapter 16, sell comparable products.

These products require third-party software packages to be useful for specific applications. Packages for morphometry and image processing are available for these modular hardware systems, many running on IBM Personal Computer AT or other popular microcomputers. In addition, you are responsible for integrating the image-processing boards, microcomputer, hardware peripherals, and software. This can require considerable computer expertise and should not be attempted without proper support personnel.

Please note that all commercial products specifically mentioned in this chapter are used as examples only. Any mention of such a product should not be taken as a positive or negative recommendation by the author.

14.8.3. Biological Measures

This chapter has described three different ways to measure immunocytochemically labeled tissue. The least-resolved measure is the field measure, often called the "region of interest" (ROI). The ROI measures are field rather than object oriented because they ignore individual objects in the field. Field measures are typically used for quantitative receptor autoradiography where the film resolution is relatively low. The primary purpose of this technique is to measure the amount of radioactivity within a region of brain to obtain estimates of bound ligand (in the case of receptor studies) or glucose utilization (in the case of the 2-deoxyglucose technique) in that region. It is rarely possible to localize the amount of radioactivity in individual cells and fibers with this technique.

Field measures of immunocytochemistry are less useful because there are such fine regional differences in labeling intensity. Averaged optical-density measures over large regions of tissue ignore these local variations and reduce the information that can be obtained. Immunoreaction product is almost always well localized within single cells and fibers and sometimes within subcellular organelles. It is therefore valuable to be able to measure these individual profiles.

Cell body measurements are easy to make because of the ease with which immunoreactive cells can be segmented. One major advantage of this approach is that numerical values can be assigned to the labeling intensity of all cells in a section. The optical density of any cell can be compared with that of any other cell. This is a relative measure but is nevertheless useful because it provides an objective criterion to distinguish labeled from unlabeled neurons. When the distribution of labeling intensities is plotted, it is often possible to identify a gap in the distribution that clearly defines labeled cells. If the gap is not obvious by visual inspection, statistical distribution analyses can be applied to the data to define the labeling threshold.

A second major advantage of the technique is the ability to obtain numeric data about the position, geometry, and optical density of cells simultaneously. This allows one to study the relationships among these variables. Correlations among cell size, labeling intensity, and locations are often revealed when analyzed statistically.

This chapter has also demonstrated how to measure fields of immunoreactive fibers. This is a less exact procedure than cell analysis because it is more difficult to segment fibers precisely. Densely packed fibers often overlap and may be partially out of focus. It is therefore usually not possible to obtain accurate measures of fiber diameter or numbers. However, field measures of fiber density provide a reasonable estimate of the numbers of fibers in a field so long as the segmentation window is held constant from one field to the next. Two field measures are useful: the total area occupied by immunoreactive fibers (above a threshold optical density) and the total length of these fibers. Total immunoreactive area estimates the density of fibers because it captures all objects whose labeling intensity is above the chosen threshold while rejecting background labeling. The total length measure reduces errors that are introduced by overlapping fibers. By thinning of the fibers, single fibers are distinguished more easily. The length measure also eliminates area effects related to the thickness of individual fibers.

The total optical density of the segmented fibers can also be examined with this technique. In most cases, differences in total optical density are closely correlated with differences in total area. However, fiber area and optical density are not always perfectly correlated. For example, the number of immunostained fibers may decrease while there is a corresponding increase in antibody labeling in the remaining fiber population. For this reason, it is useful to distinguish these two possibilities by comparing total fiber area with average optical density.

14.8.4. Biological Utility of the Measures

Quantitative immunocytochemistry can reveal transmitter levels in the CNS and detect alterations in these levels after brain lesions, after environmental modifications, and in disease states. Valid measures of labeling intensity depend critically on linear enzyme kinetics and carefully controlled reaction conditions. These factors have been discussed in detail by others (Benno et al., 1982a; Gross and Rothfeld, 1985; Pool et al., 1984; Reis et al., 1982; Schipper and Tilders, 1983; Streefkerk et al., 1975). If controlled conditions are met, the technique may accurately measure differences in antibody labeling that reflect antigen concentration in brain.

Quantitative immunocytochemistry has demonstrated differences in both the morphology and labeling intensity of subpopulations of neurons labeled by various transmitter antibodies. For example, this chapter showed that cells in the cat dorsal lateral geniculate nucleus are smaller but label more intensely than cells within the perigeniculate nucleus. The two parameters are inversely related, which suggests that there may be a constant pool of transmitter per cell. The Center's results also show that there are differences in both the size and labeling characteristics of projection and interneuron cell populations. Others have also shown quantitative differences in labeling in transmitter-specific cell populations. In the locus coeruleus, cells in the center of the nucleus were less darkly labeled by an antibody to tyrosine hydroxylase than cells in other regions (Benno et al., 1982b; Reis et al., 1982). Densely packed cells were also more lightly labeled than less densely packed cells in this structure. On the other hand, no consistent differences were found between labeling intensity and the morphology of the cells (Benno et al., 1982b). These studies demonstrate that consistent differences in antibody labeling can be shown within cell populations using relative optical-density measures that need not take account of absolute concentration of the antigen.

Measures of immunoreactive fiber labeling also reveal regional variations of biological importance. For example, the total area of immunoreactive fibers labeled by an antibody to serotonin (5-HT) is greater in the superficial gray layer than in other layers of the cat superior colliculus (Mize and Horner, 1989). This may correspond to the heavy innervation by W-type retinal ganglion cells to this region of the colliculus. Likewise, the area of GABA-antibody labeled in processes of the tree shrew lateral geniculate is greater in laminae 4,5 than in laminae 1,2. These laminae also differ in the functional properties of their cells, with the former containing off-center receptive fields and the latter on-center fields (Holdefer et al., 1988). The differences in density of GABA processes may reflect different inhibitory processing in on-center versus off-center channels. Strain differences in antibody labeling have also been demonstrated using an image analyzer. Thus, the total area of tyrosine hydroxylase and prolactin immunoreactivity is higher in BALB/cJ rats, which have higher locomotor activity, than in a CBA/J strain with low locomotor activity (Baker et al., 1985).

The image-analysis technique is also valuable for measuring alterations in antibody labeling induced by a variety of experimental manipulations. Thus, both the numbers and labeling intensity of anti-tyrosine-hydroxylase-positive cells have been shown to be increased by GM-1 ganglioside treatment in the substantia nigra (Agnati et al., 1984b). Image analysis has shown that the area of immunoreactive fibers labeled by an antibody to gonadotropin-releasing hormone is reduced by ovariectomy (Briski et al., 1983) and castration (Gross and Rothfeld, 1985) but restored by testosterone (Gross and Rothfeld, 1985). Others have shown changes in the number and labeling intensity of antibody-labeled fibers after brain lesions. Thus, the area of peptide antibody labeling was reduced in the central nucleus of the amygdala following parasagittal lesions of the substantia inominata, anterior commissure, and stria terminalis (Cassell et al., 1982). Antibody labeling to fixative-modified glutamate in the cat superior colliculus was reduced significantly by lesions that interrupt the corticotectal pathway to that structure (Hartman et al., 1988).

Although alterations in immunocytochemical labeling can be measured reliably, the functional significance of these alterations is not always clear. A reduction in antibody labeling in cell bodies could result from either reduced synthesis and storage or increased transport and release. In both cases, there would be a reduction in the amount of transmitter present in the cell body. Increases or reductions in the optical density of nerve terminals can also be ambiguous. There may be an absolute loss of fibers, as in the case of lesions of an afferent pathway, or there may simply be a reduction in the amount of transmitter labeling in fibers that remain intact. It is sometimes possible to distinguish these alternatives by comparing changes in the area of immunoreactive profiles with changes in their optical density. Thus, for example, Briski et al. (1983) showed a reduction in labeling by an antibody to gonadotropin-releasing hormone in the hypothalamus that was thought to be the result of increased synthesis and release of the hormone after ovariectomy. On the other hand, Houser et al. (1986) have shown decreases in GABA-immunoreactive puncta in the neocortex after experimentally induced epileptic seizures that appear clearly related to decreases in the amount of that transmitter in cortex.

Estimating changes in the actual concentration of neurotransmitters using quantitative immunocytochemistry is also possible when one can relate optical density to biochemical measures of concentration. Two approaches have been taken. The first uses radioimmunoassay. Reductions in antibody labeling are compared to reductions in antigen levels measured by RIA. Thus, for example, Benno et al. (1982b) and Reis et al. (1982) increased the amount of catalytically active tyrosine hydroxylase (TH) in rat locus coeruleus (LC) using reserpine. The increase in TH measured in the RIA assay (nanomoles per LC) was of the same order of magnitude as the increase in optical density produced by an antibody to TH. Likewise, Gross and Rothfeld (1985) showed correlative changes in gonadotropin-releasing hormone (GnRH) and leutinizing hormone (LH) in the hypothalamus and pituitary using RIA and immunocytochemistry. Castration reduced GnRH and increased LH, and testosterone treatment increased GnRH and decreased LH. The magnitude of the changes was similar to that measured by RIA (micrograms per milligram of wet weight tissue) and optical density using antisera directed against those hormones (Gross and Rothfeld, 1985).

In some cases, immunocytochemistry under- and overestimates changes determined by RIA assay. Thus, Briski et al. (1983) have shown that ovariectomy reduces GnRH levels to a much greater extent than it reduces the optical density of GnRH antibody-labeled fibers. This is not surprising if one considers the differences in the techniques. The RIA depends on competition in binding between exogenous and endogenous sources of antigen, whereas immunocytochemistry reflects binding of an antibody to the endogenous antigen without competition. The dynamics of binding therefore differ dramatically in the two techniques (Schipper and Tilders, 1983).

Correlations between immunoreactivity and biochemical measures have also been demonstrated using nonbiological standards. Standards have the advantage that direct comparisons can be made between the optical density of antibody-treated brain sections and standards whose antigen concentration is known. Absolute estimates of antigen concentration based on optical density are therefore possible. For example, Schipper and Tilders (1983) and Schipper et al. (1984) have shown that the concentrations of both 5-HT and corticotropin-releasing factor (CRF) are linearly related to fluorescence intensity produced by antibodies to these substances. They also used brain extracts to compare inhibition of antibody fluorescence produced by the extract with that produced by a known amount of antigen. From this immunoinhibition data they determined directly the amount of CRF per extract. Biegon and Wolff (1986) measured the amounts of acetylcholinesterase bound to brain paste standards and the optical density of these standards produced by a label for acetylcholinesterase (AChE). A standard curve was generated and used to estimate the AChE content (nanograms per milligram protein) in both the cingulate cortex and hippocampus. Consistent differences were found between different regions of the hippocampus by use of this technique.

The Center's researchers have used activated agar standards to compare the optical density produced by antibody labeling with the concentration of radiolabeled BSA-conjugated GABA measured by scintillation counts. The Center's researchers found a linear relationship between the optical density produced by the antibody and the amount of antigen coupled to the standard. It remains to be determined whether the same relationship will hold for brain sections. For amino-acid transmitters, the derived measures should be reasonable because amino acids are cross-linked to brain protein by glutaraldehyde fixation in a manner similar to the conjugated antigen (Geffard et al., 1985). The Center's researchers are currently testing this by comparing reductions in GABA and glutamate levels produced by lesions of the corticortectal pathway using HPLC and immunocytochemistry.

14.9. FOR FURTHER READING

For textbooks on computer applications in neuroanatomy, I suggest Lindsay (1977), McEachron (1986), and Mize (1985). Specific textbooks about image processing are

Gonzales and Wintz (1977), Pratt (1978), Castleman (1979), and Rosenfeld and Kak (1982).

Review chapters on densitometry and immunocytochemistry have been written by Agnati et al. (1984a), Ramm and Kulick (1985), and Reis et al. (1982).

ACKNOWLEDGMENTS. The author would like to thank Scott Davis, C. J. Jeon, Dr. Michael Hartman, and Burt Nabors for collecting much of the data reported in this chapter. Linda Horner, Tere Martin, Kathy Troughton, and Dr. Michael Gurski prepared the immunocytochemically labeled tissue and many of the figures. The author is also grateful to Drs. Robert Holdefer and Thomas Norton for letting him use the data on the tree shrew LGN. Drs. James Madl and Alvin Beitz generously provided the γ-L-glutamyl-L-glutamate antibody. James Ashberger, Sivaram Srinivassan, and Peter Gray developed some of the software used on the Center's image analyzer. John Gaylor and William Burnip of Joyce-Loebl Inc. provided needed technical advice and support. Drs. Stephen Kitai, Mel Park, and Charles Wilson offered enthusiastic and continuous support of this work. This project was funded by a Biomedical Shared Instrumentation Grant S10 RR02800 from NIH, USPHS grant EY-02973 from the National Eye Institute, and funds from the Neuroscience Center of Excellence from the State of Tennessee.

15

Fully Automatic Neuron Tracing

15.1. INTRODUCTION

Tamas Freund, a friend from Budapest, wrote me a letter in 1981 in which he said, "Joe, you should build the ultimate neuron-tracing system. In the afternoon, an anatomist will put his microscope slide in it and go to a local pub and drink a beer. The next morning when he returns to the lab, the neuron will be reconstructed, the analysis will be done, and the paper will be written. This would be the perfect neuron-tracing system." I answered Tamas' letter with "Thank you for your insight and we'll get right on it." Well, we've been getting right on it since 1974, and we're nowhere near it.

15.2. THE AUTOMATIC RECONSTRUCTION PROBLEM

Why is there such difficulty in designing a fully automatic neuron-tracing system? There are several reasons. Consider the tasks involved in following the fibers of a neuron as they meander throughout the tissue (Fig. 15-1, left).

First, the computer is very good at such details as looking at individual pixels in an image and reading their gray levels. The computer is very poor, however, at seeing the overall picture, the gestalt of an image. So to have the computer look at the overall view and recognize a neuron by its pattern deviates from the happy marriage of a computer's attention to detail and the human's superior pattern recognition skills, as you may recall from Chapter 4. Automatic tracing of neurons, or following any structure in an image automatically, is a very difficult problem, because you must write a computer program that, to a degree, duplicates the functioning of the human vision system.

To restate the problem will provide better insight. When you use a television system to capture an image, you have taken the analog video image and converted it into a digital pixel-by-pixel representation. For tracing or statistically summarizing the structures in an image, you need more than the pixel-by-pixel representation—you need a vector representation, which is a series of X,Y,Z coordinates along the border of the structure. The conversion process from a pixel representation of a structure to its vector representation is the very difficult problem. To automate the neuron tracing fully, you need to design a computer algorithm to change the pixel image into a vector image. The human visual

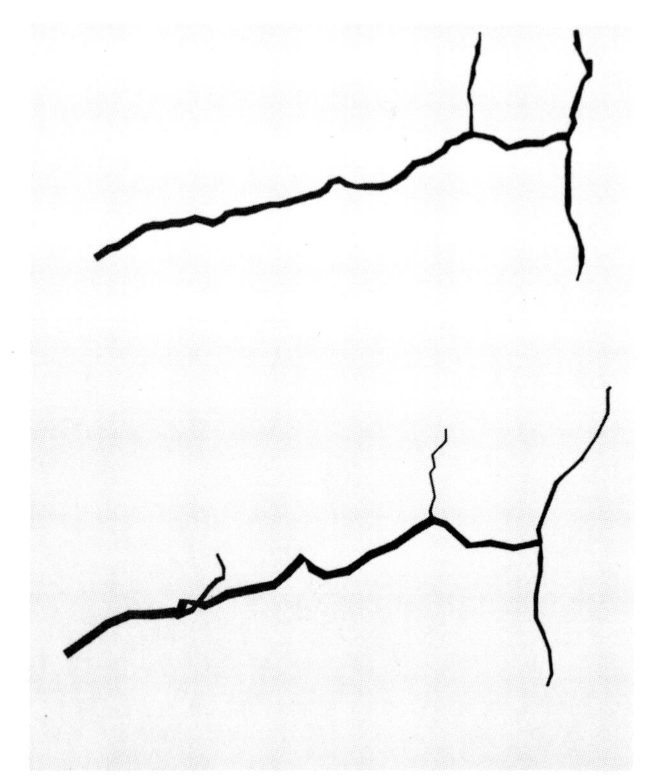

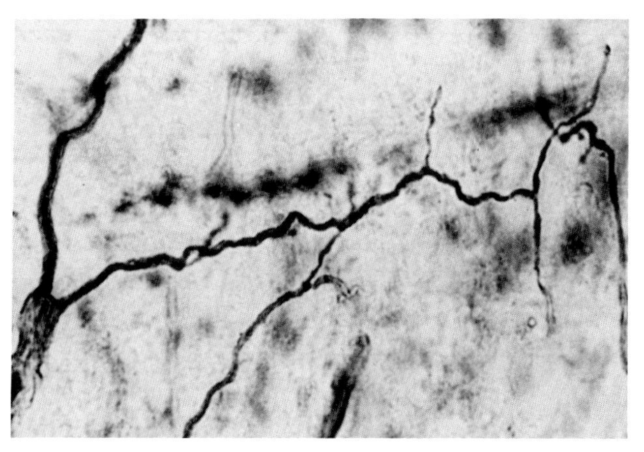

FIGURE 15-1. A piece of dendrite traced in semiautomatically and automatically. Left: The original microscope image. Center: A plot of the dendrite traced semiautomatically using the Capowski and Sedivec (1981) neuron-tracing system. Right: The same piece traced automatically using the Capowski (1983a) system. Note that a small proximal branch was missed. Reproduced from Capowski (1983a) with permission.

system performs this task with many thousands of microcomputers (neurons) operating simultaneously on the image at different levels of sophistication. At each level, the pixel data is reduced further, eventually into a vector structure.

Biological images are "noisy"; they contain unwanted elements that must be ignored. Unfortunately, the unwanted elements frequently have the same color and light intensity as the important structures, so they must be distinguished by shape and location as well as by gray level. Humans are very good at sensing this irrelevant information and quickly discarding it; computers are not.

If this is true, why do we use computers for collecting and analyzing anatomic structures? We use them to do things faster, to be more objective in what we do, to make fewer errors, to reduce tedium, and to allow us to do things that we would otherwise not be able to do. All of these characteristics apply to automatic neuron tracing.

15.3. A STANDARD OF COMPARISON

There is a standard of comparison—the well-trained technician. Suppose it would take a technician 1 hr to trace and analyze a particular neuron. Could an automatic neuron-tracing system perform as fast?

A slow but fully automated neuron-tracing system might be acceptable, for it would free the technician for other tasks. And it will become faster as newer computers are designed. The algorithm of a neuron-tracing system is first programmed on a general-purpose computer, and it may perform at a rate that is perhaps one-tenth that of a human. The algorithm will contain low-level, often-repeated calculations on which most of its time is spent. It may be possible to build special-purpose hardware to perform these lowest-level calculations, speeding them up by a factor of 100. Then the resulting system may outperform the human speed by a factor of ten.

Could the automatic neuron-tracing system perform as objectively as a person? As a person traces, he becomes tired and makes mistakes. Also, as he traces, he learns the tissue so that some of the criteria by which he makes subjective decisions (e.g., to differentiate between a spine and a small branch) do not remain constant. In short, does he perform the same way at 3 o'clock as he did at 1 o'clock? The computer does; it is constant and always makes objective decisions according to the criteria of its program.

15.4. THE CURRENT ART OF AUTOMATIC TRACING

What can be done today? It is possible to follow a dendrite where it is thick and has a high contrast with its background tissue, and with some modest accuracy, branch points and endings may be recognized. It is not yet feasible, however, to perform some of the more research-oriented tasks. For instance, automatic neuron-tracing systems fail to recognize fine details. For example, it is not possible for such a system to recognize the location of dendritic spines and varicosities. Also, when an arborization becomes very complex, especially with fine crossing fibers, no automatic reconstruction system can succeed in following the fibers. Unfortunately, such details are frequently the most scientifically important portions of a dendrite. This is particularly true when two populations of neurons are compared, as it is often the fine details that distinguish them.

For the reasons just described, automatic neuron-tracing systems are not in wide use today. Yet it is worthwhile to review two published systems to illustrate their principles.

15.4.1. Automatic Neuron Tracing with a Vidisector

The first system was built and described by Garvey et al. (1973) and by Coleman et al. (1977) and is shown in Figs. 15-2 and 15-3. This system uses a light microscope and a device called an image dissector, or vidisector, to sense the image for a computer. The image dissector is similar to a television camera in that it has a sensing tube onto whose face the image is focused. An electron beam scans the tube face and generates an electrical signal whose magnitude is proportional to the light incident on the tube face at that point. You may recall from Chapter 2 that a television camera's electron beam repeatedly scans the entire tube face in a fixed pattern called a raster. The electron beam of an image dissector does not; rather, the computer directs the electron beam to any spot on the tube face. Thus, the computer may obtain gray-level readings of any pixel by simply moving the electron beam to the point and reading the magnitude of the returned electrical signal.

When using the Garvey–Coleman system, a human operator starts the tracking process by positioning the proximal portion of a dendrite under the sensor and indicating the initial direction of the dendrite with a knob. The computer scans an arc whose center is at the starting point. A gray level for each point in the arc is read, forming a waveform of the light intensity across the dendrite (Fig. 15-3, lower left). The software looks for a dip in the gray level as the arc crosses the dendrite. It first calculates the location of the dip in the waveform, then the coordinate of the dip in the tissue, and finally moves the microscope stage to this coordinate. It then repeats the process by scanning the next arc. By

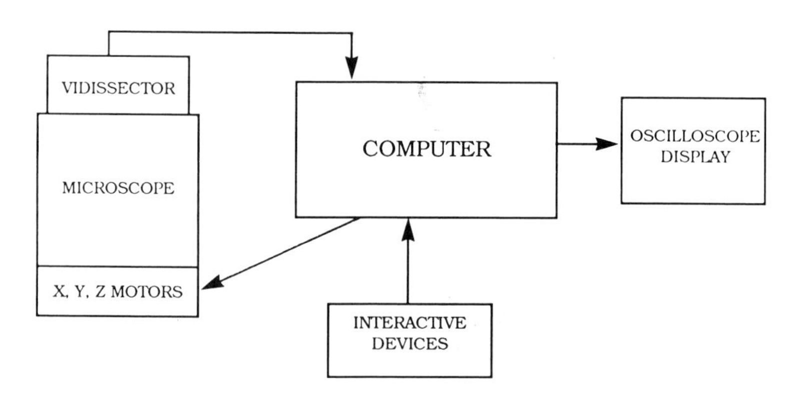

FIGURE 15-2. A simplified diagram of the hardware used in the Coleman et al. (1977) automatic neuron-tracking system. Using a vidisector (a TV-camera-like device; see text) the computer may sense the light level at selected points in the tissue mounted in the microscope. The computer program then repeatedly finds the next point along a dendrite and drives the stage and focus to that point. An oscilloscope is used to display waveforms extracted from scans across the dendrite and to display the traced neuron. Interactive devices are used to initiate the program, to start tracing, and to guide the software through tracking errors. Modified from Coleman et al. (1977) with permission.

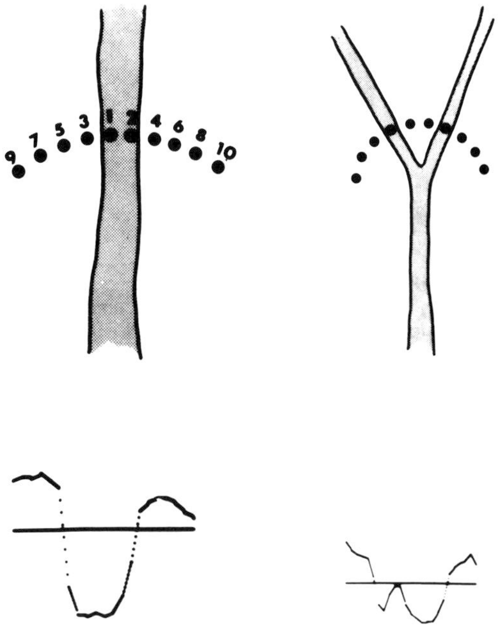

FIGURE 15-3. Scanning across a dendrite and a branch point with the Coleman et al. (1977) automatic neuron-tracing system. Upper left: The position of the points whose gray level is sampled across the dendrite. Lower left: The gray-level response of the scan. The valley in the response indicates that there is indeed a dendrite. Upper right: The sampled points at a branch point. Lower right: The gray-level response of this scan. The double valley in the response indicates that there is a branch point. Modified from Coleman et al. (1977) with permission.

repeatedly scanning an arc, sensing the dip, and moving to the dip's location, the software follows the dendrite, accumulating a list of coordinates in the computer memory that forms a vector representation of the dendrite.

When the instrument approaches a branch point, a double dip appears in the returned waveform (Fig. 15-3, lower right). The tracing software records a branch point and then arbitrarily directs the tracing down the left-hand path exiting from the branch point. When it reaches an ending, there is no dip in the response, so the software determines that, indeed, this is the end of a dendrite. Then it moves back to the branch point and traces the other fiber exiting from it.

Occasionally the system gets lost because it is unable to sense a dendrite, branch point, or ending. At this time it sounds an alarm to ask for human help to reorient it along a dendrite. One hopes that the system does not get confused so often that restarting it is more trouble than tracing the dendrites semiautomatically.

How well did this first automatic system work? Only poorly, because it examined only a small region of the image, ignoring most of the view. Although some scientific research was published using it, it was sufficiently frustrating that the laboratory abandoned it in favor of the semiautomatic tracing system described by Moyer et al. (1985).

15.4.2. Automatic Neuron Tracing with a Television Camera

The second automatic neuron-tracing system was reported by Capowski (1983a) and is illustrated in Figs. 15-4 and 15-5. He used a television digitizing system rather than an

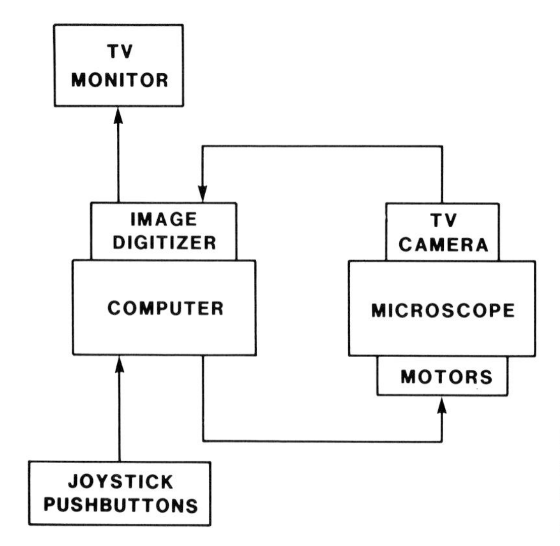

FIGURE 15-4. Block diagram of hardware in the UNC automatic neuron-tracing system. Details are given in the text. Reproduced from Capowski (1983a) with permission.

image dissector. The digitizing system captures an entire video image into a pixel memory in 1/30 sec, so that the computer can read the gray value of a pixel at any X,Y coordinate. The automatic tracking software in this system operates like the submarine in the movie *Fantastic Voyage*. Only in this case, the automatic tracker swims through the dendrites instead of through the blood vessels, always looking ahead for a target at which to aim.

Determining where to find the next point is the most difficult task for the software in the Capowski system. The task is illustrated in Fig. 15-5. As with the Garvey–Coleman system, a human operator focuses on the starting point of the dendrite, thus providing an initial X,Y,Z coordinate for the dendrite. He also provides an initial direction. Unlike the Garvey–Coleman software, however, the Capowski software then digitizes 13 images at different Z or focus values, six above, six below, and one at the current Z value. Within each image, several arcs (Fig. 15-5, top) centered about the current X,Y coordinate are generated, and the gray values of each pixel in each arc are read by the computer. In each focus plane, pixels from each arc that are in the same radial direction from the current coordinate are summed, and the computer stores the gray value in each radial direction. This is like a bicycle wheel rotated into the horizontal plane and placed so that its hub is at the current coordinate on the dendrite. The computer may look down each radial spoke from the hub and read the gray value in that direction.

The same scan of the arcs is performed at each of the 13 focal planes, and the gray levels are stored in the computer memory. This results in gray levels representing pixels on a half-cylinder directly ahead of the meandering dendrite. They form a two-dimensional image (Fig. 15-5, top inset) of the cross section of the dendrite that is ahead. In the movie analogy, they form a "target" at which to aim the swimming submarine.

The next step is to find the "bulls-eye" in this target at which to aim the dendrite tracker. After a series of image enhancements, the center and the width of the dendrite are calculated. The target point is displayed between two short black lines (Fig. 15-5, top inset) and the lengths of the lines show the thickness of the dendrite at the target point. The X,Y,Z coordinate in the tissue of this target point is then calculated, the microscope stage is moved to the point, and the coordinate is stored in the computer memory.

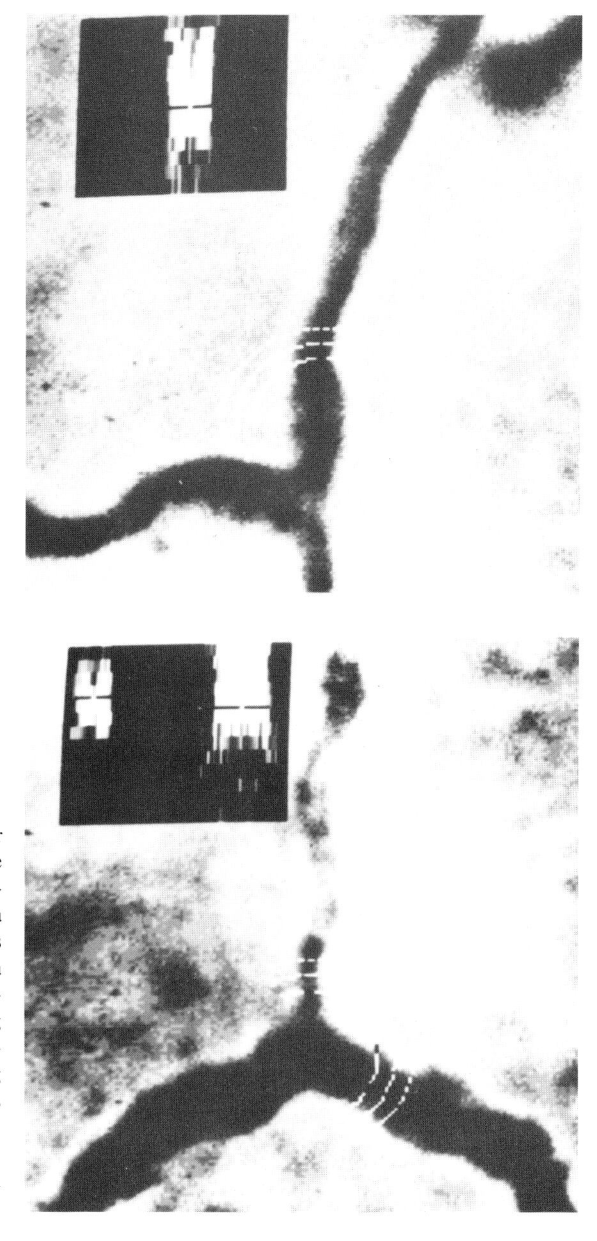

FIGURE 15-5. The television monitor image while tracing a dendrite using the Capowski (1983a) automatic neuron-tracing system. Top: From one point on a dendrite, semicircular scans at various focal planes are made. These result in a target (insert) to which to move the microscope stage for the next point along the dendrite. Bottom: When the semicircular scans cross two dendrites existing from a branch point, a double target is generated, indicating two branches, which must then be traced separately. Modified from Capowski (1983a) with permission.

When the computer looks ahead along the dendrite, if it sees two targets (Fig. 15-5, bottom inset), then it assumes a branch point. The software arbitrarily selects the left target and traces the dendrite that exits the branch point to the left. When no target can be found, the computer assumes an ending, and the dendrite is terminated. If there is a branch point from which both exiting paths have not yet been traced, the computer moves the microscope stage back to this point and begins tracing the other exiting fiber.

As shown in Fig. 15-1, this system worked well when following thick dendrites: its success rate for recognizing branch points and endings was about 67%. However, it was hopeless at recognizing such fine details as spines and swellings, and it could not follow very fine fibers. It was abandoned in 1984 because, although it formed a good computer science project, it did not promise reliable biological results for at least a decade.

15.5. FOR FUTURE WORK

In face of the difficulties described in Section 15.4, should further attempts be made to automate the tracing of structures? Yes. Progress can be made by taking advantage of recent developments in computer hardware, image enhancement software, and microscope hardware. The next sections outline some possible directions for further research in the problem.

15.5.1. Bigger and Faster Computers

Both of the automatic neuron-tracing systems just described were designed when a typical laboratory computer had about 64,000 bytes of memory. Most image-digitizing systems (Section 2.4.15b) capture images at a resolution of about 500 × 500 pixels and save a pixel in one byte of memory. Therefore a quarter megabyte of memory is required to store an image. Because of this, the tissue itself was used as the storage medium, and the gray-level data required for automatic tracing were repeatedly extracted from the tissue.

With contemporary computers, it might be possible to design an automatic tracing system using a different concept, that is, to digitize the entire slab of tissue one time. Then the data would be stored as a large number of three-dimensional pixels called "voxels" (for volume elements). All calculations would be done from the computer-stored data, and the tissue would never again be interrogated. Figure 15-6 presents this idea. A storage capacity of about 625 megabytes is needed for a typical neuron-tracing task.

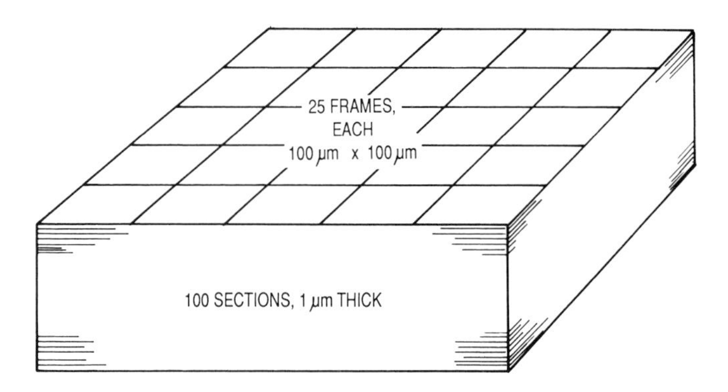

FIGURE 15-6. The enormous quantity of data in a digitized tissue slab. A slab of tissue 500 x 500 x 100 µm is divided into 2500 frames, each of size 100 µm × 100 µm. Each frame is digitized into 500 × 500 pixels of one byte each, approximately the resolution required to track dendrites automatically. This results in a total data storage of 625 megabytes.

Such a storage capacity is no longer unrealistic for a laboratory. One can purchase a gigabyte (one billion bytes) disk storage system for a laboratory minicomputer and use it to store the voxel representation of the tissue. Then the automatic tracing software must find the dendrite as it meanders throughout the voxels stored in the computer. Access to the data in this disk storage system would be hundreds of times faster than access to the original tissue and would be truly three-dimensional. Previously, moving from one pixel to another within a single digitized frame was easy; another pixel was read from the memory. But access to another pixel along the focus axis required moving a microscope focus motor and capturing a new image. With computer-stored data, moving throughout the image in any direction would be equally easy. The voxel representation of the tissue slab would allow the application of some image-enhancement techniques that previously would have been too time consuming.

15.5.2. Better Image-Enhancement Techniques

If the tissue slab is stored in a digital memory, then some image-enhancement techniques that previously have been too time consuming to apply become reasonable. These techniques are described in some detail in Chapter 13, so they are only briefly mentioned here.

The task of following a dendritic process is really the task of finding a structure that has some known parameters. You know, for example, that a neuron tree is indeed treelike with thin and repeatedly branching fibers. This is in sharp contrast to a circular structure, such as a red blood cell. You may make educated guesses about the thickness, color, and gray levels of the dendritic processes and about the color and gray level of the tissue that houses the neuron. A spatial filter may be applied to an image to enhance structures of a certain size. If you can guess at the thickness of the dendrites, then the proper spatial filter will make these dendrites stand out relative to their background, with the trade-off that thinner or thicker dendrites will be suppressed. Different spatial filters may be applied for different-thickness dendrites, so it may be possible to enhance dendrites as they meander, even as they vary in thickness.

Edge-detection schemes may be applied to enhance the contrast between the background tissue and the meandering dendrites. Here a difference in gray level between neuronal fiber and tissue background is magnified in the hope of making it easier to find the fiber in the image.

Template matching forms another opportunity. A template is a small pixel-by-pixel picture of a typical structure, such as a branch point. Branch points may be classified into several types according to their shape. Then whenever the automatic tracking software looks for the next point, it may check to see if the point matches a template, thus providing another test on whether a branch point has really been found. Templates may also be devised for varicosities, spines, and endings.

15.5.3. The Confocal Microscope

The optics of a typical light microscope are not conducive to three-dimensional work, for they are designed to maximize the depth of focus, i.e., to integrate as much information as possible along the focus axis in every image. But to digitize accurately a series of two-dimensional slices from a slab of tissue, a minimal depth of focus for each image is needed. The confocal microscope, recently introduced into the neuroscience marketplace, provides this minimal depth of focus. In the following paragraphs, the microscope is described, and the results of a brief experiment to show its promise for automatic neuron tracing are given.

A confocal laser scanning microscope, shown in Fig. 15-7, consists of a research light microscope, a laser for illuminating the specimen, a photomultiplier tube for sensing the light reflected from (or possibly transmitted through) the specimen, and an optical path to direct the laser and its reflections. The microscope is attached to an image-processing system for gathering, manipulating, and displaying the image.

Figure 15-8 shows a diagram of the light path in a confocal microscope. The laser beam is scanned across the specimen in a raster pattern as the computer rotates the two mirrors. Light that is reflected from the specimen is directed back through a small aperture to the photomultiplier tube, where it is quantified for the computer. The computer accumulates the image, pixel by pixel, requiring about 2 sec to capture a 512 × 512-pixel image. The signal-to-noise ratio of the image may be improved by having the photomultiplier tube count photons for a longer period per pixel. This, of course, lengthens

FIGURE 15-7. A confocal microscope. The box above the research light microscope contains the laser for illuminating the microscope's specimen. The laser beam is passed through the top port of the microscope and reflects from the specimen through the same port into the box where a photomultiplier tube senses it. A fiberoptic tube also delivers the beam below the stage so transmitted light may also be used to illuminate the specimen. A video digitizing board inside the computer system (lower center) captures the image and displays it on the right-hand color TV monitor. The left color TV monitor is the computer terminal. For hard copy, a 35-mm camera in the lower right may take a picture of a small monochrome TV monitor on which the microscope image is also presented.

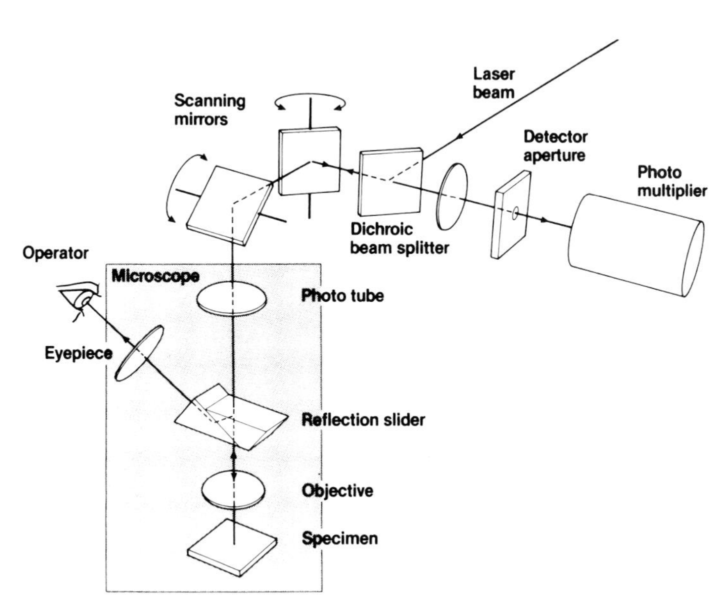

FIGURE 15-8. The beam path of a confocal microscope. The equipment outside the rectangle is located in the box above the microscope. The equipment within the rectangle is inside the light microscope. Details of the beam's flow are given in the text. Diagram courtesy of Sarastro, Inc.

the time required to capture an image. Image-processing functions are performed on the captured image, and the result is displayed on the TV monitor.

Figure 15-8 shows the light path in the microscope when the laser illuminates the specimen from above and senses the light reflected upwards from the specimen. If the laser illuminates the specimen from below the stage, a transmitted light image is captured as the computer rotates the mirrors to scan the image. Both of these techniques are used in the neuron-tracing discussions below. Note, however, that with transmitted light illumination, the advantage of extremely shallow depth of field does not exist.

15.5.3a. Confocal Scanning. A primary advantage of the confocal microscope for automatic neuron tracing is that an extremely shallow depth of field is provided. Figure 15-9 diagrams why this occurs. The result of confocal reflected light scanning is that any part of a structure that is in sharp focus will appear bright and any part that is even slightly out of focus will appear black.

15.5.3b. Confocal Microscopy Applied to Automatic Neuron Tracing. To illustrate the application of confocal microscopy to the automatic neuron-tracing problem, a brief experiment is presented. A 100-μm-thick tissue section containing Golgi-stained neurons was mounted in the microscope. With a 40× objective lens, two scans of the image were made, one with transmitted light and the other with reflected light. Both scans were captured into the frame buffer of the image processor and displayed with no further image processing. The scans are shown in Fig. 15-10.

The transmitted light image of Fig. 15-10 looks like a typical video digitized image

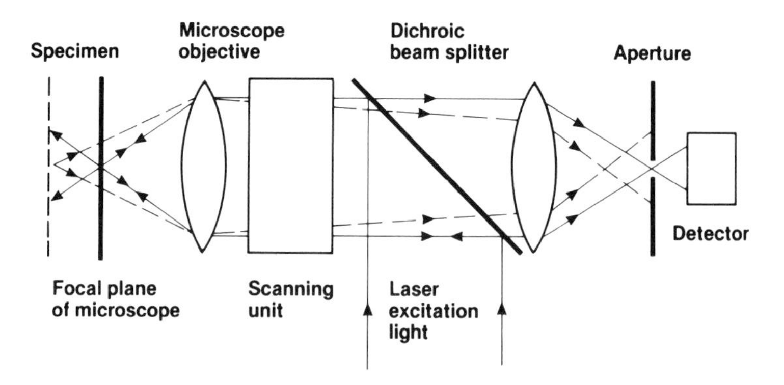

FIGURE 15-9. The confocal scanning principle. Because a laser beam is monochromatic, the microscope lenses focus it very sharply on the point in the specimen currently being scanned. Then, only light reflecting from the current spot is passed through the aperture into the photomultiplier tube. Light that is reflected from the specimen at a different focal plane cannot pass through the aperture. Thus, the depth of field is shallow, and out-of-focus portions of the specimen appear black.

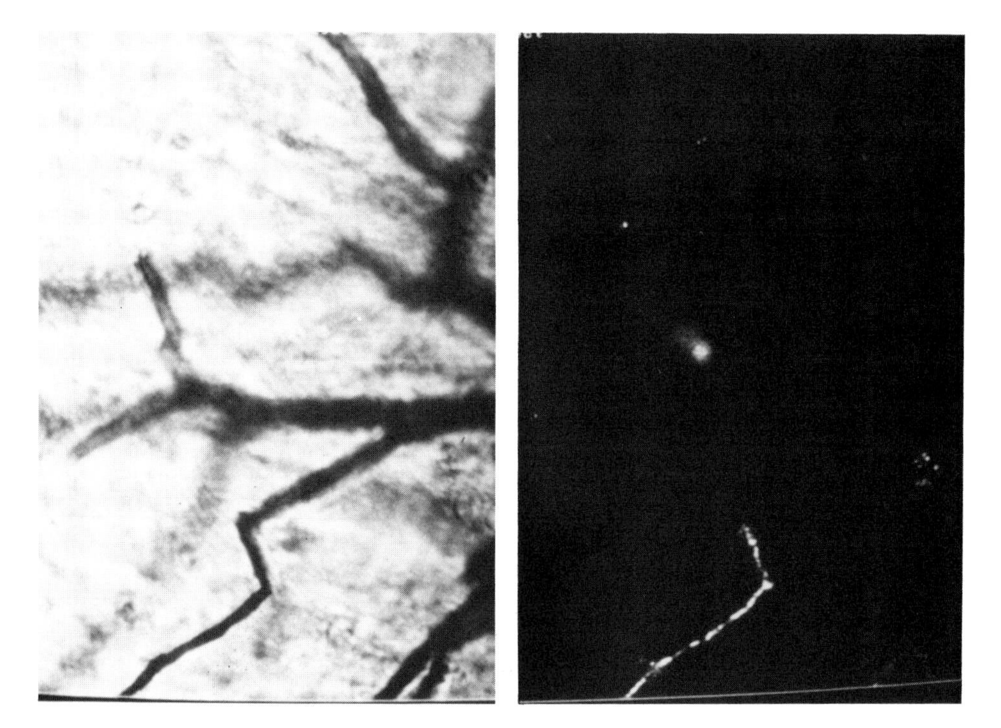

FIGURE 15-10. Images captured when scanning a Golgi-stained neuron. Left: Transmitted light was used to illuminate the specimen. Right: Reflected light was used. Further details are given in the text. The large white dot in the center of the reflected-light image is an artifact probably caused by poor illumination. Both the quality and contrast of this photograph are much poorer than they appear on the face of the CRT.

of a neuron. Note that only one part of a dendrite (in the lower center of the image) is in sharp focus, and it requires a modest human visual effort to determine this. The reflected light image of Fig. 15-10 is the exciting one. Only the portion of the dendrite that is in sharp focus is brightly illuminated, and the rest of the dendrites are all black. Furthermore, it is easy for us or for a computer to determine what is in focus. The dendritic portion that is in focus is highly visible and has high contrast to its black background.

Although the reflected light image provides superb data for a computer program, it is not a good image for a human because it does not contain the in-focus dendritic portion located in its environment. This is analogous to driving a car down the highway and being somehow restricted to view only an automobile that is leading you by exactly 100 meters. What is desired is to highlight that leading automobile so that you may see it clearly and at the same time maintain the ability to scan the environment for other objects as you desire. In the biological problem, by combining the transmitted and reflected light images, it is possible to provide both capabilities, as shown in Fig. 15-11.

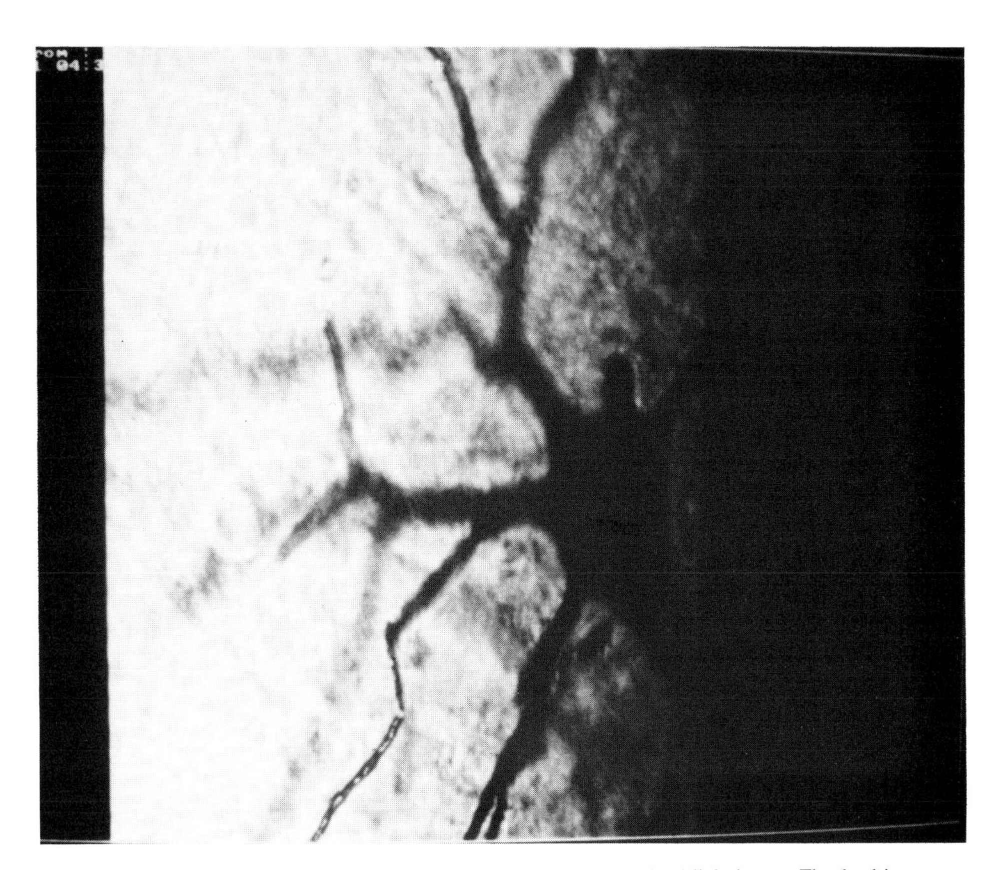

FIGURE 15-11. The reflected-light image used to enhance the transmitted-light image. The dendrites meander in and out of focus, and the sharply focused portion of the dendrite in the lower portion of the image is highlighted. For artifactual reasons, the image is scaled by one half along the X axis in comparison to Fig. 15-10. The highlighted portion of the dendrite is spectacular in bright red on the TV monitor; the quality of this photograph does not do it justice.

Images from the confocal microscope have two valuable and relevant characteristics: they show very high contrast between sharply focused and slightly out-of-focus portions of a dendrite, and they delete the unwanted background, rendering it black. A series of these images captured into a voxel representation of the tissue slab will yield the best opportunity for a computer algorithm to track dendrites successfully. Unfortunately, the imaging system of the confocal microscope requires about 2 sec to capture each image, limited by the time required to deflect the mirrors. As Fig. 15-6 shows, 2500 images are required to digitize the tissue slab containing the neuron, so initially, digitizing a series of images will demand about 1.5 hr.

15.6. CONCLUSION

Even with a confocal microscope to present images to a fast laboratory computer with a large memory, I am not sure whether an automatic neuron-tracing instrument can approach the speed of a well-trained technician using a well-designed semiautomatic neuron-tracing system. Nor am I certain about the feasibility of recognizing features such as swellings and spines or tracking the finest, most distal fibers.

I must conclude that with the current state of the art, you should regard automated neuron tracing as computer science research with only long-range benefits in neuroscience. This is not to say that developing automated systems should not be continued; rather, morphological studies that depend on automatic neuron tracers should not be designed. Similar applications, such as tracking coronary arteries and river basins as well as robot vision and recognition of manufactured parts, will benefit from this research, as neuroscience will benefit from theirs.

15.7. FOR FURTHER READING

Further details about the abilities and limitations of automatic video-based neuron-tracing systems may be found in Garvey et al. (1972, 1973), Reddy et al. (1973), Llinas and Hillman (1975), Hillman (1976), Hillman et al. (1977), Capowski (1983a), and Reuman and Capowski (1984).

A description of an automatic neuron-tracing system that uses an image dissector instead of a TV camera as the sensing device may be found in Coleman et al. (1977). Use of this system in the reconstruction of neurons has been described by Buell and Coleman (1979) and Hogan and Coleman (1981).

A description of the operation of the confocal microscope and its application to visualizing neuronal structure is available in Wallen et al. (1988).

16

Commercially Available Computer Systems for Neuroanatomy

16.1. INTRODUCTION

When a neuroanatomist surveys the market to purchase a computer system for collecting and analyzing laboratory data, he faces a dazzling array of hardware and software. One of the functions of this book is to help him communicate intelligently with the people who sell these systems.

This chapter forms yet another dazzling array, a kind of shopping mall for computer-assisted neuroanatomy. Altogether, I asked 41 companies that manufacture computer products for neuroanatomy to provide a one-page description of their products, including prices if they desired. Some companies make products (e.g., neuronal morphometry systems) that are primarily aimed at the neuroanatomy marketplace. I invited all of these companies of which I am aware to participate. Other companies make general products (e.g., television cameras) that are used in neuroanatomy as well as in other fields. Of these companies, I invited the ones who described neuroanatomy-related products in the program of the 1986 or 1987 annual meeting of the U.S. Society for Neuroscience.

After some coaxing, most of the invited companies submitted product descriptions. Their replies are presented below in alphabetical order. I have not edited their descriptions except occasionally to make them fit into the equal space that was allocated for each company. Please note that a company's presence here is not an endorsement for or against its products by any of the authors of this book.

16.2. COMPANIES AND THEIR PRODUCTS

16.2.1. American Innovision, Inc.

7750 Dagget Street, Suite 210
San Diego, CA 92111
(619) 560-9355

The American Innovision Videometric 150 image-analyzing system is designed specifically for neuroscience applications (Fig. 16-1). The image analyzer features true color

FIGURE 16-1. American Innovision Videometric 150 image-analyzing system.

thresholding that provides special image recognition and measurement that is only available on the V150 system. True color makes it simple to use multicolored images and still obtain selective measurements. Full image enhancements such as pseudocolor and remapping are built in for applications that require image processing.

Instant calculations are delivered for basic planimetry and morphometry measurements such as areas, perimeters, line lengths, ferets, shapes, points, slopes, and even centroids. Special measurements are available for grain counting and optical density.

The system is based on a hardware design to provide fast and accurate results. In addition, the V150 is open and flexible to provide answers for your specific needs. Every system has a powerful macro language that allows your experiment to be programmed into the V150. It is just like having a dedicated instrument designed for a specific experiment.

Several options are available for the system. **Automatic Feature Measurement** yields specific and accurate measurements for each object in the image that matches a model taught to the system. **3-D Rotation,** developed especially for neuroscience applications, provides solidlike modeling with rotation and reconstruction. **Autoradiography:** real-time analysis of glucose and other radio-tagged substances. **MRI:** a routine for off-site measurements and analysis of MRI imagery.

16.2.2. Analytical Imaging Concepts, Inc.

12 Woodfall
Irvine, CA 92714
(714) 857-1269

The IMAGE 3000 is a real-time image analysis system for use with an IBM AT (or compatible) microcomputer. The system includes image-processing hardware (four 512 ×

512 × 8-bit frame buffers, 60-HZ noninterlaced display, and on-board ALU), advanced morphometric and densitometric analysis software, and an optical mouse device.

Morphometric and densitometric image analysis may be performed in real time on a live video image. Object segmentation is by the full 256 gray levels with no bits being sacrificed for menus or overlays. Up to 15 different parameters may be measured automatically with data sent to a standard ASCII file for analysis. Upper and/or lower size limits may be entered for all of the measurement parameters, and logical equations may be written between parameters. The densitometric software allows the user to calibrate gray values to known densitometric references. Once the references are entered, a "best" curve fit is determined, and data are displayed in appropriate calibrated density units. A register function allows two images to be overlaid and subtracted.

Options include 3-D reconstruction, an OMDR interface, and complete real-time image-processing software developed by Universal Imaging Corporation.

16.2.3. Axon Instruments, Inc.

1429 Rollins Road
Burlingame, CA 94010
(415) 340-9988

16.2.3a. Acquisition Hardware. Axon Instruments has two different series of data acquisition hardware that operate with IBM personal computers and compatibles. The TL-1 and TL-2 series perform at 12-bit resolution at speeds of up to 1125 kHz, and the AXOLAB series performs at up to 16-bit resolution and at speeds of up to 1 MHz.

16.2.3b. Acquisition and Analysis Software. pCLAMP is a suite of programs for data acquisition and analysis. It is particularly well suited to investigations using microelectrodes, for current clamping, voltage clamping, and patch clamping. Data can be recorded in bursts with simultaneous generation of complex command waveforms. Alternatively, data can be recorded at high speed direct to a file on disk with the file size being limited only by the disk capacity. Analysis routines include histogramming, curve fitting, plotting, arithmetic on waveforms, determination of I–V curves and other X–Y parameter curves, and specialized analysis procedures for characterizing single-channel patch-clamp currents. pCLAMP supports the TL-1, TL-2, and AXOLAB types of acquisition hardware.

AXOTAPE is a program for continuous data acquisition to disk. The menu selections are as simple to use as those on a tape recorder. Record, replay, fast forward, rewind, and other functions are supported. Data are displayed in real time on standard monitors at acquisition speeds of several kilohertz, and if the display is switched off, the acquisition can proceed at speeds in excess of 30 kHz directly to disk (all times for an 8-MHz AT). Data can be recorded directly to disk up to the full capacity of the disk. When used with optical disk drives, hundreds of megabytes can be recorded nonstop.

16.2.3c. Educational Software. AXOVACS is a simulation program that graphically models the Hodgkin–Huxley equations for single or multiple channels. It is laid out as a series of lessons designed for undergraduate and graduate students. It may also be used by active researchers for modeling purposes.

16.2.3d. Third-Party Software. LPROC is a data-reduction program that operates on patch-clamp data to find all the events that meet specified amplitude and kinetic criteria. The user can create lists, define bursts, generate histograms, plot subsets of the data, signal-average events, compute spectra, and perform correlation analysis. LPROC was written by Dr. Tony Auerbach and Mr. Zhigang Xiang of the State University of New York at Buffalo.

C-SIM is a simulation program that models ionic channels and simulates the channel currents. Each channel is modeled by a kinetic model that contains up to 12 states. The transition rate constants are specified in a square matrix, so that linear, cyclic, and other models can be created. Open-channel and closed-channel fluctuations (noise) can be easily simulated by adding sufficient states to the model and setting appropriate rate constants. In addition, white noise can be added to the synthesized data. To model the responses of voltage-activated channels, the set of matrices defining the transition rate constants can be instantly changed to a new set. To model macroscopic currents, up to 100 instances of up to ten different or identical types of channels can be simultaneously simulated. C-SIM was written by Dr. Fred Sachs of the State University of New York at Buffalo.

16.2.4. Biographics, Inc.

1950 Stemmons Freeway, Suite 5001
Dallas, TX 75207
(214) 746-5041

The CARP (Computer Aided Reconstruction Package) neuroanatomy workstation provides a powerful tool for anatomists by combining image analysis with two- and three-dimensional graphic display capabilities. CARP's hierarchical data base is specifically designed for the storage and manipulation of anatomic information. With CARP, researchers may analyze photographs, projected images, video input, light microscope images, and autoradiographs in both two and three dimensions. Additional features of CARP include two- and three-dimensional quantitation and display (Fig. 16-2 and Fig. 16-3), digital densitometry, manual data entry using a graphics tablet and cursor, semi-automated entry of vectorized borders around regions from digitized images, an extensive highlight menu system for ease of use, a command language similar to both C and LISP for user programming, and a "notebook" feature for storing text with graphic information.

With the optional purchase of a microscope stepper motor stage with a *Z*-axis controller, data entry capabilities in three-dimensions from a light microscope are provided, including a cell-finding feature. CARP is available in several hardware configurations, including versions for 80386 microcomputers as well as both Sun-3 and Sun-4 workstations. Other hardware provided with each package includes a video camera, light box, high-resolution color monitor, graphics tablet, a Post-Script laser printer, a Matrox MVP-AT graphics board for the 80386 version, and both the Sun TAAC-1 applications accelerator card and the Matrox MVP-VME card for the Sun versions. Future releases of CARP will include volume-imaging capabilities for three-dimensional volume data such as CT, SPECT, PET, and MRI.

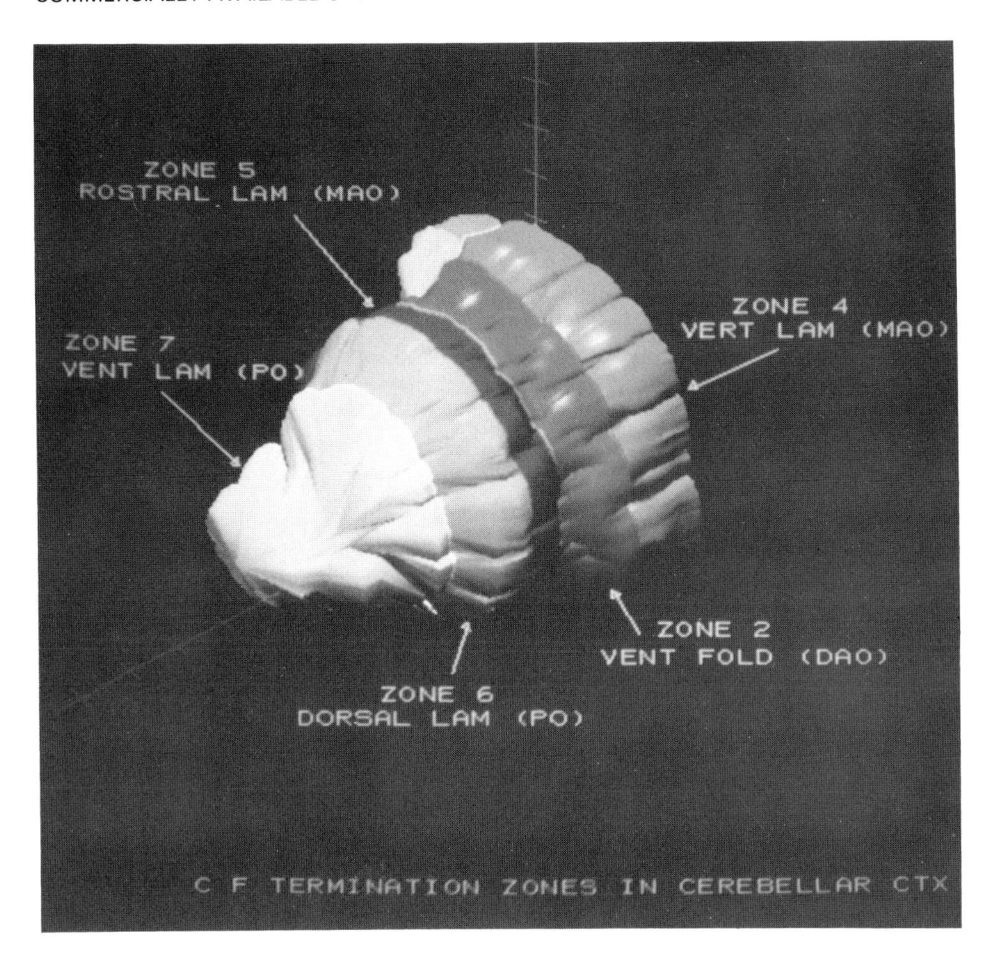

FIGURE 16-2. Biographics: Olivary projection zones onto rat cerebellum. Courtesy of Dr. Ausim Azizi, U. T. Southwestern Medical Center.

16.2.5. Bio Image—A Kodak Company

1460 Eisenhower Place
Ann Arbor, MI 48108
(800) 624-5394 (toll free)
(313) 971-7500 (in Michigan)

Bio Image supplies the life science market with a full line of machine vision image-analysis instruments for use in a wide range of applications. The Visage™ instrument family includes the Visage 2000™(Fig. 16-4), the Visage 110™, and the Visage BT™ (Fig. 16-5). The instruments are turnkey image-analysis systems featuring high-resolution image-acquisition devices, high-performance central processing units, and applications software. Visage instruments are designed to provide flexibility in accommodating the

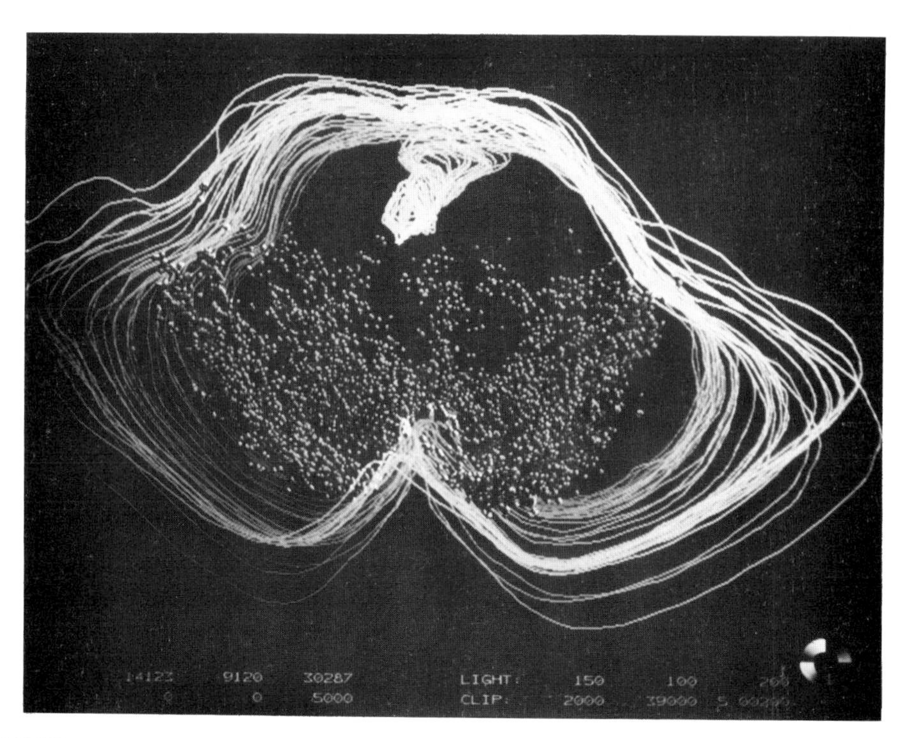

FIGURE 16-3. Biographics: Dopamine neurons in the human brainstem. Courtesy of Dr. Dwight German, U. T. Southwestern Medical Center.

user's application requirements. Capabilities include automated analysis of one- and two-dimensional electrophoretic separations that enable the user to automatically identify and quantify spots, bands, and blots; to make comparisons between images or between lanes within an image; and to create data bases for this information. The instruments accommodate wet and dry gels, autoradiograms, nitrocellulose blots, microtiter plates, TLC plates, and microscopy applications.

This technology provides a tool for studies such as the detection of protein markers for disorders and disease; pharmacology- and toxicology-related research involving protein expression and activity; studies relating to genetic expression; and research involving electrophoretic analysis of cells, tissues, and organisms.

16.2.6. Bio-Rad

Microscience Division
19 Blackstone Street
Cambridge, MA 02139
(617) 864-5820, (800) 342-2043

The Lasersharp MRC-500 Confocal Fluorescence Imaging System (Fig. 16-6) is an advanced instrument for biomedical research. The Bio-Rad Lasersharp MRC-500 is an

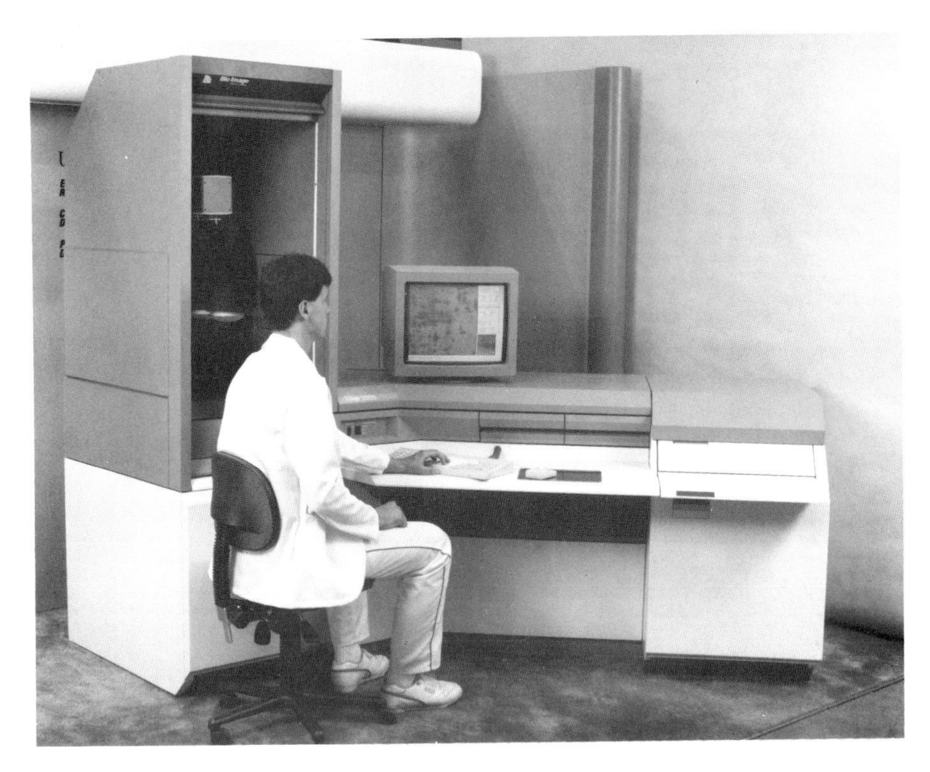

FIGURE 16-4. Bio Image: High-Speed Machine Vision Image Analysis Instrument. The Bio Image Visage 2000 system features a high-resolution 1152 × 900-pixel image monitor, a central processor with a 25-MHz clock speed enabling 4 million instructions per second (MIPS), and 8 MB of memory with data storage handled by a 260-MB Winchester drive.

optical system that converts the standard research microscope into a laser scanning confocal microscope. Designed to meet the particular needs of biomedical scientists, the MRC-500 creates an imaging system with unsurpassed contrast in reflected light or fluorescence mode.

Laser: A 25-mW air-cooled argon ion laser running in multiline mode; strongest lines 488 nm, 514 nm.

Scan system: Dual galvanometer mirror scanners; linear movement with position feedback control; variable scan angle for optical zoom; scan signal generated by dedicated microprocessor.

Detection system: Photomultiplier detector with S-20 photocathode; analogue or photon counting; preamplifier with manually variable black level and gain; switchable automatic black level and gain.

Host computer: PC/AT compatible with 80286 processor; 80287 math coprocessor (option); 80386 processor and 80387 math coprocessor (option); Microsoft MS-DOS 3.2 operating environment.

Frame store/image processor: A 768 × 512 pixel image store; images digitized to

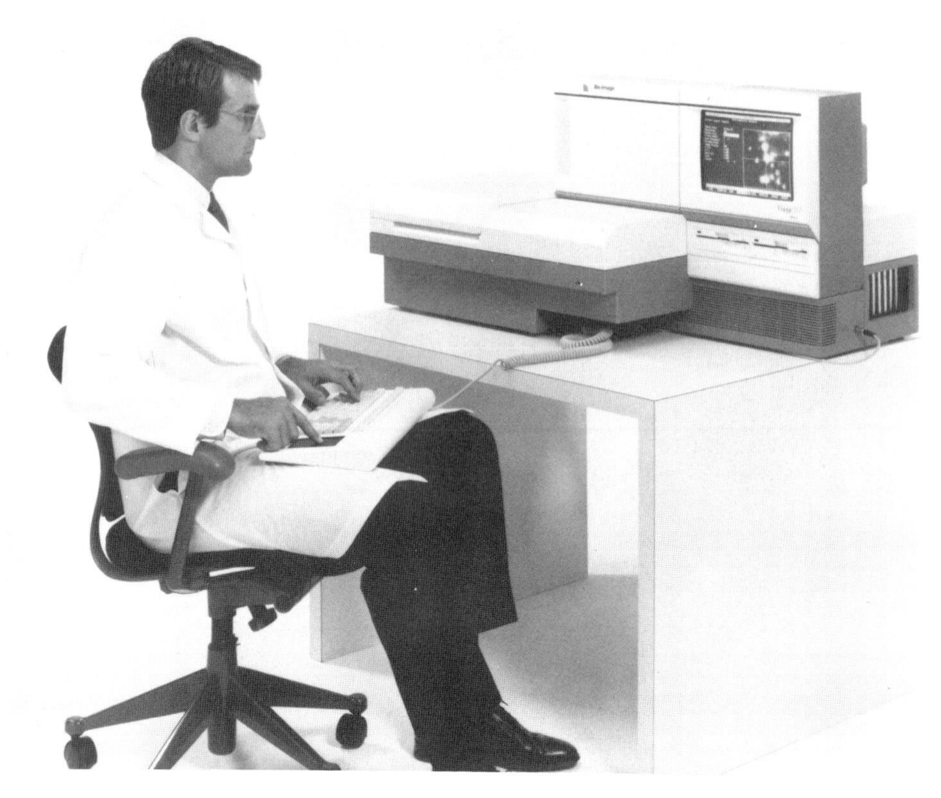

FIGURE 16-5. Bio Image: Visage BT performs fast, accurate electrophoresis pattern analysis. Bio Image Visage BT system performs one- and two-dimensional electrophoretic pattern analysis. The system is IBM PC/AT compatible and features a high-resolution monitor with pixel resolution selectable at 50 or 100 μm.

8 bits—accumulated in 16-bit words; RGB video output; three output lookup tables for pseudocolor display.

Software: Control software written in C programming language; menus for image acquisition, archiving, contrast enhancement, display subwindows, and motorized focus control (option).

16.2.7. Burleigh Instruments, Inc.

Burleigh Park
Fishers, NY 14453

The Burleigh Inchworm Micropositioning System (Figs. 16-7 and 16-8). To accommodate new techniques in the rapidly expanding fields of neuroanatomy, neuroscience, and molecular biology, new tools are required. As these fields move toward observation of smaller features, micropositioning schemes become more critical. Requirements call for a device capable of advancing in reasonably short, high-velocity, high-accelera-

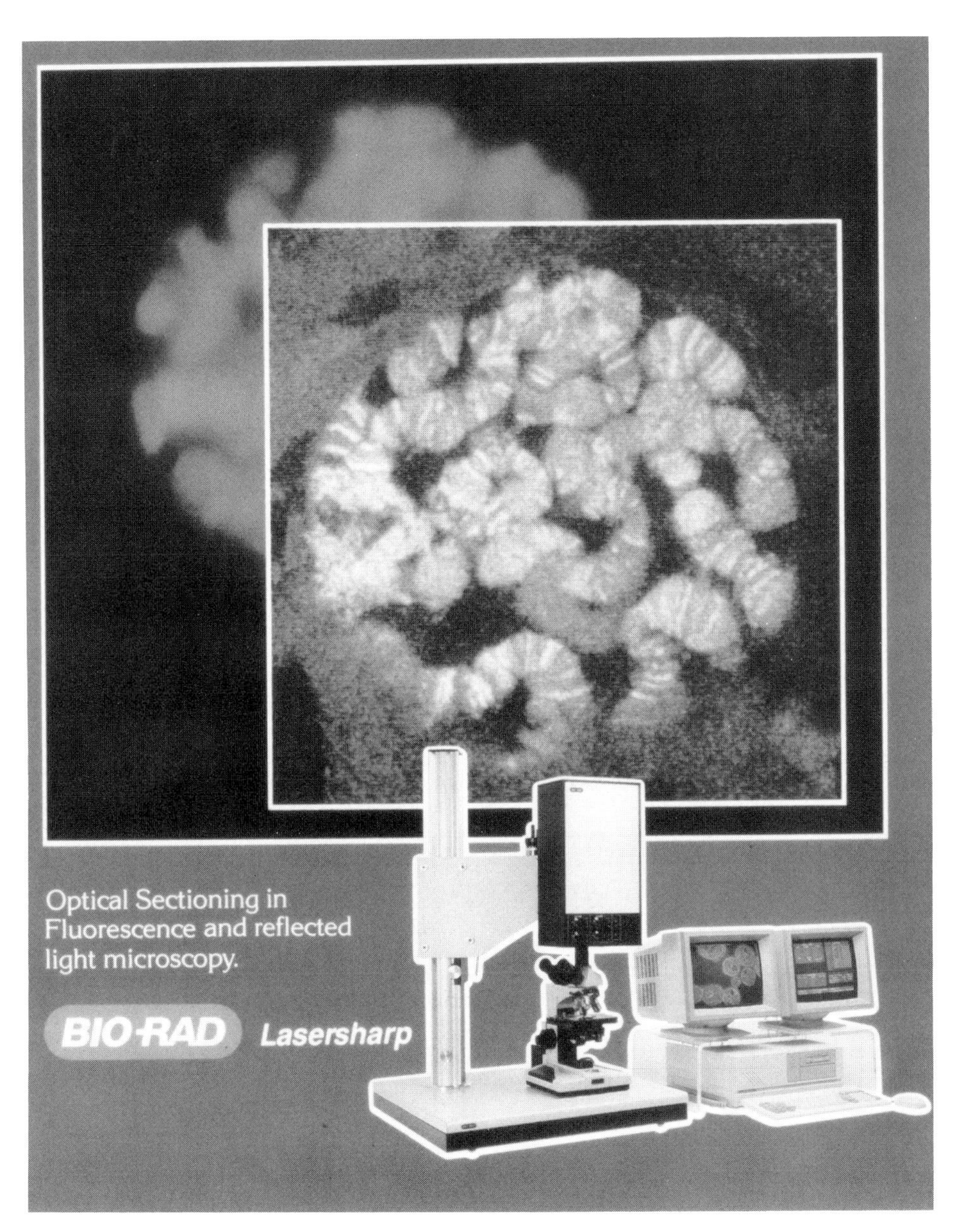

FIGURE 16-6. Bio-Rad Lasersharp MRC-500 Confocal Fluorescence Imaging System.

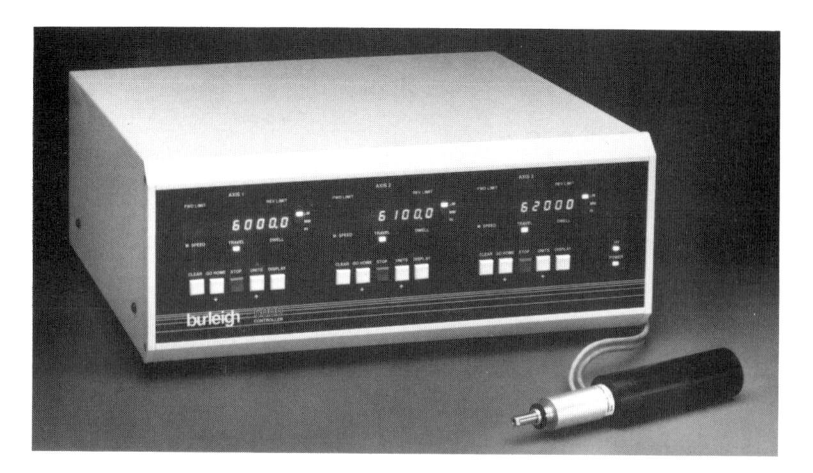

FIGURE 16-7. Burleigh Inchworm Micropositioning System.

tion/deceleration steps. Further, these requirements need to be obtainable with minimum vibration and without damage to the structures under observation.

The Burleigh Inchworm Motor provides true linear motion that is highly repeatable and has minimal backlash and mechanical drift. By patented technology, motion is created through sequential activation of three piezoelectric components. There are no rotational motions, gear boxes, hydraulic lines, or stepper-motor-related inertial problems to worry about.

The motor is capable of programmed steps down to 0.5 μm. This is achieved with an integral encoder that provides direct encoding of actual shaft motion in 0.5-μm increments. These motors have been attached to microscope stages, coarse-positioning micromanipulators, micropipettes/microelectrodes, and other probes or similar devices.

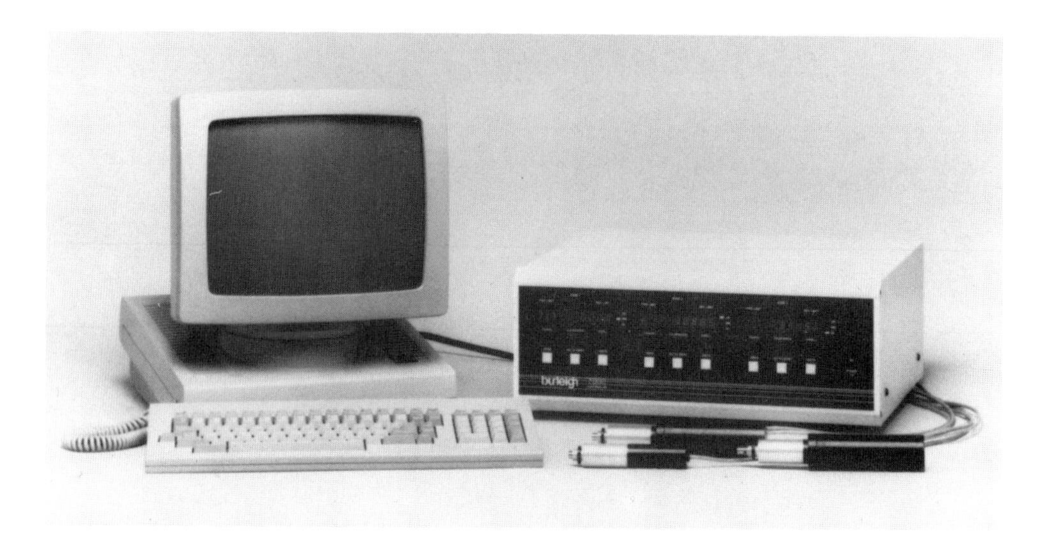

FIGURE 16-8. Burleigh Inchworm Micropositioning System.

A complete line of microprocessor-based Inchworm controllers allows the user to select an open- or closed-loop system, the level of programmability, and the method of operator interface. The user can operate these controllers either manually via joystick or handset or remotely via TTL, IEEE-488, or RS-232 commands. From the front panel of the controller the user can set step size, identify a home position, home the motor, and select the units of display (micrometers, millimeters, or inches). The handset allows the operator to select forward and reverse direction, run/jog or step mode, and motor speed.

16.2.8. Dage-MTI, Inc.

701 N. Roeske
Michigan City, IN 46360
(219) 872-5514

Dage-MTI manufactures a wide range of instrumentation cameras used in video microscopy. Our cameras are widely used as the main input detector for digital image analysis and/or image processing (Figs. 16-9 and 16-10).

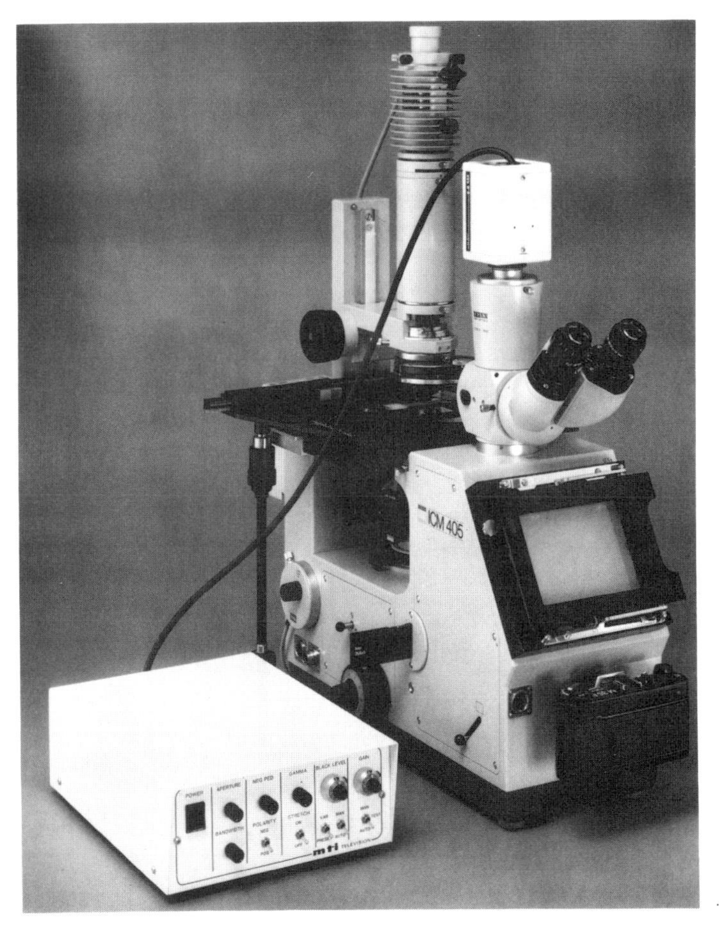

FIGURE 16-9. Dage-MTI: CCD-71 Series Camera.

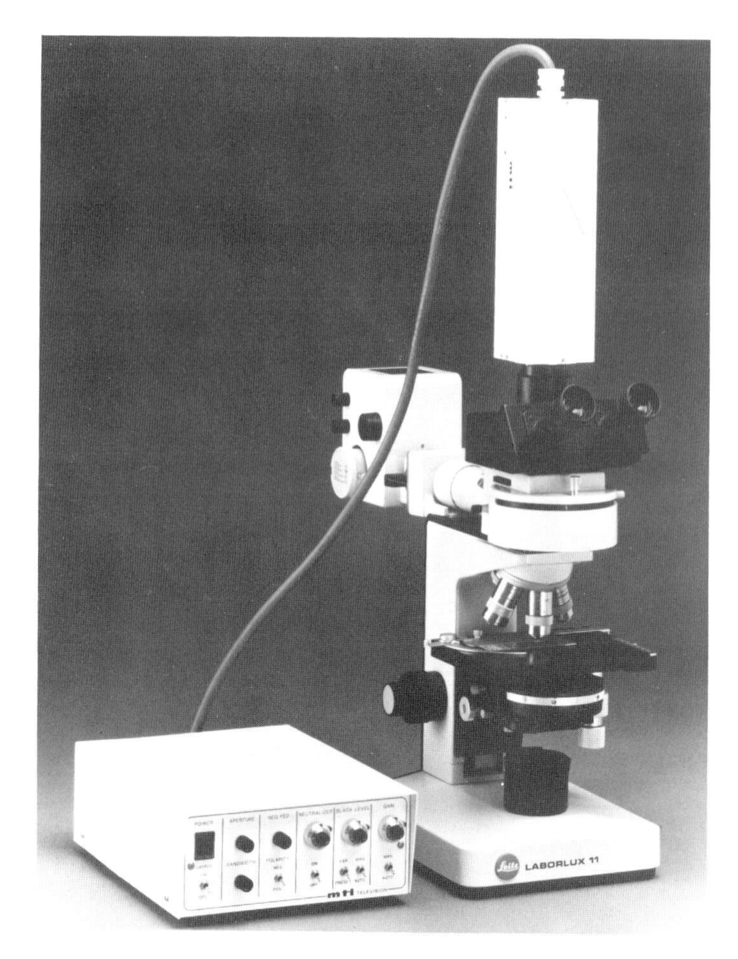

FIGURE 16-10. Dage-MTI: 70 Series Camera.

16.2.9. Dapple Systems

355 West Olive, Suite 100
Sunnyvale, CA 94086
(408) 733-3283

ImagePlus is a microcomputer-based image analyzer for the measurement and processing of images (Fig. 16-11). Automatic image analysis makes possible characterizations that would never be attempted by hand, either because it would be too time consuming to accumulate adequate statistics or because the meaningful parameters are too difficult to measure or interpret.

ImagePlus accurately measures features in images acquired from a video camera coupled to an electron or optical microscope or from a photograph. The image signal may also be the slow-scan input from an SEM. The measurements give number of features, area coverage, uniformity, relative spacing and clustering, as well as individual object

FIGURE 16-11. Dapple ImagePlus.

dimensions, location, roughness, brightness, and texture. The statistical analysis operations report mean, variance, and standard deviation. Distributions of one variable as a function of another, comparison by analysis of variance, and fitting by multiple regression facilitate presentation of the acquired data.

16.2.10. Eikonix

15 Wiggins Avenue
Bedford, MA 01730
(617) 276-7161

Eikonix, a subsidiary of Eastman Kodak Company, manufactures high-performance digital imaging camera systems. Eikonix cameras are used in a variety of application areas including medical imaging, microscopy, and neuroanatomy. Eikonix cameras provide the user with a high-resolution scanning capability that can transform electron micrographs or autoradiographs into high-resolution digital representations of neuroanatomical sections for analysis.

The Eikonix 1412 Digital Imaging Camera System (Figs. 16-12 and 16-13) is an advanced linear CCD array camera offering 4096 × 4096-pixel resolution, a dynamic range of up to 12 bits per pixel per color, and the flexibility to scan reflective and transmissive images in a broad range of formats.

16.2.11. Eutectic Electronics, Inc.

8606 Jersey Court
Raleigh, NC 27612
(919) 782-3000

Eutectic Electronics, Inc. offers two integrated software and hardware AT–MS/DOS vector graphic systems for three-dimensional reconstruction, dynamic rotation, and quantitation. Both provide an extremely high-resolution user-defined tracing mode to assure accurate and detailed structure data acquisition. Both include flexible and extensive data editing to correct for alignment and tissue variations. Both provide instant and smooth 3-D

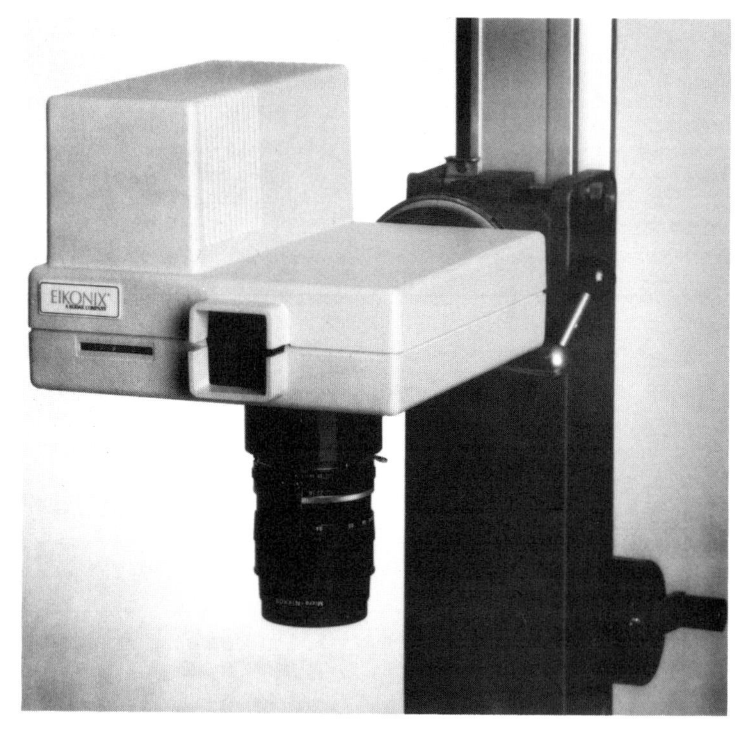

FIGURE 16-12. EIKONIX 1412 Digital Imaging Camera System. The EIKONIX 1412 Digital Imaging Camera System is an advanced linear CCD array camera offering 4096 × 4096-pixel resolution, dynamic range of up to 12 bits per pixel per color, and the flexibility to scan reflective and transmissive images in a broad range of formats.

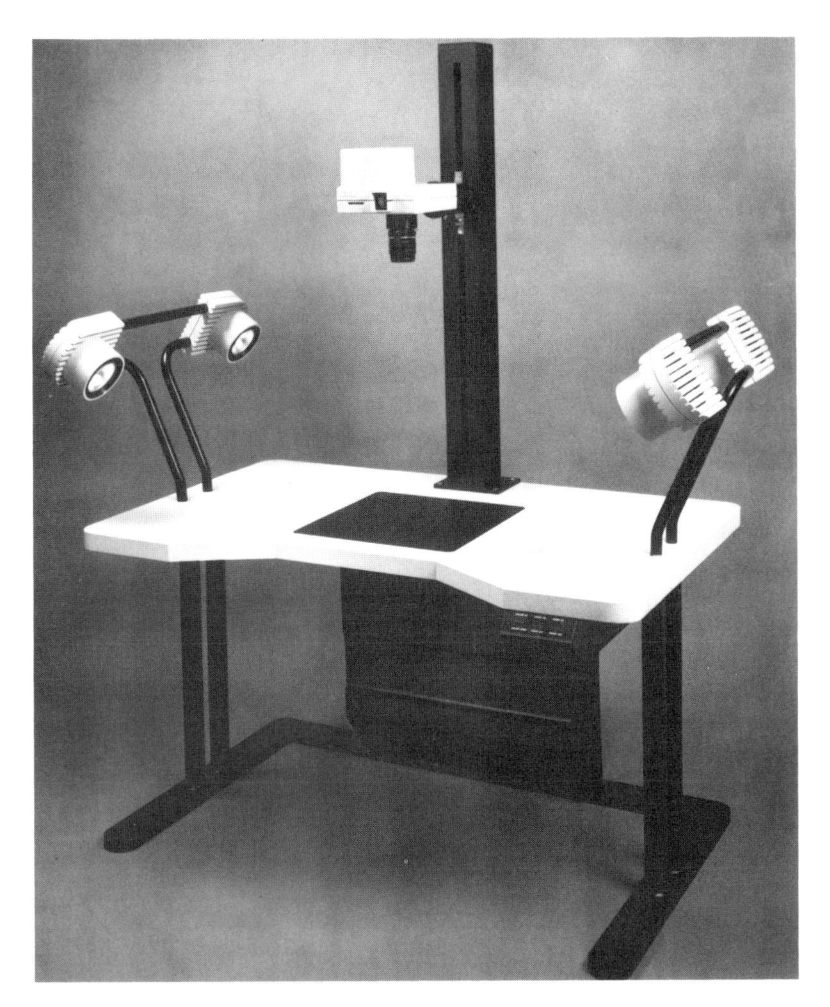

FIGURE 16-13. EIKONIX 1412 Digital Imaging Camera System and Camera Station. The EIKONIX 1412 Digital Imaging Camera System is an advanced linear CCD array camera offering 4096 × 4096-pixel resolution, dynamic range of up to 12 bits per pixel per color, and the flexibility to scan reflective and transmissive images in a broad range of formats. The optional EIKONIX Camera Station is a free-standing digitizing work surface designed for use with all EIKONIX cameras. The effective work area field is approximately 22.5 × 22.5 inches (57 × 57 cm) and offers a variety of lighting capabilities including reflective and transmissive.

rotation so that you can examine each 3-D reconstruction as if it were held in your own hand. Both provide outputs of any structure display to a multiple-color felt pen plotter.

The Neuron Tracing System (NTS) (Fig. 16-14) allows the analysis of single cells and populations. It features a high-resolution black-and-white vector graphics CRT and a motorized stage and focus. You digitize directly from the microscope using a 3-D joystick to produce an overlay to resolve 3-D position and thickness to 0.1 μm. The Serial Section Reconstruction System (SSRS) more generally applies to 3-D visualization of sequential 2-D source material. Each 2-D layer is digitized easily via a high-resolution data tablet

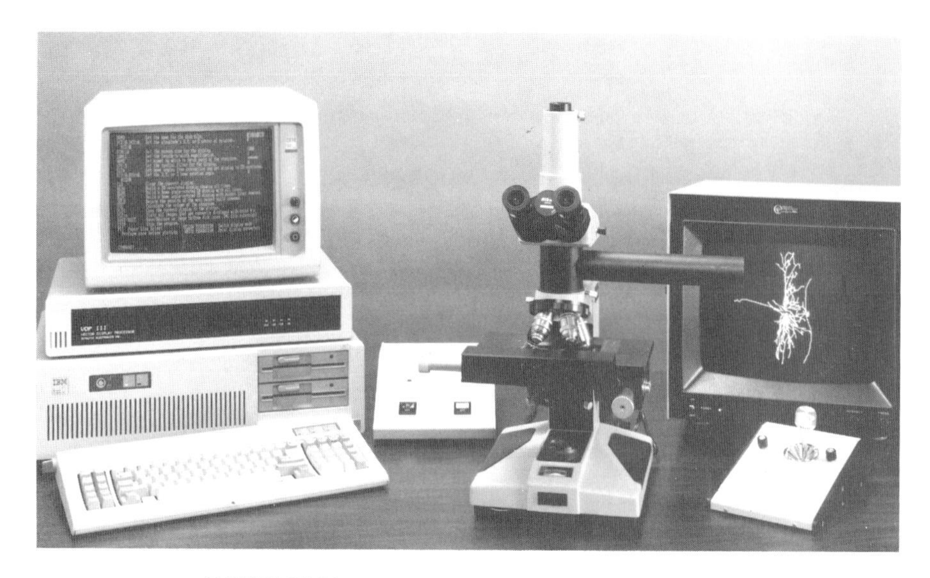

FIGURE 16-14. Eutectic Electronics Neuron-Tracing System.

and interactively displayed on a color vector graphics CRT. You can call any one or more components of the data base to the screen for 3-D rotation or 2-D hidden-line removal (Fig. 16-15). You can display and edit in the SSRS any data files that you create in the NTS. The NTS morphometery includes a wide variety of mathematical and statistical calculations so that cells are quantified based on significant classifications, e.g., trees, branches, swellings, relative distances. Analysis tables, histograms, and diagrams are output to the terminal, plotter, or printer. The SSRS analyses include volume and surface area of uniquely defined components, calculations of 2-D area, perimeter, mean diameter, and form factor for all polygons as well as 3-D linear distances and angles.

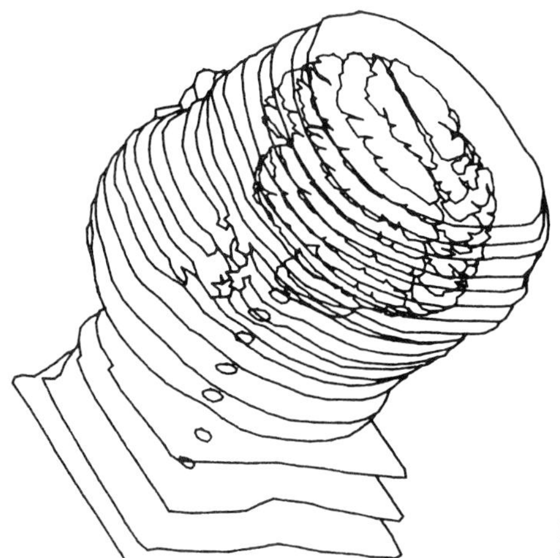

FIGURE 16-15. Eutectic Electronics: Felt-pen plot of head outline with internal brain.

16.2.12. General Imaging Corporation

P. O. Box 13455
Gainesville, FL 32604
(800) 852-9729, (904) 378-0047

General Imaging designs and manufactures a family of electronic imaging products that are fully integrated and designed to support PC-based systems. Functions provided include high-resolution image acquisition and display, image processing/analysis, optical storage, and photo-quality laser hard copy. In addition, General Imaging provides custom integration of selected imaging subsystems (manufactured by independent suppliers), contract applications engineering, and contract research and development.

GENERALMEASURE software offers image-processing (Fig. 16-16), graphics, and analysis capabilities to support the imaging systems. GENERALMEASURE modules for neuroanatomy include such programs as morphometry, densitometry and fluoremetry, and scanning, counting, and sizing.

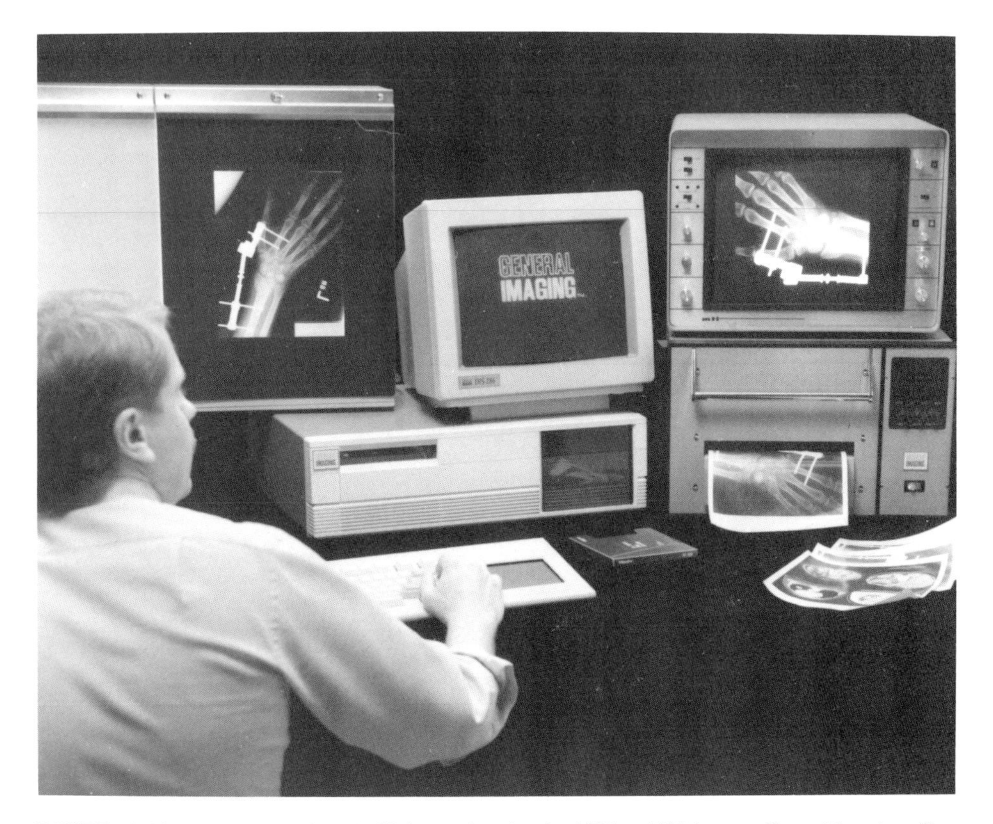

FIGURE 16-16. General Imaging: Archiving workstation for 1024 × 1024 images. General Imaging offers the DAS-1024 image archiving workstation, a turnkey, ultra-high-resolution imaging system for acquiring and archiving radiographs, documents, etc. It features an optical laser disk WORM memory that can file up to 3000 images as well as a laser printer for photo-quality images in hardcopy format.

16.2.12a. Low-Cost Imaging Systems. Standard systems, components, and software modules are relatively inexpensive. For example, the D-CAM turnkey system for converting a PC into an imaging workstation is priced at only $3995.00, including MOS-5300 solid-state camera, AVC-1 videographic display board, interface, and introductory software. Prices of high-performance imaging workstations begin at less than $14,000.00 for the DIS-286 or the portable DPS-286. Moderately priced, very-high-speed modular imaging systems, such as the DIS-386, are also in General Imaging's standard product line (Fig. 16-17).

16.2.12b. Custom Imaging Systems and Software. General Imaging also provides and supports custom systems and software for color imaging, 3-D stereoscopic imaging, 1024 × 1024 imaging, microanalysis, optical storage, and photo-quality gray-scale laser printing. Custom application engineering and programming services are also provided for single-user applications on a contract basis.

16.2.13. Image Data Systems

30 Leslie Drive
Santa Barbara, CA 93105
(805) 682-9395

A new imaging processing system tailored to the needs of the neuroresearch scientist has been developed by Image Data Systems, a Santa Barbara, CA, company. Developed in response to the biomedical research community's call for application-specific image processing, the BICS/BSAS system (Fig. 16-18) performs autoradiographic tissue slice analysis with remarkable ease, precision, and speed.

Costing less than $40,000, the system gathers information on specific chemical binding sites by digitizing the film sample and calculating appropriate statistics. Subjects under analysis are displayed in colors assigned to density levels on a high-resolution display (Fig. 16-19). Image and menu interaction is performed with the mouse and keyboard.

The imaging station, which consists of the computer, camera light table, mouse, color and monochrome displays, printer, and desk, provides a completely functional and comfortable place for gathering and analyzing data and preparing reports.

A system of overlapping menus eliminates getting lost, as can often happen when other menu techniques are used. Subjects under analysis are displayed in color on an IBM-compatible professional graphics display, and the menus are on a monochrome display. Outlining areas of interest for data gathering and point-density measurements are done with a mouse. Menu functions include the following:

1. Easy image acquisition, storage, and retrieval.
2. Effective image enhancement and annotation. Enhancement: Smoothing, edge detection, contrast stretching, three-dimensional relief in monochrome or color with various selectable resolutions. Annotation: Choose font size and character color for writing notes, comments, identification callouts, and so on.

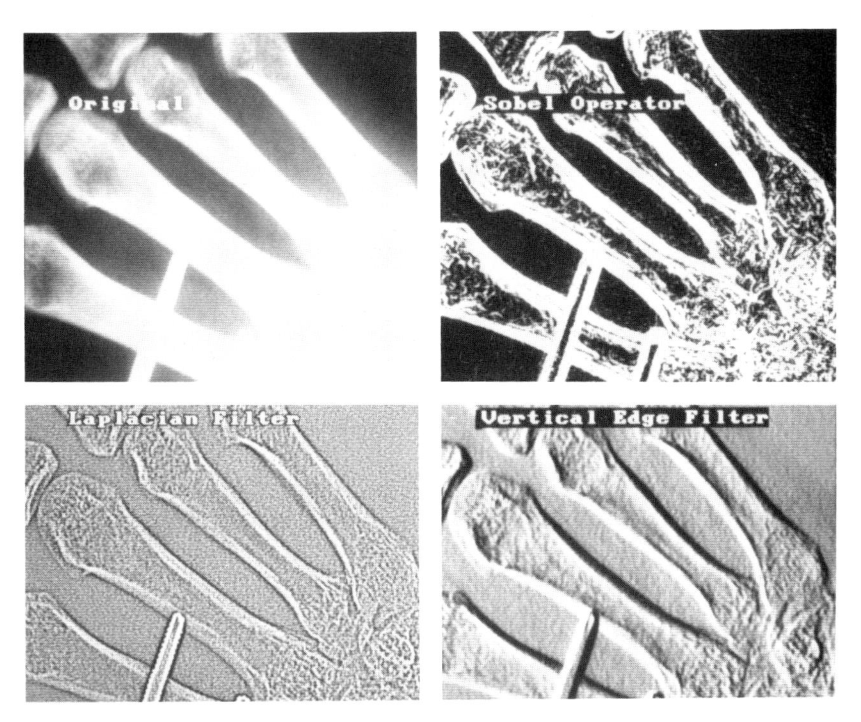

FIGURE 16-17. General Imaging: Radiograph of human hand processed by GENERALMEASURE. One of the GENERALMEASURE software modules available with General Imaging systems is GM-2300, Image Processing, which offers a powerful and comprehensive library of processing functions for neuroanatomy, such as histogram scaling, image subtraction, LUT-to-image mapping, edge filters (high pass, low pass, Roberts' edge, Sobel, etc.), and user-defined convolutions.

3. Experiment setup with titles, comments, data, and time, data base name, researcher's name, subjects, and binding sites.
4. Fast, accurate data gathering using a mouse for pointing and outlining provide area and point data in DPMs.
5. Custom formatted reports include the printing of images, header and title information, histogram, and summary or table of the statistics.
6. Data base file compatibility with SPSS, BMDP, Lotus 1-2-3, SYMPHONY, DBASE III, REFLEX, and others for statistical comparisons and analysis.

The BICS—Bio Imaging Computer System—is fully integrated with BSAS. The imaging station includes an IBM-AT compatible engineering-scientific series computer complete with math coprocessor, three megabytes of RAM, and a 20-megabyte fixed disk, CCD camera and mast, an even illumination light table (±6%), 640-line by 480-dot by 256-level digitizer and display processor, 256 color levels, mouse, color and monochrome displays, and printer.

The BICS hardware is required to run BSAS software or any imaging-processing application program from Image Data Systems. The computer system is of course compatible with IBM-AT software and fully integrated with BSAS.

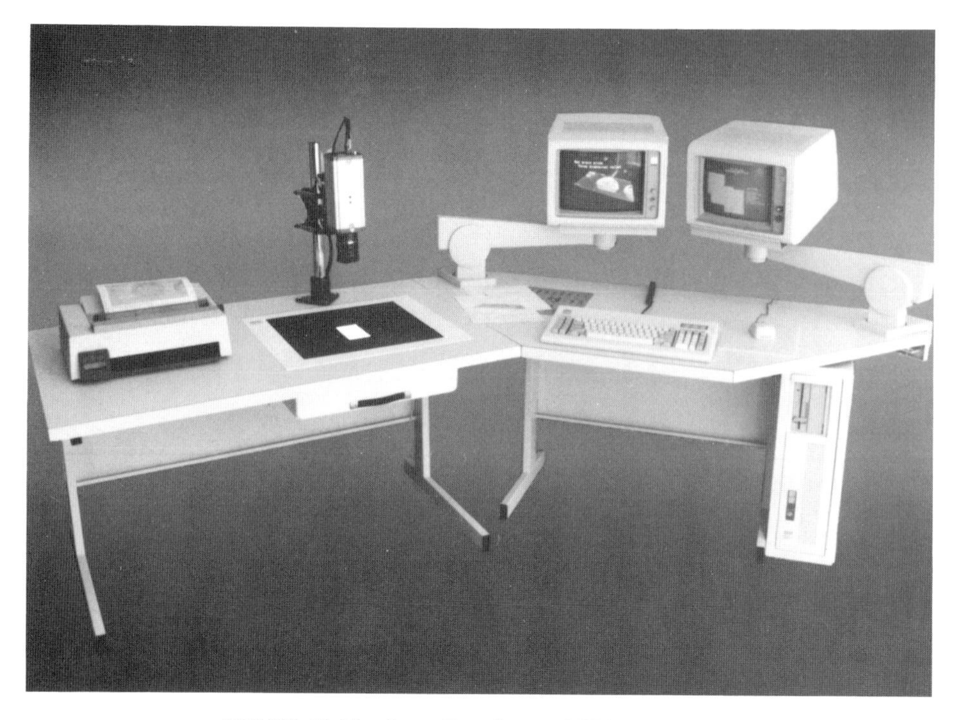

FIGURE 16-18. Image Data Systems BICS/BSAS system.

16.2.14. Imaging Research, Inc.

Brock University
St. Catherines, Ontario L2S 3A1, Canada

The MCID (the MicroComputer Imaging Device) (Fig. 16-20) is an image-analysis workstation based on the IBM/Compaq microcomputer family. The workstation includes imaging circuits, software, precision video camera, and a unique, DC-powered fluorescent illuminator (the Northern Light) with excellent stability and uniformity.

The MCID is a proven, mature tool in routine use at many research laboratories worldwide. The system offers 512 × 512-pixel resolution, 256 gray levels, pseudocolor coding, and multiple image storage. The MCID software is unusual in that all functions, including extensive facilities for data management, are provided as part of an integrated package. There are no extra-cost "modules." The MCID is available as a fully configured and installed workstation or as a cost-efficient "System Kit."

Functions for quantitative densitometry include direct reading of rates of cerebral glucose utilization, cerebral protein synthesis, and cerebral blood flow (tissue equilibration and indicator fractionation methods). The MCID can be calibrated to any standards for receptor binding autoradiography, immunocytochemistry, and other densitometric tasks. Images of specific binding and percentage specific binding can be displayed.

Grain counting and morphometric functions are available in both manual and automated modes. The MCID can simultaneously provide any combination of density, area, proportional area, perimeter, length, width, maximum chord, minimum chord, orthogonal chord, two-point distance, and count. Specialized functions are provided for cerebral blood vessel morphometry.

The MCID is acquiring new functions in response to client needs. Projects under development include three-dimensional reconstruction and solid modeling of autoradiographs and anatomic specimens.

16.2.15. INDEC Systems

1283A Mt. View-Alviso Road
Sunnyvale, CA 94089
(408) 745-1842

For researchers using IBM-XT/ATs and compatibles, INDEC offers two flexible data acquisition and analysis workstations. Basic-FASTLAB and C Lab provide turnkey applications and include source code for more tailored research needs. Both systems are easy to learn and provide menu-driven environments.

The Basic-FASTLAB package includes a set of programming tools to facilitate writing sophisticated online and offline laboratory programs. Applications for the neu-

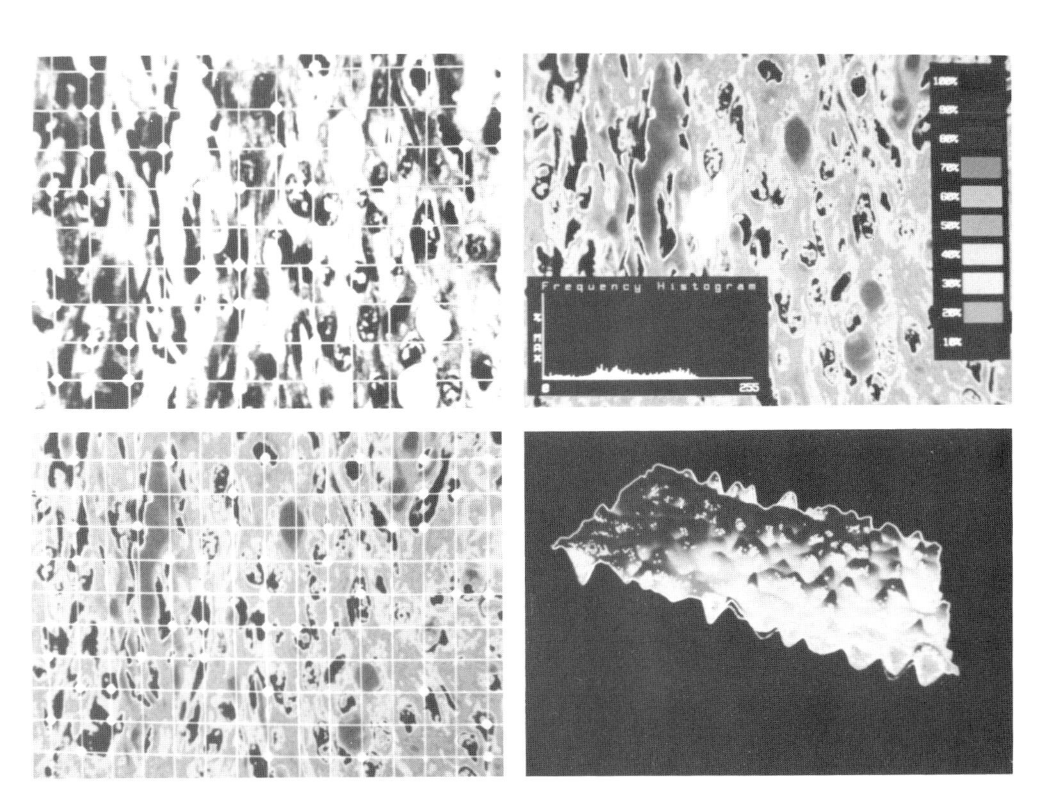

FIGURE 16-19. Image Data Systems BICS/BSAS.

FIGURE 16-20. Imaging Research, Inc. MCID (the MicroComputer Imaging Device).

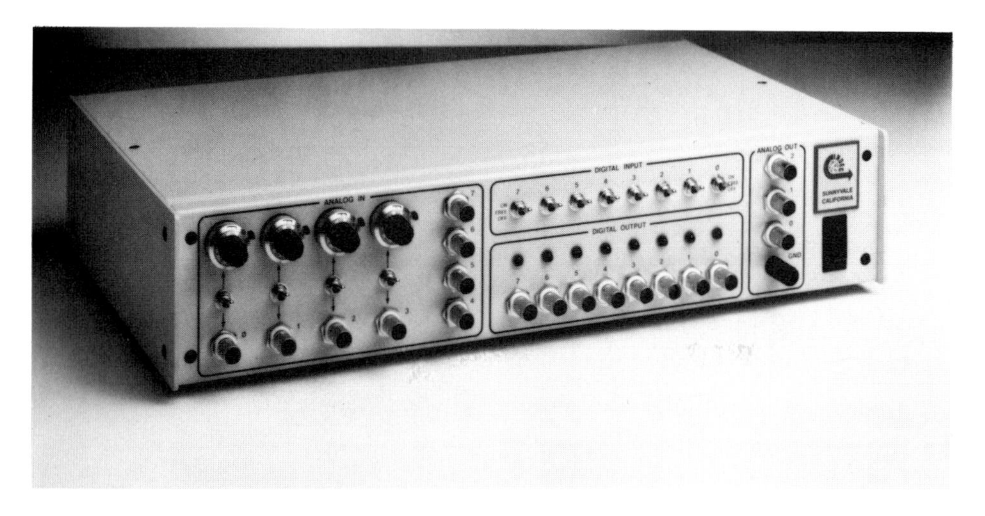

FIGURE 16-21. Indec data acquisition interface.

rophysiology laboratory include SHOCK, which turns the PC into a simple to use but elegant analog stimulator, ACQUIRE, which provides the full capabilities of simultaneous stimulation, recording, averaging display, and digital storage that are needed for online experiments, and PXY, which furnishes a powerful analysis and plotting capability.

C Lab, using C language, comes with a C-CLAMP and C-CRUNCH patch-clamp data acquisition and analysis programs. Source code is provided for C-CLAMP and C-CRUNCH. It provides a programming environment suitable for near-simultaneous measuring of up to eight separate, rapidly varying voltages. It is equipped with many high-level math functions such as fast Fourier transform, running averages, scaling routines, etc. for analyzing collected data. It includes a flexible set of graphics instructions. This allows data to be viewed on an EGA color monitor and plotted while retaining complete data resolution. There is also a family of multifunction data acquisition interfaces available (Fig. 16-21), which allow continuous recording capability. They include such features as programmable multiplexed sequencing, synchronous A/D, D/A, and parallel output, sophisticated timer/counter functions and control, general-purpose 24-bit parallel input–output, device drivers for DOS 3.*x* and OS/2 operating systems, and software making the interfaces callable from C or BASIC.

16.2.16. Jandel Scientific

65 Koch Rd.
Corte Madera, CA 94925
(415) 924-8640

SIGMA-PLOT: A scientific plotting program for IBM-PC-type computers (Fig. 16-22). Especially useful to neuroanatomists because it is so versatile and easy to use, it produces publication-quality plots and graphs with error bars, log, semilog, and linear scales. Very large data-set size. Carries out many common statistical calculations and allows you full control of axes, tic marks, label size, and position. It is very easy to learn.

SIGMA-SCAN: A digitizing tablet and software for fast and accurate measurement of anything that can be seen through a microscope or put on a digitizing tablet (Fig. 16-23). Quickly and easily measure length, area angle, X–Y position of points, slope, and count objects. Set up the worksheet to apply your own equations to the data including log and trig functions. The program will calculate while you measure. Macros allow you to set up complex sequences with a single keystroke. Ideal for measuring almost any kind of neuroanatomical data.

PC3D: Software that reconstructs complex three-dimensional objects from serial section data. Separate components can be plotted in different colors. The finished model can be rotated around any axis or converted to a stereo printout. Ideal for untangling complex nerve processes.

JAVA: Jandel Video-Analysis. Video measurement from camera or tape. Advanced image processing. Automatic counting of cells or other objects. Versatile densitometry. Automatic edge and line tracking. Digitize your strip charts or automatically measure cells.

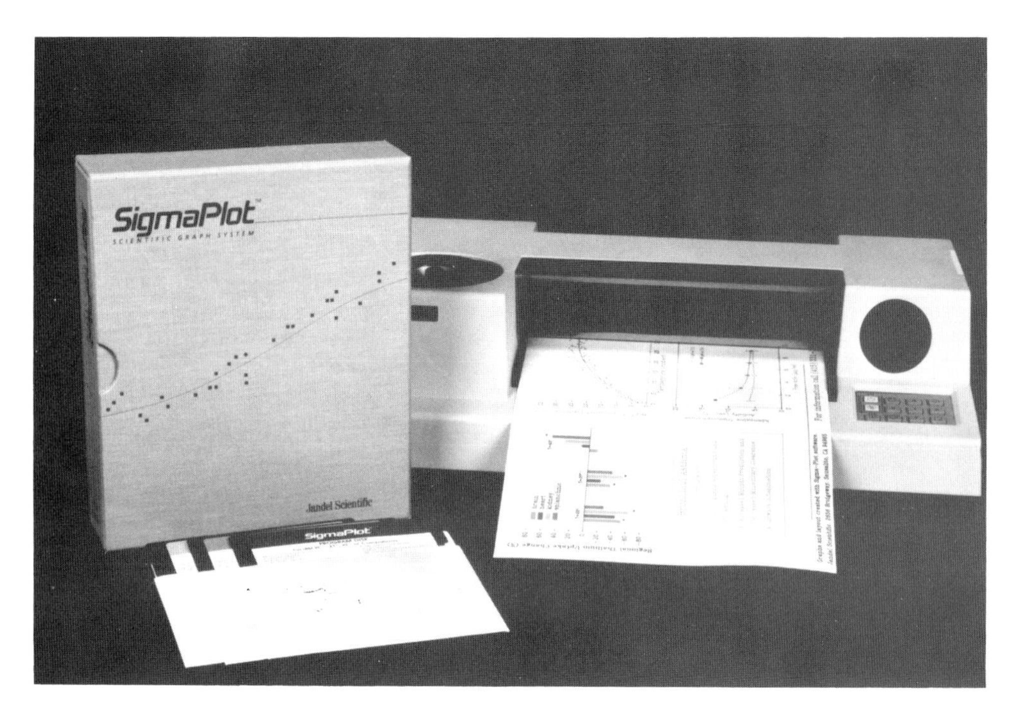

FIGURE 16-22. Jandel Scientific Sigma-Plot.

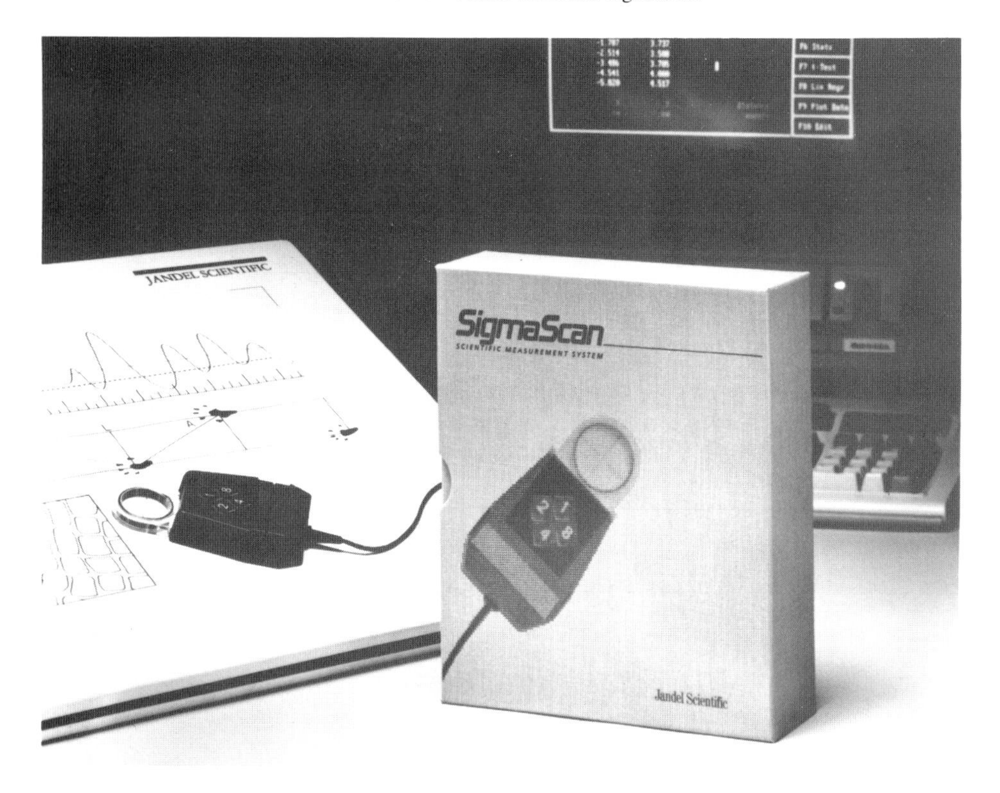

FIGURE 16-23. Jandel Scientific Sigma-Scan.

16.2.17. Joyce–Loebl

Distributed in the United States by Compix, Inc.
220 Cliff Drive, Suite #24
Pittsburgh, PA 15202
(412) 772-5277

The Joyce–Loebl range of MAGISCAN image analysis systems are being used in the fields of neuroanatomy and neurobiology in universities and commercial research centers across the world. The MAGISCAN (Figs. 16-24 and 16-25) is a sophisticated image-analysis system combining a fast architecture with flexible and mature software. Image capture from light microscopes in all illumination modes, including dual-stained fluorescence, and from electron microscopes to macro lenses allows analysis of specimens from complete brain sections to intracellular details. Animated capture and display can be used for voltage-sensitive dyes, intracellular ion concentration measurements, animated morphology, and motility. Applications software packages are focused on autoradiography, both for high-resolution grain counting and at low magnification for receptor binding studies. Three-dimensional reconstruction and animation software of full gray-level images increases the information that can be viewed in its own perspective. Even simple camera lucida using video microscopy can be enhanced by adding quantitative stain intensity measurements, profiles, and graphics. The comprehensive software libraries of Microsoft PASCAL/C callable routines available with MAGISCAN encourages the development of expansions to existing programs or new dedicated programs tailored to the individual researcher.

FIGURE 16-24. Joyce–Loebl MAGISCAN.

FIGURE 16-25. Joyce–Loebl MAGISCAN.

16.2.18. Media Cybernetics

8484 Georgia Avenue, Suite 200
Silver Spring, MD 20910
(301) 495-3305

Media Cybernetics, a leading software developer in the microcomputer-based image-processing industry, markets IMAGE-PRO II to the medical research community. IMAGE-PRO II is a powerful, comprehensive image-processing software system for the microcomputer. Operating on 286/386 machines with imaging boards from over 20 leading hardware manufacturers, IMAGE-PRO's advanced processing functions have earned it the reputation as the foremost software in its class. IMAGE-PRO II is available as menu-driven interactive software for the end user or as a subroutine library for the development of custom image-processing software.

IMAGE-PRO II contains over 100 functions including histogram analysis, frame averaging, spatial filtering, line length, area, and perimeter measurement, color mapping, Boolean and mathematical operations, particle counting and analysis, and complete graphics and text overlay for image annotation. With IMAGE-PRO II, medical researchers can capture images from a video source such as a camera or VCR, enhance and analyze these images, and store them in digital form. IMAGE-PRO II is installed in over 1000 labs worldwide, providing solutions in CT/MRI analysis, grain counting, autoradiography, densitometry, and radiology. IMAGE-PRO II is priced between $1000 and $3000. Extensive documentation is provided with the product.

16.2.19. Microscience, Inc.

31101 18th Ave. S.
Federal Way, WA 98003
(206) 941-7976

Microscience is a leading microcomputer-based supplier and developer of image-analysis software. The firm was established by Dr. M. A. Weissman in 1984 and today has over 150 installations in six countries. It counts among its customers organizations such as Ciba Geigy, Bristol Meyers, Howard Hughes Medical Institute, UCLA, Johns Hopkins, Fred Hutchinson Cancer Research Center, Texas A & M, Northwestern, University of Washington, USC, and many other leading companies, universities, hospitals, laboratories, and research institutions.

IMAGEMEASURE (Figs. 16-26 and 16-27), Microscience's high-powered, low-cost image-analysis system, consists of 14 different software modules, each developed with different applications in mind. These applications range from autoradiography (automated grain counting) to densitometry to morphometry to serial section analysis and more.

FIGURE 16-26. Microscience IMAGEMEASURE.

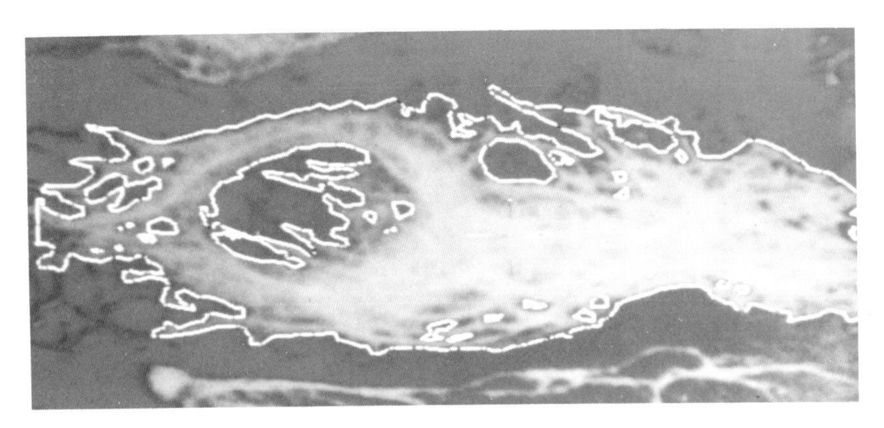

FIGURE 16-27. Image produced by the Microscience IMAGEMEASURE.

A complete set of utility routines including basic image processing, a data base for the storage and retrieval of both data and images, and statistical analysis is provided as part of every IMAGEMEASURE system.

The hardware components of an IMAGEMEASURE system are an IBM/PC-XT-AT or a Compaq 386-compatible computer, a video digitizer board, an instrumentation-quality video camera, a video monitor, and mouse or digitizer tablet. Microscience can provide a complete system based on an XT-type computer for under $10,000 or any combination of the above components necessary to work with your existing equipment.

16.2.20. Olympus Corporation

Scientific Products Group
4 Nevada Drive
Lake Success, NY 11042
(516) 488-3880

The Cue series of image-analysis systems include the Cue-2, Cue-3, and Cue-4 (Fig. 16-28). The Cue-2 offers significant advantages for an economical system. Advantages include control of the displayed measurement parameters, line profiles, versatile image editing, shading, correction, and automatic measurement of over 30 parameters.

The Cue-3 offers the same versatility as the Cue-2 and includes analysis by true color and of true color, according to the standard LAB format.

The Cue-4 is an advanced-performance research instrument that includes powerful hardware to provide greater speed of operation, increased statistical and analytical capabilities, and user definable spatial filtering.

Fast Fourier transforms for image manipulation are integrated into a flexible and easy to learn and use system.

Each of these systems can capture images at TV frame rates and enhance and analyze

FIGURE 16-28. Olympus Cue Image Analysis System.

the image to provide morphological measurements of three-dimensional data with push-button ease.

16.2.21. Quantex Corporation

252 N. Wolfe Road
Sunnyvale, CA 94086
(408) 733-6730

The Quantex Model QX7 is a real-time digital image-processing system used by researchers as a video image-enhancement, processing, and analysis tool. It is commonly used in conjunction with a microscope (light, fluorescent, confocal, etc.) and video camera. The system is managed by an IBM PC/AT-compatible 80286-based computer and features a Motorola 32-bit/16-MHz coprocessor board with an optional 68881 floating point math coprocessor. A Sun Microsystems workstation can also be utilized as a host computer. At the time of this writing, IBM PS/2 systems as well as other 80386-based systems do not have widespread proliferation in the marketplace, although their use is currently being explored by Quantex. The system's hardware provides up to 12 image buffers and processes at 12 MHz with 12-bit resolution.

The QX7 addresses the general image-processing and analysis needs of the neuroscientist by providing real-time image enhancement and offline analysis. Real-time image-processing features include continuous averaging and/or summing with simultaneous background subtraction. Real-time spatial filtering for edge enhancement is also available. The system has specific software for fura-2 as well as other image-ratioing applications. pH analysis software is currently being developed.

The system price range is $35–50,000.

16.2.22. R & M Biometrics, Inc.

5611 Ohio Avenue
Nashville, TN 37209
(615) 350-7866

BQ System IV is a moderately priced hardware/software system for image analysis and morphometric research. The system is IBM PC compatible and has easy to use "pull-down" menus for collecting data from the digitizing tablet and optional video camera overlay. The modular user-interactive programs provide accurate measurements, data listing, editing, and storage, statistics, frequency distribution, correlation, calculations, and tutorials.

Accessories to BQ System IV include color camera and monitor display, pseudocolor image enhancement, 3-D serial reconstruction and rotation, bone morphometry, pseudocolor image enhancement, and *XY* stage encoder. The Meg IV Automated Image Analysis accessory performs automated particle counts, integrated optical density measurements, and area and perimeter calculations on video images that can be edited and enhanced. The Meg IV microdensitometry accessory, which provides extensive image-enhancement capabilities, performs microdensitometry and video counts.

Bioquant software is also available for Apple II users.

16.2.23. Sarastro, Inc.

Suite A-103, Benjamin Fox Pavilion
Jenkintown, PA 19046
(215) 884-3345

PHOIBOS 1000 (Fig. 16-29) is a confocal scanning microscope system. It consists of an optical microscope fitted with fluorescence equipment, an electronically controlled scanning unit with built-in or external laser, a buffer image memory, and an instrument computer for 2-D or 3-D image processing.

With PHOIBOS 1000, the field of view seen in the eyepiece of the microscope is digitized by laser scanning. Scanning may be performed at different depths in the specimen in order to "slice" it optodigitally. The images scanned can be viewed instantly on a TV monitor. But they are also stored in digital form on disk or magnetic tape for processing and subsequent 2-D or 3-D presentation.

The PHOIBOS 1000 system offers exciting new ways to extract, visualize, and quantify microscopic information in practically every field where microscopy is presently used.

Specifications:
Image size: Up to 1024 × 1024 pixels
Photometric resolution: 256 gray levels
Resolution (a^t = 488 nm, NA = 1.3)**:**
  Lateral resolution (two-point resolution)**:** 0.25 μm
  Depth resolution (section thickness, FWHM)**:** 1 μm
Built-in laser: 5-mW 488-nm air-cooled argon-ion laser

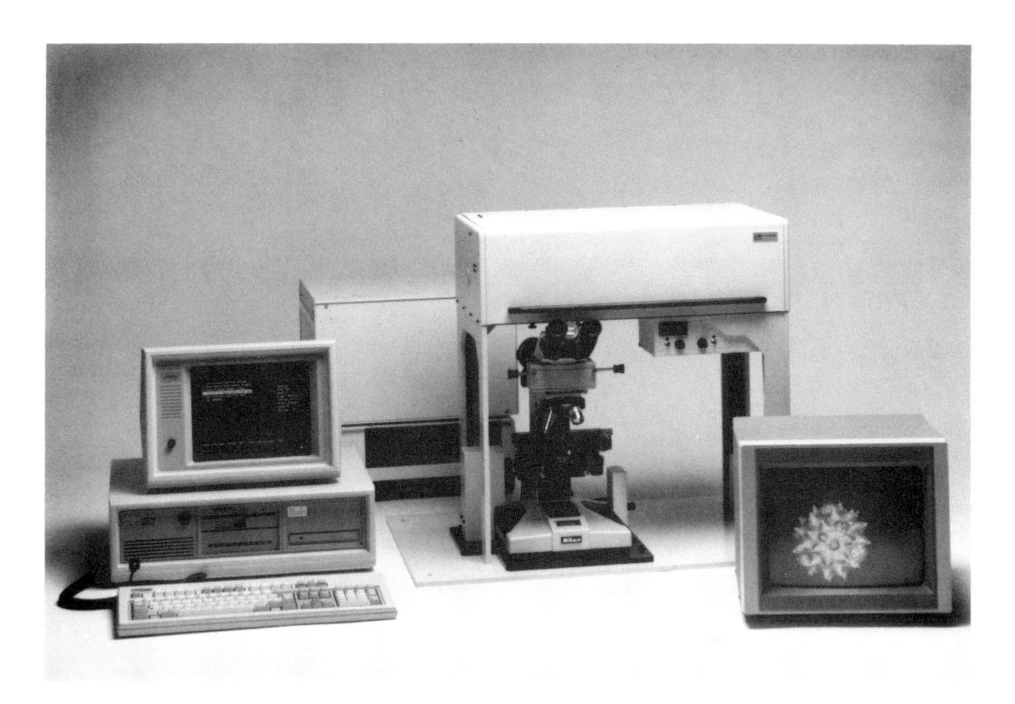

FIGURE 16-29. Sarastro PHOIBOS 1000 Laser Scanning Microscope. The image on the display is a 3-D reconstruction of a pollen from *Metalisa pungens*. The specimen is provided by P.O. Karis, University of Stockholm.

Scanning speed:

Picture size	Time
256 × 256	5.1 sec
512 × 512	10.2 sec
1024 × 1024	20.4 sec

Instrument computer:
Compaq Deskpro 286 running DOS (2-D model)
Compaq Deskpro 386 running XENIX-386 (3-D model)

16.2.24. Southern Micro Instruments

120 Interstate North Parkway East, Suite 308
Atlanta, GA 30339
(800) 241-3312 (toll free)
(404) 956-0343 (in Georgia)

SMI, an industry leader in the field of image analysis, manufactures a family of IBM-compatible software entitled MICROCOMP. Software includes planar morphometry, densitometry, particle analysis, grain counting, and 3-D reconstruction.

16.2.24a. Planar Morphometry. The MICROCOMP-PM package is a manually operated system incorporating a live video image with a computer-generated graphics overlay. The user employs an electromagnetic mouse to create a graphic trace of an object, and MICROCOMP software generates a variety of data for display, storage, or printing. A wide variety of geometric, distance, and counting parameters are available.

16.2.24b. Grain Counting. MICROCOMP's automated grain-counting software takes the guesswork out of deciding how many grains appear within the field of interest. Measurements include count of grains, population density, area, percentage area of grains, total and average intensity of grains, and more.

16.2.24c. Particle Analysis. Our particle-analysis software will automatically count discrete objects in an area of interest. Image manipulation includes object cut, paint, and shade correction. Measurements include area, perimeter, hole area, center of gravity, feret diameter, form factor, and number.

16.2.24d. Densitometry. Measurements include cross-sectional intensity profiles, frequency histograms, single-pixel intensity, average pixel intensity, integrated intensity, average/area/percentage area within an area of interest, image arithmetic, contrast enhancement, image negation and alignment, image smoothing, shade correction, background subtraction, and nonlinear intensity calibration.

16.2.25. SPEX Industries, Inc.

3880 Park Ave.
Edison, NJ 08820
(201) 549–7144

Changes in the intracellular concentration of cations are basic to many physiological events, including neurotransmission. Fully automated SPEX Cation Measurement Systems (Fig. 16-30) facilitate study of calcium, pH, and other cations in living cells via a fluorometric technique that employs fluorogenic dyes with known affinity for a particular cation. Fluorometric analysis is the fastest, least biologically disruptive method for acquiring these data from cells in suspension or on the microscope stage.

SPEX Cation Measurement Systems apply the potential of dual-excitation or dual-emission fluorogenic probes with outstanding ease and accuracy. In large measure, this convenience results from the SPEX DM3000CM spectroscopy computer, a PC/AT-compatible unit that automates all experiments via dedicated software. The DM3000CM controls the scanning monochromators, light chopper, and PMT high-voltage sources, and it offers precise management of experimental parameters.

Full computer automation is also a feature of SPEX IM Digital Imaging Systems designed for use with microsampling SPEX cation measurement instruments. Imaging locates site-specific biochemical events within a cell and distinguishes heterogeneous cellular behavior in an apparently uniform sample. Similarly, SPEX Cation Measurement

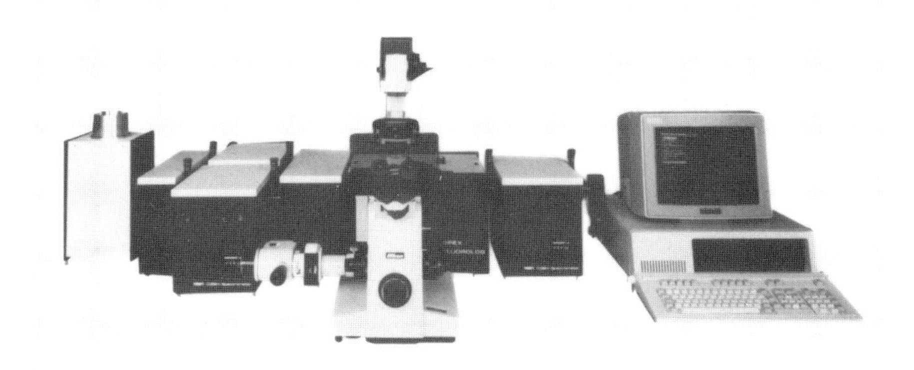

FIGURE 16-30. SPEX Cation Measurement System.

and IM Imaging Systems can automatically monitor ionic changes associated with cell-to-cell communication.

16.2.26. Technology Resources, Inc.

P.O. Box 120294
Nashville, TN 37212
(615) 292-1436

The MicroScan Gel Analysis System (Fig. 16-31) is a powerful and versatile computerized analysis system for quantitating polypeptides resolved on one-dimensional and two-dimensional gels. The system is designed to quantitate wet or dry stained gels, autoradiograms, and a variety of other media that require analysis of intensity information. The main analysis program automatically determines the number, positions, volumes, amplitudes, and shapes of all polypeptides present in a user-specified area of the gel. A very powerful set of gel comparison programs perform automatic conformal mapping of one gel image onto another (translation, rotation, and nonlinear scaling), so that gels run on different gradients can be quantitatively compared.

The IMAGEMASTER Image Analysis System is a powerful, sophisticated automatic image-analysis system for rapidly analyzing and extracting quantitative information from images of all kinds. The software supports grain counting, densitometric measurements, 3-D reconstructions, and a variety of geometric and stereological measurements.

Software for both systems is fully integrated. It is possible to process files scanned by the MicroScan digital imaging camera with the IMAGEMASTER software, and vice versa. There is virtually no image-analytic operation that cannot be performed by combining these two systems. Hardware consists of an AT-style computer, high-resolution RGB image monitor, EGA command monitor, and a three-button mouse. The MicroScan System includes a proprietary solid-state photodiode array equipped with a Nikon mount 55-mm macro lens and an intensity-stabilized light table. Hardware options and custom programming are available.

FIGURE 16-31. Technology Resources, Inc. MicroScan Gel Analysis System.

16.2.27. Universal Imaging Corporation

600 North Jackson Street
Media, PA 19063
(215) 891-0333

Universal Imaging Corporation was formed in 1983 in response to a growing need for software suited for research environments. Today, UIC is a leader in the field of image processing, developing systems tailored for basic research. By working closely with scientists on the leading edge of imaging technology, UIC has been uniquely responsive to requests for software both powerful enough to satisfy the needs of sophisticated users and easy enough to operate by the casual user. In addition, UIC prides itself on its high level of user support. By providing customers with a direct "help" hotline to our support staff, one never has to wait long for answers to technical questions.

UIC's commitment to the basic sciences has allowed the development of the IMAGE-1/AT (Figs. 16-32 and 16-33), an image-processing package unequaled in ease of use and flexibility. Unlike packages that restrict a user's options in order to make the software easy to use, IMAGE-1/AT encourages experimentation by allowing access to all parameters of a given function if needed. At the same time, a powerful journaling mechanism permits often used series of commands to be activated in a single keystroke. Such journals can then be added to the main menu to be activated later like any internal function. Options allow control of laboratory equipment, shutters, and filter wheels for quantitative fluorescence microscopy.

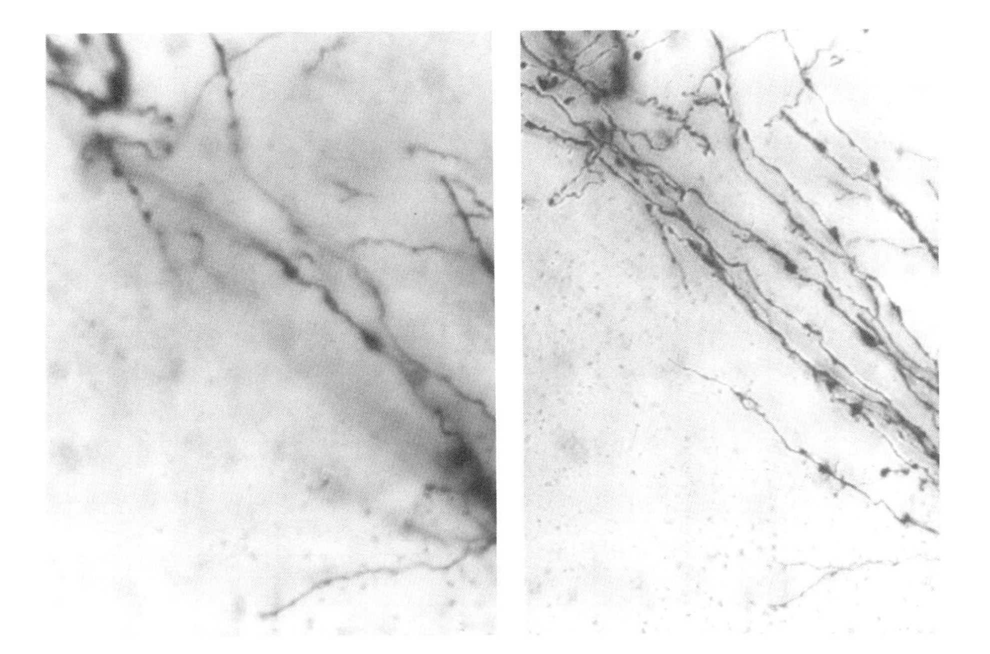

FIGURE 16-32. Universal Imaging: A before/after pair showing the effects of enhanced depth of field processing. The before image is a normal, unprocessed image showing the relatively shallow depth of field (5 μm) obtained with a normal light microscope under brightfield illumination. The after image shows a greatly enhanced depth of field using an image plane combination algorithm. This resultant image was created by stacking 10 focal planes, 5 μm apart, hence creating a 50-μm-thick result. This image was then sharpened and contrast enhanced so as to bring out fine details. The result is striking, allowing visualization of fine structures through the entire thickness of the specimen.

Some of the more powerful IMAGE-1/AT functions are as follows: real-time image averaging, live background subtraction, jumping average, automatic contrast enhancement, histogram calculation and display, region of interest processing, arithmetic and logic functions, image sharpening, user-definable convolutions, motion detection, 3-D intensity profiles, image ratioing, line brightness scans, text and graphic overlays, real-time zoom and roam, fast archival disk storage, optical disk support, easily modifiable menus and journals, which allow the user to run whole experiments at the touch of only a few keys.

Universal Imaging Corporation IMAGE-1/AT Image Processing System costs $16,900.

16.2.28. Wild Leitz U.S.A., Inc.

24 Link Drive
Rockleigh, NJ 07647
(201) 767-1100

The Leitz Microvid (Fig. 16-34) is a component option for use with the Diaplan, Orthoplan, and Aristoplan microscopes. The Microvid contains a miniature television

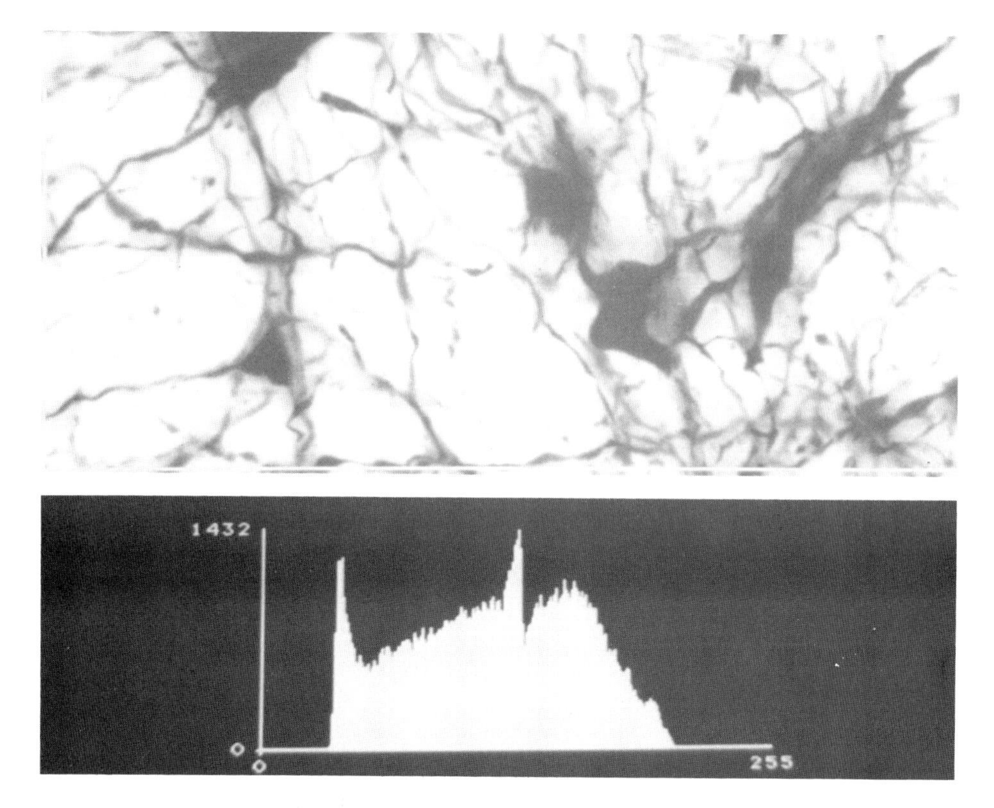

FIGURE 16-33. Universal Imaging: This picture shows an image created using the same method. We then made a measurement of the brightness distributions in the image and drew the graph of this on the image.

monitor. The monitor's screen image is mixed with the microscopic image for simultaneous viewing through the binocular tube. Several interesting possibilities result from such an arrangement:

1. For interactive image analysis, the Microvid (plus mouse) is a very attractive alternative to the digitizing tablet with camera lucida or the light pen. Since the tracing is performed on the "live" image, the criterion "color" can be fully explored, and the undiminished resolution of the microscope optics is available.
2. The Multicon is a "data back" par excellence, capable of copying any kind of alphanumeric or graphic information anywhere onto the micrographic image. This might include micrometer markers, reticles for length or area measurements, or histograms. Time measurements of dynamic events and velocity are additional examples of applications.

All information can be displayed in reversed video, as white lines on a black background. The severe reduction of image contrast, typical of camera lucidas that mix the white paper's image with that of the object, is thus avoided.

The Microvid is compatible with the IBM PC/AT with CGA or Hercules graphics card.

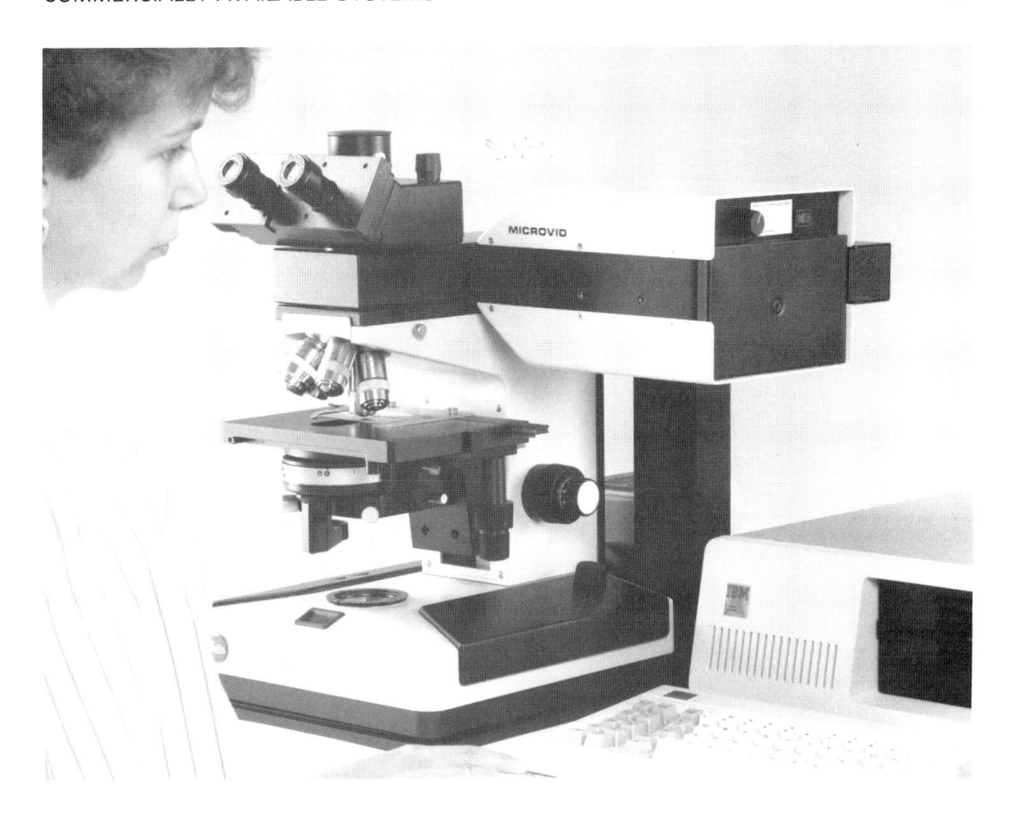

FIGURE 16-34. Wild Leitz Microvid.

References

Afshar, F., and E. Dykes (1982). A three-dimensional reconstruction of the human brain stem. *J. Neurosurg.* **57**:491–495.

Afshar, F., and E. Dykes (1984). Computer-generated three-dimensional visualization of the trigeminal nuclear complex. *Surg. Neurol.* **22**:189–196.

Afshar, F., E. Dykes, and E. S. Watkins (1983). Three-dimensional stereotactic anatomy of the human trigeminal nerve nuclear complex. *Appl. Neurophysiol.* **46**:147–153.

Agard, D. A. (1984). Optical sectioning microscopy: Cellular architecture in three dimensions. *Annu. Rev. Biophys. Bioeng.* **13**:191–219.

Agnati, L. F., K. Fuxe, F. Benfenati, I. Zini, M. Zoli, L. Fabbri, and A. Harfstrand (1984a). Computer-assisted morphometry and microdensitometry of transmitter-identified neurons with special reference to the mesostriatal dopamine pathway. I. Methodological aspects. *Acta Physiol. Scand. [Suppl.]* **532**:5–36.

Agnati, L. F., K. Fuxe, L. C. M. Goldstein, G. Toffano, L. Giardino, and M. Zoli (1984b). Computer-assisted morphometry and microdensitometry of transmitter-identified neurons with special reference to the mesostriatal dopamine pathway. II. Further studies on the effects of the GM_1 ganglioside on the degenerative and regenerative features of mesostriatal dopamine neurons. *Acta Physiol. Scand. [Suppl.]* **532**:37–44.

Alexander, G., and R. J. Schwartzman (1984). Quantitative computer analysis of autoradiographs utilizing a charge-coupled device solid-state camera. *J. Neurosci. Methods* **12**:29–36.

Allen, R. D., and N. S. Allen (1983). Video-enhanced microscopy with a computer frame memory. *J. Microsc.* **20**:3–17.

Altar, C. A., J. N. Joyce, and J. F. Marshall (1986). Functional organization of dopamine and serotonin receptors in the rat forebrain. In: *Quantitative Receptor Autoradiography* (C. A. Boast, E. W. Snowhill, and C. A. Altar, eds.). New York: Alan R. Liss, pp. 53–78.

Ameil, M., J. F. Delattre, B. Cordobes, and J. B. Flament (1984). Computerized reconstruction of an anatomical structure based on digitized sections. *Anat. Clin.* **5**:261–264.

Amthor, F. R. (1985). Quantitative analysis and reconstruction of retinal ganglion cells using a color graphics computer. In: *The Microcomputer in Cell and Neurobiology Research* (R. R. Mize, ed.). New York: Elsevier, pp. 135–153.

Andrews, H. C., and B. R. Hunt (1977). *Digital Image Restoration.* Englewood Cliffs, NJ: Prentice-Hall.

Andrews, H. P., R. D. Snee, and M. H. Sarner (1980). Graphical display of means. *Am. Stat.* **34**:195–199/

Antal, M., R. Kraftsik, G. Székely, and H. van der Loos (1986). Distal dendrites of frog motor neurons: A computer-aided electron microscopic study of cobalt-filled cells. *J. Neurocytol.* **15**:303–310.

Argus Communications (1981). *Trees of Europe through the Four Seasons.* Harlow, Essex, UK: Argus Communications.

Arkin, M. S., and R. F. Miller (1986). Golgi-like staining of ganglion cells using transported HRP. *ARVO Abstr.* **27**(3):96.

Arkin, M. S., and R. F. Miller (1988). Mudpuppy retinal ganglion cell morphology revealed by an HRP impregnation technique which provides Golgi-like staining. *J. Comp. Neurol.* **270**(2):185–208.

Artwick, B. A. (1980). *Microcomputer Interfacing.* Englewood Cliffs, NJ: Prentice Hall.

Artwick, B. A. (1984). *Applied Concepts in Microcomputer Graphics*. Englewood Cliffs, NJ: Prentice Hall.
Augustine, J. R., T. Huntsberger, and M. Moore (1985). Computer-aided reconstructive morphology of the baboon abducens nucleus. *Anat. Rec.* **212:**210–217.
Axen, R., J. Porath, and S. Ernback (1967). Chemical coupling of peptide and proteins to polysaccharides by means of cyanogen halides. *Nature* **214:**1302–1304.
Badler, N. I., K. H. Maanoocherhri, and G. Walters (1987). Articulated figure positioning by multiple contraints. *IEEE Comput. Graphics Appl.* **7:**28–38.
Baker, H., A. F. Sved, L. W. Tucker, S. M. Alden, and D. J. Reis (1985). Strain differences in pituitary prolactin content: Relationship to number of hypothalamic dopamine neurons. *Brain Res.* **358:**16–26.
Batschelet, E. (1981). *Circular Statistics in Biology*. London: Academic Press.
Becker, L. E., D. L. Armstrong, and F. Chan (1986). Dendritic atrophy in children with Down's syndrome. *Ann. Neurol.* **20:**520–526.
Bellinger, D. L., and W. J. Anderson (1987). Postnatal development of cell columns and their associated dendritic bundles in the lumbosacral spinal cord of the rat. I. The ventrolateral cell column. *Dev. Brain Res.* **35:**55–67.
Benno, R. H., L. W. Tucker, T. H. Joh, and D. J. Reis (1982a). Quantitative immunocytochemistry of tyrosine hydroxylase in rat brain. I. Development of a computer-assisted method using the peroxidase–antiperoxidase technique. *Brain Res.* **246:**225–236.
Benno, R. H., L. W. Tucker, T. H. Joh, and D. J. Reis (1982b). Quantitative immunocytochemistry of tyrosine hydroxylase in rat brain. II. Variations in the amount of tyrosine hydroxylase among individual neurons of the locus coeruleus in relationship to neuronal morphology and topography. *Brain Res.* **246:**237–247.
Beran, R. J. (1968). Testing for uniformity on a compact homogeneous space. *J. Appl. Prob.* **5:**177–195.
Berry, M., and R. Flinn (1984). Vertex analysis of Purkinje cell dendritic trees in the cerebellum of the rat. *Proc. R. Soc. Lond. [Biol.]* **221:**321–348.
Biegon, A., and M. Wolff (1986). Quantitative histochemistry of acetylcholinesterase in rat and human brain postmortem. *J. Neurosci. Methods* **16:**39–45.
Bigbee, J. W., J. C. Kosek, and L. F. Eng (1977). Effects of primary antiserum dilution on staining of antigen-rich tissues with the peroxidase antiperoxidase technique. *J. Histochem. Cytochem.* **25:**443–447.
Blackstad, T. W., K. K. Osen, and E. Mugnaini (1984). Pyramidal neurones of the dorsal cochlear nucleus: A Golgi and computer reconstruction study in cat. *Neuroscience* **13:**827–854.
Blaisdel, G. (1986). Voltage-sensitive dyes reveal a modular organization in monkey striate cortex. *Nature* **321:** 579–585.
Boast, C. A., E. W. Snowhill, and C. A. Altar (1986). *Quantitative Receptor Autoradiography*. New York: Alan R. Liss.
Boivie, J. G., G. Grant, and H. Ulfendahl (1968). The *X–Y* recorder used for mapping under the microscope. *Acta Physiol. Scand.* **74:**A1–A2.
Bok, S. T. (1959). *Histonomy of the Cerebral Cortex*. Amsterdam: Elsevier.
Borges, S., and M. Berry (1978). The effects of dark rearing on the development of the visual cortex of the rat. *J. Comp. Neurol.* **180:**277–300.
Born, G. (1883). Die Plattenmodellirmethode. *Arch. Mikrosk. Anat.* **22:**584–599.
Boyle, P. J. R., and D. G. Whitlock (1974). The application of a computer controlled microscope to autoradiographs of nerve tissue. *DECUS Proc.* **1**(2)**:**95–99.
Boyle, P. J. R., and D. G. Whitlock (1977). A computer-controlled microscope as a device for evaluating autoradiographs. In: *Computer Analysis of Neuronal Structures* (R. D. Lindsay, ed.). New York: Plenum Press, pp. 133–148.
Braverman, M. S., and I. M. Braverman (1986). Three-dimensional reconstruction of objects from serial sections using a microcomputer graphics system. *J. Invest. Dermatol.* **86:**290–294.
Bregman, B. S., and W. L. R. Cruce (1980). Normal dendritic morphology of frog spinal motoneurons: A Golgi study. *J. Comp. Neurol.* **193**(4)**:**1035–1045.
Briarty, L. G., J. Patrick, J. Fisher, and P. H. Jenkins (1982). Microscopy, morphology and microcomputers. *Acta Stereol.* **82:**227–234.
Briski, K. P., B. L. Baker, and A. K. Christensen (1983). Effect of ovariectomy on the hypothalamic content of immunoreactive gonadotropin-releasing hormone in the female mouse as revealed by quantitative immunocytochemistry and radioimmunoassay. *Am. J. Anat.* **166:**187–208.
Brooks, F. P., Jr. (1975). *The Mythical Man-Month*. Reading, MA: Addison-Wesley.

Brooks, F. P., Jr. (1987). No silver bullet: Essence and accidents of software engineering. *Computer* **20**(4):10–19.

Brown, C. (1977). Neuron orientations: A computer application. In: *Computer Analysis of Neuronal Structures* (R. D. Lindsay, ed.). New York: Plenum Press, pp. 177–188.

Brown, P. B., ed. (1976). *Computer Technology in Neuroscience.* Washington, DC: Hemisphere.

Brown, P. B., G. R. Busch, and J. Whittington (1979). Anatomical changes in cat dorsal horn cells after transection of a single dorsal root. *Exp. Neurol.* **64**:453–468.

Buell, S. J., and P. D. Coleman (1979). Dendritic growth in the aged human brain and failure of growth in senile dementia. *Science* **206**:854–856.

Buell, S. J., and P. D. Coleman (1981). Quantitative evidence for selective dendritic growth in normal human aging but not in senile dementia. *Brain Res.* **214**:23–41.

Buskirk, D. R. (1978). Computer analysis of dendritic morphology. *Brain Theor. Newsl.* **3**:184–186.

Cahan, L. D., and B. T. Trombka (1975). Computer graphics three-dimensional reconstruction of thalamic anatomy from serial sections. *Comput. Prog. Biomed.* **5**:91–98.

Cajal, S. R. (1906). *Studien uber die Hirnrinde des Menschen.* Leipzig: Verlag von Johann Ambrosius Barth.

Cajal, S. R. (1909). *Histologie du Systeme Nerveux de l'Homme et des Vertebres.* Paris: Maloine.

Cajal, S. R. (1984). *The Neuron and the Glial Cell* (Translated and edited by J. de la Torre and W. C. Gibson). Springfield, IL: Charles C. Thomas.

Calvet, M. C., and J. Calvet (1984). Computer assisted analysis of HRP-labelled and Golgi-stained Purkinje neurons. *Prog. Neurobiol.* **23**:251–272.

Calvet, M. C., J. Calvet, and R. Camacho-Garcia (1985). The Purkinje cell dendritic tree: A computer-aided study of its development in the cat and in culture. *Brain Res.* **331**:235–250.

Cameron, W. E., D. B. Averill, and A. J. Berger (1985). Quantitative analysis of the dendrites of cat phrenic motoneurons stained intracellularly with horseradish peroxidase. *J. Comp. Neurol.* **230**:91–101.

Capel, P. J. A. (1974). A quantitative immunofluorescence method based on the covalent coupling of protein to sepharose beads. *J. Immunol. Methods* **5**:165–178.

Capowski, J. J. (1973). A general purpose 3D physical modeling program. *Comput. Graphics* **7**(3):24–28.

Capowski, J. J. (1976). Characteristics of neuroscience computer graphics displays and a proposed system to generate those displays. *Comput. Graphics* **10**(2):257–261.

Capowski, J. J. (1977). Computer-aided reconstruction of neuron trees from several serial sections. *Comput. Biomed. Res.* **10**:617–629.

Capowski, J. J. (1978a). The neuroscience display processor. *Computer* **11**(11):48–58.

Capowski, J. J. (1978b). The neuroscience display processor model 2. *DECUS Proc.* **5**(2):763–766.

Capowski, J. J. (1979). The modeling, display, and analysis of nerve cells. *DECUS Proc.* **6**(2):739–742.

Capowski, J. J. (1983a). An automatic neuron reconstruction system. *J. Neurosci. Methods* **8**(4):353–364.

Capowski, J. J. (1983b). The neuroscience display processor model 3. *DECUS Proc.* **10**(2):167–170.

Capowski, J. J. (1985). The reconstruction, display, and analysis of neuronal structure using a computer. In: *The Microcomputer in Cell and Neurobiology Research* (R. R. Mize, ed.). New York: Elsevier, pp. 85–109.

Capowski, J. J. (1987). A new comprehensive neuron reconstruction system. system. *Neurosci.* **22**(supp.): S369.

Capowski, J. J. (1988). Computer graphics display processor for generating dynamic refreshed vector images. U.S. Patent 4,736,330.

Capowski, J. J. (1988b). Anatomical measurement and analysis. In: *Microcomputers in Physiology: A Practical Approach* (P. J. Fraser, ed.). Oxford: IRL Press, pp. 95–127.

Capowski, J. J., and W. L. R. Cruce (1979). How to configure a computer-aided neuron reconstruction and graphics display system. *Comput. Biomed. Res.* **12**(6):569–587.

Capowski, J. J., and W. L. R. Cruce (1981). The personnel needs of a neuroscientific computer center. *SIGBIO Newsletter* special issue **March:**110–128.

Capowski, J. J., and E. M. Johnson (1985). A simple hidden line removal algorithm for serial section reconstruction. *J. Neurosci. Methods* **13**:145–152.

Capowski, J. J., and M. Réthelyi (1978). Computer analysis of the distribution of synaptic elements of Golgi-stained axon trees. *Brain Theory Newsl.* **3**:179–183.

Capowski, J. J., and M. Réthelyi (1982). Neuron reconstruction using a Quantimet image analyzing computer. *Acta Morphol. Sci. Hung.* **30**:243–251.

Capowski, J. J., and S. A. Schneider (1985). A simple motor controller for computer-assisted microscopy. *J. Neurosci. Methods* **13:**97–102.

Capowski, J. J., and M. J. Sedivec (1981). Accurate computer reconstruction and graphics display of complex neurons utilizing state-of-the-art interactive techniques. *Comput. Biomed. Res.* **14:**518–532.

Capowski, J. J., M. J. Sedivec, and L. M. Mendell (1986). An illustration of spinocervical tract cells and their computer reconstruction. *J. Neurosci.* **6**(3)**:**front cover.

Carr, D. B., W. L. Nicholson, R. J. Littlefield, and D. L. Hall (1986). Interactive color display methods for multivariate data. In: *Statistical Image Processing and Graphics* (E. J. Wegman and D. J. DePriest, eds.). New York: Marcel Dekker.

Casale, E. J. (1988). *Anatomy and Physiology of Corticospinal Tract Neurons in the Rat and Cat.* Doctoral Dissertation, University of North Carolina.

Cassell, M. D., N. J. Mankovich, T. S. Gray, and T. H. Williams (1982). Computer-assisted image analysis of the distributions of peptidergic terminals in the central nucleus of the amygdala: A preliminary study. *Peptides* **3:**283–290.

Castleman, K. R. (1979). *Digital Image Procesing.* Englewood Cliffs, NJ: Prentice Hall.

Chatfield, C., and A. J. Collins (1980). *Introduction to Multivariate Analysis.* London: Chapman and Hall.

Chawla, S. D., L. Glass, and J. W. Procter (1981). Three-dimensional reconstruction of disseminated cancer modules. *Cancer Biochem. Biophys.* **5:**153–161.

Chawla, S. D., L. Glass, S. Friewald, and J. W. Procter (1982). An interactive computer graphic system for 3-D steroscopic reconstruction from serial sections: Analysis of metastatic growth. *Comput. Biol. Med.* **12**(3)**:** 223–232.

Claiborne, B. J., D. G. Amaral, and W. M. Cowan (1986). A light and electron microscopic analysis of the mossy fibers of the rat dentate gyrus. *J. Comp. Neuol.* **246**(4)**:**435–458.

Cochran, W. G. (1977). *Sampling Techniques,* 3rd ed. New York: John Wiley & Sons.

Coleman, P. D., and D. G. Flood (1987). Neuron numbers and dendritic extent in normal aging and Alzheimer's disease. *Neurobiol. Aging* **8:**521–545.

Coleman, P. D., C. F. Garvey, J. H. Young, and W. Simon (1977). Semiautomatic tracing of neuronal processes. In: *Computer Analysis of Neuronal Structures* (R. D. Lindsay, ed.). New York: Plenum Press, pp. 91–109.

Coleman, P. D., D. G. Flood, M. C. Whitehead, and R. C. Emerson (1981). Spatial sampling by dendritic trees in visual cortex. *Brain Res.* **214:**1–21.

Colonnier, M. (1964). The tangential organization of the visual cortex. *J. Anat.* **98:**327–344.

Comans, P. E., I. S. McLennan, R. F. Mark, and I. A. Hendry (1988). Mammalian motoneuron development: Effect of peripheral deprivation on motoneuron numbers in a marsupial. *J. Comp. Neurol.* **270**(1)**:**111–120.

Conover, W. J. (1980). *Practical Nonparametric Statistics,* 2nd ed. New York: John Wiley & Sons.

Conover, W. J., and R. L. Iman (1981). Rank transformations as a bridge between parametric and non-parametric statistics. *Am. Stat.* **35:**124–133.

Coombs, G. H., L. Tetley, V. A. Moss, and K. Vickerman (1986). Three dimensional structure of the leishmania amastigote as revealed by computer-aided reconstruction from serial sections. *Parasitology* **92:** 13–23.

Cornelisse, J. T. W. A., and T. J. T. P. van den Berg (1984). Profile boundary length can be overestimated by as much as 41% when using a digitizer tablet. *J. Microsc.* **136**(3)**:**341–344.

Cowan, W. M., and D. F. Wann (1973). A computer system for the measurement of cell and nuclear sizes. *J. Microsc. (Oxf.)* **99:**331–448.

Cruce, W. L. R., and S. L. Steusse (1987). Three-dimensional neuron reconstruction: A retrospective view of ten years' use in the anatomical laboratory. *Neurosci.* **22**(suppl)**:**S393.

Cruz-Orive, L.-M., and E. B. Hunziker (1986). Stereology for anisotropic cells: Application to growth cartilage. *J. Microsc. (Oxf.)* **143:**47–80.

Cullheim, S., J. W. Fleshman, L. L. Glenn, and R. E. Burke (1987a). Membrane area and dendritic structure in type-identified triceps surea alpha motoneurons. *J. Comp. Neurol.* **255:**68–81.

Cullheim, S., J. W. Fleshman, L. L. Glenn, and R. E. Burke (1987b). Three-dimensional architecture of dendritic trees in type-identified α-motoneurons. *J. Comp. Neurol.* **255:**82–96.

Curcio, C. A., and K. R. Sloan (1981). A computer system for combined neuronal mapping and morphometry. *J. Neurosci. Methods* **4:**267–276.

Curcio, C. A., and K. R. Sloan (1986). Computer-assisted morphometry using video-mixed microscopic images and computer graphics. *Anat. Rec.* **214:**329–337.

Dann, J. F., and E. H. Buhl (1987). Retinal ganglion cells projecting to the accessory optic system in the rat. *J. Comp. Neurol.* **262:**141–158.

Davis, B. J. (1985). The electronic pantograph: Amplifier couples microscope stage to *X–Y* plotter. *Brain Res. Bull.* **15:**533–536.

DePasquale, J. D., and C. S. Izzard (1987). Evidence for an actin-containing cytoplasmic precursor of the focal contact and the timing of incorporation of vinculin at the focal contact. *J. Cell Biol.* **105:**2803–2809.

De Ruiter, J. P., and H. B. M. Uylings (1987). Morphometric and dendritic analysis of fascia dentata granule cells in human aging and senile dementia. *Brain Res.* **402:**217–229.

DeVoogd, T. J., F. L. F. Chang, M. K. Floeter, M. J. Jencius, and W. T. Greehough (1981). Distortions induced in neuronal quantification by camera lucida analysis: Comparisons using a semiautomated data acquisition system. *J. Neurosci. Methods* **3:**284–294.

Dierker, M. L. (1976a). An algorithm for the alignment of serial sections. In: *Computer Technology in Neuroscience* (P. B. Brown, ed.). Washington, DC: Hemisphere, pp. 131–133.

Dierker, M. L. (1976b). An algorithm for the display and manipulation of lines in three dmensions. In: *Computer Technology in Neuroscience* (P. B. Brown, ed.). Washington, DC: Hemisphere, pp. 139–151.

Dinopoulos, A., J. G. Parnavelas, H. B. M. Uylings, and C. G. Van Eden (1988). Morphology of neurons in the basal forebrain nuclei of the rat: A Golgi study. *J. Comp. Neurol.* **272:**461–474.

Dunn, R. F., O'Leary, D. P., and Kumley, W. E. (1975). Quantitative analysis of micrographs by computer graphics. *J. Microsc. (Oxf.)* **105:**205–213.

Dunn, R. F., D. P. O'Leary, and W. E. Kumley (1977). Online computerized analysis of peripheral nerves. In: *Computer Analysis of Neuronal Structures* (R. D. Lindsay, ed.). New York: Plenum Press, pp. 111–132.

DuVarney, D., and R. C. DuVarney (1985). A computer-based video microscope. In: *The Microcomputer in Cell and Neurobiology Research* (R. R. Mize, ed.). New York: Elsevier, pp. 233–246.

Dykes, E., and F. Afshar (1982). Computer generated three dimensional reconstructions from serial sections. *Acta Sterol.* **82:**289–296.

Dykes, E., and J. G. Clement (1980). The construction and application of an *X–Y* coordinate plotting microscope. *J. Dent. Res.* **59:**1800.

Eayrs, J. T. (1955). The cerebral cortex of normal and hypothyroid rats. *Acta Anat.* **25:**160–183.

Eidelberg, E., and F. Davis (1977). An improved electronic pantograph. *J. Histochem. Cytochem.* **25:**1016–1018.

Ellias, S. A., and J. K. Stevens (1980). The dendritic varicosity: A mechanism for electrically isolating the dendrites of cat retinal amacrine cells? *Brain Res.* **196:**365–372.

Ellisman, M., R. Ranganathan, T. Deerinck, S. Young, R. Terry, and S. Mirra (1987). Neuronal fibrillar cytoskeleton and endomembrane system organization in Alzheimer's disease. In: *Alterations in the Neuronal Cytoskeleton in Alzheimer's Disease* (G. Perry, ed.). New York: Plenum Press, pp. 61–73.

Eysel, U. T., L. Peichl, and H. Wassle (1985). Dendritic plasticity in the early postnatal feline retina: Quantitative characteristics and sensitive period. *J. Comp. Neurol. B242:*134–145.

Fahle, M. (1988). The double microscope: Its use in three-dimensional reconstruction from serial sections. *J. Neurosci. Methods* **23:**95–99.

Falck, B., N. A. Hillarp, G. Thieme, and A. Thorpe (1962). Fluorescence of catecholamines and related compounds condensed with formaldehyde. *J. Histochem. Cytochem.* **10:**348–354.

Falen, S. W., and D. S. Packard, Jr. (1982). Computer-assisted stereoscopic reconstruction of biological tissues. *Proc. Natl. Comput. Graphics Assn.* **2:**995–1003.

Feingold, E., S. B. Seshadri, and O. J. Tretiak (1986). Hardware and software design consideration in engineering and image processing workstation: Autoradiographic analysis with DUMAS and the BRAIN autoradiograph analysis software package. In: *Functional Mapping in Biology and Medicine: Computer Assisted Autoradiography* (D. L. McEachron, ed.). Basel: Karger, pp. 175–203.

Fisher, J. B., and H. Honda (1977). Computer simulation of branching pattern and geometry in terminalia (combretaceae), a tropical tree. *Bot. Gaz.* **138**(4)**:**377–384.

Flaming, D. G. (1982). A short review of current laboratory microcomputer systems and practice. *J. Neurosci. Methods* **51:**1–6.

Flaming, D. G. (1985). CP/M and programming languages. In: *Microcomputers in the Neurosciences* (G. A. Kerkut, ed.). Oxford: Clarendon Press, pp. 142–151.

Flood, D. G., S. J. Buell, G. J. Horwitz, and P. D. Coleman (1987). Dendritic extent in human dentate gyrus granule cells in normal aging and senile dementia. *Brain Res.* **402:**205–216.

Foley, J. D., and A. van Dam (1982). *Fundamentals of Interactive Computer Graphics.* Reading, MA: Addison-Wesley.

Foley, J. D., and V. L. Wallace (1974). The art of natural graphic man-machine conversation. *Proc. IEEE* **62:** 462–471.

Foley, J. D., V. L. Wallace, and P. Chan (1984). The human factors of computer graphics interaction techniques. *IEEE Comput. Graphics Appl.* **4**(11)**:**13.

Foote, S. L., S. E. Loughlin, P. S. Cohen, F. E. Bloom, and R. B. Livingston (1980). Accurate three-dimensional reconstruction of neuronal distributions in brain: Reconstruction of the rat nucleus locus coeruleus. *J. Neurosci. Methods* **3:**159–173.

Forbes, D. J., and R. W. Petry (1979). Computer-assisted mapping with the light microscope. *J. Neurosci. Methods* **1:**77–94.

Ford-Holevinski, T. S., T. A. Dahlberg, and B. W. Agranoff (1986). A Microcomputer-based image analyzer for quantitating neurite outgrowth. *Brain Res.* **368:**339–346.

Fram, E. K. (1985). 3-D reconstruction: Seeing beyond the surface. *PC Magazine* **Aug 20:**170–174.

François, C., J. Yelnik, and G. Percheron (1987). Golgi study of the primate substantia nigra. II. Spatial organization of dendritic arborizations in relation to the cytoarchitectonic boundaries and to the striatonigral bundle. *J. Comp. Neurol.* **265:**473–493.

Frasch, A. C. C., M. E. Itoiz, and R. L. Cabrini (1978). Microspectrophotometric quantitation of the di-aminobenzidine reaction for histochemical demonstration of cytochrome oxidase activity. *J. Histochem. Cytochem.* **26:**157–162.

Freeman, J., and R. S. Meltzer (1983). CARTOS revives biological approach from turn of century. *Comput. Graphics News* **May/June:**17–18.

Freiherr, G. (1987). Image combining microscope resource. *Research Resources Reporter. U.S. Heath and Human Services* **12**(12)**:**13–14.

Freire, M. (1986). An inexpensive and interactive microcomputer system for codifying Golgi-impregnated neuronal morphology. *J. Neurosci. Methods* **16:**103–117.

Frenkel, K. A. (1988). The art and science of visualizing data. *Commun. ACM* **31**(2)**:**110–122.

Gage, S. H. (1941). *The Microscope.* Ithaca, NY: Comstock.

Gallistel, C. R., and O. Tretiak (1985). Microcomputer systems for analyzing 2-deoxyglucose autoradiographs. In: *The Microcomputer in Cell and Neurobiology Research* (R. R. Mize, ed.). New York: Elsevier, pp. 379–408.

Gambino, D. R., L. T. Malmgren, and R. R. Gacek (1985). Three-dimensional computer reconstruction of the neuromuscular junction distribution in the human posterior cricoarytenoid muscle. *Laryngoscope* **95**(5)**:** 556–560.

Garvey, C. F., J. Young, W. Simon, and P. D. Coleman (1972). Semiautomatic dendrite tracking and focusing by computer. *Anat. Rec.* **172:**314.

Garvey, C. F., J. Young, W. Simon, and P. D. Coleman (1973). Automated three-dimensional dendrite tracking system. *Electroencephalogr. Clin. Neurophysiol.* **35:**199–204.

Gaunt, P. N., and W. A. Gaunt (1978). *Three-Dimensional Reconstruction in Biology.* Tunbridge Wells: Pitman.

Gdowski, G. T., W. D. Eldred, and H. F. Voight (1987). A simple device for the computer quantification of depth measurements in thick light microscope sections. *J. Neurosci. Methods* **20:**249–260.

Geffard, M., A. M. Henrich-Rock, H. Dulluc, and P. Seguela (1985). Antisera against small neurotransmitter-like molecules. *Neurochem. Int.* **7:**403–413.

Geisow, M. J., and A. N. Barrett, eds. (1983). *Computing in Biological Sciences.* Amsterdam: Elsevier Biomedical Press.

Gentile, A. M., and E. Harth (1978). The alignment of serial sections by spatial filtering. *Comput. Biomed. Res.* **11:**537–551.

German, D. C., D. S. Schlusselberg, and D. J. Woodward (1983). Three-dimensional computer reconstruction of midbrain dopaminergic neuronal populations: From mouse to man. *Neural Transm.* **57:**243–254.

Geröcs, K., M. Réthelyi, and B. Halasz (1986). Quantitative analysis of dendritic protrusions in the medial preoptic area during postnatal development. *Dev. Brain Res.* **26:**49–57.

Giloi, W. K. (1978). *Interactive Computer Graphics; Data Structures, Algorithms, Languages*. Englewood Cliffs, NJ: Prentice-Hall.

Glaser, E. M. (1981). A binary identification system for use in tracing and analyzing dichotomously branching dendrite and axon systems. *Comput. Biol. Med.* **11**:17–19.

Glaser, E. M. (1982). Snell's law: The bane of computer microscopists. *J. Neurosci. Methods* **5**:201–202.

Glaser, E. M., and N. T. McMullen (1984). The fan-in projection method for analyzing dendrite and axon systems. *J. Neurosci. Methods* **12**:37–42.

Glaser, E. M., and H. van der Loos (1965). A semiautomatic computer microscope for the analysis of neuronal morphology. *IEEE Trans. Biomed. Eng.* **12**:22–31.

Glaser, E. M., and H. van der Loos (1980). Computer microscope apparatus and method for superimposing an electronically-produced image from the computer memory upon the image in the microscope's field of view. U.S. Patent 4,202,037.

Glaser, E. M., and H. van der Loos (1981). Analysis of thick brain sections by obverse–reverse computer microscope: Application of a new, high clarity Golgi–Nissl stain. *J. Neurosci. Methods* **4**:117–125.

Glaser, E. M., H. van der Loos, and M. Gissler (1978). Preferred tangential orientation and spatial order in dendritic fields of cat auditory cortex: A computer–microscope study of Golgi-stained material. *Soc. Neurosci. Abstr.* **4**:210.

Glaser, E. M., M. Gissler, and H. van der Loos (1979a). An interactive camera lucida computer–microscope. *Soc. Neurosci. Abstr.* **5**:1697.

Glaser, E. M., H. Van der Loos, and M. Gissler (1979b). Tangential organization and spatial order in dendrites of cat auditory cortex: A computer microscope study of Golgi-impregnated material. *Exp. Brain Res.* **36**: 411–431.

Glaser, E. M., M. Tagamets, N. T. McMullen, and H. van der Loos (1983). The image-combining computer microscope—an interactive instrument for morphometry of the nervous system. *J. Neurosci. Methods* **8**: 17–32.

Glasser, S., J. Miller, N. G. Xuong, and A. Selverston (1977). Computer reconstruction of invertebrate nerve cells. In: *Computer Analysis of Neuronal Structures* (R. D. Lindsay, ed.). New York: Plenum Press, pp. 21–58.

Glenn, L. L., and R. E. Burke (1981). A simple and inexpensive method for 3-dimensional visualization of neurons reconstructed from serial sections. *J. Neurosci. Methods* **4**:127–134.

Gonzales, R. C., and P. W. Wintz (1977). *Digital Image Processing*. Reading, MA: Addison-Wesley.

Goochee, C., W. Rasband, and L. Sokoloff (1979). Computerized densitometry and color coding of ^{14}C-deoxyglucose autoradiographs. *Ann. Neurol.* **7**:359–370.

Graham, R. C., and M. J. Karnovsky (1966). The early stages of absorption of injected horseradish peroxidase in the proximal tubules of mouse kidney: Ultrastructural cytochemistry by a new technique. *J. Histochem. Cytochem.* **14**:291–302.

Gras, H. (1984). A "hidden line" algorithm for 3D-reconstruction from serial sections—An extension of the NEUREC program package for a microcomputer. *Comput. Prog. Biomed.* **18**:217–226.

Gras, H., and F. Killman (1983). NEUREC—a program package for 3-D reconstruction from serial sections using a microcomputer. *Comput. Prog. Biomed.* **17**:145–156.

Graveland, G. A., R. S. Williams, and M. DiFiglia (1985). A Golgi study of the human neostriatum: Neurons and afferent fibres. *J. Comp. Neurol.* **234**:317–333.

Greenberg, M., J. Stevens, and S. Ellias (1985). Highly irregular shapes of normal type C axons: Serial EM study. *Soc. Neurosci. Abstr.* **11**:184.4.

Greenough, W. T., and F. R. Volkmar (1973). Pattern of dendritic branching in occipital cortex of rats reared in complex environments. *Exp. Neurol.* **40**:491–504.

Greenough, W. T., C. S. Carter, C. Steerman, and T. J. De Voogd (1977). Sex differences in dendritic patterns in hamster preoptic area. *Brain Res.* **126**:63–72.

Gross, D. S., and J. M. Rothfeld (1985). Quantitative immunocytochemistry of hypothalamic and pituitary hormones: Validation of an automated, computerized image analysis system. *J. Histochem. Cytochem.* **33**: 11–20.

Gundersen, H. J. G., and R. Osterby (1981). Optimizing sampling efficiency of stereological studies in biology: Or "Do more less well!" *J. Microsc. (Oxf.)* **121**:65–73.

Gupta, M., T. M. Mayhew, K. S. Bedi, A. K. Sharma, and F. H. White (1982). Inter-animal variation and its influence on the overall precision of morphometric estimates based on nested sampling designs. *J. Microsc. (Oxf.)* **131:**147–154.

Haberly, L. B. (1983). Structure of the piriform cortex of the opossum. I. Description of neuron types with Golgi methods. *J. Comp. Neurol.* **213:**163–187.

Haberly, L. B., and J. M. Bower (1982). Graphical methods for three-dimensional rotation of complex axonal arborizations. *J. Neurosci. Methods* **6:**75–84.

Hansen, W. J. (1971). User engineering principles for interactive systems. In: *Proceedings of the Fall Joint Computer Conference.* Montvale, NJ: AFIPS Press.

Harding, E. F. (1971). The probabilities of rooted tree-shapes generated by random bifurcation. *J. Appl. Prob.* **3:**44–77.

Harms, H., and H. M. Aus (1984). Comparison of digital focus criteria for a TV microscope system. *Cytometry* **5:**236–243.

Harpring, J. E., J. C. Pearson, J. R. Norris, and B. L. Mann (1985). Subclassification of neurons in the ventrobasal complex of the dog: Quantitative Golgi study using principal component analysis. *J. Comp. Neurol.* **242:**230–246.

Harris, K. M., and J. K. Stevens (1988). Study of dendritic spines by serial electron microscopy and three-dimensional reconstructions. In: *Intrinsic Determinants of Neuronal Form and Function* (R. J. Lasek and M. M. Black, eds.). New York: Alan R. Liss, pp. 179–199.

Hartman, M. L., A. J. Beitz, J. E. Madl, and R. R. Mize (1988). Glutamate-like antibody staining in the cat superior colliculus is reduced by visual decortication. *Invest. Ophthalmol. Vis. Sci. [Suppl].* **29:**32.

Hengstenberg, R., H. Bulthoff, and B. Hengstenberg (1983). Three-dimensional reconstruction and stereoscopic display of neurons in the fly visual system. In: *Functional Neuroanatomy* (N. J. Strausfeld, ed.). Berlin: Springer-Verlag, pp. 183–205.

Henson, O. W., and M. M. Henson (1986). Morphometric analysis of cochlear structures in the mustached bat, *Pteronotus parnellii parnellii.* In: *3rd International Symposium on Animal Sonar Systems. Helsingor, Denmark.* New York: Plenum Press, pp. 301–305.

Hibbard, L. S., and R. A. Hawkins (1984). Three-dimensional reconstruction of metabolic data from quantitative autoradiography of rat brain. *Am. J. Physiol.* **247:**E412–E419.

Hibbard, L. S., and R. A. Hawkins (1988). Objective image alignment for three-dimensional reconstruction of digital autoradiograms. *J. Neurosci. Meth.* **26:**55–74.

Hibbard, L. S., J. S. McGlone, D. W. Davis, and R. A. Hawkins (1987). Three-dimensional representation and analysis of brain energy metabolism. *Science* **236:**1641–1646.

Hillman, D. E. (1976). A tridimensional reconstruction computer system for neuroanatomy. *Comput. Med.* **5**(6)**:**1–2.

Hillman, D. E. (1979). Neuronal shape parameters and substructures as a basis of neuronal form. In: *The Neuroscience Fourth Study Program* (F. O. Schmitt and F. G. Worden, eds.). Cambridge, MA: MIT Press, pp. 477–498.

Hillman, D. E., R. Llinas, and M. Chujo (1977). Automatic and semiautomatic analysis of nervous system structure. In: *Computer Analysis of Neuronal Structures* (R. D. Lindsay, ed.). New York: Plenum Press, pp. 73–90.

His, W. (1880). *Anatomie Menschlicher Embryonen.* Leipzig: Vogel.

Hitchcock, P. F., and S. S. Easter, Jr. (1986). Retinal ganglion cells in goldfish: Qualitative classification into four morphological types, and a quantitative study of the development of one of them. *J. Neurosci.* **6**(4)**:** 1037–1050.

Hofman, M. A., A. C. Laan, and H. B. M. Uylings (1986). Bivariate linear models in neurobiology: Problems of concept and methodology. *J. Neurosci. Methods* **18:**103–114.

Hogan, R. N., and P. D. Coleman (1981). Experimental hyperphenylalaninemia: Dendritic in motor cortex of rat. *Exp. Neurol.* **74:**218–233.

Holdefer, R. N., T. T. Norton, and R. R. Mize (1988). Laminar organization and ultrastructure of GABA-immunoreactive neurons and processes in the dorsal lateral geniculate nucleus of the tree shrew (*Tupaia belangeri*). *Vis. Neurosci.* **1:**189–204.

Hollingworth, T., and M. Berry (1975). Network analysis of dendritic fields of pyramidal cells in neocortex and Purkinje cells in the cerebellum of the rat. *Phil. Trans. R. Soc. Lond. [Biol.]* **270:**227–264.

Honda, H., and J. B. Fisher (1978). Tree branch angle: Maximizing effective leaf area. *Science* **199:**888–890.

Horsfield, K. (1980). Are diameter, length and branching ratios meaningful in the lung? *J. Theor. Biol.* **87:**773–784.

Horsfield, K., and M. J. Woldenberg (1986). Comparison of vertex analysis and branching ratio in the study of trees. *Respir. Physiol.* **65:**245–256.

Horwitz, B. (1981). Neuronal plasticity: How changes in dendritic architecture can affect the spread of postsynaptic potentials. *Brain Res.* **224:**412–418.

Houser, C. R., A. B. Harris, and J. E. Vaughn (1986). Time course of the reduction of GABA terminals in a model of focal epilepsy: A glutamic acid decarboxylase immunocytochemical study. *Brain Res.* **383:**129–145.

Hsu, S. M., L. Raine, and H. Fanger (1981). A comparative study of the peroxidase–antiperoxidase method and an aviden–biotin complex method for studying polypeptide hormones with radioimmunoassay antibodies. *Am. J. Clin. Pathol.* **75:**734–738.

Huijsmans, D. P. (1983). Closed 2D contour algorithms for 3D reconstruction. In: *Eurographics '83 Conference Proceedings* (P. J. W. ten Hagen, ed.). Amsterdam: Elsevier, pp. 157–168.

Huijsmans, D. P., W. H. Lamers, J. A. Los, J. Smith, and J. Strackee (1984). Computer-aided three-dimensional reconstruction from serial sections: A software package for reconstruction and selective image generation for complex topologies. In: *Eurographics '84 Conference Proceedings.* (K. Bo and H. Tucker, eds.). Amsterdam: Elsevier, pp. 3–13.

Huijsmans, D. P., W. H. Lamers, J. A. Los, and J. Strackee (1986). Toward computerized morphometric facilities: A review of 58 software packages for computer-aided three-dimensional reconstruction, quantification and picture generation from parallel serial sections. *Anat. Rec.* **216:**449–470.

Hull, C. D., J. P. McAllister, M. S. Levine, and A. M. Adinolfi (1981). Quantitative developmental studies of feline neostriatal spiny neurons. *Dev. Brain Res.* **1:**309–332.

Inoué, S. (1986). *Video Microscopy.* New York: Plenum Press.

Ipiña, S. L. (1987). A multivariate phenetic approach to neuronal nuclei; resemblances across species, with examples from three regions of cerebral cortex of a mammal, a bird and a reptile. *J. Theor. Biol.* **126:**105–124.

Ipiña, S. L., and A. Ruiz-Marcos (1986). Dendritic structure alterations induced by hypothyroidism in pyramidal neurons of the rat visual cortex. *Dev. Brain Res.* **29:**61–67.

Ipiña, S. L., A. Ruiz-Marcos, F. Escobar del Rey, and G. Morreale de Escobar (1987). Pyramidal cortex cell morphology studied by multivariate analysis: Effects of neonatal thyroidectomy, ageing and thyroxine substitution therapy. *Dev. Brain Res.* **37:**219–229.

Ireland, W., J. Heidel, and E. Uemura (1985). A mathematical model for the growth of dendritic trees. *Neurosci. Lett.* **54:**243–249.

Jacobs, J. R., and J. K. Stevens (1986). Changes in the organization of the neuritic cytoskeleton during nerve growth factor-activated differentiation of the PC12 cells: A serial electron microscopic study of the development and control of neurite shape. *J. Cell Biol.* **103:**895–906.

Jarvis, R. S., and A. Werritty (1975). Some comments on testing random topology stream network models. *Water Resources Res.* **11:**309–318.

Jarvis, R. S., and M. J. Woldenberg (1984). *River Networks. Benchmark Papers in Geology/80.* Stroudsberg, PA: Hutchinson Ross.

Jimenez, J., A. Santisteban, J. M. Carazo, and J. L. Carrascosa (1986). Computer graphic display method for visualizing three-dimensional biological structures. *Science* **232:**1113–1115.

Johnson, E. M., and J. J. Capowski (1983). A system for the three-dimensional reconstruction of biological structures. *Comput. Biomed. Res.* **16:**79–87.

Johnson, E. M., and J. J. Capowski (1985). Principles of reconstruction and three-dimensional display of serial sections using a computer. In: *The Microcomputer in Cell and Neurobiology Research* (R. R. Mize, ed.). New York: Elsevier, pp. 249–263.

Jones, E. G. (1975). Varieties and distribution of non-pyramidal cells in the somatic sensory cortex of the squirrel monkey. *J. Comp. Neurol.* **160:**205–268.

Jones, S. C., and D. Lu (1988). The evaluation of quantitative autoradiogram processing systems for cerebrovascular research. *J. Neurosci. Methods* **24**(1)**:**11–25.

Jones, W. H., and D. B. Thomas (1962). Changes in the dendritic organization of neurons in the cerebral cortex following deafferentiation. *J. Anat.* **96:**375–381.

Juraska, J. M. (1984). Sex differences in developmental plasticity in the visual cortex and the hippocampal dentate gyrus. In: *Sex Differences in the Brain, Progress in Brain Research,* Vol. 61 (G. J. de Vries, J. P. C. de Bruin, H. B. M. Uylings, and M. A. Corner, eds.). Amsterdam: Elsevier, pp. 205–214.

Juraska, J. M., W. T. Greenough, C. Elliot, K. J. Mack, and R. Berkowitz (1980). Plasticity in adult rat visual cortex: An examination of several cell populations after differential rearing. *Behav. Neurol. Biol.* **29:**157–167.

Kapps, C., and L. Mays (1978). An inexpensive system for digitizing pictoral information. *DECUS Proc.* **5**(2): 735–749.

Kastschenko, N. (1886). Methode zur genauen rekonstruktion kleinerer makroskopischer gegenstande. *Arch. Anat. Physiol. Abt.* **1:**388–394.

Kater, S. B., C. S. Cohan, G. A. Jacobs, and J. P. Miller (1986). Image intensification of stained, functioning, and growing neurons. In: *Optical Methods in Cell Physiology* (P. de Weer and B. M. Salzberg, eds.). New York: John Wiley & Sons, pp. 31–50.

Katz, L. C. (1987). Local circuitry of identified projection neurons in cat visual cortex brain slices. *J. Neurosci.* **7:**1223–1249.

Katz, L., and C. Levinthal (1972). Interactive computer graphics and representation of complex biological structures. *Ann. Rev. Biophys. Bioeng.* **1:**465–504.

Keppel, G. (1982). *Design and Analysis. A Researcher's Handbook,* 2nd ed. Englewood Cliffs, NJ: Prentice-Hall.

Kernell, D. (1966). Input resistance, electrical excitability and size of ventral horn cells in cat spinal cord. *Science* **152:**1637–1640.

Kimura, O., E. Dykes, and R. W. Fearnhead (1977). The relationship between the surface area of the enamel crowns of human teeth and that of the dentine–enamel junction. *Arch. Oral Biol.* **22:**677–683.

Kinnamon, J. C., T. A. Sherman, and S. D. Roper (1988). Ultrastructure of mouse vallate taste buds: III. Patterns of synaptic connectivity. *J. Comp. Neurol.* **270:**1–10.

Kropf, N., I. Sobel, and C. Levinthal (1985). Serial section reconstruction using CARTOS. In: *The Microcomputer in Cell and Neurobiology Research* (R. R. Mize, ed.). New York: Elsevier, pp. 265–292.

Kujoory, M. A., B. H. Mayall, an M. L. Mendelsohn (1973). Focus-assist device for a flying-spot microscope. *IEEE Trans. Biomed. Eng.* **20:**126–132.

Kusinitz, M. (1987). 3-D nerve images made easy. *New Med. Sci.* **1**(4)**:**5–7.

Lear, J. L., K. Mido, J. Plotnick, and R. Muth (1986). High-performance digital image analyzer for quantitative autoradiography. *J. Cereb. Blood Flow Metab.* **6:**625–629.

Leontovich, T. A. (1973). Methodik zur quantitativen Beschreibung subcorticaler Neurone. *J. Hirnforsch.* **14:** 59–87.

Leuba, G., and L. J. Garey (1984a). Orientation of dendrites in the lateral geninculate nucleus of the monkey. *Exp. Brain Res.* **56:**369–376.

Leuba, G., and L. J. Garey (1984b). Development of dendritic patterns in the lateral geniculate nucleus of the monkey: A quantitative Golgi study. *Dev. Brain Res.* **16:**285–299.

Leventhal, A. G., and J. D. Schall (1983). Structural basis of orientation sensitivity of cat retinal ganglion cells. *J. Comp. Neurol.* **220:**465–475.

Levinthal, C., and R. Ware (1972). Three dimensional reconstruction from serial sections. *Nature* **236:**207–210.

Levinthal, C., E. R. Macagno, and C. Tountas (1974). Computer-aided reconstruction from serial sections. *Fed. Proc.* **33**(12)**:**2336–2340.

Levinthal, F., E. R. Macagno, and C. Levinthal (1976). Anatomy and development of identified cells in isogenic organisms. *Cold Spring Harbor Symp. Quant. Biol.* **XL:**321–331.

Lieth, E. (1987). *Neuronal–Glial Interactions in CNS Development.* Doctoral Dissertation, University of North Carolina at Chapel Hil.

Light, A. R., and A. M. Kavookjian (1988). Morphology and ultrastructure of physiologically identified substantia gelatinosa (lamina II) neurons with axons that terminate in deeper dorsal horn laminae (III–IV). *J. Comp. Neurol.* **267:**172–189.

Light, A. R., and E. R. Perl (1979). Spinal termination of functionally identified primary afferent neurons with slowly conducting myelinated fibers. *J. Comp. Neurol.* **186**(2)**:**133–150.

Lindsay, R. D. (1977a). Computer analysis of neuronal structures. In: *Computers in Biology and Medicine* (G. P. Moore, ed.). New York: Plenum Press, pp. 71–79.

Lindsay, R. D. (1977b). The video computer microscope and A.R.G.O.S. In: *Computer Analysis of Neuronal Structures* (R. D. Lindsay, ed.). New York: Plenum Press, pp. 1–19.

Lindsay, R. D. (1977c). Tree analysis of neuronal processes. In: *Computer Analysis of Neuronal Structures* (R. D. Lindsay, ed.). New York: Plenum Press, pp. 149–164.

Lindsay, R. D. (1977d). Neuronal field analysis using Fourier series. In: *Computer Analysis of Neuronal Structures* (R. D. Lindsay, ed.). New York: Plenum Press, pp. 165–175.

Lindsay, R. D., and A. B. Scheibel (1976). Quantitative analysis of dendritic branching pattern of granular cells from human dentate gyrus. *Exp. Neurol.* **52:**295–310.

Litzinger, B. E. (1988). VGA-compatible issues perplex board buyers. *Comput. Technol. Rev.* **8**(3)**:**1–6.

Llinas, R., and D. E. Hillman (1975). A multipurpose tridimensional reconstruction computer system for neuroanatomy. In: *Golgi Centennial Symposium Proceedings* (M. Santini, ed.). New York: Raven Press, pp. 71–79.

Lopresti, V., E. R. Macagno, and C. Levinthal (1973). Structure and development of neuronal connections in isogenic organisms: Cellular interactions in the development of the optic lamina of *Daphnia. Proc. Natl. Acad. Sci. U.S.A.* **70**(2)**:**433–437.

Lund, J. S. (1973). Organisation of neurons in the visual cortex, area 17, of the monkey, *Macaca mulatta. J. Comp. Neurol.* **147:**455–496.

Macagno, E. R. (1978). Mapping synaptic sites between identified neuronsin leech CNS by means of 3-D computer reconstructions from serial sections. *Brain Theory Newsl.* **3**(3/4)**:**186–189.

Macagno, E. R., V. Lopresti, and C. Levinthal (1973). Structure and development of neuronal connections in isogenic organisms: Variations and similarities in the optic system of *Daphnia magna. Proc. Natl. Acad. Sci. U.S.A.* **70:**57–61.

Macagno, E. R., C. Levinthal, C. Tountas, R. Bornholdt, and R. Abba (1976). Recording and analysis of 3-D information from serial section micrographs: The CARTOS system. In: *Computer Technology in Neuroscience* (P. B. Brown, ed.). Washington, DC: Hemisphere, pp. 97–112.

MacDonald, N. (1983). *Trees and Networks in Biological Models.* Chichester: John Wiley & Sons.

Madri, J. A., and K. W. Barwick (1983). Use of avidin–biotin complex in an ELISA system: A quantitative comparison with two other immunoperoxidase detection systems using keratin antisera. *Lab. Invest.* **48:** 98–107.

Mannen, H. (1966). Contribution to the quantitative study of the nervous tissue. A new method for measurement of the volume and surface area of neurons. *J. Comp. Neurol.* **126:**75–89.

Mannen, H. (1975). Reconstruction of axonal trajectory of individual neurons in the spinal cord using Golgi-stained serial sections. *J. Comp. Neurol.* **159:**357–374.

Mannen, H., and Y. Sugiura (1975). Reconstruction of neurons of dorsal horn proper using Golgi-stained serial sections. *J. Comp. Neurol.* **168:**303–312.

March, S. C., I. Parikh, and P. Cuatrecasas (1974). A simplified method for cyanogen bromide activation of agarose for affinity chromatography. *Anal. Biochem.* **60:**149–152.

Mardia, K. V. (1975). Statistics of directional data. *J. R. Stat. Soc.* **B37:**349–393.

Marino, T. A., P. Nong Cook, L. T. Cook, and S. J. Dwyer III (1980). The use of computer imaging techniques to visualize cardiac muscle cells in three dimensions. *Anat. Rec.* **198:**537–546.

Marx, J. L. (1976). Computers: Helping to study nerve cell structure, *Science* **193:**565–608.

McEachron, D. (1986). *Function Mapping in Biology and Medicine: Computer Assisted Autoradiography.* Basel: Karger.

McEachron, D. L., C. R. Gallistel, J. L. Eilbert, and O. J. Tretiak (1988) The analytic and functional accuracy of a video densitometry system. *J. Neurosci. Methods* **25**(1)**:**63–74.

McGrath, R. A. (1985). Prodesign II: CAD on a budget. *Comput. Graphics World* **8**(12)**:**69–72.

McInroy, J. L., and J. J. Capowski (1977). A graphics subroutine package for the neuroscience display processor. *Comput. Graphics* **11**(1)**:**1–12.

McKanna, J. A. (1985). Micros applied to neuroanatomy: Computer-aided morphometry. In: *Microcomputers in the Neurosciences* (G. A. Kerkut, ed.). Oxford: Clarendon Press, pp. 152–201.

McKanna, J. A., and V. A. Casagrande (1985). Computerized radioautographic grain counting. In: *The Microcomputer in Cell and Neurobiology Research* (R. R. Mize, ed.). New York: Elsevier, pp. 356–373.

McKenzie, J. D., Jr., and B. A. Vogt (1976). An instrument for light microscopic analysis of three-dimensional neuronal morphology. *Brain Res.* **111:**411–415.

McMullen, N. T., E. M. Glaser, and M. Tagamets (1984). Morphometry of spine-free nonpyramidal neurons in rabbit auditory cortex. *J. Comp. Neurol.* **222**:383–395.

McMullen, N. T., B. Goldberger, C. M. Suter, and E. M. Glaser (1988). Neonatal deafening alters nonpyramidal dendrite orientation in auditory cortex: A computer microscope study in the rabbit. *J. Comp. Neurol.* **267**:92–106.

Mendelsohn, M. L., and B. H. Mayall (1972). Computer-oriented analysis of human chromosomes. III. Focus. *Comput. Biol. Med.* **2**:137–150.

Mercer, R. R., and J. D. Crapo (1987). Three-dimensional reconstruction of the rat acinus. *J. Appl. Physiol.* **63** (2):785–794.

Mercer, R. R., J. M. Laco, and J. D. Crapo (1987). Three-dimensional reconstruction of alveoli in the rat lung for pressure–volume relationships. *J. Appl. Physiol.* **62**(4):1480–1487.

Miletic, V., and H. J. Tan (1987). Morphology of Golgi-impregnated neurons in the cat and rat nucleus submedius. *Soc. Neurosci. Abstr.* **13**:329.2.

Millar, D. A., and E. D. Williams (1982). A step-wedge standard for the quantification of immunoperoxidase techniques. *Histochem. J.* **14**:609–620.

Miller, J. P., and G. A. Jacobs (1984). Relationships between neuronal structure and function. *J. Exp. Biol.* **112**:129–145.

Mize, R. R. (1983a). A computer electron microscope plotter for mapping spatial distributions in biological tissues. *J. Neurosci. Methods* **8**:183–195.

Mize, R. R. (1983b). A microcomputer system for measuring neuron properties from digitized images. *J. Neurosci. Methods* **9**:105–113.

Mize, R. R. (1984). Computer applications in cell and neurobiology: A review. *Int. Rev. Cytol.* **90**:83–124.

Mize, R. R. (1985a). Morphometric measurement using a computerized digitizing system. In: *The Microcomputer in Cell and Neurobiology Research* (R. R. Mize, ed.). New York: Elsevier, pp. 177–215.

Mize, R. R. (1985b). A microcomputer plotter for use with light and electron microscopes. In: *The Microcomputer in Cell and Neurobiology Research* (R. R. Mize, ed.). New York: Elsevier, pp. 112–133.

Mize, R. R. (1985c). Computer applications in neuroscience research. In: *Modern Neuroanatomical Methods, Short Course Syllabus* (H. J. Karten, ed.). Washington, DC: Society for Neuroscience, pp. 94–111.

Mize, R. R. (1987). Morphometry of antibody stained cells and synapses using an image analyzer. *Neuroscience [Suppl.]* **22**:S202.

Mize, R. R., and L. H. Horner (1984). Retinal synapses of the cat medial interlaminar nucleus and ventral lateral geniculate nucleus differ in size and synaptic organization. *J. Comp. Neurol.* **224**:579–590.

Mize, R. R., and L. H. Horner (1989). Origin, distribution, and morphology of serotonergic inputs to the cat superior colliculus: A quantitative light and electron microscope immunocytochemistry study. *Exp. Brain Res.* **75**:83–98.

Mize, R. R., R. F. Spencer, and L. H. Horner (1986). Quantitative comparison of retinal synapses in the dorsal and ventral parvicellular C, laminae of the cat dorsal lateral geniculate nucleus. *J. Comp. Neurol.* **248**:57–73.

Mize, R. R., R. N. Holdefer, and L. B. Nabors (1988). Quantitative immunocytochemistry using an image analyzer. I. Hardware, image processing, and data analysis. *J. Neurosci. Methods* **26**(1):1–23.

Moens, P. B., and T. Moens (1981). Computer measurements and graphics of three-dimensional cellular ultrastructure. *J. Ultrastruct. Res.* **75**:131–141.

Moraff, H. (1976). Laboratory programming: How can we really get it done? In: *Computer Technology in Neuroscience* (P. B. Brown, ed.). Washington, DC: Hemisphere, pp. 623–648.

Moreton, R. B. (1985). Choosing a microcomputer system. In: *Microcomputers in the Neurosciences* (G. A. Kerkut, ed.). Oxford: Clarendon Press, pp. 29–89.

Moyer, A., V. Moyer, and P. D. Coleman (1985). An inexpensive PC based system for quantification of neuronal processes. *Soc. Neurosci. Abstr.* **11**:261.18.

Mrzljak, L., H. B. M. Uylings, I. Kostovic, and C. G. Van Eden (1988). Prenatal development of neurons in the human prefrontal cortex: I. A qualitative Golgi study. *J. Comp. Neurol.* **271**:355–386.

Mungai, J. M. (1967). Dendritic patterns in the somatic sensory cortex of the cat, *J. Anat.* **101**:403–418.

Nabors, L. B., E. Songu-Mize, and R. R. Mize (1988). Quantitative immunocytochemistry using an image analyzer. II. Concentration standards for transmitter immunocytochemistry. *J. Neurosci. Methods* **26**:25–34.

Nelson, A. C. (1986). Computer-aided microtomography with true 3-D display in electron microscopy. *J. Histochem. Cytochem.* **34**(1):57–60.

Nierzwicki-Bauer, S. A., D. L. Balkwill, and S. E. Stevens, Jr. (1983). Use of a computer-aided reconstruction system to examine the three-dimensional architecture of cyanobacteria. *J. Ultrastruc. Res.* **84:**73–82.

Norman, D. A. (1981). The trouble with UNIX. *Datamations* 27:(12)**:**139.

Odhner, N. (1911). Eine neue graphische Methode zur Rekonstruktion von Schnittserien in schrager Stellung. *Anat. Anz.* **39:**273–281.

Overdijk, J., H. B. M. Uylings, K. Kuypers, and A. W. Kamstra (1978). An economical, semi-automtic system for measuring cellular tree structures in three dimensions, with special emphasis on Golgi-impregnated neurons. *J. Microsc.* **114**(3)**:**271–284.

Palacios, J. M., D. L. Niehoff, and M. J. Kuhar (1981). Receptor autoradiography with tritium-sensitive film: Potential for computerized densitometry. *Neurosci. Lett.* **25:**101–105.

Paldino, A. M. (1979). A novel version of the computer microscope for the quantitative analysis of biological structures: Application to neuronal morphology. *Comput. Biomed. Res.* **12:**413–431.

Paldino, A., and E. Harth (1977a). A measuring system for analyzing neuronal fiber structure. In: *Computer Analysis of Neuronal Structures* (R. D. Lindsay, ed.). New York: Plenum Press, pp. 59–71.

Paldino, A., and E. Harth (1977b). A computerized study of Golgi-impregnated axons in rat visual cortex. In: *Computer Analysis of Neuronal Structures* (R. D. Lindsay, ed.). New York: Plenum Press, pp. 189–207.

Park, D. (1985). Does Horton's law of branch length apply to open branching systems? *J. Theor. Biol.* **112:** 299–313.

Parnavelas, J. G., and H. B. M. Uylings (1980). The growth of non-pyramidal neurons in the visual cortex of the rat: A morphometric study. *Brain Res.* **193:**373–382.

Patterson, H. A., W. B. Warr, and A. J. Kleinman (1976). A mapping device for attachment to the light microscope. Technical note. *Brain Res.* **102:**323–328.

Pavlidis, T. (1982). *Algorithms for Graphics and Image Processing.* Rockville, MD: Computer Science Press.

Pearlstein, R. A., and R. L. Sidman (1986). Computer graphics presentation modes for biologic data. *Anal. Quant. Cytol. Histol.* **8**(2)**:**89–95.

Pearson, J. C., J. R. Norris, and C. H. Phelps (1985). Subclassification of neurons in the subthalamic nucleus of the lesser bushbaby *Galago senegalensis*: A quantitative Golgi study using principal components analysis. *J. Comp. Neurol.* **238:**323–339.

Percheron, G. (1979). Quantitative analysis of dendritic branching. I. Simple formulae for the quantitative analysis of dendritic branching. *Neurosci. Lett.* **14:**287–293.

Perkins, W. J., and R. J. Green (1982). Three-dimensional reconstruction of biological sections. *J. Biomed. Eng.* **4:**37–43.

Poler, S. M., S. Akeson, and D. G. Flaming (1985). Selection of hardware and software for laboratory microcomputers. In: *The Microcomputer in Cell and Neurobiology Research* (R. R. Mize, ed.). New York: Elsevier, pp. 47–82.

Pool, C. W., S. Madlener, P. C. Diegenbach, A. A. Sluiter, and P. Van der Sluis (1984). Quantification of antiserum retivity in immunocytochemistry: Two new methods for measuring peroxidase activity on antigen-coupled beads incubated according to an immunocytoperoxidase method. *J. Histochem. Cytochem.* **32:** 921–928.

Porath, J., R. Axen, and S. Ernback (1967). Chemical coupling of proteins to agarose. *Nature* **215:**1491–1492.

Pratt, W. K. (1978). *Digital Image Processing.* New York: Plenum Press.

Prentice, M. J. (1978). On invariant tests of uniformity for directions and orientations. *Ann. Stat.* **6:**169–176.

Prothero, J. S., and J. W. Prothero (1982). Three-dimensional reconstruction from serial sections. I. A portable microcomputer-based software package in FORTRAN. *Comput. Biomed. Res.* **15:**598–604.

Prothero, J. S., and J. W. Prothero (1986). Three-dimensional reconstruction from serial sections IV. The reassembly problem. *Comput. Biomed. Res.* **19:**361–373.

Prothero, J. W., A. Tamarin, and R. Pickering (1973). Morphometrics of living specimens. A methodology for the quantitative three-dimensional study of growing microscopic embryos. *J. Microscp.* **101**(1)**:**31–58.

Prothero, J. S., M. Riggins, A. Lindsay, R. Harris, and J. W. Prothero (1985). Three-dimensional reconstruction from serial sections. III. AUTOSCAN, a software package in FORTRAN for semiautomated photomicrography. *Comput. Biomed. Res.* **18:**132–136.

Pullen, A. H. (1982). A structured program in BASIC for the analysis of peripheral nerve morphometry. *J. Neurosci. Methods* **5:**103–120.

Purves, D., and J. W. Lichtman (1985). Geometrical differences among homologous neurons in mammals. *Science* **228:**298–302.

Putnam, B. W. (1987). *RS-232 Simplified.* Englewood Cliffs, NJ: Prentice-Hall.

Radermacher, M., and J. Frank (1983). Representation of three-dimensionally reconstructed objects in electron microscopy by surfaces of equal density. *J. Microsc.* **136**(1):77–85.

Radermacher, M., T. Wagenknecht, A. Verschoor, and J. Frank (1987a). Three-dimensional reconstruction from a single-exposure, random conical tilt series applied to the 50S ribosomal subunit of *Escherichia coli*. *J. Microsc.* **146**(2):113–136.

Radermacher, M., T. Wagenknecht, A. Verschoor, and J. Frank (1987b). Three-dimensional structure of the large ribosomal subunit from *Escherichia coli*. *EMBO J.* **6**(4):1107–1114.

Rakic, P., L. J. Stensas, E. P. Sayre, and R. L. Sidman (1974). Computer-aided three-dimensional reconstruction and quantitative analysis of cells from serial electron microscopic montages of foetal monkey brain. *Nature* **250**:31–34.

Rall, W. (1959). Branching dendritic trees and motorneurons membrane resistivity. *Exp. Neurol.* **1**:491–527.

Rall, W. (1977). Core conductor theory and cable properties of neurons, In: *Handbook of Physiology*, Section I: *The Nervous System*, Vol. 1, part 1 (J. M. Brookhart and V. B. Mountcastle, Section eds.; E. R. Kandel, Vol. ed.). Bethesda: American Physiological Society, pp. 39–97.

Ramm, P., and J. H. Kulick (1985). Principles of computer-assisted imaging in autoradiographic densitometry. In: *The Microcomputer in Cell and Neurobiology Research* (R. R. Mize, ed.). New York: Elsevier, pp. 311–334.

Ramm, P., J. H. Kulick, M. P. Stryker, and B. J. Frost (1984). Video and scanning micordensitometer-based imaging systems in autoradiographic densitometry. *J. Neurosci. Methods* **11**:89–100.

Reddy, D. R., W. J. Davis, R. B. Ohlander, and D. J. Bihary (1973). Computer analysis of neuronal structure. In: *Intracellular Staining in Neurobiology* (S. B. Kater and C. Nicholson, eds.). New York: Springer-Verlag, pp. 227–253.

Reed, D. J., R. Gold, and D. R. Humphrey (1980). A simple computerized system for plotting the locations of cells of specified sizes in a histological section. *Neurosci. Lett.* **20**:233–236.

Reis, D. J., R. H. Benno, L. W. Tucker, and T. H. Joh (1982). Quantitative immunocytochemistry of tyrosine hydroxylase in brain. In: *Cytochemical Methods in Neuroanatomy* (V. Chan-Palay and S. L. Palay, eds.). New York: Alan R. Liss, pp. 205–228.

Réthelyi, M. (1981). The modular construction of the neuropil in the substantia gelatinosa of the cat's spinal cord. A computer aided analysis of Golgi-specimens. *Acta Morphol. Acad. Sci. Hung.* **29**(1):1–18.

Réthelyi, M., and J. J. Capowski (1977). The terminal arborization pattern of primary afferent fibers in the substantia gelationosa of the spinal cord in the cat. *J. Physiol. (Paris)* **73**:269–277.

Reuman, S. R., and J. J. Capowski (1984). Automated neuron tracing using the Marr–Hildreth zerocrossing technique. *Comput. Biomed. Res.* **17**:93–115.

Ritz, L. A., and J. D. Greenspan (1985). Morphological features of lamina V neurons receiving nociceptive input in cat sacrocaudal spinal cord. *J. Comp. Neurol.* **238**:440–452.

Rodieck, R. W., K. F. Binmoeller, and J. Dineen (1985). Parasol and midget ganglion cells of the human retina. *J. Comp. Neurol.* **233**:115–132.

Rogers, W. T. (1986). Digital microscopy for neurobiology research. *College Report, The DuPont Company* Sept/Oct.

Rose, P. K., S. A. Kierstead, and S. J. Vanner (1985). A quantitative analysis of the geometry of cat motoneurons innervating neck and shoulder muscles. *J. Comp. Neurol.* **239**:89–107.

Rose, R. D., and D. Rohrlich (1987). Counting sectioned cells via mathematical reconstruction. *J. Comp. Neurol.* **263**:365–386.

Rosenfeld, A., and A. C. Kak (1982). *Digital Image Processing*, 2nd ed. New York: Academic Press.

Rosenthal, B. M., and W. L. R. Cruce (1984). Contralateral motoneuron dendritic changes induced by transection of frog spinal nerves. *Exp. Neurol.* **85**:565–573.

Rosenthal, B. M., and W. L. R. Cruce (1985). The dendritic extent of motoneurons in frog brachial spinal cord: A computer reconstruction of HRP-filled cells. *Brain Behav. Evol.* **27**:106–114.

Royer, S. M., and J. C. Kinnamon (1988). Ultrastructure of mouse foliate taste buds: Synaptic and nonsynaptic interactions between taste cells and nerve fibers. *J. Comp. Neurol.* **270**:11–24.

Ruigrok, T. J. H., A. Crowe, and H. J. Ten Donkelaar (1985). Dendrite distribution of identified motoneurons in the lumbar spinal cord of the turtle *Pseudemys scripta elegans*. *J. Comp. Neurol.* **238**:275–285.

Ruiz-Marcos, A. (1983). Mathematical models of cortical structures and their application to the study of pathological conditions. In: *Ramon y Cajal's Contribution to the Neurosciences* (S. Grisolia, C. Guerri, F. Samson, S. Norton, and F. Reinoso-Suarez, eds.). Amsterdam: Elsevier, pp. 209–222.

Ruiz-Marcos, A., and S. L. Ipiña (1986). Hypothyroidism affects preferentially the dendritic densities on the more superficial region of pyramidal neurons of the rat cerebral cortex. *Dev. Brain Res.* **28:**259–262.

Ruiz-Marcos, A., and F. Valverde (1970). Dynamic architecture of the visual cortex. *Brain Res.* **19:**25–39.

Sadler, M., and M. Berry (1983). Morphometric study of the development of Purkinje cell dendritic trees in the mouse using vertex analysis. *J. Microsc. (Oxf.)* **131:**341–354.

Sasaki, S., J. K. Stevens, and N. Bodick (1983). Serial reconstruction of microtubular arrays within dendrites of the cat retinal ganglion cell: The cytoskeleton of a vertebrate dendrite. *Brain Res.* **259:**193–206.

Schall, J. D., and A. G. Leventhal (1987). Relationships between ganglion cell dendritic structure and retinal topography in the cat. *J. Comp. Neurol.* **257:**149–159.

Schipper, J., and F. J. H. Tilders (1983). A new technique for studying specificity of immunocytochemical procedures: Specificity of serotonin immunostaining. *J. Histochem. Cytochem.* **31:**12–18.

Schipper, J., T. R. Werkman, and F. J. H. Tilders (1984). Quantitative immunocytochemistry of corticotropin-releasing factor CRF: Studies on nonbiological models and on hypothalamic tissues of rats after hypophysectomy, adrenalectomy and dexamethasone treatment. *Brain Res.* **293:**111–118.

Schwaber, J. S., W. T. Rogers, K. Satoh, and H. C. Fibiger (1987). Distribution and organization of cholinergic neurons in the rat forebrain demonstrated by computer-aided data acquisition and three-dimensional reconstruction. *J. Comp. Neurol.* **263:**309–325.

Sedivec, M. J., J. J. Capowski, J. Ovelmen-Levitt, and L. M. Mendell (1982). Changes in dendritic organization of spinocervical tract neurons following partial chronic deafferentation. *Soc. Neurosci. Abstr.* **8:**217.1.

Sedivec, M. J., J. J. Capowski, and L. M. Mendell (1986). Morphology of HRP-injected spinocervical tract neurons: Effect of dorsal rhizotomy. *J. Neurosci.* **6**(3)**:**661–672.

Seldon, H. L., and D. G. Von Keyserlingk (1978). Preferential orientations of nerve processes in cat and monkey cortex. *J. Anat.* **126:**65–86.

Shantz, M. J. (1976). A minicomputer-based image analysis system. In: *Computer Technology in Neuroscience* (P. B. Brown, ed.). Washington, DC: Hemisphere, pp. 113–129.

Shantz, M. J. (1980). Human cortex reconstruction modeled with 70,000 polygons (30,000 polygon portion). Slide 42 of the 1979 Core System Slide Set. *Comput. Graphics* **13**(4)**:**242–246.

Shantz, M. J., and G. D. McCann (1978). Computational morphology: Three-dimensional computer graphics for electron microscopy. *IEEE Trans. Biomed. Eng.* **25:**99–103.

Shay, J. (1975). Economy of effort in electron microscope morphometry. *Am. J. Pathol.* **81:**503–512.

Shneiderman, B. (1987). *Designing the User Interface: Strategies for Effective Human–Computer Interaction.* Reading, MA: Addison-Wesley.

Sholl, D. A. (1953). Dendritic organization in the neurons of the visual and motor cortices of the cat. *J. Anat.* **87:**387–406.

Sholl, D. A. (1956). *The Organization of the Cerebral Cortex.* London: John Wiley & Sons.

Shreve, R. L. (1966). Statistical law of stream numbers. *J. Geol.* **74:**17–37.

SIGGRAPH (1983). Conference proceedings. *Comput. Graphics* **17**(3)**:**i.

Simons, D. J., and T. A. Woolsey (1984). Morphology of Golgi–Cox-impregnated barrel neurons in rat SmI cortex. *J. Comp. Neurol.* **230:**119–132.

Sing, R. L. A., and E. D. Salin (1985). Programming languages for the laboratory. In: *The Microcomputer in Cell and Neurobiology Research* (R. R. Mize, ed.). New York: Elsevier, pp. 25–45.

Sinha, U. K., L. I. Terr, F. R. Galey, and F. H. Linthicum (1987). Computer-aided three-dimensional reconstruction of the cochlear nerve root. *Arch. Otolaryngol. Head Neck Surg.* **113:**651–655.

Sivapragasam, S., J. G. Clement, and E. Dykes (1982). A three-dimensional assessment of dental asymmetry in human maxillary first premolar teeth. *Acta Stereol.* **82:**297–304.

Slepecky, N., H. Larsen, and C. Angelborg (1984). Computerized reconstruction of the regional blood flow in the rodent cochlea. *Hearing Res.* **15:**95–101.

Smart, J. S. (1969). Topological properties of channel networks. *Bull. Geol. Soc. Am.* **80:**1757–1774.

Smit, G. J., and H. B. M. Uylings (1975). The morphometry of the branching pattern in dendrites of the visual cortex pyramidal cells. *Brain Res.* **87:**41–53.

Smit, G. J., H. B. M. Uylings, and L. Veldmaat-Wansink (1972). The branching pattern in dendrites of cortical neurons. *Acta Morphol. Neerl. Scand.* **9:**253–274.

Smith, R. G. (1987). MONTAGE: A system for three-dimensional reconstruction by personal computer. *J. Neurosci. Methods* **21:**55–69.

Sobel, I., C. Levinthal, and E. R. Macagno (1980). Special techniques for the automatic computer reconstruction of neuronal structures. *Annu. Rev. Biophys. Bioeng.* **9:**347–362.

Sokal, R. R., and F. J. Rohlf (1981). *Biometry,* 2nd ed. San Francisco: W. H. Freeman.

Somogyi, J., J. J. Capowski, F. Zsuppán, and L. Dobransky (1987). Data collecting computer terminal for 3D stick model reconstruction of neuronal structure. *Neurosci.* **22**(suppl)**:**S378.

Spacek, J., and M. Hartmann (1983). Three-dimensional analysis of dendritic spines. *Anat. Embryol.* **167:**289–310.

Spacek, J., and A. R. Lieberman (1974). Ultrastructure and three-dimensional organization of synaptic glomeruli in rat somatosensory thalamus. *J. Anat.* **117**(3)**:**487–516.

Speck, P. T., and N. J. Strausfeld (1983). Portraying the third dimension in neuroanatomy. In: *Functional Neuroanatomy* (N. J. Strausfeld, ed.). Berlin: Springer-Verlag, pp. 156–182.

Steffen, H., and H. Van der Loos (1980). Early lesions of mouse vibrissal follicles: their influence on dendrite orientation in the cortical barrelfield. *Exp. Brain Res.* **40:**419–431.

Stephens, M. A. (1964). The testing of unit vectors for randomness. *J. Am. Stat. Assoc.* **59:**160–167.

Stephens, M. A. (1966). Statistics connected with the uniform distribution: Percentage points and application to testing for randomness of directions. *Biometrika* **53:**235–240.

Stephens, M. A. (1969). Multi-sample tests for the Fisher distribution for directions. *Biometrika* **56:**169–181.

Sternberger, L. A. (1979). *Immunocytochemistry,* 2nd ed. New York: John Wiley & Sons.

Stevens, C. F., and A. N. Van den Pol (1982). Appendix to Van den Pol and Cassidy (1982). *J. Comp. Neurol.* **204:**65–98.

Stevens, J. K., and J. Trogadis (1984). Computer-assisted reconstruction from serial electron micrographs: A tool for the systematic study of neuronal form and function. *Adv. Cell. Neurobiol.* **5:**341–369.

Stevens, J. K., T. L. Davis, N. Friedman, and P. Sterling (1980a). A systematic approach to reconstructing microcircuitry by electron microscopy of serial sections. *Brain Res. Rev.* **2:**265–293.

Stevens, J. K., B. A. McGuire, and P. Sterling (1980b). Toward a functional architecture of the retina: Serial reconstruction of adjacent ganglion cells. *Science* **207:**317–319.

Stevens, J. K., R. Jacobs, and M. L. Jackson (1984). Rings of cross-striated fibrils within the cat cone pedicle: A computer-assisted serial EM analysis. *Invest. Ophthal. Vis. Sci.* **25:**201–208.

Stevens, J. K., J. Trogadis, and J. R. Jacobs (1988). Development and control of axial neurite form: A serial electron microscopic analysis. In: *Intrinsic Determinants of Neuronal Form and Function* (R. J. Lasek and M. M. Black, eds.). New York: Alan R. Liss, pp. 115–145.

Strahler, A. N. (1952). Hypsometric (area-altitude) analysis of erosial topography. *Bull. Geol. Soc. Am.* **63:** 1117–1142.

Streefkerk, J. G., and A. M. Deelder (1975). Serodiagnostic application of immunohistoperoxidase reactions on antigen-coupled agarose beards. *J. Immunoglobul. Methods* **7:**225–236.

Streefkerk, J. G., and M. van der Ploeg (1973). Quantitative aspects of cytochemical peroxidase procedures investigated in a model system. *J. Histochem. Cytochem.* **21:**715–722.

Streefkerk, J. G., M. van der Ploeg, and P. van Duijn (1975). Agarose beads as matrices for proteins in cytophotometric investigations of immunohistoperoxidase procedures. *J. Histochem. Cytochem.* **23:**243–250.

Street, C. H., and R. R. Mize (1983). A simple microcomputer-based three-dimensional serial section reconstruction system (MICROS). *J. Neurosci. Methods* **7:**359–375.

Street, C. H., and R. R. Mize (1985). An algorithm for removing hidden lines in serial section reconstructions using MICROS. In: *The Microcomputer in Cell and Neurobiology Research* (R. R. Mize, ed.). New York: Elsevier, pp. 293–308.

Sugiura, Y., E. Schrank, and E. R. Perl (1985). Central terminal distribution of unmyelinated afferent fibers. *Soc. Neurosci. Abstr.* **11:**35.6.

Sugiura, Y., C. L. Lee, and E. R. Perl (1986). Central projections of identified, unmyelinated (C) afferent fibers innervating mammalian skin. *Science* **234:**358–361.

Sundsten, J. W., and J. W. Prothero (1983). Three-dimensional reconstruction from serial sections: II. A microcomputer-based facility for rapid data collection. *Anat. Rec.* **207:**665–671.

Sutherland, I. E. (1965). SKETCHPAD: *A Man–Machine Graphical Communication System.* Cambridge, MA: MIT Lincoln Lab. Abridged version Baltimore: Spartan Books.

Sutherland, I. E., R. F. Sproull, and R. A. Schumacker (1974). A characterization of ten hidden-surface algorithms. *Comput. Surv.* **6:**1–55.

Tamamaki, N., K. Abe, and Y. Nojyo (1988). Three-dimensional analysis of the whole axonal arbors originating from single CA2 pyramidal neurons in the rat hippocampus with the aid of a computer graphic technique. *Brain Res.* **452:**255–272.

Tayrien, M. W., and R. Loy (1984). Computer-assisted image analysis to quantify regional and specific ligand binding: Up regulation of [^{3}H]QNB and [^{3}H]WB4101 binding in denervated hippocampus. *Brain Res. Bull.* **13:**743–750.

Ten Hoopen, M., and H. A. Reuver (1970). Probabilistic analysis of dendritic branching patterns of cortical neurons. *Kybernetik* **6:**176–188.

Thompson, R. P., Y. M. Wong, and T. F. Fitzharris (1983). A computer graphic study of cardiac truncal septation. *Anat. Rec.* **206:**207–214.

Tieman, D. G., and R. K. Murphey (1985). A computer-assisted video technique for preparing pictures of intracellularly filled, whole-mounted neurons in the cricket. *Soc. Neurosci. Abstr.* **11:**184.5.

Tieman, D. G., R. K. Murphey, J. T. Schmidt, and S. B. Tieman (1986). A computer-assisted video technique for preparing high resolution pictures and sterograms from thick specimens. *J. Neurosci. Methods* **17:**231–245.

Tieman, S. B., and H. V. B. Hirsch (1982). Exposure to lines of only one orientation modifies dendritic morphology of cells in the visual cortex of the cat. *J. Comp. Neurol.* **211:**353–362.

Toga, A. W., and T. L. Arnicar (1985). Image analysis of brain physiology. *Comput. Graph. Appl.* **5:**20–25.

Toga, A. W., and T. L. Arnicar-Sulze (1987). Digital image reconstruction for the study of brain structure and function. *J. Neurosci. Methods* **20:**7–21.

Tolivia, J., D. Tolivia, and M. Alvarez-Uria (1986). A three-dimensional reconstruction program for personal computers. *J. Neurosci. Methods* **17:**55–62.

Tömböl, T., M. Madarasz, and J. Martos (1983). Quantitative aspects in the analysis of the synaptic architecture of thalamic sensory relay nuclei. *Acta Biol. Hung.* **34**(2–3)**:**275–301.

Triller, A., and H. Korn (1986). Variability of axonal arborizations hides simple rules of construction: A topological study from HRP intracellular injections. *J. Comp. Neurol.* **253:**500–513.

Tumosa, N., S. B. Tieman, and D. G. Tieman (1989). Binocular competition affects the patter and intensity of ocular activation columns in the visual cortex of cats. *Vis. Neurosci.* (in press).

Ulfhake, B., and J.-O. Kellerth (1981). A quantitative light microscopic study of the dendrites of cat spinal α-motoneurons after intracellular staining with HRP. *J. Comp. Neurol.* **202:**571–584.

Unnerstall, J. R., D. L. Niehoff, M. J. Kuhar, and J. M. Palacios (1982). Quantitative receptor autoradiography using [^{3}H]Ultrofilm: Application to multiple benzodiazepine receptors. *J. Neurosci. Methods* **6:**59–73.

Upfold, J. B., M. S. R. Smith, and M. J. Edwards (1987). Three-dimensional reconstruction of tissue using computer-generated images. *J. Neurosci. Methods* **20:**131–138.

Usson, Y., S. Torch, and G. Douret d'Aubigny (1987). A method for automatic classification of large and small myelinated fibre populations in peripheral nerves. *J. Neurosci. Methods* **20:**237–248.

Uylings, H. B. M., and H. K. P. Feirabend (1983). Golgi staining methods, In: *Manual of Neuroanatomical Staining Methods* (H. B. M. Uylings, E. Marani, A. H. M. Lohman, and G. Vrensen, eds.). Amsterdam: Netherlands Neuroanatomists Verhaart Meetings Publication, pp. 9.1–9.42 (in Dutch).

Uylings, H. B. M., and G. J. Smit (1975). Three-dimensional branching structure of pyramidal cell dendrites. *Brain Res.* **87:**55–60.

Uylings, H. B. M., G. J. Smit, and W. A. M. Veltman (1975). Ordering methods in quantitative analysis of branching structures of dendritic trees. *Adv. Neurol.* **12:**247–254.

Uylings, H. B. M., K. Kuypers, M. C. Diamond, and W. A. M. Veltman (1978a). Effects of differential environment on plasticity of dendrites of cortical pyramidal neurons in adult rats. *Exp. Neurol.* **62:**658–677.

Uylings, H. B. M., K. Kuypers, and W. A. M. Veltman (1978b). Environmental influences on neocortex in later life. In: *Maturation of the Nervous System, Progress in Brain Research,* Vol. 48 (M. A. Corner, R. E. Baker, N. E. van de Poll, D. F. Swaab, and H. B. M. Uylings, eds.). Amsterdam: Elsevier, pp. 261–274.

Uylings, H. B. M., J. G. Parnavelas, H. Walg, and W. A. M. Veltman (1980). The morphometry of the branching pattern of developing non-pyramidal neurons in the visual cortex of rats. *Mikroskopie* **37**(Suppl.)**:**220–224.

Uylings, H. B. M., R. W. H. Verwer, J. van Pelt, and J. G. Parnavelas (1983). Topological analysis of dendritic growth at various stages of cerebral development. *Acta Stereol.* **2**(1)**:**55–62.

Uylings, H. B. M., A. Ruiz-Marcos, and J. van Pelt (1986a). The metric analysis of three-dimensional dendritic tree patterns: A methodological review. *J. Neurosci. Methods* **18:**127–151.

Uylings, H. B. M., C. G. Van Eden, and M. A. Hofman (1986b). Morphometry of size/volume variables and comparison of their bivariate relations in the nervous system under different conditions. *J. Neurosci. Methods* **18:**19–37.

Uylings, H. B. M., M. A. Hofman, and M. A. H. Matthijssen (1987). Comparison of bivariate linear relations in biological allometry research. *Acta Stereol.* **6**(Suppl. III)**:**467–472.

Valverde, F. (1970). The Golgi method. A tool for comparative structural analyses. In: *Contemporary Research Methods in Neuroanatomy* (W. J. H. Nauta and S. O. E. Ebbeson, eds.). Berlin: Springer-Verlag, pp. 12–31.

Van Dalen, J. P. R., W. Knapp, and J. S. Ploem (1973). Microfluorometry on antigen–antibody interaction in immunofluorescence using antigens covalently bound to agarose beads. *J. Immunol. Methods* **2:**383–392.

Van den Pol, A. N., and J. R. Cassidy (1982). The hypothalamic arcuate nucleus of the rat—a quantitative Golgi analysis. *J. Comp. Neurol.* **204:**65–98.

Van der Loos, H. (1959). [*Dendro–dendritic Connections in the Cerebral Cortex.*] Doctoral dissertation, University of Amsterdam (in Dutch).

Vandesande, F. (1979). A critical review of immunocytochemical methods for light microscopy. *J. Neurosci. Methods* **1:**3–23.

Van Duijn, P., E. Pascoe, and M. van der Ploeg (1967). Theoretical and experimental aspects of enzyme determination in a cytochemical model system of polyacrylamide films containing alkaline phosphatase. *J. Histochem. Cytochem.* **15:**631–645.

Van Pelt, J., and R. W. H. Verwer (1983). The exact probabilities of branching patterns under terminal and segmental growth hypothesis. *Bull. Math. Biol.* **45:**269–285.

Van Pelt, J., and R. W. H. Verwer (1984a). New classification methods of branching patterns, *J. Microsc. (Oxf.)* **136:**23–34.

Van Pelt, J., and R. W. H. Verwer (1984b). Cut trees in the topological analysis of branching patterns. *Bull. Math. Biol.* **46:**283–294.

Van Pelt, J., and R. W. H. Verwer (1985). Growth models (including terminal and segmental branching) for topological binary trees. *Bull. Math. Biol.* **47:**323–336.

Van Pelt, J., and R. W. H. Verwer (1986). Topological properties of binary trees grown with order-dependent branching probabilities. *Bull. Math. Biol.* **48:**197–211.

Van Pelt, J., and R. W. H. Verwer (1987). Mean centrifugal order as a measure for branching pattern topology. *Acta Stereol.***6:**393–397.

Van Pelt, J., R. W. H. Verwer, and H. B. M. Uylings (1986). Application of growth models to the topology of neuronal branching patterns. *J. Neurosci. Methods* **18:**153–165.

Vaughn, J. E., R. P. Barber, and T. J. Sims (1988). Dendritic development and preferential growth into synaptogenic fields: A quantitative study of Golgi-impregnated spinal motor neurons. *Synapse* **2:**69–78.

Veen, A., and L. D. Peachey (1977). TROTS: A computer graphics system for three-dimensional reconstruction from serial sections. *Comput. Graphics* **2:**135–150.

Verwer, R. W. H., and J. van Pelt (1983). A new method for the topological analysis of neuronal tree structures. *J. Neurosci. Methods* **8:**335–351.

Verwer, R. W. H., and J. Van Pelt (1985). Topological analysis of binary tree structures when occasional multifurcations occur. *Bull. Math. Biol.* **47:**305–316.

Verwer, R. W. H., and J. Van Pelt (1986). Descriptive and comparative analysis of geometrical properties of neuronal tree structures. *J. Neurosci. Methods* **18:**179–206.

Verwer, R. W. H., and J. Van Pelt (1987). Multifurcations in topological trees: Growth models and comparative analysis. *Acta Stereol.* **6:**399–404.

Verwer, R. W. H., J. Van Pelt, and H. B. M. Uylings (1985). A simple statistical test for the vertex ratio using Monte Carlo simulation. *J. Neurosci. Methods* **14:**137–142.

Verwer, R. W. H. J. Van Pelt, and A. J. Noest (1987). Parameter estimation in topological analysis of binary tree structures. *Bull. Math. Biol.* **49:**363–378.

Villa, A. E. P., M. Bruchez, G. M. Simm, and S. Jeandrevin (1987). A computer-aided three-dimensional reconstruction of brain structures using high level computer graphics. *Int. J. Biomed. Comput.* **20:**289–302.

Voyvodic, J. T. (1987). Development and regulation of dendrites in the rat superior cervical ganglion. *J. Neurosci.* **7:**904–912.

Wallen, P., K. Carlsson, A. Liljeborg, and S. Grillner (1988). Three-dimensional reconstruction of neurons in the lamprey spinal cord in whole-mount, using a confocal laser scaning microscope. *J. Neurosci. Methods* **24:**91–100.

Wann, D. F. (1976). Counting high contrast closed objects in biological images using a 525-line raster scan television camera and a minicomputer. In: *Computer Technology in Neuroscience* (P. B. Brown, ed.). Washington, DC: Hemisphere, pp. 135–137.

Wann, D. F., T. A. Woolsey, M. L. Dierker, and W. M. Cowan (1973). An on-line digital computer system for the semiautomatic analysis of Golgi-impregnated neurons. *IEEE Trans. Biomed. Eng.* **20:**233–247.

Wann, D. F., J. L. Price, W. M. Cowan, and M. A. Agulnek (1974). An automated system for counting silver grains in autoradiographs. *Brain Res.* **81:**31–58.

Ware, R. W., and V. LoPresti (1975). Three-dimensional reconstruction from serial sections. In: *International Review of Cytology*, Vol. 40 (G. H. Bourne and J. F. Danielli, eds.). New York: Academic Press, pp. 325–440.

Waters, J. R., and A. J. Chester (1987). Optimal allocation in multivariate, two-stage sampling designs. *Am. Stat.* **41:**46–50.

Watson, G. S., and E. J. Williams (1956). On the construction of significance tests on the circle and the sphere. *Biometrika* **43:**344–352.

Webb, W. W. (1986). Light microscopy—a modern renaissance. *Ann. N.Y. Acad. Sci.* **483:**387–391.

Wegman, E. J., and D. J. DePriest, eds. (1986). *Statistical Image Processing and Graphics*. New York: Marcel Dekker.

Weinstein, M., and K. R. Castleman (1971). Reconstructing three-D specimens from two-D section images. *Proc. Soc. Photo-opt. Instrum. Eng.* **26:**131.

Werner, C., and J. S. Smart (1973). Some new methods of topologic classification of channel networks. *Geogr. Anal.* **5:**271–295.

West, M. J. (1985). Neuroanatomical modeling with CADCAM. *Soc. Neurosci. Abstr.* **11:**184.6.

Willey, T. J., R. L. Schultz, and A. H. Gott (1973). Computer graphics in three dimensions for perspective reconstruction of brain ultrastructure. *IEEE Trans. Biomed. Eng.* **20:**288–291.

Williams, F. G., and R. Elde (1982). A microcomputer-aided system for the graphic reproduction of neurohistochemical maps. *Comput. Prog. Biomed.* **15:**93–102.

Williams, R. S., and S. Matthijsse (1983). Morphometric analysis of granule cell dendrites in the mouse dentate gyrus. *J. Comp. Neurol.* **215:**154–164.

Williams, R. S., and S. Matthijsse (1986). Age-related changes in Down syndrome brain and the cellular pathology of Alzheimer disease. *Prog. Brain Res.* **70:**49–67.

Wind, G., V. K. Dvorak, and J. A. Dvorak (1986). Computer graphic modeling in surgery. *Orthop. Clin. North Am.* **17**(4)**:**657–668.

Wind, G., R. W. Finley, and N. M. Rich (1988). Three-dimensional computer graphics modeling of ballistic injuries. *J. Trauma* **28**(1)**:**S16–S20.

Winslow, J. L., M. Bjerknes, and H. Cheng (1987). Three-dimensional reconstruction of biological objects using a graphics engine. *Comput. Biomed. Res.* **20**(6)**:**583–602.

Woolsey, T. A., and M. L. Dierker (1978). Computer-assisted recording of neuroanatomical data. In: *Neuroanatomical Research Techniques* (R. T. Robertson, ed.). New York: Academic Press, pp. 47–85.

Woolsey, T. A., and M. L. Dierker (1982). Morphometric approaches to neuroanatomy with emphasis on computer-assisted techniques. In: *Cytochemical Methods in Neuroanatomy* (V. Chan-Palay and S. L. Palay, eds.). New York: Alan R. Liss, pp. 69–91.

Yaegashi, H., T. Takahashi, and M. Kawasaki (1987). Microcomputer-aided reconstruction: A system designed for the study of 3-D microstructure in histology and histopathology. *J. Microsc.* **146**(1)**:**55–65.

Yelnik, J., G. Percheron, J. Perbos, and C. François (1981). A computer-aided method for the quantitative analysis of dendritic arborizations reconstructed from several serial sections. *J. Neurosci. Methods* **4:**347–364.

Yelnik, J., G. Percheron, C. François, and Y. Burnod (1983). Principal component analysis: A suitable method for the 3-dimensional study of the shape, dimensions and orientation of dendritic arborizations. *J. Neurosci. Methods* **9:**115–125.

Yelnik, J., C. François, G. Percheron, and S. Heyner (1987). Golgi study of the primate substantia nigra. I. Quantitative morphology and topology of nigral neurons. *J. Comp. Neurol.* **265:**455–472.

Young, S. L., E. K. Fram, and B. L. Craig (1985). Three-dimensional reconstruction and quantitative analysis of rat lung type II cells: a computer-based study. *Am. J. Anat.* **174:**1–14.

Young, S. L., S. Royer, P. M. Groves, and J. C. Kinnamon (1987). Three-dimensional reconstructions from serial micrographs using the IBM PC. *J. Electron Microsc. Tech.* **6:**207–217.

Zsuppán, F. (1984). A new approach to merging neuronal tree segments traced from serial sections. *J. Neurosci. Methods* **10:**199–204.

Zsuppán, F. (1985). A computer reconstruction system for biological macro- and microstructures traced from serial sections. *Acta Morphol. Hung.* **33**(1–2)**:**33–44.

Zsuppán, F. (1987). The methodology of entering neuronal structure into a computer. *Neurosci.* **22**(suppl)**:** S393.

Zsuppán, F., and M. Réthelyi (1985). Approximation of missing sections in computer reconstruction of serial EM pictures. *J. Neurosci. Methods* **15:**203–212.

Selected Reading

BOOKS AND SPECIAL JOURNAL VOLUMES

Bourne, J. R. (1981). *Laboratory Minicomputing*. New York: Academic Press.

Brooks, F. P., Jr. (1975). *The Mythical Man-Month*. Reading, MA: Addison-Wesley.

Brown, P. B., ed. (1976). *Computer Technology in Neuroscience*. Washington, DC: Hemisphere.

Castleman, K. R. (1979). *Digital Image Processing*. Englewood Cliffs, NJ: Prentice - Hall.

Foley, J. D., and A. van Dam (1982). *Fundamentals of Interactive Computer Graphics*. Reading, MA: Addison-Wesley.

Fraser, P. J., ed. (1988). *Microcomputers in Physiology: A Practical Approach*. Oxford: IRL Press.

Gage, S. H. (1941). *The Microscope*. Ithaca, NY: Comstock.

Gaunt, P. N., and W. A. Gaunt (1978). *Three-Dimensional Reconstruction in Biology*. Tunbridge Wells: Pitman.

Gonzales, R. C., and P. W. Wintz (1977). *Digital Image Processing*. Reading, MA: Addison-Wesley.

Inoué, S. (1986). *Video Microscopy*. New York: Plenum Press.

Karshmer, A. I., and M. A. Arbib, eds. (1978). *Computers in the Neurosciences*. Special issue of the *Brain Theory Newletter* **3**(3–4). Amherst, MA: Center for Systems Neuroscience.

Kerkut, G. A., ed. (1985). *Microcomputers in the Neurosciences*. Oxford: Clarendon Press.

Lasek, R. J., and M. M. Black, eds. (1988). *Intrinsic Determinants of Neuronal Form and Function*. New York: Alan R. Liss.

Lindsay, R. D., ed. (1977). *Computer Analysis of Neuronal Structures*. New York: Plenum Press.

McEachron, D. (1986). *Function Mapping in Biology and Medicine: Computer Assisted Autoradiography*. Basel: Karger.

Mize, R. R., ed. (1985). *The Microcomputer in Cell and Neurobiology Research*. New York: Elsevier.

Newman, W. M., and R. F. Sproull (1979). *Principles of Interactive Computer Graphics*, 2nd ed. New York: McGraw-Hill.

Pratt, W. K. (1978). *Digital Image Processing*. New York: John Wiley & Sons.

Rosenfeld, A., and A. C. Kak (1982). *Digital Imaging Processing*, 2nd ed. New York: Academic Press.

Sholl, D. A. (1956). *The Organization of the Cerebral Cortex*. London: John Wiley & Sons.

Somlyo, A. P., ed. (1986). Recent advances in electron and light optical imaging in biology and medicine. *Ann. N.Y. Acad. Sci.* **483**:387–391.

Uylings, H. B. M. (1977). *A Study on Morphometry and Functional Morphology of Branching Structures, with Applications to Dendrites in Visual Cortex of Adult Rats under Different Environmental Conditions*. Amsterdam: Kaal's Printing House.

Uylings, H. B. M., R. W. H. Verwer, and J. van Pelt, eds. (1988). *Morphometry and Stereology in Neurosciences*. Special issue of *Journal of Neuroscience Methods* **18**(1–2). Amsterdam: Elsevier.

PERIODICALS

Computer Applications in the Biosciences. Oxford: IRL Press.
Computers in Biology and Medicine. Oxford: Pergamon Press.
Computers and Biomedical Research. San Diego: Academic Press.
Computers and Medicine. Glencoe, IL: Medical Group News.
Computer Methods and Programs in Biomedicine. Amsterdam: Elsevier.
IEEE Transactions on Biomedical Engineering. Piscataway, NJ: Institute of Electrical and Electronic Engineers.
International Journal of Biomedical Computing. Limerick: Elsevier.
Journal of Electrophysiological Techniques. New York: Pergamon Press.
Journal of Microscopy. Oxford: Blackwell Scientific Publications.
Journal of Neuroscience Methods. Amsterdam: Elsevier.
SIGBIO (Special Interest Group on Biomedical Computing). New York: Association for Computing Machinery.
SIGCHI (Special Interest Group on Computer–Human Interaction). New York: Association for Computing Machinery.
SIGGRAPH (Special Interest Group on Computer Graphics). New York: Association for Computing Machinery.

Index

Note: The suffixes g, f, n, and t denote glossary, figure, footnote, and table, respectively.